CONCRETE

BRE Building Research Series

Volume 1

Practical Studies from the Building Research Establishment

CONCRETE

THE CONSTRUCTION PRESS

LANCASTER LONDON NEW YORK

The Construction Press Ltd,
Lancaster, England.

A subsidiary company of Longman Group Ltd, London.
Associated companies, branches and representatives
throughout the world.

Published in the United States of America by
Longman Inc, New York.

First published 1978

ISBN 0 904406 37 7

Printed in Great Britain at The Pitman Press, Bath

Preface

This book and its companion volumes bring together in a bound and edited form research papers originating from the British Building Research Establishment (BRE) and originally issued singly in a series of Current Papers. Until now this rich source of information has never been collated and published in traditional bound format and in consequence many potential readers have not been aware of, or had ready access to, the valuable data contained in the series.

These volumes do not contain every paper relating to each subject that has ever been issued by the BRE. To have included them all would have resulted in unmanageable volumes of enormous size! More importantly, to have included them all would have meant preserving in a permanent reference format for an international readership many papers which were intended to be of relevance only to a local British readership or which, important though they were at the time, are not of lasting significance.

The volumes have therefore been carefully compiled so as to contain all those papers which are considered to be of long-term value and genuine international interest issued during the five year span 1973-1978. As such they are permanent, convenient and practical reference works which will be well used by both research workers and building practitioners. Each volume has been carefully subject indexed and, for the benefit of Overseas readers, the contents are published in French, German and Spanish as well as English.

ACKNOWLEDGEMENT

We have pleasure in acknowledging the co-operation of both the Building Research Establishment and Her Majesty's Stationery Office in granting us permission to publish the Current Papers in this volume.

Contents

Table des matières

Inhaltsverzeichnis

Table de materias

Materials for concrete

High magnesia cements 1: curing at 50 °C as a measure of volume stability (CP 27/74)

M.E. Gaze and M.A. Smith

Summary

Present Standard Specifications limit the total amount of magnesia permitted in Portland cement in order to avoid the long-term volume instability which can occur if too much free magnesia (periclase) is present. If these limits could be raised it would be possible to use a wider range of raw materials, for example dolomitic limestones, for cement manufacture. One method proposed for utilization of high-magnesia Portland cements involves the addition of active siliceous admixtures such as pulverized fuel ash or blast furnace slag. Such blended cements have been shown to be sound by means of the ASTM autoclave test for volume stability; however, it has now become necessary to verify the applicability of this test to such blended 'stabilized' cements. As a preliminary step towards this end the expansions and compressive strengths of high magnesia cements, with and without a siliceous admixture (pulverized fuel ash), have been determined following the ASTM autoclave test and prolonged storage in water at 50°C.

At 50°C maximum expansions were achieved by cements without pfa addition within six months whereas those containing pfa required about one year. Addition of pfa reduced expansion both at 50°C and in the autoclave. X-ray analysis showed periclase to be fully hydrated both after autoclaving and after one year at 50°C. Although after storage at 50°C Portland and Portland/pfa cements develop similar compressive strengths, in the autoclave Portland cements give strengths much below those given by Portland/pfa cements, which are themselves similar to the strengths obtained at 50°C.

The higher strength given by the Portland/pfa cements in the autoclave is a sufficient departure from the behaviour of Portland cements to raise some doubts about the applicability of the ASTM test to such blended high-magnesia cements since it may consequently give unduly optimistic results. In contrast the conditions of the 50°C test are closer to normal and should give results similar to those obtained at ordinary temperatures after much longer curing times. The 50°C test can thus be used with reasonable confidence to evaluate and verify the results of the ASTM test obtained on blended high-magnesia cements and is preferable as a means of studying such cements. There is however no evidence to suggest that the ASTM test should not continue to be used as a basis of comparison between Portland cements.

The validity of storage at 50°C as a measure of volume stability can itself be ascertained only by comparison with the long-term behaviour of specimens stored at normal temperatures (e.g. 20°C): such tests are in progress at the Building Research Station.

Introduction

It is well known that the presence of magnesia can cause the delayed expansion of set mortars and concretes and because of this possibility, most Standard Specifications for Portland cements place a limit on the total MgO content allowed. The British standard[1] allows 4%, the ASTM[2] 5%, total magnesia. This restricts the use of materials for cement manufacture to those which contain little magnesia. The use of Portland cements containing amounts of magnesia greater than 5% is of particular interest to countries such as India, which, unlike Great Britain, are deficient in good quality limestone but have extensive deposits of dolomitic limestones which could be utilized if higher limits (10–15%) were to be allowed.

Rosa[3,4] showed that the addition of certain 'active' siliceous materials which include ground granulated slag, pulverised fuel ash (pfa) and natural pozzolans will reduce expansion in the ASTM autoclave test and enable otherwise unsound cements containing up to 15% MgO to pass the test. He reported that concrete made from such stabilized cements has remained sound for more than ten years. In contrast Dolezsai and Szatura[5] reported unsoundness in stabilized cement specimens stored at ordinary temperatures and concluded that the autoclave test is not suitable for this type of mixed cement.

In view of these discrepancies it was decided to carry out an investigation at the Building Research Station into the correlation between the autoclave expansions of high magnesia cements and their expansions under ordinary conditions. The present report describes the results of some preliminary tests in which the expansions of high magnesia cements, with and without a siliceous admixture (pulverized fuel ash), have been accelerated by curing at 50°C.

Background

Magnesia may be present in cement clinker in solution in the glassy phase and crystalline compounds, or as free 'hard-burned', crystalline MgO (periclase). It is only in the latter state that it is capable of causing expansion. Since the periclase is hard-burned, its hydration to $Mg(OH)_2$ takes place only slowly. This reaction is accompanied by a volume increase of more than 100%, leading to possible disruption of the mortar or concrete long after it has hardened. Gille[6] showed that the expansive effect of periclase depends on the particle size and that, weight for weight, coarse particles cause more expansion than fine particles.

'Unsoundness' in a cement, including that due to magnesia, can be shown by the ASTM autoclave test.[7] This involves autoclaving neat cement specimens for three hours at 295 ± 5 psi (2·0 MN/m^2), 24 hours after preparation. The temperatures developed in the autoclave are of the order of 216°C. Under the test conditions most or all of the free magnesia is fully hydrated, and a cement giving an expansion in excess of 0·8% is considered to be unsound. It has been found[8] that cements containing up to 5% MgO usually show autoclave expansions below the ASTM limit; above this amount, cements show a rapid increase in expansion (Figure 1). Majumdar and Rehsi[9] showed that expansions

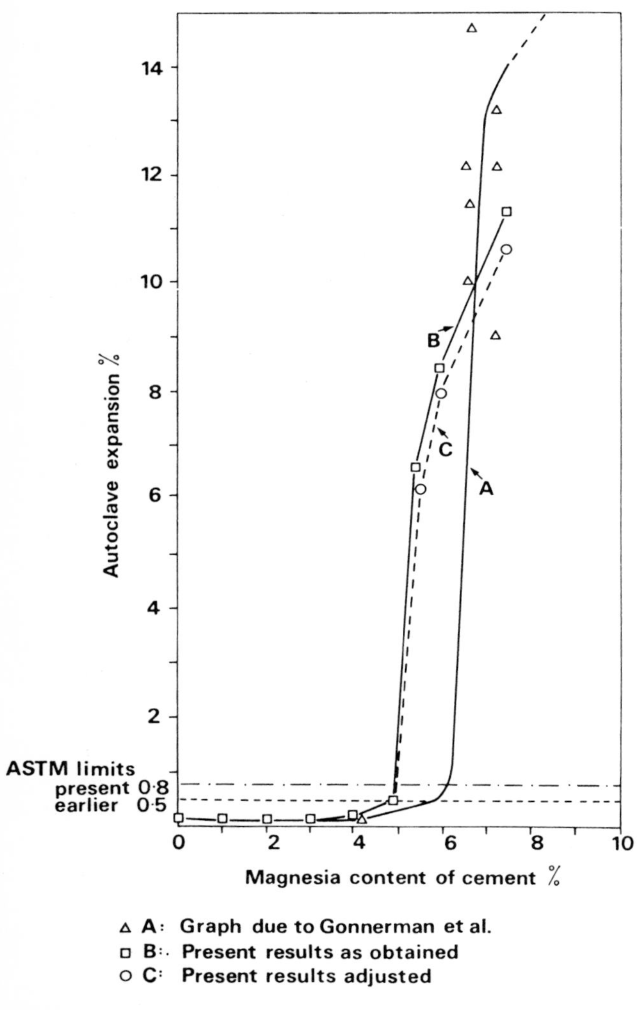

Figure 1: Relationship between MgO content and autoclave expansion of ordinary Portland cement (after Rehsi and Majumdar 9)

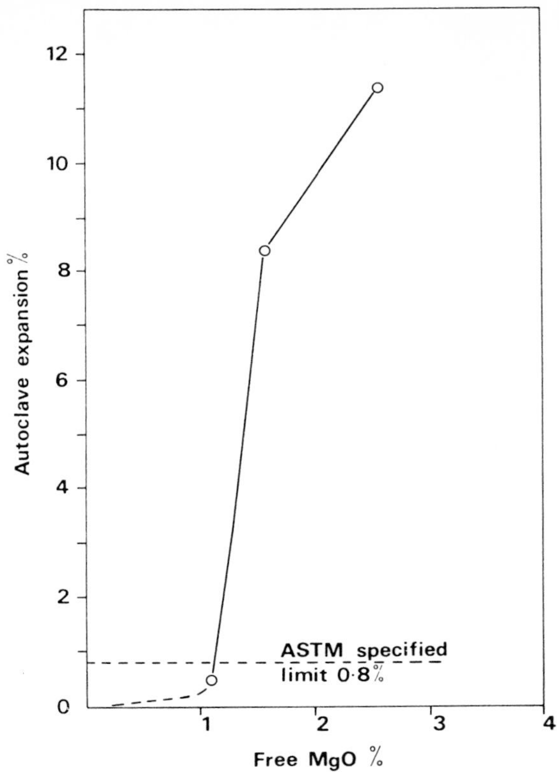

Figure 2: Relationship between 'free' MgO content and autoclave expansion of cement

in excess of ASTM limits could be associated with as little as 1·2% free periclase (Figure 2).

Following Rosa's work[3,4] Majumdar and Rehsi[9] confirmed that cements containing up to 15% MgO could be stabilized in the autoclave test and investigated the mechanism involved. Rehsi[10] has since established tests to determine the long-term volume stability of concretes made with stabilized high-magnesia cements. Dolezsai and Szatura[5] studied the behaviour of cements containing 11% and 14% MgO with admixtures of the type recommended by Rosa, both in the autoclave and after 5½ years' water storage at room temperature. The admixtures reduced expansion under both conditions, but the effect was insufficient at room temperature and many mixtures, rated as sound by the autoclave test, showed signs of unsoundness in the latter test. Dolezsai and Szatura concluded that the autoclave test is unsuitable for predicting the durability of high magnesia cements containing admixtures, but considered it to be still valid for high magnesia cements alone.

Present work

A major part of the work at present in progress at the Building Research Station has consisted of the preparation of high-magnesia clinkers and their use in blended stabilized cements to produce concretes for long-term studies at ordinary temperatures, in order to correlate the results with the results of accelerated tests for volume stability, such as the ASTM autoclave test and storage at 50°C as described below.

Although experimental concretes are required to give a general idea of the behaviour of stabilized high-magnesia cements in practice, this will be influenced by factors such as mix design, aggregate, method of preparation etc., which are unrelated to magnesia content. Consequently, neat cement paste specimens have also been included in the present work in order to eliminate most of the variables and to provide a basis for direct correlation with the results from the autoclave test.

The main problem with respect to obtaining such a correlation is the time involved for the expansion of specimens stored at ordinary temperatures. White[11, 12] has indicated that periods of up to 40 years may be required. It was therefore decided to try the effect of prolonged water storage of neat cement specimens at a temperature a little above normal, so that earlier results might be obtained. The temperature had to be high enough to increase the rate of magnesia hydration without significantly altering the nature of the hydration reactions of the base Portland cement. Thus the results obtained would be closely representative of those obtainable at ordinary temperatures. As a preliminary approach 50°C was chosen, since up to to about 60°C the only expected effects of increasing the curing temperature are increases in the rates of hydration and strength development and slight changes in the composition of the tobermorite gel. However, tests are in progress, at both higher and lower temperatures, in order to confirm the validity of these assumptions.

Majumdar and Rehsi[9] have stated that stabilization of high-magnesia cements by reactive silica in the autoclave is due to the formation of a high-strength matrix of 11 Å tobermorite which resists the expansive force of hydrating magnesia. They assumed that a similar strength mechanism operates at ambient temperatures; however, the possibility exists that sufficiently high strengths are not obtainable at ordinary temperatures. If this is so, autoclave results for stabilized cements could be too optimistic. It has been reported[13] that steam curing of Portland cements at temperatures up to 100°C results in the formation of the normal high strength calcium silicate hydrates but at 100–190°C low strength αC_2S hydrate is formed. Autoclave conditions, whilst not identical to the latter, seem sufficiently close for formation of the low-strength hydrate to occur. If low strength is created in a Portland cement matrix in the autoclave, this must be considered an essential feature of the test and any change from a low strength condition could affect the validity of the results. It was decided to investigate the magnitude of such effects, if any, by comparing the compressive strengths of Portland cement and OPC/pfa mixes (without magnesia) after autoclave treatment and after curing for one week at 50°C. It was thought that the latter condition should develop a fair amount of strength in both cements without unduly favouring the strength development of one matrix over the other, and thus provide a fair, if arbitrary, basis for comparison with autoclave strengths.

Experimental procedures

Materials used

(*a*) *Periclase*

Magnesium carbonate (GPR) was decarbonated at 1300°C and ground to pass a 200-mesh sieve. It was then hard burned by heating at 1 500°C and resieved to pass a 44 μm mesh sieve (ASTM No. 325).

The preparation of periclase was designed to give particles at the coarser end of the spectrum investigated by Gille.[6] This enabled large expansions to be obtained for small magnesia additions so that the lime/silica ratios of the mixes, with an addition of 40% pfa, could be kept fairly constant (1·21–1·25) and close to that (1·25) found by Majumdar and Rehsi[9] to give the most favourable result in the autoclave test.

(*b*) *Portland cement*

An ordinary Portland cement was used for the tests with surface area 341 m²/kg. The chemical analysis is given in Table 1.

(*c*) *Pulverized fuel ash*

The chemical analysis of the pulverized fuel ash used is given in Table 1. The surface area was 381 m²/kg. X-ray analysis showed the presence of quartz, iron oxide and mullite. Pfa contains 'active' amorphous silica which will not be shown by X-ray analysis. The amount of SiO_2 shown in Table 1 includes both quartz and amorphous silica.

Table 1. Chemical analysis

	Portland Cement wt.% (B.R.S. No. 753)	Pulverized fuel ash wt.%
SiO_2	20·60	48·12
Al_2O_3	5·15	25·83
TiO_2	0·33	1·18
Fe_2O_3	2·50	9·72
CaO	65·25	2·61
MgO	1·05	1·90
Na_2O	0·32	1·54
K_2O	0·34	3·55
Mn_2O_3	0·07	0·09
P_2O_5	0·21	—
SO_3	2·60	—
C	—	2·09
LOI	1·75	4·23
Total	100·17	100·86

The pfa was shown to have pozzolanic activity by means of the Rio-Fratini test (ISO Recommendation No. 863), and also that this activity was sufficient for it to be classified as a good pozzolan, by means of the accelerated curing test developed by F. M. Lea.[14]

Mixes tested

Two sets of mixed cements were prepared, one containing periclase and Portland cement and the the other periclase, Portland cement, and pfa as a stabilizer. The OPC/MgO mixes contained 2, 3, 4, 5% MgO with Portland cement to 100%. The OPC/MgO/pfa mixes contained the same amounts of periclase and 40% pfa with Portland cement to 100%. In order to complete the autoclave expansion curve for pfa cements, mixes containing 0, 0·5 and 1·0% added periclase were also made. The mixes are listed in Table 2.

Table 2. Mixes prepared

Mix No.	Added periclase wt.%	Pfa content wt.%	OPC content wt.%	Total MgO content	Lime/silica ratio
2	2·0	—	98·0	3·03	3·17
2A	2·0	40	58·0	3·37	1·25
3	3·0	—	97·0	4·02	3·17
3A	3·0	40	57·0	4·36	1·23
4	4·0	—	96·0	5·01	3·17
4A	4·0	40	56·0	5·35	1·22
5	5·0	—	95·0	6·00	3·17
5A	5·0	40	55·0	6·34	1·21
1A	1·0	40	59·0	2·38	1·26
0·5A	0·5	40	59·5	1·88	1·27
Control	Nil	40	60·0	1·39	1·27

It should be noted that:

(a) The total magnesia figure given in Table 2 includes the added periclase plus the MgO present in the cement and pfa. No free magnesia was detected in the X-ray analysis of the two latter and it is thus concluded that the magnesia is present in an inactive form. Consequently the only expansive MgO present is the added periclase.

(b) The highest total MgO content is 6·34%, therefore strictly speaking these are not 'high magnesia' cements. However, all the cements contain a greater proportion of free magnesia than would be expected in clinkers with the same analytical total. In addition, the added periclase is probably coarser than clinker MgO and this will add to the expansive effect. Thus these cements show expansions more characteristic of clinkers with much higher total MgO contents.

(c) The lime/silica ratios of all the mixes with pfa, are similar and close to that (1·25) shown by Majumdar and Rehsi[9] to give the best results in the autoclave test.

(d) Corresponding cements with and without pfa contain the same amount of added periclase and will thus be subjected to the same potential expansive force. In contrast, for a normally prepared cement, the addition of pfa would act as a diluent and effectively reduce the potential for expansion irrespective of a specific stabilizing action.

Preparation and testing

(*a*) *Autoclave test*

The tests described below were carried out on 13 mm neat cement cubes according to the modified ASTM procedure described by Rehsi and Majumdar.[15] This method is identical except for the size of test specimen. Sufficient material of each mix to make four 13 mm test cubes was weighed out mixed with 26–27% by weight of distilled water, and placed in brass moulds. The moulds were filled in two layers and were then stored for 24 hours over water at 18°C and 100% humidity. After this time the cubes were demoulded, measured and autoclaved (180 mins at 2 MN/m^2, 216°C), according to the ASTM procedure. After cooling, the cubes were removed from the autoclave, remeasured and their percentage expansion calculated.

(*b*) *Water storage at 50°C*

After the autoclave test results for the mixes had been obtained, further quantities of the mixes were prepared. Eight cubes were made from each mix as described above for the autoclave test and were placed in plastic containers. These were stored in a water bath at 50°C. At suitable intervals four cubes, chosen at random, from each set of eight were removed from the bath, surface dried, measured, and then replaced.

(*c*) *Compressive strength tests*

Compressive strength tests were made on 13 mm neat cement paste cubes made from three ordinary Portland cements (including the one used in the other tests) and corresponding 60% OPC/40% pfa mixes, after autoclaving in accordance with the ASTM test and after one week's storage in water at 50°C. The storage period of one week was chosen following a preliminary test which showed that after this time an OPC/pfa mix developed strength similar to that obtained after autoclaving.

(*d*) *X-ray analysis*

X-ray analysis was carried out by means of a powder focusing camera on the mixes containing 2 and 5% added periclase, i.e. the lowest and highest contents. This analysis was carried out after autoclave treatment and after 16 and 52 weeks water storage at 50°C.

Results and discussion

Autoclave test

The results of the autoclave test are given in Table 3 together with the results for water storage at 50°C. Figure 3 shows the expansion in relation to MgO content for the pfa mixes. The OPC/pfa/MgO mixes all show a marked decrease in expansion, compared with the corresponding OPC/MgO mix, and the pfa has clearly had a 'stabilizing action'. As the MgO content is the same for the mixes with and without pfa, no dilution of the 'expansive-force' is involved. Although only the expansion of the 2% MgO mix was reduced to below the ASTM limit of 0·8%, it is reasonable to describe all the cements as having been 'stabilized against volume expansion in the autoclave test'. The description 'stabilized' is applied in the remainder of this report to all cements in which the expansion of the pfa mix is less than that of the corresponding mix without pfa under any particular test regime and does not imply that it passes the ASTM autoclave test limit of 0·8%.

Table 3. Expansion results (Autoclave and 50°C)

Mix	Added periclase wt.%	Pfa wt. %	Expansion in autoclave %	Expansion after 52 weeks in water at 50°C%
2	2·0	—	11·00	0·15
2A	2·0	40	0·44	0·10
3	3·0	—	~62	1·30
3A	3·0	40	3·30	1·15
4	4·0	—	~77	3·05
4A	4·0	40	8·20	2·05
5	5·0	—	*	4·60
5A	5·0	40	15·90	2·85
1A	1·0	40	0·27	—
0·5A	0·5	40	0·23	—
Control	Nil	40	0·15	Nil

*Mixes 3 and 4 were near disintegration after autoclave treatment, consequently mix 5 was not autoclaved.

Water storage tests

The results of the water storage tests are shown graphically in Figures 4 and 5. The expansions achieved after 52 weeks are listed in Table 3. The results show that:

(a) For OPC/MgO cements the rate of expansion is relatively fast and, after approximately 20 weeks' storage, reaches a more or less constant value for a given amount of MgO. In contrast, the pfa cements show a slow continuous rise up to 52 weeks. These differences in expansion rates agree with a suggestion by Rosa that pfa delays magnesia hydration.

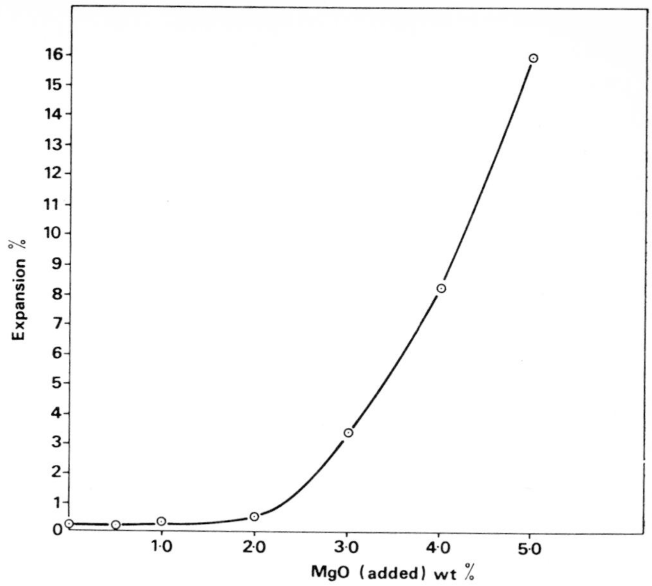

Figure 3: Autoclave expansion of mixes with pfa

It can be seen from Figure 4 that the difference in expansion between an OPC/MgO cement and the corresponding cement containing pfa, does not reach a constant value until very late ages. At earlier ages the effect of the stabilizer appears greater than final results justify and is most marked at ages where the unstabilized cement is approaching its maximum (Figure 5). This indicates that at ordinary temperatures the beneficial effects of pfa may be overestimated unless tests are continued for very long ages.

(b) For each pair of mixes, the one containing pfa shows the lower final expansion. This indicates that pfa acts as a stabilizer outside of autoclave conditions, but the effect is not as beneficial as suggested by the autoclave results. Table 3 shows a reduction in autoclave expansion on addition of pfa much in excess of 100% for autoclave conditions, whereas reduction at 50°C is of the order of 30%. In the case of the 2 and 3% MgO mixes, the differences lie within the limits of experimental error.

(c) In all cases the expansion at 50°C is less than the autoclave expansion. This was not unexpected, since the ASTM test involves far from normal conditions, however the autoclave results for pfa cements are only a few times greater than the 50°C results, but many times greater for the cements without stabilizer.

Compressive strength

The results of the compressive strength tests are given in Table 4. After one week at 50°C both the Portland and Portland/pfa cements had developed considerable strength, the Portland cement strengths being somewhat higher. Compared with this arbitrary standard, autoclave treatments increased the Portland/pfa strength by an average of 16% and decreased that of the Portland cement alone by 68%.

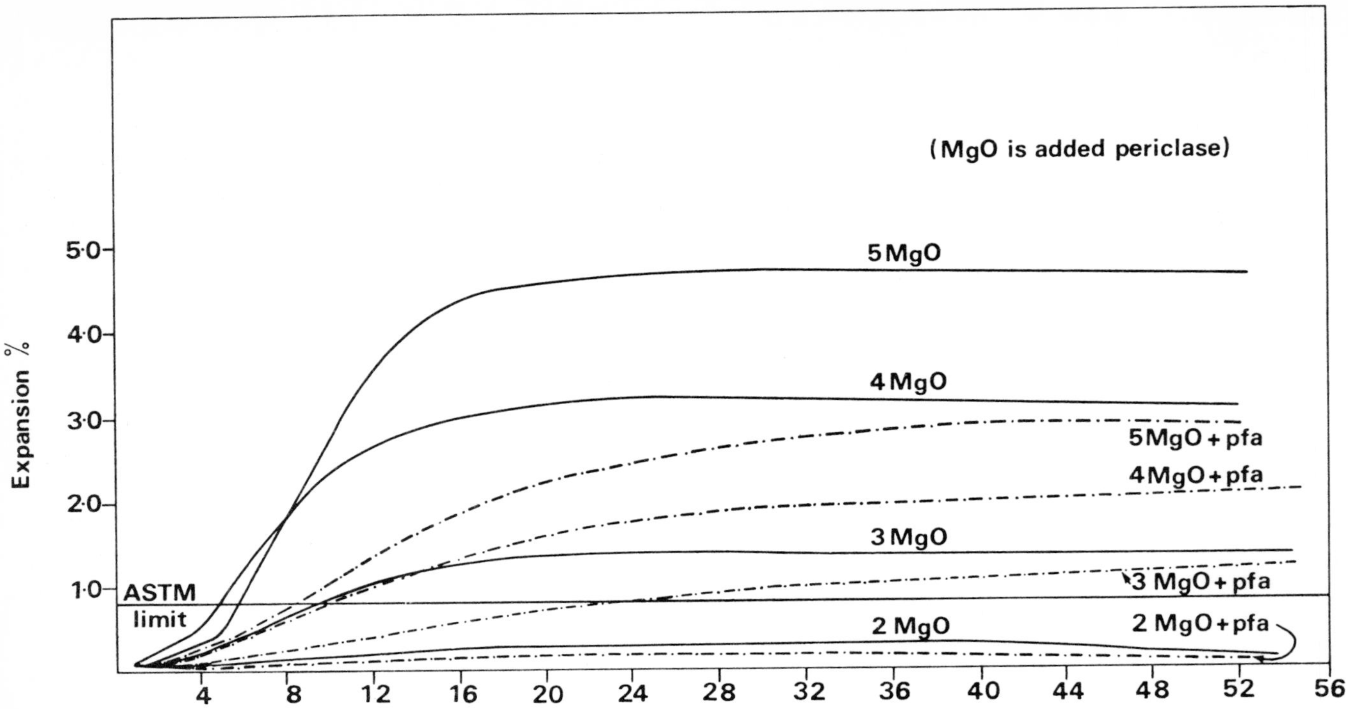

Figure 4: Expansion of mixes in 50°C water bath

Table 4. Compressive strengths MN/m².

Cement	After one week water at 50°C	After autoclave treatment	Average strength difference $\left(\frac{50°C—autoclave}{50°C}\right)\times 100$
OPC 753	15·1	4·0	
OPC 1	11·6	4·2	−68%
OPC 2	11·2	3·8	
OPC 753/pfa 60 40	12·2	13·9	
OPC 1/pfa 60 40	8·4	10·4	+16%
OPC 2/pfa 60 40	9·6	10·8	

It has been mentioned earlier that the strength developed by an OPC/pfa mix after one week at 50°C was chosen as standard, since it was close to the strength of a similar mix after autoclave treatment. Continued hydration at 50°C resulted in only a small increase in strength, so it would seem that the strength developed by the conditions of the autoclave test and the best strength obtainable at 50°C are of the same order.

The pass mark for the ASTM test was primarily determined on the basis of the behaviour of Portland cements, thus the weakening effect must be incorporated as one of the test conditions. It has been shown that the OPC/pfa matrix maintains its strength under these conditions. Consequently it seems likely that autoclave test results for this type of cement could be unduly optimistic.

The magnitude of the weakening effect probably

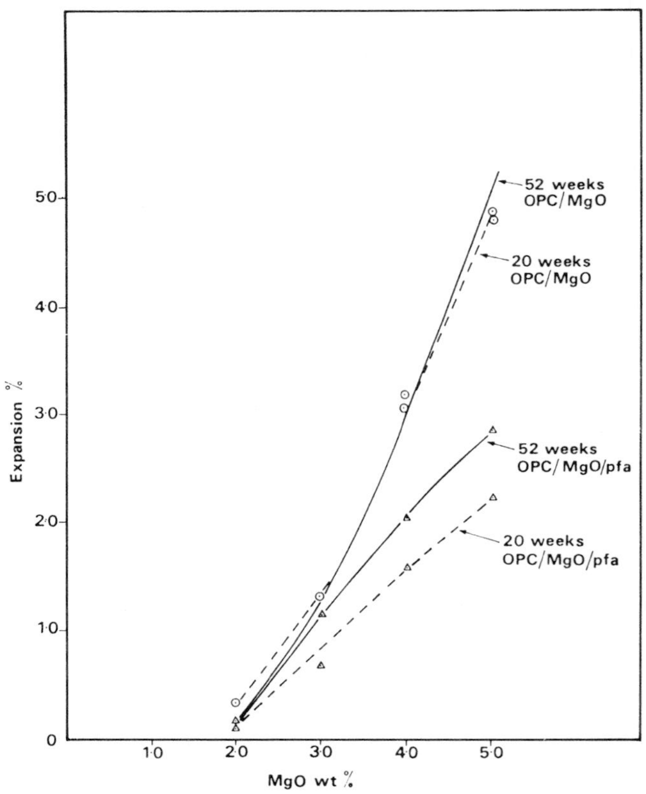

Figure 5: Expansion of mixes at 20 and 52 weeks (water storage 50°C)

explains the marked differences in expansion found between OPC/MgO mixes at 50°C and in the autoclave since under the latter condition the magnesium hydroxide is expanding against a low strength matrix and thus causes much more damage.

X-ray results

The results in Table 5 show that complete hydration of the magnesia was achieved in the autoclave and after storage in water at 50°C for 52 weeks.

Table 5. X-ray results

Mix	After autoclave treatment	After 16 weeks in water 50°C	After 52 weeks in water 50°C
2, 5	Unhydrated cement $Ca(OH)_2$ $Mg(OH)_2$	Unhydrated cement Ettringite $Ca(OH)_2$ $Mg(OH)_2$ MgO	Unhydrated cement Ettringite $Ca(OH)_2$ $Mg(OH)_2$
2A, 5A	As above + pfa compounds 11 Å tobermorite No $Ca(OH)_2$	As above + pfa compounds $Ca(OH)_2$ faint	As above + pfa compounds $Ca(OH)_2$ faint

Summary of the main results

(1) Final expansions can be obtained for high magnesia cements with and without pfa admixtures within a period of about one year by continuous water storage at 50°C.

(2) During this period all MgO is hydrated.

(3) The addition of pfa delays MgO expansion.

(4) Pfa has a stabilizing action of its own independent of dilution effects, both in the autoclave and at 50°C, but this effect is relatively less under the latter condition.

(5) Portland/pfa cements give higher strength than Portland cements after autoclaving.

Conclusions

The main object of the present preliminary study was to see if prolonged water storage at 50°C was a reasonable test to employ in the testing of high magnesia cements. The results indicate that this is so. It has been shown that complete hydration of magnesia can be achieved at 50°C and this fact, coupled with the shape of the expansion curves, shows that final expansions for magnesia cements can be obtained by the use of this test. Although maximum expansion was not reached at 50°C until after 52 weeks storage for pfa cements and 20 weeks for those without pfa addition, this period of time is short compared with the times required at normal temperatures (10–40 years).

Portland/pfa cements gave higher strengths than Portland cements in the autoclave, and this is sufficient departure from the behaviour of Portland cements to raise some doubts about the applicability of the ASTM test to high-magnesia Portland/pfa cements, since it may consequently give unduly optimistic results. It is therefore thought that prolonged storage at 50°C is preferable to the ASTM test as a means of studying the volume stability of Portland/pfa cements, as it is much closer to the conditions to be encountered in practice. There is however no evidence to suggest that the ASTM test should not continue to be used as a basis of comparison between various Portland cements.

In view of the limited data available, the fact that these were obtained from mixes with added free MgO rather than from high MgO clinkers, and the absence of long-term data at normal temperatures with which comparison can be made, it is too early to consider what limits might be established for the 50°C test as indicating long-term stability.

Further research

Further tests are in progress on: (1) the effect of different curing temperatures on the time taken to achieve maximum expansion; (2) the behaviour of mixes with different admixtures; (3) cements made from clinkers containing up to 13% MgO with various admixtures; (4) the correlation of results obtained at 50°C, or other chosen temperatures, with the long-term behaviour of specimens stored at normal temperatures.

Acknowledgement

The work described has been carried out as part of the research programme of the Building Research Establishment of the Department of the Environment and is published by permission of the Director.

REFERENCES

1. British Standard 12, Part 2, 1971.
2. American Society for Testing and Materials, ASTM designation C150–71. Standard specification for Portland cement. 1971 Book of ASTM Standards, Part 9. pp. 135–140.
3. ROSA, J. Raumbestandige Zemente mit hohem MgO-Gehalt (Sound cements with high MgO content). Zement-Kalk-Gips. Vol. 18, No. 9, September 1965. pp. 460–470.
4. ROSA, J. The mechanism of the stabilisation of high magnesia Portland cement. Proceedings of eighth Conference on the Silicate Industry, Budapest, 1966. Budapest, Akadamiai Kiado, 1966. pp. 263–273.
5. DOLEZSAI, K. and SZATURA, L. Epitoanyag, 1970 22(b), 208–12.
6. GILLE, F. Untersuchungen uber das Magnesia—Treiben von Portlandzement (Investigations on expansion tendencies of Portland cement due to MgO). Zement-Kalk-Gips. Vol. 5, No. 5, May 1952. pp. 142–151.
7. American Society for Testing and Materials, ASTM Designation C151–68. Autoclave expansion of Portland cement. 1971 Book of ASTM Standards, Part 9. pp. 141–143.
8. GONNERMAN, H. F., LERCH, W. and WHITESIDE, T. M. Investigations of the hydration expansion characteristics of Portland cements. Chicago, Portland Cement Association, June 1953. pp. 168. Research Department Bulletin 45.
9. MAJUMDAR, A. J. and REHSI, S. S. Mechanism of stabilisation of high magnesia Portland cements by reactive silica under autoclave conditions. Magazine of Concrete Research. Vol. 21, No. 68, 1969, pp. 141–150.
10. REHSI, S. S. and GARG, S. K. Room temperature hydration of high magnesia cement containing fly ash. (To be published. Indian Concrete Journal.)
11. WHITE, A. H. Volume changes of Portland cement as affected by chemical composition and aging. Procs. Amer. Soc. Test. Mat. Vol. 28, Part II, 1928. pp. 398–431.
12. WHITE, A. H. and KEMP, H. S. Long time volume changes of Portland cement bars. Proc. Amer. Soc. Test. Mat. Vol. 42, 1942. pp. 727–749.
13. VERBECK, G. and COPELAND, L. E. Paper SP 32–1. Proc. Menzel symposium on high pressure steam curing. 1969. (American Concrete Institute, SP 32, Detroit, 1972). pp. 1–13.
14. LEA, F. M. The testing of pozzolanic cements. Cement Technology. January/February 1973. pp. 21–25.
15. REHSI, S. S. and MAJUMDAR, A. J. The use of small specimens for measuring autoclave expansion of cements. Magazine of Concrete Research. Vol. 19, No. 61, 1967. pp. 243–246.

High magnesia cements 2: the effect of hydraulic and non-hydraulic admixtures on expansion (CP 60/74)

M.E. Gaze and **M.A. Smith**

Summary

Present standard specifications limit the total amount of magnesia permitted in Portland cement in order to avoid the long-term volume instability which can occur if too much free magnesia (periclase) is present. If these limits could be raised it would be possible to use a wider range of raw materials, for example dolomitic limestones, for cement manufacture. One method proposed for utilization of high-magnesia Portland cements involves the addition of active siliceous admixtures such as pulverized fuel ash (pfa) or ground granulated blast-furnace slag.

In the present investigation the expansion of mixes of high-magnesia cement with hydraulic admixtures (pfa and granulated slag) and non-hydraulic admixtures (quartz and trachyte) in the ASTM autoclave test and after prolonged storage in water at 50°C, have been compared. Under autoclave conditions the non-hydraulic admixtures reduced the expansion of high-magnesia cement, but at 50°C had no stabilizing action. In contrast the hydraulic materials, pfa and slag, had a distinct and similar stabilizing action at 50°C, although the autoclave result for the slag was not as good as that for the quartz and trachyte. It is suggested that the pfa and slag may act to reduce expansion by increasing the tensile strength of the cement matrix.

Introduction

Present standard specifications limit the total amount of magnesia permitted in Portland cement in order to avoid the long-term volume instability of set mortars and concretes which can occur if too much free magnesia (periclase) is present. The British Standard[1] allows 4%, the ASTM[2] 5%, total magnesia. If these limits could be raised it would be possible to use a wider range of raw materials for cement manufacture. One method proposed for the utilization of high-magnesia Portland cements involves the addition of 'active' siliceous materials such as ground granulated blast-furnace slag, pulverized fuel ash (pfa) and natural pozzolans. Rosa[3,4] showed that such blended cements containing up to 15% MgO would pass the ASTM autoclave test for unsoundness[5] and that concrete made from such 'stabilized' cements has remained stable for more than ten years.

The ASTM test involves the autoclaving of neat cement specimens for three hours at 2 MN/m² (295 ± 5 psi). The temperatures developed in the autoclave are of the order of 216°C. Under the test conditions most or all of the free magnesia is fully hydrated, and a cement giving an expansion in excess of 0·8% is considered to be unsound.

In an earlier paper[6] the authors have explained their reasons for doubting the applicability of this test to blended cements containing pfa. It was found that pfa cements developed much higher strengths in the autoclave than Portland cements, for which the test was originally devised, and that consequently results for pfa cements could be too optimistic. An alternative method for laboratory investigation of high magnesia cements was explored using 'synthetic' cements, prepared by mixing hard-burned magnesia with an ordinary Portland cement, which involved the prolonged storage of cement specimens in water at 50°C. Under these conditions complete magnesia hydration and final expansions were achieved within a period of about one year.

As an extension of this work it was decided to compare the stabilizing properties of both 'active' and non-hydraulic siliceous materials. A series of mixtures of a 'synthetic' high magnesia cement blended with various admixtures has been subjected to the ASTM autoclave test and to the 50°C test. The admixtures tested were pulverized fuel ash (a pozzolan), and ground granulated blast-furnace slag (a latent hydraulic cement), active materials of the type used by Rosa,[3,4] with ground trachyte and pure quartz as non-hydraulic additives. The quartz was ground to three different surface areas. Trachyte is a fine-grained intermediate igneous rock. Earlier unpublished work[7] carried out at the Building Research Station had shown that ground trachyte enabled a high magnesia cement to pass the autoclave test although shown to be non-hydraulic by both the ISO chemical method[8] and the accelerated curing test due to F. M. Lea.[9] A summary of this work[7] is given in the next section prior to the description of the present investigation.

Trachyte as a stabilizer in the ASTM autoclave test

A high-magnesia Portland cement was prepared by

Table 1. Chemical analysis of clinker used in tests with trachyte (wt%)

CaO	64·20
SiO_2	21·19
Al_2O_3	3·64
Fe_2O_3	3·37
MgO	5·46
TiO_2	0·46
P_2O_5	0·10
Na_2O	0·75
K_2O	0·04
Mn_2O_3	0·05
Acid soluble SO_3	0·35
Loss to 1 000°C	0·58
Total	100·19
Free CaO	1·5

Note: The clinker was shown by both microscopy and X-ray analysis to contain uncombined MgO.

Table 2. Results of ASTM autoclave test applied to cement/trachyte mixtures

Proportion of trachyte wt %	Expansion %
Nil	cubes disrupted
5	cubes disrupted
10	0·71
15	0·42
20	0·43
25	0·45
30	0·56
35	0·40
40	0·50

Notes: 1. ASTM limit on expansion, 0·8%.
2. Surface area of cement = 366 m^2/kg.
Surface area of trachyte = 367 m^2/kg.

grinding a cement clinker containing 5·5 wt% MgO (see Table 1 for full chemical analysis) with 4% gypsum. When subjected to the ASTM autoclave test using the modified small-scale procedure of Rehsi and Majumdar[10] it was found to be unsound. Portions of the cement were blended with various amounts of a finely ground trachyte (see Table 5 for chemical analysis) and each mixture was tested for volume stability in the autoclave test. The results of these tests are given in Table 2. Stabilization was readily achieved and a blended cement was subsequently prepared containing 85 wt% Portland cement and 15 wt% trachyte and tested in accordance with BS 12 : 1958. The blended cement fulfilled the performance requirements of the Standard (Table 3).

The trachyte was subjected to both the ISO test[8] for pozzolanic activity and the accelerated curing test due to F. M. Lea.[9] The trachyte failed to show any pozzolanic activity in either test. The

Table 3. Results of BS 12 : 1958 tests on high-magnesia cement stabilized by the addition of 15% trachyte

Compressive strength		
Vibrated mortar cubes MN/m^2	1 day	8·1
	3 days	22·7
	7 days	37·1
	28 days	52·4
Setting times		
	initial	4h 10min
	final	6h 40min
Surface area		368 m^2/kg

detailed results of the accelerated curing test are given in Table 4.

These results show clearly that a material without hydraulic properties at ordinary temperatures can act as a stabilizer in the autoclave and provide a blended cement which passes standard specifications for strength. It is obvious that addition of an inert material to a high magnesia cement may reduce expansion simply by dilution of the total magnesia content, irrespective of any specific stabilizing action. With a cement prepared from a high-magnesia clinker, these two effects leading to reduced expansion are not distinguishable. The use of 'synthetic' magnesia cements, however, enables each cement/stabilizer mixture to contain the same amount of magnesia and eliminates the dilution effect. Thus the present work offered the chance of investigating any specific stabilizing properties which trachyte might possess.

Experimental work

The high-magnesia cement used in the tests was prepared by mixing hard-burned magnesia with a commercial Portland cement. The method of preparation enabled cements with admixtures to contain the same amount of magnesia as the control cement without admixture and therefore to be subjected to the same 'expansive force'.

A magnesia addition (4%) was chosen that would give substantial expansion, since only a comparative result between the various admixtures was sought and this seemed to offer the best chance of detecting any significant differences in their behaviour. An admixture can be identified as a 'stabilizer' if it results in an expansion which is less than that of the control without admixture. All the mixes contained the same amount of admixture (40 wt%). The earlier work[7] had shown this quantity to give the best autoclave result for the pfa used, and use of a constant admixture content gives some possibility of ranking the stabilizers in order of activity per unit weight. In practice, as can be seen from the results given for the trachyte above, the

Table 4. Results of accelerated curing test

Material under test	Compressive strength results (MN/m^2)		
	Cured at 18°C	Accelerated cure (50°C)	Difference
Cement A	40·1	45·0	4.9
40% trachyte: 60% cement A	18·8	23·1	4·3
Cement B	36·9	43·0	6·1
40% trachyte: 60% cement B	21·8	24·8	3·0

Note: (a) Test is carried out on vibrated mortar cubes containing either 20 or 40 wt% pozzolan with OPC to 100% and treated as follows:
(1) 7 days at 18°C
(2) 5 days at 18°C, 46 hours at 50°C, 2 hours at 18°C.

(b) The strength differences obtained between the two conditions are a measure of the activity of the pozzolan portion. For a 40 wt% substitution the strength difference for a good pozzolan should be at least 14 MN/m^2 (2 000 psi).

(c) Cements A and B are ordinary Portland cements given the same treatment. It can be seen that the strength differences for the trachyte mixes are less than for the Portland cements alone, thus showing that trachyte has no pozzolanic activity.

Table 5. Materials used and chemical analyses

Material	Surface area m_2/kg	Chemical analysis wt%														Notes
		CaO	SiO_2	Al_2O_3	MgO	Fe_2O_3	SO_3	Na_2O	K_2O	TiO_2	Mn_2O_3	P_2O_5	C	LOI	Total	
Magnesia	Ground to pass 44 μm sieve															Made from GPR magnesium carbonate. Hard burned at 1 500°C
Ordinary Portland cement	340	65·25	20·60	5 15	1·05	2·50	2·60	0·32	0·34	0·33	0·21	—	1·75	1.75	100·17	
Pulverized fuel ash	380	2·61	48·12	25·83	1·90	9·72	—	1·54	3·55	1·18	0·09	—	2·09	4·23	100·86	Good pozzolanic properties
Trachyte	575	0·77	63·25	17·15	0·20	4·50	—	6·46	4·56	0·16	0·21	—	—	1·70	98·96	No hydraulic properties
Ground granulated blast-furnace slag	315	39·77	32·70	14·60	4·99	0·20	2·79	0·38	1·37	—	1·14	0·12	—	—	98·36	$+F^{-}$0·3%
Quartz	A 420 B 565 C 880															99·95% SiO_2

optimum addition (i.e. minimum quantity to provide sufficient reduction in expansion) will vary with admixture and magnesia content.

Materials used and mixes tested
The materials used in the investigation are listed in Table 5 together with the chemical analyses of the Portland cement, trachyte, pfa and slag. A control mix was prepared by adding 4% of the hard-burned magnesia to the Portland cement. The other mixes contained 4% MgO, 40% admixture and 56% Portland cement (Table 6).

Preparation and testing
13 mm neat cement cubes were prepared according to the modified ASTM autoclave test procedure

described by Rehsi and Majumdar.[10] Twelve cubes of each mix were prepared and stored for 24 hours over water at 18°C and 100% humidity. After this time the cubes were measured, four of each were subjected to the autoclave test and the remainder were stored under water at 50°C. At suitable intervals the 50°C cubes were remeasured until final expansions were achieved.

Table 6. Mixes prepared, autoclave expansion and final expansions at 50°C

Mixes (wt %)	Autoclave expansion %	Final expansion at 50°C %
96 OPC/4 MgO	approx. 60	3·4
40 pfa/56 OPC/4 MgO	8·6	2·1
40 slag/56 OPC/4 MgO	approx. 32	1·7
40 Trachyte/56 OPC/4 MgO	approx. 23	3·9
40 Quartz A/56 OPC/4 MgO	14·9	4·0
40 Quartz B/56 OPC/4 MgO	14·6	3·0
40 Quartz C/56 OPC/4 MgO	14·5	3·6

Results and discussion

The results of the autoclave test are given in Table 6 together with the final expansions obtained following curing at 50°C in water. Figure 1 shows the expansion of the mixes at 50°C as a function of time. In the autoclave the ranking of admixtures in terms of decreasing stabilizing ability was pfa, quartz (variation of the surface area of the quartz had no effect on the autoclave expansion), trachyte and slag, whereas in the 50°C test the ranking order was slag, pfa, quartz B (565 m^2/kg), quartz C (880 m^2/kg), trachyte, quartz A (420 m^2/kg). There is thus a significant difference between the two sets of results. The results obtained at 50°C show only the two materials of known hydraulicity at ordinary temperatures (pfa and slag) to have marked stabilizing action although the autoclave identifies all the admixtures as stabilizers. At 50°C the pfa and slag mixes have similar expansions; in the autoclave they give least and most expansion respectively.

As mentioned earlier, it was suggested in the previous paper that the autoclave results obtained with pfa might overestimate its stabilizing action. On the basis of present results it is equally reasonable to consider the autoclave as underestimating the stabilizing action of slag under ordinary conditions.

The fact that the two non-hydraulic materials (quartz and trachyte) have some stabilizing action under autoclave conditions is not completely unexpected, since under the high temperatures and pressures of the test even 'inert' quartz acquires hydraulic properties. The results obtained at 50°C on these materials are more difficult to interpret. The quartz sample of intermediate fineness (565 m^2/kg) had a small stabilizing action although the other two quartz samples (420 and 880 m^2/kg) and the trachyte had no such action, and resulted in a small

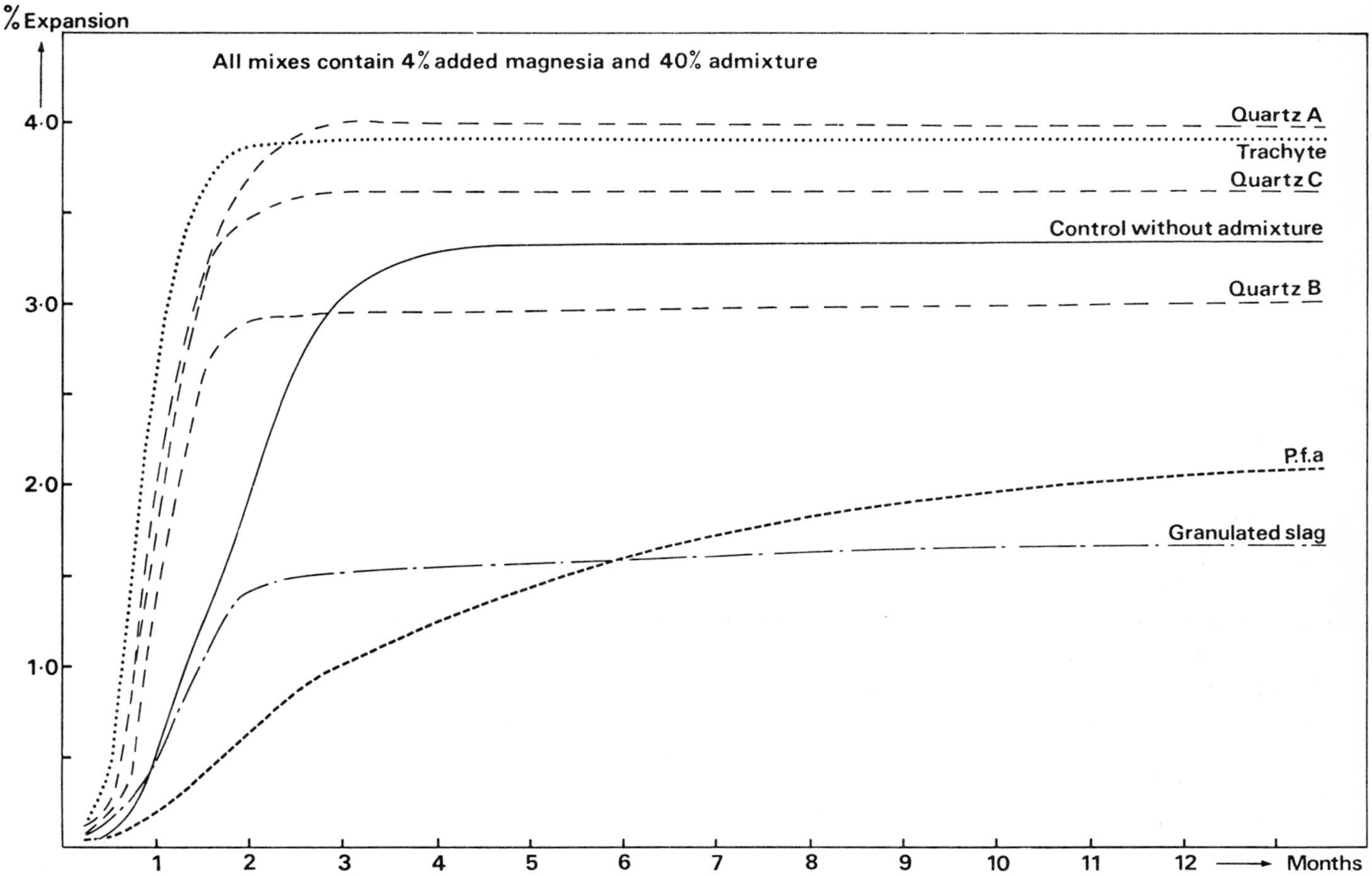

Figure 1: Expansion of mixes at 50°C.

increase in expansion. During the grinding of the quartz it was noticed that the finest sample showed a tendency towards agglomeration of particles. Consequently it may have acted as a coarser material than that of intermediate fineness which would then have acted as the finest in practice.

Whatever the explanation for the differences in stabilizing ability, results for the other non-hydraulic admixtures were better than might be expected for a 40% substitution of allegedly inert materials, as the amount of OPC available to constrain expansion was reduced by 40%, while the MgO content remained the same as for the control.

An explanation of this may lie in the work of Lyubimova *et al.*,[11] who showed that at ordinary temperatures the compressive strength of mixes of OPC with finely ground quartz falls as the amount of quartz is increased, but for additions between 20 and 50%, tensile strength is equal to or slightly higher than that of the Portland cement alone. The hydration of MgO to $Mg(OH)_2$ results in a solid volume increase of over 100%. This expansion at points within the set cement paste must exert a very large tensile force against the matrix. If there are pores within the paste into which the newly formed magnesium hydroxide can expand this process can occur without damage. As the size of the available pores decreases, the force required to occupy these will increase, and if this force exceeds the tensile strength of the matrix, it is reasonable to assume that small pores will remain unoccupied, whilst the full expansive force is directed against the matrix, resulting in micro-cracking and expansion. Thus any improvement in the tensile strength of the matrix should enable more of the smaller pores to be occupied and reduce overall expansion. If this explanation is correct, it implies that the particular fineness and quantity of the quartz of intermediate fineness raised the tensile strength of the mix above that of the control and resulted in some reduction in expected expansion. A similar process occurring with the other two quartz samples and the trachyte was insufficient to stabilize, but resulted in less expansion than might otherwise have been expected. Specimens at present under investigation are not suitable for tensile strength measurements, so this explanation must remain a speculation for the moment; however, it may be mentioned that both slag and pfa are materials which can improve tensile strength. Tensile strength would not be the only factor influencing expansion in this case, since the size and distribution of the available pores and propensity for self-healing of micro-cracks would also be expected to affect the results.

Finally, it has been shown that pure quartz acts as a stabilizer in the autoclave. The hydraulic properties of pozzolanic materials such as pfa are heavily dependent on their content of amorphous silica, but most also contain a proportion of crystalline silica (quartz). A poor pozzolan may contain a large proportion of quartz and a small amount of amorphous silica, which will behave in the same way under autoclave conditions, but differently under normal conditions. It seems advisable to ensure that a pozzolan intended for use as a stabilizer is of proven good quality at ordinary temperatures.

Conclusions

Under autoclave conditions non-hydraulic materials such as quartz or trachyte are reactive and their use as admixtures with high-magnesia cements results in reduced expansion. In contrast, curing at 50°C in water provides no useful stabilization, so the results of the autoclave test on such materials could be misleading.

Only the materials of known hydraulic activity, the pfa and granulated slag, were shown to have definite stabilizing action at 50°C. A pozzolan intended for use as a stabilizer should be assessed by standard tests to ensure that its activity is adequate under ordinary conditions. The autoclave test may underestimate the stabilizing properties of granulated slag under normal conditions of cure.

It is suggested that under normal conditions stabilizers may act by improving the tensile strength of the cement matrix. In this context the size and distribution of pores would also be of significance.

Acknowledgement

The work described has been carried out as part of the research programme of the Building Research Establishment of the Department of the Environment and this paper is published by permission of the Director.

REFERENCES

1. British Standard 12, Part 2, 1971.
2. American Society for Testing and Materials, ASTM designation C150-71 Standard Specification for Portland cement. 1971 Book of ASTM Standards Part 9. pp 135-140.
3. ROSA, J. Raumbestandige Zemente mit hohem MgO-Gehalt (Sound cements with high MgO content). Zement-Kalk-Gips Vol. 18, No. 9, Sept 1965. pp. 460–470.
4. ROSA, J. The mechanism of the stabilization of high magnesia Portland cement. Proceedings of eighth Conference on the Silicate Industry. Budapest, 1966. Budapest, Akadamiai Kiado, 1966. pp. 263–273.
5. American Society for Testing and Materials, ASTM Designation C151–68 Autoclave expansion of Portland cement. 1971 Book of ASTM Standards Part 9. pp. 141–143.
6. GAZE, M. E. and SMITH, M. A. High magnesia cements I: Curing at 50°C as a measure of volume stability. Cement Technology, Vol. 4, No. 6, Nov–Dec 1973. pp. 224–236.
7. GUTT, W. and SMITH, M. A. (unpublished work).
8. International Standards Organization. Recommendation No. 863. (British Standard 4550 Part 2: 1970. Chemical Tests).
9. LEA, F. M. The testing of pozzolanic cements. Cement Technology, Vol. 4, No. 1, Jan–Feb 1973. pp. 21–25.
10. REHSI, S. S. and MAJUMDAR, A. J. The use of small specimens for measuring autoclave expansion of cements. Magazine Concrete Research Vol. 19, No. 61, 1967. pp. 243–246.
11. LYUBIMOVA, T. YU, KUDRYAUTSEVA, N. L. and MELENT'EVA, G. G. Kinetics of formation and properties of crystallization structures arising during the hardening of cements in presence of finely ground quartz. Doklady Akademii Nauk SSSR (Proceedings of USSR Academy of Sciences) 1972. Vol. 202. Section 2. pp. 399–402.

Studies of phosphatic Portland cements (CP 95/74)

W. Gutt and M.A. Smith

INTRODUCTION

At the 5th International Symposium on the Chemistry of Cement, Gutt[1] gave an account of researches into the manufacture of Portland cement from phosphatic raw materials. He reported, inter alia, high temperature phase equilibrium studies in the system $CaO-SiO_2-P_2O_5$ which have provided a fundamental basis for understanding the role of phosphate. The special importance of fluorides as mineralisers in the manufacture of phosphatic-cements was elucidated. Gutt and Smith[2] in 1971 described studies of the use of phosphogypsum, the by-product of wet process phosphoric acid manufacture, in the combined cement/sulphuric acid process and reviewed the difficulties arising from the presence of phosphate and fluoride in the phosphogypsum. Gutt and Nurse[3] in their principal paper to this symposium have reviewed phase equilibrium studies carried out since the Tokyo symposium on phosphate and fluoride containing systems. The present paper contains the results of studies at the Building Research Station on the use of phosphatic raw materials completed since the 5th Symposium.

ASSESSMENT OF PHOSPHATIC RAW MATERIALS FOR CEMENT MANUFACTURE

Three sets of phosphatic raw materials have been assessed for suitability for Portland cement manufacture:

(1) limestone and clay from West Africa for use in the conventional limestone/clay process producing clinkers of 'normal' composition apart from phosphate and fluoride;

(2) limestone and shale from Uganda for use in the limestone/clay process producing clinkers with high iron content;

(3) phosphogypsums for use in the cement/sulphuric acid process.

CLAY/LIMESTONE PROCESS: EXPERIMENTAL METHODS

West African raw materials

The raw materials investigated were two limestones, with P_2O_5 contents of 1.7 and 6.7% (in this paper % = weight per cent), and a clay containing 0.36% P_2O_5. The full chemical analyses are given in Table 1.

In the first stage of the investigation a series of eighteen clinkers containing 2.2 to 6.0% P_2O_5 and having lime factors, according to the Lea and Parker formula ($CaO/(2.80\ SiO_2 + 1.18\ Al_2O_3 + 0.65\ Fe_2O_3)$), in the range 1.00 to 1.06 were made in order to determine the effect of varying P_2O_5 content on lime combination. The Lea and Parker formula is not strictly applicable to phosphatic clinkers as it makes no allowance for the effects of P_2O_5 on phase equilibria but it provided a useful means of comparison between clinkers and for proportioning raw materials. The clinkers were ignited for 30 minutes at 1450°C. The full list of clinkers prepared in this way and their chemical compositions calculated from the proportioning of the raw materials are given in Table 2. The results of chemical analysis for free CaO (all the clinkers), P_2O_5 and fluoride (clinkers with up to 4% P_2O_5) are given in Table 3 and the compound composition calculated from the proportioning of the raw meals in Table 4. Graphs of free CaO against total CaO and LF for each set of clinkers are shown in Figure 1. The clinkers were also examined by X-ray analysis and microscopy to determine their phase composition (Table 5).

The main object of the second series of clinkers was to produce sufficient cement for compressive strength tests to be carried out in accordance with BS 12:1958. All clinkers were made with LF = 1.02 since the first series of tests (see below) had shown this to correspond to the optimum CaO content, irrespective of P_2O_5 contents (Figure 1). Clinkers were produced with P_2O_5 contents (calculated) of 2.2, 2.5, 3.0 and 3.5%. In order to assess the likely effects of

SO_3 in the clinkers, which would in practice arise from the fuel oil, clinkers were also prepared containing 2.5% P_2O_5, 1.0% SO_3; 3.5% P_2O_5, 1.0% SO_3; 3.5% P_2O_5, 2.0% SO_3 (calculated). The calculated chemical compositions and technical characteristics of the clinkers are given in Table 2. The clinkers were ignited at 1450°C for 30 minutes. The clinkers were each analysed for free CaO, P_2O_5, F (and SO_3 when appropriate) and examined by X-ray analysis and microscopy. Cements were prepared by grinding the clinkers, with sufficient gypsum to give a total SO_3 content of 1.75%, to about 280 m^2/kg surface area and compressive strengths determined on 1:3 mortar specimens made and tested in accordance with BS12:1958. The results are shown in Figures 6 and 7. Other tests in accordance with BS12:1958, eg setting time, were made on a selective basis (Table 6). Three of the prepared cements (see below) were examined by continuous conduction calorimetry together with a cement made from the Ugandan raw materials.

Table 1 Chemical analyses of raw materials

Material	West African			Ugandan		
	Limestone A	Limestone B	Clay	Limestone M152	Limestone M153	Shale M151
	wt%	wt%	wt%	wt%	wt%	wt%
SiO_2	3.06	2.72	50.74	3.55	7.31	61.51
CaO	52.35	53.14	7.23	48.85	46.02	1.98
MgO	0.45	0.29	5.62	1.20	0.92	1.23
Fe_3O_4	-	-	-	2.71	-	-
Fe_2O_3	0.44	0.76	6.22	3.29	4.00	7.65
Al_2O_3	1.05	1.53	12.91	1.40	16.55	0.72
Mn_2O_3	0.01	0.04	0.23	0.30	0.31	0.73
TiO_2	0.06	0.08	0.87	0.20	0.33	0.18
P_2O_5	1.69	6.69	0.36	2.65	1.85	0.25
SO_3	0.09	0.22	0.07	0.45	1.43	0.04
Na_2O	0.06	0.12	0.16	0.22	0.59	1.96
K_2O	0.07	0.05	0.66	0.34	1.09	2.56
F	0.21	0.83	0.17	0.23	0.34	0.08
Cl	-	-	0.04	-	-	-
S^{2-}	-	-	-	0.05	0.12	-
SrO	0.04	0.04	-	0.44	0.59	0.03
BaO	-	-	-	-	-	0.11
H_2O	(0.69)	(0.96)	(8.72)	-	-	-
LOI (1000°C)	40.14	34.79	14.63	35.27	33.72	4.86
O_2 equiv for F	-0.09	-0.35	-0.07	-0.10	-0.14	-
TOTAL	99.63	100.95	99.84	100.37	99.88	99.72
F/P_2O_5	0.12	0.12	0.47	0.09	0.18	0.32

Results are reported on dried at 110°C basis

Table 2 West African raw materials: Calculated chemical composition of clinkers

Clinker No	Chemical analysis (wt %)											Technical moduli					
	P_2O_5	SiO_2	CaO	Al_2O_3	Fe_2O_3	MgO	K_2O	Na_2O	SO_3	F	Total	LF	LFN	SM	FM	SR	Liquid at 1400°C
First series																	
RM 1	2.15	20.37	65.25	5.51	2.58	2.45	0.30	0.12	0.13	0.31	99.17	1.00	1.01	2.52	2.14	3.20	24.36
2	2.16	20.09	65.66	5.44	2.54	2.42	0.30	0.12	0.13	0.31	99.17	1.02	1.03	2.52	2.14	3.27	24.06
3	2.18	19.82	66.05	5.38	2.51	2.39	0.30	0.12	0.13	0.31	99.19	1.04	1.05	2.51	2.14	3.33	23.80
4	2.19	19.56	66.43	5.31	2.48	2.36	0.29	0.12	0.13	0.31	99.18	1.06	1.07	2.51	2.14	3.39	23.54
9	3.00	19.62	65.54	5.42	2.55	2.34	0.29	0.13	0.15	0.41	99.45	1.04	1.06	2.46	2.13	3.34	24.01
10	3.50	19.50	65.22	5.44	2.57	2.31	0.29	0.14	0.16	0.47	99.60	1.04	1.07	2.43	2.12	3.35	24.13
5	4.00	19.91	64.11	5.60	2.66	2.33	0.29	0.14	0.18	0.54	99.75	1.00	1.04	2.41	2.11	3.22	24.80
6	4.00	19.64	64.52	5.53	2.62	2.30	0.28	0.14	0.18	0.54	99.75	1.02	1.06	2.41	2.11	3.29	24.52
7	4.00	19.37	64.91	5.47	2.59	2.28	0.28	0.14	0.18	0.54	99.76	1.04	1.08	2.40	2.11	3.35	24.25
8	4.00	19.11	65.30	5.40	2.56	2.25	0.28	0.14	0.18	0.54	99.76	1.06	1.10	2.40	2.11	3.42	23.99
12	4.00	18.87	65.67	5.34	2.53	2.22	0.28	0.14	0.18	0.53	99.76	1.08	1.12	2.40	2.11	3.48	23.73
13	5.00	19.39	63.90	5.58	2.66	2.24	0.28	0.16	0.20	0.66	100.07	1.02	1.08	2.35	2.10	3.30	24.77
14	5.00	19.13	64.29	5.52	2.63	2.21	0.28	0.16	0.20	0.66	100.08	1.04	1.10	2.35	2.10	3.36	24.50
15	5.00	18.88	64.67	5.45	2.60	2.18	0.27	0.15	0.20	0.66	100.06	1.06	1.12	2.34	2.10	3.43	24.24
16	6.00	19.14	63.28	5.63	2.71	2.17	0.27	0.17	0.23	0.78	100.38	1.02	1.09	2.30	2.08	3.31	25.01
17	6.00	18.88	63.67	5.57	2.68	2.14	0.27	0.17	0.23	0.78	100.39	1.04	1.12	2.29	2.08	3.37	24.75
18	6.00	18.63	64.04	5.50	2.64	2.12	0.27	0.16	0.23	0.78	100.37	1.06	1.14	2.29	2.08	3.44	24.49
Second series																	
RM 20	2.16	20.09	65.66	5.44	2.54	2.42	0.30	0.12	0.13	0.31	99.17	1.02	1.03	2.52	2.14	3.27	24.06
22	2.50	20.01	65.45	5.46	2.56	2.40	0.30	0.13	0.14	0.35	99.30	1.02	1.03	2.50	2.14	3.27	24.16
21	3.00	19.88	65.14	5.48	2.58	2.37	0.29	0.13	0.15	0.41	99.43	1.02	1.04	2.47	2.13	3.28	24.28
23	3.50	19.76	64.83	5.51	2.60	2.34	0.29	0.14	0.16	0.47	99.60	1.02	1.05	2.44	2.12	3.28	24.40
28	2.46	19.80	64.84	5.40	2.53	2.37	0.29	0.13	1.14	0.34	99.30	1.02	1.03	2.50	2.13	3.27	23.92
24	3.44	19.56	64.23	5.45	2.57	2.31	0.29	0.14	1.16	0.47	99.62	1.02	1.05	2.44	2.12	3.28	24.16
25	3.38	19.37	63.61	5.40	2.55	2.28	0.28	0.14	2.16	0.46	99.63	1.02	1.05	2.44	2.12	3.28	23.93

Notes:
1 LF = CaO/ (2.8 SiO_2 + 1.18 Al_2O_3 + 0.65 Fe_2O_3)
2 LFN = CaO/ (2.8 SiO_2 + 1.65 Al_2O_3 + 0.35 Fe_2O_3 - 1.0 P_2O_5)
3 SM = SiO_2/ (Fe_2O_3 + Al_2O_3)
4 FM = Al_2O_3/Fe_2O_3
5 SR = CaO/SiO_2
6 Liquid at 1400°C = 2.95 Al_2O_3 + 2.2 Fe_2O_3 + K_2O + Na_2O + 2.0

Table 3 West African raw materials: Results of chemical analysis of clinkers

Clinker No	LF	Analysis of clinker (wt %)					F/P_2O_5	
		Calculated P_2O_5	Experimental				Calculated	Experimental
			P_2O_5	F	SO_3	Free CaO		
First series								
RM 1	1.00	2.15	2.25	0.29	ND	NIL	0.144	0.129
2	1.02	2.16	2.26	0.23	ND	0.04	0.144	0.102
3	1.04	2.18	2.27	0.30	ND	0.6	0.142	0.132
4	1.06	2.19	2.27	0.32	ND	0.8	0.142	0.140
9	1.04	3.00	3.05	0.44	ND	0.2	0.137	0.144
10	1.04	3.50	3.57	0.47	ND	0.3	0.134	0.132
5	1.00	4.00	4.00	0.44	ND	NIL	0.134	0.110
6	1.02	4.00	4.00	0.54	ND	NIL	0.134	0.135
7	1.04	4.00	4.02	0.59	ND	0.4	0.134	0.147
8	1.06	4.00	4.05	0.53	ND	ND	0.134	0.131
12	1.08	4.00	ND	ND	ND	1.2	-	-
13	1.02	5.00	ND	ND	ND	NIL	0.132	-
14	1.04	5.00	ND	ND	ND	0.4	0.132	-
15	1.06	5.00	ND	ND	ND	0.8	0.132	-
16	1.02	6.00	ND	ND	ND	0.03	0.130	-
17	1.04	6.00	ND	ND	ND	0.5	0.130	-
18	1.06	6.00	ND	ND	ND	ND	0.130	-
Second series								
20	1.02	2.15	2.14	0.24	<0.05	0.5	0.144	0.112
22	1.02	2.50	2.49	0.22	0.06	0.2	0.142	0.088
21	1.02	3.00	2.97	0.31	<0.05	0.3	0.118	0.104
23	1.02	3.50	3.38	0.35	0.12	0.2	0.101	0.104
28	1.02	2.46	2.43	0.38	0.68	0.6	0.143	0.156
24	1.02	3.44	3.35	0.35	0.97	0.2	0.102	0.105
25	1.02	3.38	3.42	0.43	1.38	0.1	0.103	0.126

ND - Not determined

Table 4 Calculated compound composition of clinker made from West African materials

Clinker No	P_2O_5 content	Bogue - no allowance for P_2O_5					Nurse - allowing for P_2O_5					
		C_4AF	C_3A	C_3S	C_2S	free CaO	C_4AF	C_3A	C_3S	PSS	C_2S	free CaO
First series												
RM 1	2.15	8.36	10.92	74.84	5.88	0.00	8.17	10.68	48.44	32.09	0.00	0.62
2	2.16	8.25	10.79	79.37	1.59	0.00	8.06	10.55	47.12	32.27	0.00	2.00
3	2.18	8.14	10.66	80.33	0.00	0.87	7.95	10.42	45.85	32.44	0.00	3.35
4	2.19	8.03	10.53	79.25	0.00	2.19	7.85	10.29	44.62	32.59	0.00	4.65
9	3.00	8.31	10.78	80.06	0.00	0.84	8.05	10.44	32.60	44.65	0.00	4.25
10	3.50	8.42	10.86	79.89	0.00	0.83	8.11	10.46	24.60	52.02	0.00	4.80
5	4.00	8.75	11.21	74.02	6.03	0.00	8.38	10.74	18.75	59.41	0.00	2.72
6	4.00	8.64	11.07	78.56	1.74	0.00	8.28	10.61	17.66	59.40	0.00	4.05
7	4.00	8.52	10.94	79.72	0.00	0.82	8.17	10.48	16.61	59.38	0.00	5.35
8	4.00	8.42	10.81	78.61	0.00	2.17	8.07	10.36	15.57	59.36	0.00	6.65
12	4.00	8.31	10.68	77.59	0.00	3.42	7.96	10.24	14.62	59.33	0.00	7.84
13	5.00	8.85	11.22	78.09	1.83	0.00	8.39	10.64	1.73	74.08	0.00	5.16
14	5.00	8.74	11.09	79.38	0.00	0.79	8.28	10.52	0.70	74.05	0.00	6.44
15	5.00	8.63	10.96	78.30	0.00	2.10	8.18	10.40	-0.28	74.02	0.00	7.69
16	6.00	9.07	11.39	77.62	1.92	0.00	8.51	10.68	-14.13	88.68	0.00	6.26
17	6.00	8.96	11.25	79.03	0.00	0.76	8.40	10.56	-15.14	88.65	0.00	7.53
18	6.00	8.85	11.12	77.95	0.00	2.07	8.30	10.44	-16.11	88.61	0.00	8.76
Second series												
RM 20	2.16	8.25	10.79	79.37	1.59	0.00	8.06	10.55	47.12	32.27	0.00	2.00
22	2.50	8.32	10.84	79.23	1.61	0.00	8.10	10.56	41.70	37.26	0.00	2.38
21	3.00	8.42	10.92	79.00	1.66	0.00	8.16	10.58	33.67	44.66	0.00	2.94
23	3.50	8.53	10.99	78.78	1.70	0.00	8.22	10.59	25.65	52.04	0.00	3.50
28	2.46	8.31	10.82	79.50	1.37	0.00	8.09	10.54	41.95	36.98	0.00	2.44
24	3.44	8.52	10.97	79.24	1.28	0.00	8.21	10.57	25.99	51.63	0.00	3.60
25	3.38	8.52	10.97	79.08	1.44	0.00	8.21	10.57	26.44	51.26	0.00	3.52

Notes: 1 Compound compositions derived from direct application of the relevant equations recalculated to total 100%. No allowance made for SO_3.

2 See text for details of equations used in calculations.

Table 5 West African raw materials
Results of microscopic point counting of polished sections

RM No	P_2O_5 wt%	Weight % of main phases		
		Alite	Belite	Interstitial
20	2.16	60	24	16
2	2.16	63	23	14
9	3.0	59	31	10
10	3.5	54	33	12
7	4.0	45	43	12
13	5.0	32	59	9
16	6.0	23	68	9

Table 6 **Results of tests in accordance with BS 12:1958**

Cement		Test: Sulphate content of cement (wt % SO_3)	Test: Surface area of cement (m^2/kg)	Test: Setting times (19°C): Initial (min)	Test: Setting times (19°C): Final (min)	Test: Soundness (Le Chatelier) (mm expansion)
West African	RM 20	1.75	285	ND	ND	ND
	22	1.75	294	340	505	2
	21	1.75	294	390	600-700	1
	23	1.87	284	>600	ND	1
	28	1.75	289	285	495	2
	24	2.08	299	>600	ND	1
	25	1.75	282	>600	ND	0
Ugandan	UM 28	3.0	365	>540	>600	2
	26	3.0	380	300	600	1
	34	3.0	350	480	600	1
	31	3.0	390	>600	>600	Sample disintegrated on boiling
BS 12 requirements for OPC		$\ngtr$ 3.0 $\ngtr$ 2.5	$\nless$ 220	$\nless$ 45	$\ngtr$ 600	$\ngtr$ 10

ND - Not determined

Raw materials from Uganda

The Uganda Cement Industry has produced good quality cement at Tororo from phosphatic raw materials since 1952. The industry is based on the exploitation of a phosphatic limestone deposit which had previously been considered as unsuitable for cement manufacture due to the detrimental effect of the phosphate. Early work by Nurse[4] established that 2.5% P_2O_5 was the maximum level of phosphate which could be tolerated in cements made from these materials. This finding was supported and consolidated by means of detailed phase equilibrium studies by Gutt[5,6] and was confirmed by extensive practical experience in Uganda[1]. In order to alleviate partly the deleterious effect of phosphate, fluorspar is added to the Tororo raw meal. The practical experience of UCI, which takes into account possible adverse effects due to fluoride itself, has shown that the optimum fluoride content for their clinkers is approximately 0.85% F. Similarly the optimum gypsum content for the ground cement has been established by UCI as equivalent to a total SO_3 content of about 3%.

Known reserves of limestone of sufficiently low phosphate content are however limited and the Building Research Station has carried out work using a limestone from an alternative source to assess whether it can be utilised to prolong economic cement production at Tororo. The main problems associated with the use of this limestone deposit stem from the higher level of P_2O_5 and the unexpectedly high magnetite (Fe_3O_4) content.

The Ugandan raw materials studied consisted of two limestones with P_2O_5 contents of 1.85 and 2.6% P_2O_5 respectively and a shale with P_2O_5 content of 0.25%. The full chemical analyses are given in Table 1. Two series of clinkers were prepared: (1) to determine the maximum combinable lime and (2) for testing to BS12:1958.

For the first series of clinkers these raw materials and reagent grade calcite (99.9% pure) were used to prepare clinkers with P_2O_5 in the range 2.0 to 3.4%. Fluorspar was added to each raw meal so that the fired clinkers contained approximately 0.85% F. For convenience the Lea and Parker formula for lime saturation was again used for proportioning the raw materials. The clinkers were ignited for 30 minutes at 1400°C and were analysed for CaO and in some cases for P_2O_5 and fluoride (Table 7). They were also examined by X-ray analysis and microscopy to determine their compound composition (Table 9). The calculated chemical and compound compositions of the clinkers are given in Tables 8 and 9 respectively. The free CaO content of the clinkers in relation to total CaO content and P_2O_5 content are shown in Figure 2.

The main object of the second series of preparations was to produce sufficient cement at each P_2O_5 content for compressive strength tests, setting times and soundness determinations to be carried out, in accordance with BS12:1958. Cement clinkers with P_2O_5 contents (calculated) of

2.0, 2.5, 3.0 and 3.3% were prepared using the raw meal compositions considered to be the most suitable as a result of the first series of tests; ie those giving 0.5% free CaO in the clinkers. The compositions of the raw meals and the results of selective chemical analysis of the clinkers are given in Table 7. The calculated chemical compositions and technical moduli, based on the raw meal analyses without the addition of fluorspar, are given in Table 8. The desired fluoride level of 0.85% F for the clinkers was achieved by addition of CaF_2 to the raw meal. The clinkers were examined by reflected light microscopy and by X-ray analysis. The phase compositions as calculated by applying the Bogue formula (with no allowance made for P_2O_5) and the Nurse formula (allowance made for P_2O_5) are given in Table 9 and are compared with the results obtained from the microscopic point counts. The clinker (UM 26) containing approximately 2.5% P_2O_5 was fully analysed and the results are compared with the analyses calculated from the raw material compositions in Table 10.

For the tests to BS12:1958 the clinkers were ground with sufficient gypsum to give 3.0% SO_3 in the cement to a surface area of 350-390 m^2/kg. It should be noted that 3.0% SO_3 is in slight excess of the BS12 limit of 2.5% SO_3 for total sulphur for cements such as these containing less than 7% by weight of C_3A. The results of compressive strength tests are shown in Figure 8; the results of setting time and soundness tests are given in Table 6. The compressive strength was determined on 70.6 mm 1:3 mortar cubes at 1, 3, 7 and 28 days. Considerable difficulty was experienced in demoulding the mortar cubes at 24 hours due to their low strengths and the cubes made with cement containing 2.5% P_2O_5 were removed at 48 hours (except for 2 which were tested at 24 hours) and the cubes made using cements with 2.95 and 3.33% P_2O_5 were demoulded at 3 days, having been stored in moist air.

Table 7 Ugandan raw materials: Results of chemical analysis of clinkers

Clinker UM No	Lime factor (LF)	Analysis of clinker (wt %)					
		Calculated		Experimental			
		P_2O_5	F	P_2O_5	F	SO_3	Free CaO
First series							
27	1.04	2.00	0.85	–	–	–	0.25
24	1.02	2.46	0.85	–	–	–	0.00
25	1.06	2.49	0.85	2.44	0.80	2.21	0.42
33	1.04	3.00	0.85	–	–	–	0.66
32	1.05	3.00	0.85	–	–	–	0.83
30	1.04	3.35	0.85	–	–	–	1.15
29	1.06	3.36	0.85	–	–	–	2.10
Second series							
28	1.04	2.00	0.85	2.00	0.80	1.60	0.27
26	1.06	2.49	0.85	2.44	1.08	1.56 (S^{2-} = 0.14)	0.39
34	1.04	3.00	0.85	2.95	0.90	1.23 (S^{2-} = 0.02)	0.73
31	1.03	3.34	0.85	3.33	0.87	0.73	0.13

Table 8 Ugandan raw materials: calculated chemical composition of clinkers

Clinker No	Chemical analyses (based on raw meal without fluorspar addition)													Technical moduli						
UM	P_2O_5	SiO_2	CaO	Al_2O_3	Fe_2O_3	MgO	K_2O	Na_2O	SO_3	F	TiO_2	SrO	Mn_2O_3	Total	LF	LFN	SM	FM	SR	Liquid at 1400°C
First series																				
27	2.00	18.34	62.51	4.33	5.55	1.19	1.60	0.96	1.52	0.37	0.45	0.63	0.38	99.83	1.04	1.07	1.85	0.78	3.41	28.74
24	2.46	18.03	60.64	4.11	6.30	1.38	1.78	1.04	1.88	0.46	0.51	0.79	0.46	99.82	1.02	1.06	1.73	0.65	3.36	30.17
25	2.49	17.51	61.29	3.96	6.28	1.38	1.77	1.03	1.90	0.46	0.50	0.79	0.46	99.82	1.06	1.11	1.71	0.63	3.50	29.67
33	3.00	17.33	61.06	4.19	8.02	1.60	1.28	0.82	1.09	0.36	0.52	0.64	0.35	100.26	1.04	1.11	1.41	0.52	3.52	33.69
32	3.00	17.21	61.22	4.15	8.00	1.60	1.28	0.82	1.11	0.37	0.52	0.65	0.36	100.26	1.05	1.12	1.42	0.52	3.56	33.54
30	3.35	17.05	61.13	4.30	9.15	1.74	0.95	0.68	0.57	0.30	0.52	0.55	0.29	100.57	1.04	1.12	1.27	0.47	3.59	36.18
29	3.36	16.79	61.45	4.22	9.16	1.75	0.94	0.67	0.57	0.30	0.52	0.56	0.29	100.57	1.06	1.14	1.25	0.46	3.66	35.96
Second series																				
28	2.00	18.34	62.51	4.33	5.55	1.19	1.60	0.96	1.52	0.37	0.45	0.63	0.38	99.83	1.04	1.07	1.85	0.78	3.41	28.74
26	2.49	17.51	61.29	3.96	6.28	1.38	1.77	1.03	1.90	0.46	0.50	0.79	0.46	99.82	1.06	1.11	1.71	0.63	3.50	29.67
34	3.00	17.33	61.06	4.19	8.02	1.60	1.28	0.82	1.09	0.36	0.52	0.64	0.35	100.26	1.04	1.10	1.42	0.52	3.52	33.69
31	3.34	17.18	60.96	4.33	9.15	1.74	0.95	0.68	0.57	0.30	0.52	0.55	0.29	100.56	1.03	1.10	1.27	0.47	3.54	36.28

Table 9 Ugandan raw materials: calculated and experimental phase compositions of second series of clinkers

Clinker UM No	Content (wt %)	Calculated Compositions											Experimental Compositions				
		Bogue formula - No allowance for P_2O_5					Nurse formula - Allowing for P_2O_5						Phases identified by microscopic point counting of polished sections (wt%)				Free CaO
		C_4AF	C_3A	C_3S	C_2S	Free CaO	C_4AF	C_3A	C_3S	PSS	C_2S	Free CaO	Alite	C_2S (etched blue)	Inter-stitial	C_2S (unetched)	
28	2.00	18.40	2.29	76.98	0.0	2.33	18.00	2.24	44.17	30.93	0.0	4.66	63	22	15	-	0.3
26	2.44	21.21	-0.17	74.94	0.0	4.02	20.63	-0.17	33.69	38.94	0.0	6.91	66	16	16	2	0.4
25	2.44	21.21	-0.17	74.94	0.0	4.02	20.63	-0.17	33.69	38.94	0.0	6.91	61	25	13	2	-
34	2.95	26.64	-2.76	72.96	0.0	3.16	25.78	-2.66	24.31	45.98	0.0	6.59	38	38	22	2	0.8
31	3.33	30.08	-4.43	71.54	0.0	2.81	29.02	-4.28	18.23	50.44	0.0	6.59	18	50	30	3	1.1

Notes: 1 Compound compositions derived by direct application of the relevant equations, before fluorspar addition, and recalculated to total 100%. No allowance made for SO_3.

2 Phases etched by HF vapour. C_3S (Alite) brown; C_2S, blue or unetched; Interstitial, white.

Table 10 Calculated and experimental chemical analyses of clinker UM 26 made from Ugandan raw materials

	Calculated	Experimental
SiO_2	17.37	17.83
CaO	61.37	61.40
Al_2O_3	3.93	4.05
Fe_2O_3	6.23	6.52
MgO	1.37	1.31
K_2O	1.76	1.36
Na_2O	1.02	0.79
SO_3	1.89	1.56
S^{2-}	-	0.14
F	0.85	1.08
TiO_2	0.50	0.40
SrO	0.78	-
Mn_2O_3	0.46	0.46
P_2O_5	2.47	2.44
LOI *	-	1.07
Less O_2 equiv of F	-0.35	-0.45
TOTAL	99.65	99.96

* Loss on ignition

CLAY/LIMESTONE PROCESS: PHASE COMPOSITION AND COMBINATION

West African materials

The first series of tests showed that clinkers with lime factor (LF) as high as 1.06 could be made with an acceptable free CaO content ie not more than 1.0%. The results of analysis for free CaO were consistent with the clinkers being fully reacted, ie the free CaO detected did not arise from inhomogeneity of the raw meal.

The 'breakpoints' in the graphs of free lime against LF and total CaO (Figure 1) for clinkers with P_2O_5 contents of 2.2, 4.0, 5.0 and 6.0 all occur at LF = 1.02 and at LFs above 1.02 there is a progressive increase in uncombined CaO. The form of the four graphs is very similar and a free CaO content of 1.0% is reached for each at about LF = 1.06. The CaO content of clinkers of constant LF however decreases with increasing P_2O_5 content, so that the total combinable lime falls from about 65.6% for 2.16% P_2O_5 to 63.3% for 6.0% P_2O_5 (ie approximately 0.6% combined CaO per 1.0% P_2O_5) (see Figure 3) but this reduction is simply one of dilution. The calculated difference in CaO content of clinkers with LFs 1.02 and 1.06 is 0.8% (variation of the P_2O_5 content does not alter this figure) which is in close agreement with the average free CaO content (0.9%) of clinkers with LF = 1.06 indicated by Figures 1a, b, c and d. The average experimental value is 0.6%. Thus in practice the Lea and Parker formula could provide a satisfactory proportioning formula to obtain minimum free lime.

The success of the Lea and Parker formula which makes no allowance for P_2O_5 or F is however fortuitous since the microscopic examination (Table 5) revealed a significant deviation with increasing P_2O_5 content from the compound composition predicted by the Bogue equations, (equations 1 to 4 below) although not to the extent predicted by the Nurse equations (equations 5 to 7 below). The main phases identified by X-ray analysis were alite, β-C_2S, $\alpha' C_2S$ and calcium aluminoferrite solid solution (close to C_6A_2F). It is almost certain that C_3A was also present but in quantities below those detectable by X-ray analysis.

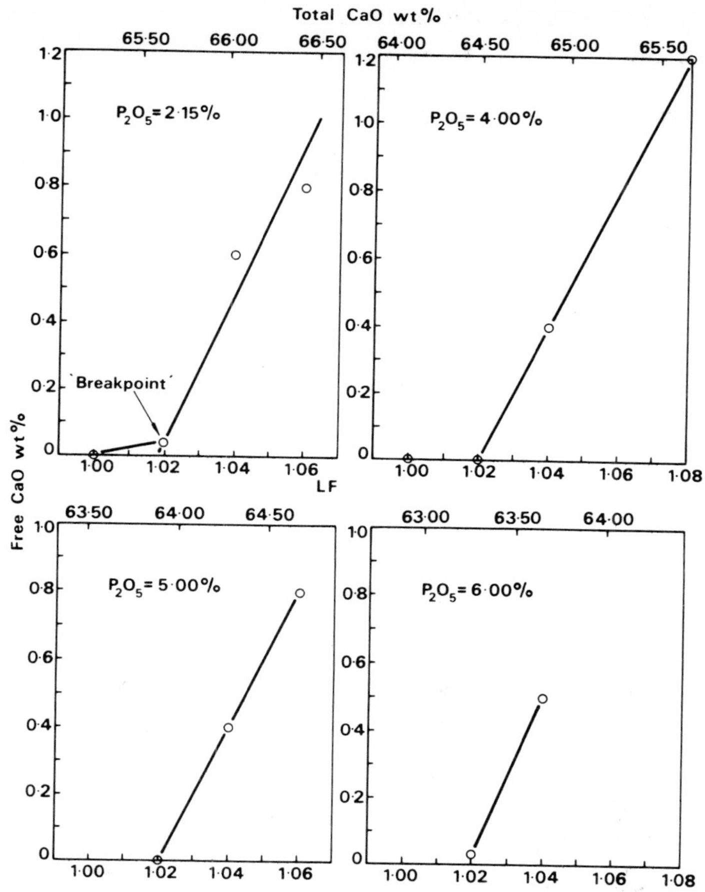

Figure 1 West African raw materials: variation of free CaO with total CaO and lime factor

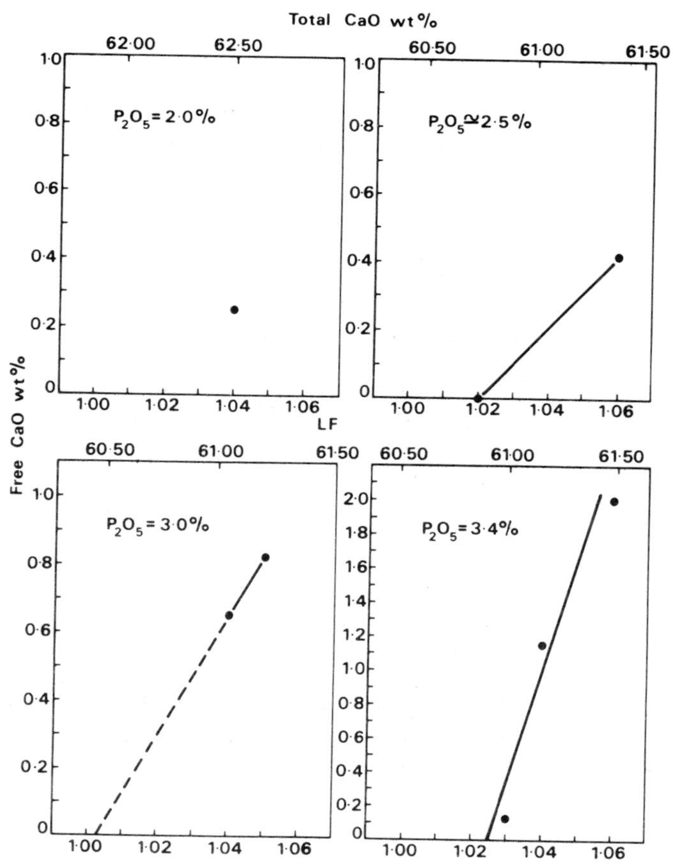

Figure 2 Ugandan raw materials: variation of free CaO with total CaO and lime factor

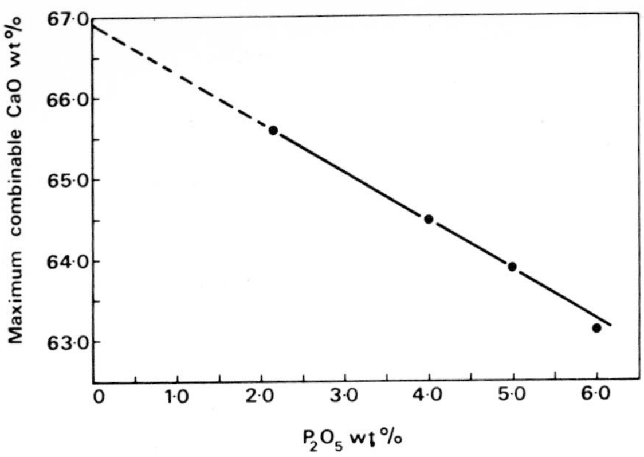

Figure 3 West African raw materials: relation of maximum combinable CaO to P_2O_5 content of clinkers

Bogue equations:

$C_4AF = 3.04\ Fe_2O_3$ (1)

$C_3A = 2.65\ Al_2O_3 - 1.69\ Fe_2O_3$ (2)

$C_3S = 4.07\ CaO - 7.60\ SiO_2 - 6.72Al_2O_3 - 1.43\ Fe_2O_3$ (3)

$C_2S = 8.60\ SiO_2 - 3.07\ CaO + 5.10\ Al_2O_3 + 1.08\ Fe_2O_3$ (4)

Nurse equations:

$C_3S = 4.07\ CaO - 7.60\ SiO_2 - 6.72\ Al_2O_3 - 1.43\ Fe_2O_3 - 9.9\ P_2O_5$ (5)

$C_2S = 8.60\ SiO_2 - 3.07\ CaO + 5.10\ Al_2O_3 + 1.08\ Fe_2O_3 - 3.4\ P_2O_5$ (6)

$PSS = 14.3\ P_2O_5$ (7)

These equations can only be applied to clinkers with $LF \leqslant 1.00$. If LF is greater than 1.00, then $C_2S = 0.0$, and $C_3S = 3.80\ SiO_2$ and Free CaO = CaO- (CaO combined in other phases)- 2.80 SiO_2 where SiO_2 = total SiO_2 if P_2O_5 is not present or (SiO_2-SiO_2 in PSS) if P_2O_5 is present. For clinkers with LF = 1.02 (see Figure 4) the variation in the experimentally determined compound composition with P_2O_5 content can be expressed by the equations (8 to 10) set out below:

Alite	$= 83.292 - 10.128\ P_2O_5$	(8)
Belite	$= -1.877 + 11.875\ P_2O_5$	(9)
Interstitial phase	$= 18.584 - 1.719\ P_2O_5$	(10)

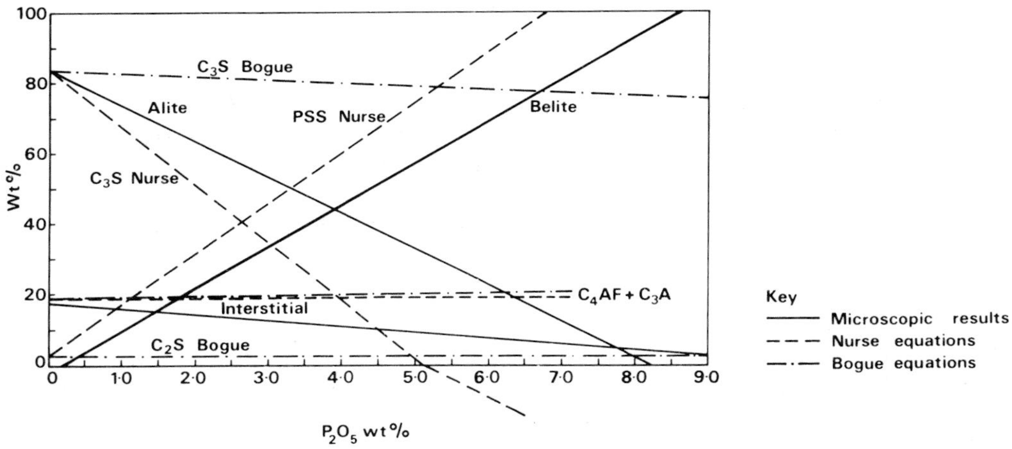

Figure 4 West African raw materials: variation of phase composition of clinkers with LF = 1.02

The negative intercept for belite indicates that some free CaO should be present in a clinker with nil P_2O_5 if it were possible to make such a clinker from these raw materials. The reduction in interstitial content with increasing P_2O_5 content suggests some combination of Al_2O_3, Fe_2O_3 or both, in the phosphatic belite phase. Further evidence for this can be deduced as follows:

1 From the analyses of the clinkers with 2.16-6.0% P_2O_5 it is possible, from the variation with P_2O_5 content, to arrive at a composition for a clinker with nil% P_2O_5 of 20.626% SiO_2, 66.999% CaO, 5.333% Al_2O_3, 2.445% Fe_2O_3 in terms of the four major oxides.

2 If the standard Bogue equations are now applied, and the results of this calculation recalculated to total 100 on the assumption of equal distribution of all the minor components among the phases the following phase composition is obtained: 7.79% C_4AF, 10.48% C_3A, 80.29% C_3S, 1.61% C_2S. This is tolerably close to the composition predicted from the microscopic point count (which is 81.76% alite, 18.24% interstitial) correcting for the apparently negative content of belite. Major inaccuracies are therefore unlikely to occur if it is assumed that the alite is in fact pure C_3S.

3 On this assumption the interstitial phase at 0.0% P_2O_5 would have the overall composition 30.644% Al_2O_3, 14.050% Fe_2O_3, 54.736% CaO, 0.570% SiO_2.

4 Considering now a clinker with 8.0% P_2O_5, applying equations (8) to (10) (giving 2.264% alite, 92.905% belite, 4.831% interstitial) and assuming (a) that the minor components are equally distributed between all phases, (b) that the composition of the interstitial phase is constant, (c) alite = 'pure C_3S'; it can be shown that the belite phase at this point has the composition 8.617% P_2O_5, 20.041% SiO_2, 64.269% CaO, 4.768% Al_2O_3, 2.273% Fe_2O_3 or in molar terms $C_{1.146}\,S_{0.333}\,P_{0.061}\,A_{0.047}\,F_{0.015}$ or $C_{2.000}\,S_{0.581}\,P_{0.106}\,(A, F)_{0.107}$.

If SiO_2 is substituted for the P_2O_5, Al_2O_3 and Fe_2O_3 this is seen to be equivalent to $C_{1.986}S$ and that the formula of the solid solution can be expressed in general terms as $C_2S_{(1-4x)}P_x(A, F)_x$, ie there is a 1:1 substitution of Al or Fe with P in the C_2S lattice. The solid solution can also be represented as being on the hypothetical joint C_2S-$C_8P(A, F)$. A large number of assumptions have been made to arrive at this attractively simple formula and the existence of this solid solution has not been proved. However, the indirect evidence for some form of combined solid solution of P_2O_5, Al_2O_3 and Fe_2O_3 seems reasonably strong. An investigation of the clinkers by electron microprobe analysis has been started together with an examination of mixes made from pure reagents in the appropriate part of the system CaO-SiO_2-Al_2O_3-Fe_2O_3-P_2O_5 to test the new hypothesis. Taking into account the proved existence[4,5,6] of solid solutions on the join C_2S-C_3P it is likely that the solid solutions found in practice will show a more complex substitution of P_2O_5, Al_2O_3 and Fe_2O_3 than can be represented by the formula $C_2S_{(1-4x)}P_x(A, F)_x$.

In the absence of knowledge of any end member to the $C_2S_{(1-4x)}P_x(A, F)_x$ series it is not possible to derive even tentative alternative equations to those derived by Nurse for predicting phase composition on the basis of the C_2S-C_3P solid solution series. The problem is complicated by the probability that the Al:Fe ratio will correspond, within certain limits no doubt, to the Al:Fe ratio of the clinker. A thorough study of the relevant parts of the system CaO-SiO_2-Al_2O_3-Fe_2O_3-P_2O_5 is necessary before such equations can be prepared. The presence of fluoride and other minor components may also influence the nature of the solid solution formed.

Ugandan raw materials

Figures 2a, b, c and d show the relationship of uncombined (free) CaO and total CaO for the first series of clinkers. The analytical free CaO results are consistent with the clinkers being fully reacted, ie the free lime detected does not arise from inhomogeneity of the raw meal. As for the West African raw materials total lime combination corresponds with lime factors of 1.00 to 1.02 showing that for practical purposes the standard formula of Lea and Parker may be used for proportioning these particular phosphatic raw materials.

The main phases identified by X-ray analysis were alite, $\alpha'C_2S$ and C_4AF solid solution. The alite X-ray diffraction pattern was very similar to that of the trigonal polymorph of C_3S, a finding typical of cement clinkers made from the phosphatic fluorinated Ugandan raw meal. Two forms of $\alpha'C_2S$ were detected, (a) with a pattern similar to the α'm pattern given by potassium stabilised $\alpha'C_2S$, (b) with a pattern similar to that of α'_L stablised with phosphate[7,8]. The first form was probably present only in small quantities and the intensities of the X-ray diffraction patterns were weak compared with the moderately strong patterns of the P_2O_5 stablished $\alpha'C_2S$. A correlation exists between the X-ray diffraction data and the quantitative point count results summarised in Table 9: it is possible that the irregular shaped C_2S phase which etched blue in HF vapour and the rounded phase which remained unetched, represent respectively, the phosphate stabilised $\alpha'C_2S$ and the potassium stabilised $\alpha'C_2S$, although the latter could be $C_{12}A_7$ or a related phase.

The microscopic examination showed progressive decrease in C_3S content and corresponding increase in C_2S as the P_2O_5 level was raised from 2.0 to 3.3% P_2O_5. In contrast to the finding for the West African clinkers however, the interstitial content increased sharply with increased P_2O_5 content (due to the association of Fe_2O_3 and P_2O_5 in the high P_2O_5 limestone) and the rate of fall of C_3S content is much faster (about 37% C_3S per 1.0% P_2O_5) indicating significant solution of silicates in the interstitial phase (see Figure 5). The microscopic results can be represented by the equations 11 to 13.

Alite	= 146.335–36.946 P_2O_5	(11)
Belite	= –33.784+24.984 P_2O_5	(12)
Interstitial	= –12.551+11.962 P_2O_5	(13)

Unlike the equations derived for the West African clinkers, these equations can only be applied over the limited P_2O_5 range of the clinkers examined. If clinkers with less than 2% P_2O_5 were made it is probable that the alite contents would be fairly close to those predicted by the Bogue equations, ie would be higher than those predicted by the Nurse equations. Such behaviour would comply with the finding of Welch and Gutt[5] that the deliberate addition of calcium fluoride to clinkers containing 2.35% P_2O_5 resulted in an increase in alite content. This finding has been confirmed in later work for clinkers of similar composition but clinkers with higher P_2O_5 content (above about 3%) have not been examined so that it is not possible to say whether this first increase in C_3S is usually followed by the rapid decline in C_3S content encountered with the Ugandan clinkers. The marked increase of Fe_2O_3 with P_2O_5 must also be taken into account as there will be increased liquid formation at clinkering temperatures as P_2O_5 is increased leading to solution of SiO_2, and probably also P_2O_5 and their preservation in a glassy phase if there is rapid cooling. The phase relations governing the clinkers are clearly extremely complex and the system CaO–SiO_2–P_2O_5–Al_2O_3–Fe_2O_3–CaF_2 needs a thorough and careful investigation before general rules and guidelines can be established.

Electron probe analysis of clinker UM28 containing total of 0.85% F gave the following approximate analysis for fluorine: C_3S=0.5, C_2S=0.2, Matrix 0.2%.

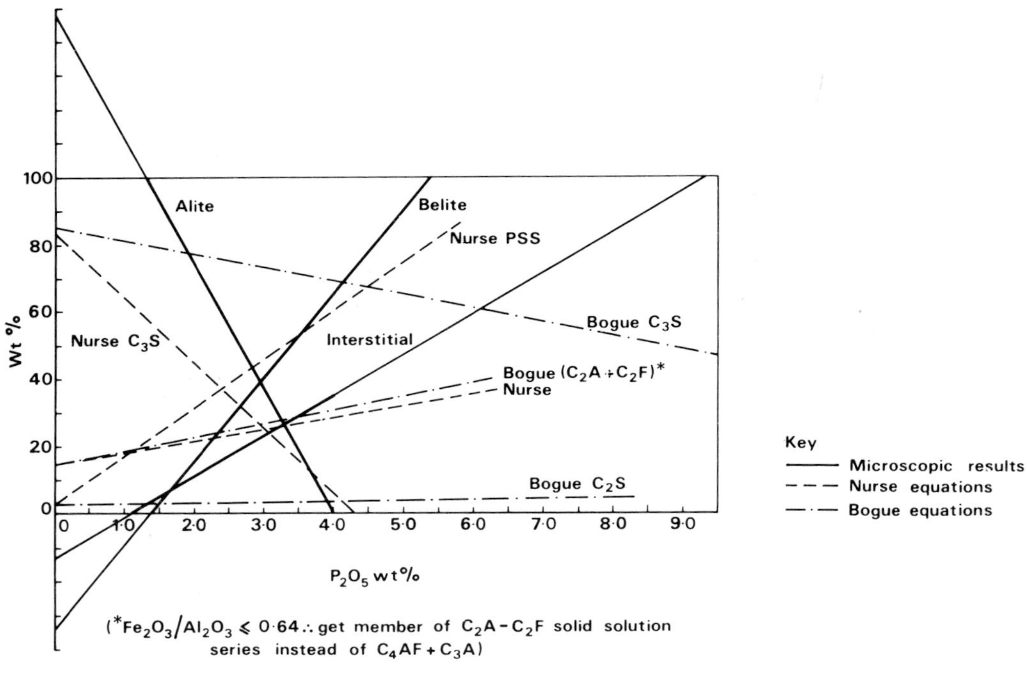

Figure 5 Ugandan raw materials: variation of phase composition of clinkers

CLAY/LIMESTONE PROCESS: RESULTS OF COMPRESSIVE STRENGTH TESTS

West African raw materials

The microscopic examination of the second series of clinkers made from the West African raw materials showed the same decrease of C_3S content with P_2O_5 as was seen in the first series of clinkers. The clinkers were all well burnt and there was reasonable agreement between calculated and experimentally determined P_2O_3 contents. The results of chemical analysis for P_2O_5, F, SO_3 and free CaO are given in Table 3; the results for surface area, setting times and soundness appropriate to BS12:1958 are given in Table 6. The results of compressive strength tests are shown in Figures 6 and 7.

All but one of the cements gave 28-day strengths comparable to normal UK ordinary Portland cements but only those with 2.5% P_2O_5 or less gave completely satisfactory results at other ages of test. Although the cement containing 2.5% P_2O_5 failed marginally the 3-day strength requirements (15 MN/m^2) of BS12:1958, the strength could have been raised to a more satisfactory level by finer grinding to about 340 m^2/kg which is more typical of British OPCs. Similarly the strengths of the cements with 3.0% P_2O_5 could be improved by finer grinding. All

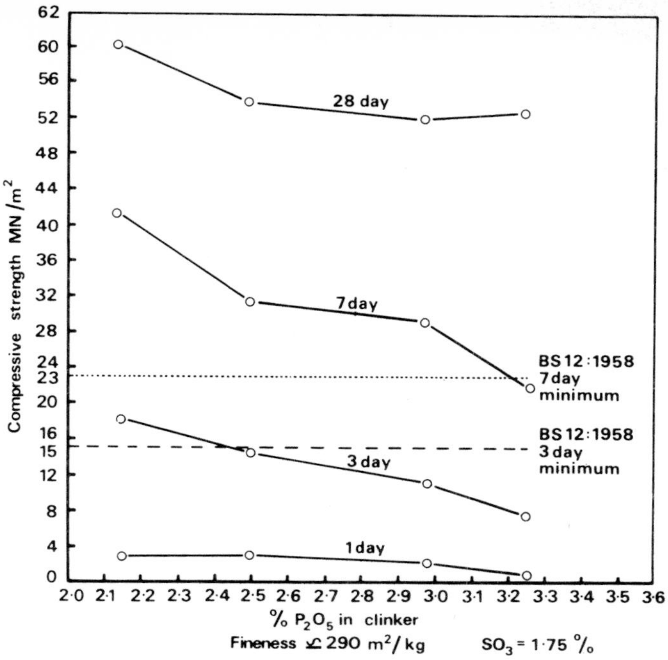

Figure 6 Effects of P_2O_5 on compressive strength of West African cements. 70.6mm mortar cubes tested to BS:12:1958

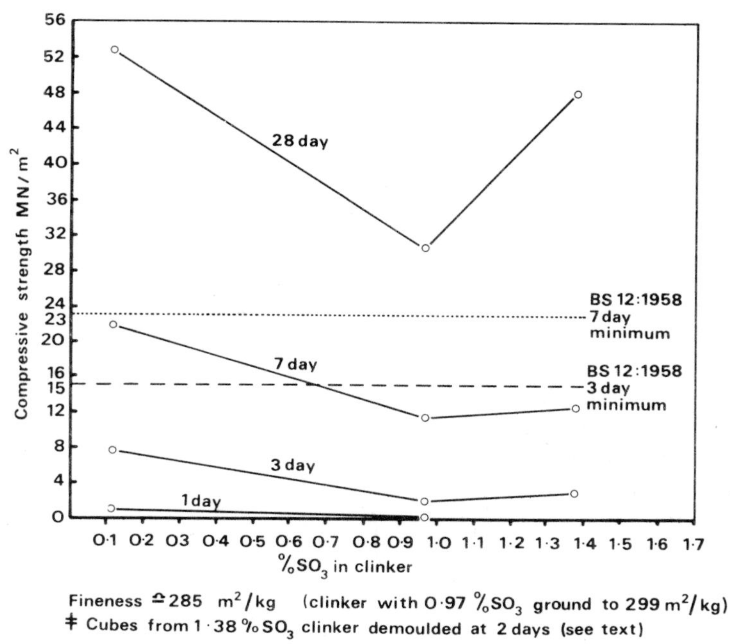

Figure 7 Effect of SO_3 on co compressive strength of West African cements containing 3.4% P_2O_5 70.6mm mortar cubes tested to BS:12:1958

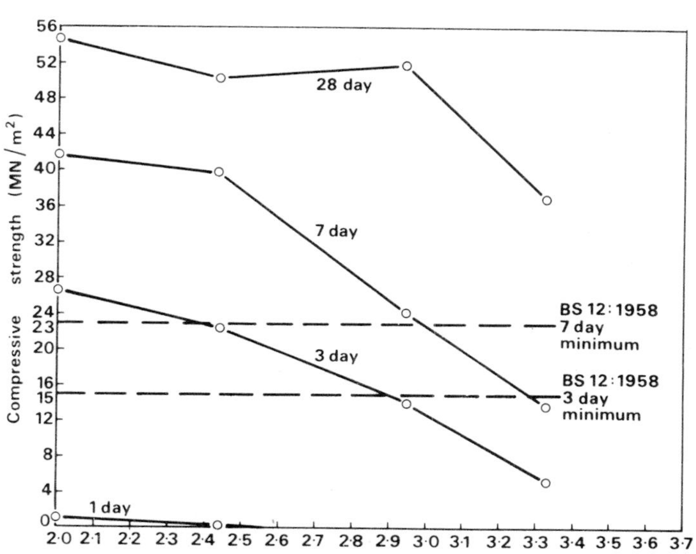

Figure 8 Effect of P_2O_5 on compressive strength of cements made from Ugandan raw materials. 70.6mm mortar cubes tested to BS:12:1958

Note. Typical British OPC (348 m^2/kg) gives 12, 29, 42, 54 MN/m^2 at 1, 3, 7, 28 days respectively.

cements containing more than 2.5% P_2O_5 failed the 3-day strength requirements and that containing 3.4% P_2O_5 also failed at 7 days.

The presence of deliberately-added SO_3 in the clinker produced a markedly deleterious effect upon the compressive strength of the cements, particularly at early ages of test. The presence of 0.7% SO_3 in clinker containing 2.5% P_2O_5 produced a lowering of strength of about 40% at 3 days. These results appear to be in line with the known sensitivity of fluorinated phosphatic clinkers to SO_3 [1,5].

All the cements gave the extremely low strength at 24 hours (in all cases less than 3 MN/m^2) characteristic of phosphatic-fluorinated cements. These low strengths at 24 hours were accompanied by initial setting times of not less than $5\frac{1}{2}$ hours and in the case of cements with more than 3.0% P_2O_5 no set was obtained before the 10-hour limit for the final set. The final set of cements with 3.0% or less P_2O_5 was about 10-hours. No attempt was made to determine the amount of gypsum which should be added to give the optimum strength results and setting times The deleterious effects (see Figure 7) of the 'SO_3' added to some of the clinkers does not necessarily indicate that increasing the amount of gypsum added during the grinding would produce a lowering of strength. The effect of sulphate ions combined in the clinker and originating from the raw material or fuel is not always comparable to the effect of sulphate added during grinding of the fired clinker [9].

The expansion of all the cements in the Le Chatelier test was satisfactory, however, the cements with the higher P_2O_5 contents showed a remarkably large expansion during the initial curing at 19°C.

Ugandan raw materials

The effect of phosphate on the compressive strength of cements with a constant level of 3.0% SO_3 is illustrated in Figure 8 which shows the decrease in strength caused by increased P_2O_5 content, there being a near-linear relationship at most ages of test. Cements containing 2.00 and 2.44% P_2O_5 satisfied the BS12 strength requirements whereas the cement with 2.9% P_2O_5 marginally failed the 3-day limit but just passed at 7 days. The cement with 3.3% P_2O_5 failed to satisfy, by a sizeable margin, both the 3-day and 7-day limits although a substantial recovery in strength did occur at later ages. The two outstanding features, characteristic of fluorinated phosphatic cements, were the extremely low one-day strengths and the complete recovery to normal strength which had occurred at 28 days; all but one cement (P_2O_5 = 3.33%) gave 28-day results comparable to typical British OPCs. This reduction in strength with increasing P_2O_5 is in line with the changes in phase composition (Table 9, Figure 5). As the C_3S content of the clinkers decreased from above 60% to below 20%, with corresponding increases in the weakly hydraulic $\alpha'C_2S$ and the interstitial, the 28-day strength, for example, dropped from ~20 MN/m^2 to 6 MN/m^2.

The only cement to satisfy the requirement of a final setting time of less than 10 hours was that containing 2.44% P_2O_5. The cements with 2.00 and 3.33% P_2O_5 failed to show an initial set within 9-10 hours. This inevitably led to the problems encountered in demoulding mortar cubes for strength testing. However with the exception of the 2.0% P_2O_5 cement, the setting times increased with P_2O_5 content as experienced in practice by the UCI. The test of setting time was carried out at 18°C and a higher temperature of curing (eg 30°C), permissible where the cement is to be used under tropical conditions as in Uganda, would produce lower setting times.

The expansion of the three cements containing ~2.0, 2.5 and 3.0% P_2O_5 in the Le Chatelier test was satisfactory, however, the cements with the higher P_2O_5 contents expanded several millimetres during the initial curing in water at 19°C. The cement with 3.3% P_2O_5 completely disintegrated on boiling the specimen in water due to the very slow rate of set.

These results are also in line with and substantiate the findings from the cement production at Tororo over the last few years where cements containing up to 2.9% P_2O_5 have satisfied British Standards requirements.

CEMENT/SULPHURIC ACID PROCESS

The work at the Building Research Station on the use of by-product calcium sulphate from the manufacture of phosphoric acid has been reported in detail elsewhere [2]. The paper reviewed earlier work on the combined process for manufacture of cement and sulphuric acid and the problems likely to arise from the use of phosphogypsum in the process, due to presence of phosphate and fluoride, were identified. The results of successful laboratory and pilot scale experiments in which clinkers with up to 1.5% P_2O_5 were prepared were reported. The results agreed generally with the known effects of P_2O_5 and fluorides on cement quality. Cements were obtained giving good strength at later ages but with a tendency towards low strength at 1 and 3 days. The inevitable presence of small quantities of residual sulphate in the clinkers tended to reinforce the adverse effects of phosphate and fluoride and some cements were obtained which failed to give useful strength even at 28 days.

It was concluded that Portland cement can be made from phosphogypsum and clay by the cement/sulphuric acid process. The essential requirements are that phosphate must be kept below 2.5% P_2O_5 and that fluorine should be at an optimum level which must be established for each assemblage of raw materials. Since sulphate residual in the clinker influences its cementing quality, this also must be controlled as indeed is the case when anhydrite is the raw material. The magnesia level has to be ascertained. A small amount, up to 2% is likely to be necessary otherwise the formation of C_3S might be hindered by the combined action of aluminium and sulphate[10]. It is self-evident that the use of phosphogypsum from the new processes of phosphoric acid manufacture - in which hemihydrate or gypsum of low phosphate content is produced - is preferable, but phosphogypsums from the older processes with up to 1.3% P_2O_5 can be used. To obtain the best results the lime saturation factor should be kept as high as is practicable. Relatively low strengths at one day are to be expected. The fluoride present in the phosphogypsum will act beneficially as a mineraliser as long as it is not present in excess and in some cases where the fluoride level is low, and the phosphate level high, CaF_2 addition may be beneficial.

CONDUCTION CALORIMETRIC STUDIES

The heat evolution during hydration of a number of phosphatic cements has been studied using a calorimeter of the type developed by Forrester[11] and described previously[9].

Three cements prepared from the West African materials (RM 22, 23, 24) and one prepared from the Ugandan (UM11) were hydrated in the conduction calorimeter (temperature of hydration 25°C, water/cement = 0.6) and the rates of heat evolution monitored for three days. The results are shown in Figure 9, and it is apparent that considerable differences were observed. The

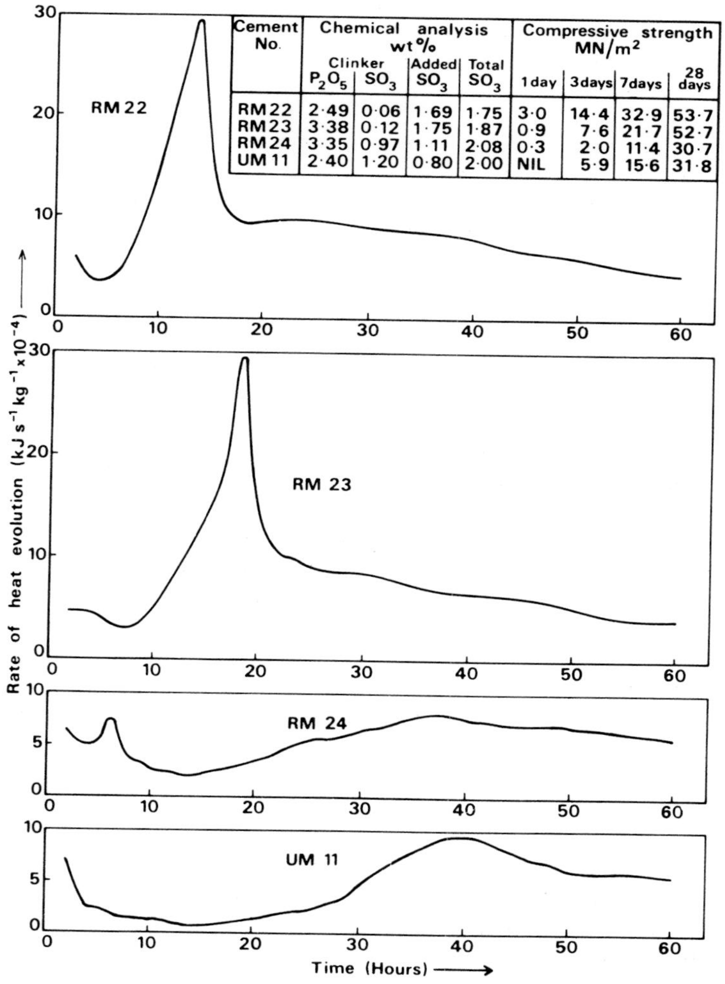

Cement No.	Chemical analysis wt %				Compressive strength MN/m²			
	Clinker P_2O_5	Clinker SO_3	Added SO_3	Total SO_3	1 day	3 days	7 days	28 days
RM22	2·49	0·06	1·69	1·75	3·0	14·4	32·9	53·7
RM23	3·38	0·12	1·75	1·87	0·9	7·6	21·7	52·7
RM24	3·35	0·97	1·11	2·08	0·3	2·0	11·4	30·7
UM 11	2·40	1·20	0·80	2·00	NIL	5·9	15·6	31·8

Figure 9 Rates of heat evolution for phosphatic cements

phosphate, clinker SO_3 and added SO_3 (as gypsum) contents of the four cements, together with the results of compressive strength tests, are given in the table attached to Figure 9.

The two cements with clinker SO_3 contents of about 1% (RN 24 and UM 11) do not show the sharp peak which appears with the two cements having very little clinker SO_3, although there are signs of this peak beginning to show in RM 24 before being brought to a halt. It is not clear whether this peak is due to the acceleration of alite or belite hydration by added gypsum, or to the interaction of gypsum with the interstitial phases, although microscopic examination of the clinkers suggested that the C_3A contents were low. In RM 24 and UM 11 the whole hydration process is severely retarded and modified compared with a normal OPC.

The effect of phosphate can be seen by comparing RM 22 (2.49% P_2O_5) with RM 23 (3.38% P_2O_5). The increase in phosphate level results in a retardation of the appearance of the heat evolution peak.

All four cements gave low compressive strengths at early ages this being particularly noticeable in RM 24 and UM 11 which have no sharp heat evolution peaks. RM 22 and RM 23 recover to give good strengths at 28 days while RM 24 and UM 11 do not recover strength to the same extent. It should be remembered, however, that cement UM 11 was prepared from entirely different raw materials to the other three, has very different chemical and physical characteristics, and the strengths are estimated from the results obtained on 12.7 mm 1:3 mortar cubes.

DISCUSSION AND CONCLUSIONS

Studies of the manufacture of phosphatic Portland cements made since the 5th International Symposium have fully confirmed the earlier conclusions [1, 2, 4] that the maximum amount of P_2O_5 that can be tolerated in the cement, if adequate compressive strength is to be obtained, is approximately 2.5%. They have also reafirmed the important combined effects of phosphate, fluoride and sulphate in the clinker on cement quality. While the new results confirm that the phase equilibria in phosphatic cements can in many instances be explained by analogy with the systems CaO-SiO_2-P_2O_5 and CaO-SiO_2-P_2O_5-CaF_2 they also suggest the possibility that solid solutions of dicalcium silicate containing P_2O_5, Al_2O_3 and Fe_2O_3 in combination can be formed in some clinkers. An investigation has commenced of the subsystem C_2S-Al_2O_3-Fe_2O_3-P_2O_5 to explore the possible existence of such solid solutions of C_2S in the pure system and to examine in particular the hypothesis that these are of the generalised form $C_2S_{(1-4x)}P_x(A, F)_x$ - C_3P.

As explained above there are at present insufficient data on which to base formulae to calculate the phase composition of clinkers, although the deviation of actual composition from that predicted by the Bogue and Nurse equations indicates the need for improved formulae. It is, however, apparent from the unique mineralogical characteristics of the high Fe_2O_3 Ugandan clinkers that a single set of equations is unlikely to be sufficient to predict the phase composition of all phosphatic clinkers, particularly when significant quantities of fluoride are present. Nor are there sufficient data on which to base new lime saturation and proportioning formulae. The need for such modified formulae however is less urgent since the standard Lea and Parker formula proved useful in proportioning two very different sets of raw materials. This applicability of the simple formula is, however, fortuitous since as the phosphate content increases, the Bogue formulae predict increasingly false mineralogical compositions.

In practical terms the present work on diverse phosphatic raw materials has: (1) confirmed the need to control the P_2O_5 content of cement to below 2.5% and to control fluoride and residual SO_3 levels very closely, and (2) shown that the Lea and Parker lime saturation factor can provide fortuitously a practical basis for proportioning phosphatic raw materials although this factor makes no allowance for the effects of P_2O_5 and fluoride phase equilibria.

SUMMARY

Earlier studies of high temperature phase equilibria in the system CaO-SiO_2-P_2O_5 have provided a fundamental basis for understanding the role of phosphate in the manufacture of Portland cement from phosphatic raw materials. The special importance of fluorides as mineralisers in the manufacture of phosphatic cements was also elucidated. Phosphatic raw materials of diverse origin have recently been assessed for cement manufacture and the results have confirmed that the maximum practical level of P_2O_5 which can be tolerated in Portland cement, if cement of acceptable quality is to be obtained, is about 2.5 per cent. Above this limit, setting times become unacceptably long and early strength (1 and 3 days) very low, although strength at later ages (28 days) may remain acceptable. New results also indicate that for a wide range of phosphatic raw materials the Lea and Parker lime saturation factor ($CaO/(2.8SiO_2 + 1.18Al_2O_3 + 0.65Fe_2O_3)$) should serve as a practical guide to raw meal proportioning despite the fact that it makes no allowance for the effects of P_2O_5 and fluoride on high temperature phase equilibria. This applicability of the LSF is fortuitous since the compound composition predicted by the related Bogue equations is incorrect.

Mineralogical examination of clinkers suggests the existence of dicalcium silicate solid solutions with P_2O_5, Al_2O_3 and Fe_2O_3 in combination and the possibility that these are of the general form $(CaO)_2\ (SiO_2)_{1-4x}\ (P_2O_5)_x\ (Al_2O_3,\ Fe_2O_{3x}$, ie substitution of P^{5+} $(Al, Fe)^{3+}$ for $2Si^{4+}$, is being investigated.

The heat evolution of phosphatic cements has been followed using continuous conduction calorimetry.

ACKNOWLEDGEMENTS

The authors would like to thank G J Osborne and J D Matthews for their help in this investigation.

REFERENCES

1 **Gutt, W.** Manufacture of Portland cement from phosphatic raw materials, Proc 5th International Symposium on the Chemistry of Cement, Tokyo 1968, Vol 1, pp 93-105. Cement Association of Japan, Tokyo 1970.

2 **Gutt, W and Smith, M A.** The use of phosphogypsum as a raw material in the manufacture of Portland cement. Cement Technology, Vol 2, 1971: No 2, pp 41-50, and No 3, pp 91-93 and 100.

3 **Gutt, W and Nurse, R W.** Phase composition of Portland cement clinker. Basic paper presented at the 6th International Symposium on the Chemistry of Cement, Moscow 1974. (To be BRE Current Paper CP 96/74)

4 **Nurse, R W.** The effect of phosphate on the constitution and hardening of Portland cement. Journal of Applied Chemistry (London), Vol 2, No 12, 1952, pp 708-716.

5 **Welch, J H and Gutt, W.** The effect of minor components on the hydraulicity of the calcium silicates. Proc 4th International Symposium on the Chemistry of Cements, Washington 1960, Vol 1, pp 59-68. US Department of Commerce, National Bureau of Standards, Monograph 43, Washington 1962.

6 **Gutt, W.** High temperature phase equilibria in polycomponent systems. London University 1966 (PhD thesis).

7 **Gutt, W and Osborne, G J.** The effect of potassium on the hydraulicity of dicalcium silicate. Cement Technology, Vol 1, No 4, 1970, pp 121-5.

8 **Midgley, H G.** The polymorphism of calcium orthosilicate. Supplementary paper presented at the 6th International Symposium on the Chemistry of Cement, Moscow 1974.

9 **Smith, M A and Matthews, J D.** Conduction calorimetric studies of the effect of sulphate on the hydration reactions of Portland cement. Cement and Concrete Research, Vol 4, 1974, pp 45-55, 851.

10 **Gutt, W and Smith, M A.** Studies of the role of calcium sulphate in the manufacture of Portland cement clinker. Trans British Ceram Society, Vol 67, No 10, 1968, pp 487-509. (BRS Current Paper CP 89/68.)

11 **Forrester, J A.** A conduction calorimeter for the study of cement hydration. Cement Technology, Vol 1, No 3, 1970, pp 95-99.

The phase composition of Portland cement clinker (CP 96/74)

W. Gutt and R.W. Nurse

INTRODUCTION

Since Lea and Parker reported on the phase equilibria of the system CaO-SiO_2-Al_2O_3-Fe_2O_3 in the 1930s there has been a progressive delineation of the systems relating to cement manufacture with increasing emphasis during recent years on phase equilibria involving minor components and on the detailed polymorphism and solid solutions of the major cement minerals. The early workers in cement chemistry had very practical aims in seeking to establish the basic phase equilibria of Portland cement clinker: they wanted to know the relation between the chemical composition and the properties of Portland cement, and they wanted criteria to guide the proportioning of cement raw materials. Current work on phase equilibria should still be judged in relation to these two basic aims. The problems of composition and proportioning are increasingly concerned with cements made from low-grade or impure raw materials and phase equilibria hitherto of only minor significance, for example those involving sulphates or fluorides, are gaining in importance with the introduction of special cements based upon such complex compounds as $3(CaO.Al_2O_3).CaSO_4$ and $11CaO.7Al_2O_3.CaF_2$.

For the period reviewed, attention has been paid to the detailed phase equilibria of the solid solutions found in clinker rather than to the main systems themselves. In the case of the silicate phases this is complicated by the complex polymorphism. The polymorphism of C_3S was reviewed at the Tokyo Symposium by Guinier and Regourd[1] who described six forms. Uchikawa and Tsukiyama[2] have since suggested that there are ten polymorphs. The existence of high and low α' forms of C_2S is established, but some doubt exists as to whether there is more than one β modification.

A number of workers have attempted, with at best limited success, to use the better knowledge of solid solution limits and actual composition of the phases in Portland cement clinker, to improve upon the Bogue equations for calculating the compound composition of clinker.

THE SYSTEM CaO-SiO_2

Trömel, Fix and Heinke[3] have produced a new version of the high-lime region of the system CaO-SiO_2. They confirmed the low temperature dissociation of C_3S into C_2S and CaO, and suggested that, as found by the earliest workers, C_3S decomposes in the same way at higher temperatures. This is in disagreement with Welch and Gutt[4] and Gutt[5] who studied the liquidus and found that C_3S melted incongruently to CaO and liquid and that three compositions studied melted to yield C_3S in contact with liquid. Trömel et al did not study the liquidus but held C_3S and neighbouring compositions for long periods in the high temperature X-ray camera, suggesting C_2S and CaO as decomposition products. In fact, however, their X-ray data showed very little or zero CaO, although very strong reflections for C_2S developed. Since the cubic CaO gives very strong X-ray reflections and for full decomposition some 24.6 per cent by weight of CaO should form, this is a serious discrepancy, which the authors explain by solid solution of CaO in C_2S. The maximum solubility according to their own data is 6 per cent by weight of CaO, so that this cannot fully explain the results. (Unless otherwise stated, in this paper all percentages given are by weight.) It seems possible that during the long heating period the composition of the specimen is changed, CaO being volatilised from the very exposed X-ray specimens forming C_2S, and that smaller amounts of CaO condense on the cooler part of the specimen holder. The boundaries of the C_3S field in ternary systems[5,6,7] such as CaO-SiO_2-P_2O_5 and CaO-C_2S-CaF_2, support the conclusions of Welch and Gutt that C_3S melts incongruently. The authors find that the solubility of CaO in αC_2S at 1850°C is 6 per cent and at 1350°C $\alpha' C_2S$ takes up 1 per cent of CaO. The solid solution of CaO in C_2S has also been studied[8] by centrifugal separation of the constituents of fused mixtures of CaO and SiO_2. Further confirmation of the low-temperature decomposition of C_3S was obtained by Butt, Timashev and Kanshanskii[9] using single crystals.

Woermann, Eysel and Hahn[10] failed to find any evidence for solid solutions of C_3S in the system $CaO-SiO_2$.

TRICALCIUM SILICATE AND ITS SOLID SOLUTIONS

In a series of five reports Woermann et al[10] delineate and discuss the phase field of alite in the systems $CaO-MgO-Al_2O_3-SiO_2$ and $CaO-Al_2O_3-Fe_2O_3-SiO_2$. It is pointed out that since C_3S is unstable below about 1200°C, and the inversion temperatures of all six main polymorphs are below this temperature, then the solubilities found refer to the rhombohedral high temperature form. In the systems studied, this form was not stabilised at room temperature, but the forms of lower symmetry were stabilised by quenching. No solid solution was found in the pure system $CaO-SiO_2$.

In the ternary system $CaO-Al_2O_3-Fe_2O_3$ the limit of solid solution of Fe ions at 1550°C was found to be 1.1 per cent as compared with 1.05 per cent found by Fletcher[11]. Half the Fe ions substituted for Si and the other half for Ca.

Three types of solid solutions were discovered in the system $CaO-Al_2O_3-SiO_2$; Al ions substituting for Si and Ca, or occurring interstitially. The distribution of ions between these sites depends on the Ca available . The maximum solubility of Al_2O_3 in C_3S was 1.0 per cent, and was independent of temperature.

The solid volume representing the C_3S solid solution in the system $CaO-MgO-Al_2O_3-SiO_2$ at 1550°C widens out towards the MgO apex[10], showing an increase in the Ca incorporated as the MgO increases. The combined limits of solid solution are unchanged at 2 per cent MgO and 1 per cent Al_2O_3, in agreement with Midgley and Fletcher[12]. Temperature has no effect on the Al_2O_3 solubility, but MgO is increasingly soluble with rising temperature. In the system $CaO-MgO-Fe_2O_3-SiO_2$ a small amount of Ca is incorporated into the C_3S lattice for higher MgO contents. The extent of solid solution is 2 per cent MgO and 1.1 per cent Fe_2O_3. Al and Fe ions occupy similar lattice positions in C_3S solid solutions in the system $CaO-Al_2O_3-Fe_2O_3-SiO_2$ and the solubilities are interdependent. Thus the presence of 0.5 per cent Fe_2O_3 reduces the solubility of Al_2O_3 from 2.0 per cent to 1.6 per cent.

In addition to this work on the main cement clinker systems there has been continued effort on those involving various minor components. Studies of systems containing phosphate, sulphates and fluoride are discussed separately below. Considerable interest has been shown in the system $3CaO.SiO_2-3CaO.GeO_2$ because a complete series of solid solutions is formed[13], and the germanate undergoes the same polymorphic transformations as the silicate. The germanate melts incongruently at 1880°C and decomposes into $2CaO.GeO_2$ and CaO below 1335°C (Shirvinskaya, Grebenshchikov and Toropov[14]).

The maximum solubility of Cr in C_3S is 0.9 per cent calculable as Cr_2O_3 (Butt, Timashev and Malozhon[15]). Yellow crystals obtained at 600°C in air contained chiefly Cr^{6+}; in air at 1500°C green crystals were formed containing a mixture of Cr^{6+} and Cr^{3+}, while in argon at 1800°C the blue solid solution was mainly Cr^{3+}.

The solubilities at 1500-1600°C of Mn_2O_3 and TiO_2 are 0.7 per cent and 0.6 per cent respectively[16].

The limit of solubility of BaO in C_3S at 1600°C is 1.5 mol per cent (Kurdowski and Wollast[17]) and any higher quantity causes partial decomposition of C_3S into C_2S and CaO. In the equilibrium state at 1600°C the C_2S formed contains approximately 1 mol per cent BaO.

The solid solution limits determined in phase equilibrium studies may be compared with those obtained by electron-microprobe analysis of alite in Portland cement clinker. For example Midgley[18] obtained 1.1 per cent Al_2O_3, 1.4 per cent Fe_2O_3, 0.9 per cent MgO, 0.3 per cent Na_2O and 0.1 per cent K_2O.

DICALCIUM SILICATE AND ITS SOLID SOLUTIONS

Midgley et al[19] have made an extensive study of the polymorphism of dicalcium silicate. Mineralogical investigations by high temperature X-ray diffraction (XRD) and by differential thermal analysis (DTA) of pure calcium orthosilicate (Ca_2SiO_4) have shown that, on heating, the polymorphic forms encountered are: γ, room temperature to 711°C; α'_m, 711 to 979°C; α'_L, 979 to 1177°C; α'_H, 1177 to 1447°C; α above 1447°C. On cooling the changes are α to α'_H to α'_L to α'_m and the α'_m is supercooled to 676°C below which the metastable form exists to room temperature.

The unit cell dimensions were determined as:

γ: a = 5.082, b = 11.225, c = 6.780Å, space group Pbmn

α'_m: a = 5.42, b = 6.85, c = 9.50Å, β = 91.3°, space group P2, $P2_1$ or Pm

α'_L : a = 11.35, b = 18.45, c = 6.76Å, space group Pmcm

α'_H : a = 5.60, b = 9.32, c = 6.90Å, space group Pmcm.

The system Ca_2SiO_4-$CaNaPO_4$ proved a suitable system to study the polymorphism of calcium orthosilicate stabilised at room temperature by substituting ions. The β polymorph has been stabilised at room temperature for 30 years, by 0.5 to 10 per cent $CaNaPO_4$; at 12.5 per cent $CaNaPO_4$ there is a mixture of forms, while between 15 and 25 per cent the α'_L polymorph occurs; at 30 to 70 per cent the α'_H polymorph is stable. In all of the forms an increase in the substituting ions alters the unit cell dimensions without changing the symmetry of the polymorph[19].

The monoclinic α'_m polymorph of calcium orthosilicate, Ca_2SiO_4, was found in the pure system CaO-SiO_2 between 711° and 979°C, in the system Ca_2SiO_4-$CaNaPO_4$ and in Ca_2SiO_4 stabilised with 3.6 per cent K_2O. All have very similar unit cells, for example in the pure sample at 900°C, a = 5.42, b = 6.85, c = 9.50Å, β = 91.3°; probable space groups P2, $P2_1$ or Pm (Midgley et al[19]).

It is claimed on crystallochemical grounds[19] using XRD that bredigite, and T-phase formerly identified as a separate compound[5, 20, 21], are α'_L polymorphs of calcium orthosilicate. This discrepancy is discussed further on page 4. Bredigite has a unit cell of: a = 10.98, b = 18.60, c = 6.88Å; T-phase a = 10.72, b = 18.41, c = 6.64Å; α'_L a = 11.18, b = 18.94, c = 6.84Å; all with space group Pmcm. It has also been shown that sodium and phosphorus produce room temperature α'_L stable polymorphs[19].

Potassium, sodium, vanadium, chromium, manganese, P_2O_5 and SO_3 at 0.83, 1.75, 2.20, 1.24, 0.88, 0.08 and 5.0 per cent respectively will completely stabilise the beta polymorph. Aluminium, ferric iron, magnesium and titanium fail to stabilise the beta polymorph[19].

Niesel and Thormann[22] distinguish between chemical and physical stabilisation of polymorphic forms and moreover note that prolonged annealing in the α'_H stability range is required if the $\beta - \gamma$ transformation is to develop fully. They also pay particular attention to the effect of particle size; fine grained C_2S giving inversion on cooling to a β_L form which did not further invert to γ (Lehmann, Niesel and Thormann[23]). Similar ideas are expressed by Chromy[24].

Eremin et al[25] found the maximum solubility of Na_2O, Fe_2O_3, TiO_2 and MgO in C_2S to be 0.6, 1.8, 0.75 and 1.0 per cent respectively. Gutt and Osborne[26] found it necessary to quench preparations containing 3.6 per cent K_2O ($KC_{23}S_{12}$) in order to obtain the α'_L form for hydraulicity tests. Otherwise the β form was obtained.

Fletcher[27] reported on the composition of belite in a Portland cement clinker (OPC) and in a sulphate-resisting Portland cement clinker (SRPC). The two analyses were Al_2O_3 1.7 per cent, Fe_2O_3 0.8 per cent, MgO 0.4 per cent, K_2O 0.3 per cent, Na_2O 0.3 per cent, TiO_2 0.3 per cent, Al_2O_3/Fe_2O_3 2.12, for the OPC, and Al_2O_3 1.1 per cent, Fe_2O_3 1.5 per cent, MgO 0.2 per cent, K_2O 0.5 per cent, Na_2O 0.3 per cent, TiO_2 0.2 per cent, Al_2O_3/Fe_2O_3 0.73, for the SRPC. The ratio of Al_2O_3 to Fe_2O_3 in the belite grain was found to correspond closely with the ratio of Al_2O_3 to Fe_2O_3 in the clinkers (1.90 for OPC and 0.67 for SRPC). The polymorphic form of the belite was not identified.

Midgley and Bennett[28] used electron-microprobe analysis to determine the composition of natural larnite (βC_2S) and bredigite ($\alpha'_L C_2S$) co-existing at Scawt Hill, Northern Ireland. The results of their analyses are given in Table 1. The most significant features of the results are the high magnesia content of the bredigite and the presence of minor amounts of P_2O_5 and SO_3 in both minerals.

Gutt and Smith[29] determined the composition of βC_2S and $\alpha' C_2S$ (no distinction was made between α'_L and α'_H forms) in the systems CaO-SiO_2-SO_3 and CaO-SiO_2-Al_2O_3-SO_3 and also determined the SO_3 content of alite in clinkers with various sulphate contents. They showed both βC_2S and $\alpha' C_2S$ on the hypothetical join C_3S-SO_3 to contain about 6 per cent of SO_3. The βC_2S was formed on the lime-rich side of the C_2S-$CaSO_4$ join and the $\alpha' C_2S$ on the silica-rich side of this join. In the presence of Al_2O_3, the βC_2S was found to contain about 3 per cent of SO_3 and the $\alpha' C_2S$ to contain about 5.5 per cent of SO_3. In a synthetic clinker containing 3.5 per cent of SO_3 the C_2S was found to contain 3.25 per cent of SO_3 and 3.6 per cent of Al_2O_3, and the Al_2O_3 content of the alite phase was increased from 0.7 per cent in a clinker without SO_3 to 1.50 per cent in the clinker with SO_3. Examination of plant clinkers showed the sulphur content

Table 1 Analysis of larnite and bredigite (Midgley and Bennett[28])

	Larnite (wt per cent ± one standard deviation)	Bredigite (wt per cent ± one standard deviation)
CaO	65.5±1.8	59.5±3.0
SiO_2	34.0±2.1	34.3±2.9
MgO	0.05	6.4
Al_2O_3	0.02	0.06
K_2O	0.037±0.015	0.083±0.015
Na_2O	0.250±0.06	0.18±0.06
TiO_2	0.029±0.019	0.001
MnO	0.074±0.028	0.31
Fe_2O_3	0.011±0.003	0.26
P_2O_5	0.48±0.07	0.50
SO_3	0.057±0.005	0.50
BaO	nd	nd
FeO	-	-

nd = not determined

of the alite phase to be related to the total sulphur content of the clinker. The highest value of sulphur recorded in the alite phase was 5.20 per cent SO_3 in a clinker containing 10 per cent total SO_3 (4.6 per cent of SO_3 and 2.16 per cent of S^{2-}). However both in the synthetic clinkers and in the plant clinkers the sulphate was generally concentrated in the institial phases.

THE SYSTEM $CaO-Al_2O_3$

The end member of the ferrite solid solution series, C_2A, has been synthesised at a pressure of 25 000 bars[30]. It is probably isostructural with brown-millerite, C_4AF. The starting materials were $C_{12}A_7$ and C_3A. Imlach, Dent-Glasser and Glasser[31] have further investigated the stability of $C_{12}A_7$. They find that at temperatures between 1200°C and the melting point, $C_{12}A_7$ contains 0.07-0.10 per cent by weight excess oxygen. The excess oxygen may be derived from an oxygen-containing atmosphere or from traces of water in inert atmospheres. Thus the existence of oxygen or water concentrations as an independent variable explains the ternary behaviour of this compound noted by previous workers[32].

Tarte[33] studied the isomorphic replacements in C_3A by diffractometry and found the following limiting compositions, in mol per cent: substitution Al-Fe 10, Ca-Mg 2, CaAl-NaSi 6, CaAl-KSi 5.

THE SYSTEM $CaO-MgO-Al_2O_3-SiO_2$

Biggar and O'Hara have published extensively about this system. The high magnesia regions have been dealt with from the point of view of refractories[34, 35]. Of more importance to cement technology is a study[36] of bredigite and the quaternary compound C_6A_4MS. Biggar identifies bredigite with C_5MS_3, the T-phase of Gutt[20], and does not consider it to be a polymorph of C_2S. In his analysis of the situation Biggar does not consider the subdivision of the α' series of solid solutions into α'_H and α'_L. Midgley[19] considers that both bredigite and T-phase are α'_L polymorphs of C_2S. On the other hand Gutt[5, 21] in his system $C_2S-C_3MS_2$ shows a definite field for C_5MS_3, which co-exists with an α' form of C_2S. This $\alpha' C_2S$ is not identified as α'_H or α'_L, but if Midgley is correct it is presumably α'_H. A decision cannot be made on the present data, and more work is required to follow the α'_H - α'_L transition from pure C_2S into the binary $C_2S-C_3MS_2$.

Biggar considers a number of crystallisation paths for cement clinker and suggests the possibility of both C_5MS_3 and C_6A_4MS occurring in clinker under conditions of fractional crystallisation.

More recently[37, 38] the same authors described melilite crystallisation in the quaternary system. A polythermal section of part of the plane $C_3A-MgO-SiO_2$ is presented in which bredigite (C_5MS_3) is shown as forming the phase assemblages bredigite-C_2S-melilite, melilite-bredigite-merwinite, C_2S-spinel-bredigite-melilite and spinel-merwinite-bredigite-melilite. These assemblages confirm Gutt's binary assemblages of C_5MS_3-merwinite and C_5MS_3-C_2S. It would appear to be undesirable to use the name bredigite for both C_5MS_3 and $\alpha'_L C_2S$, even if as seems likely they form a solid solution series. Gutt[33] found that merwinite metastabily but persistently crystallises from some melilite compositions in the systems C_2AS-C_2MS_2 and C_3MS_2-C_2MS_2-C_2AS.

The composition of the quaternary phase originally thought to be C_6A_4MS has been determined by Midgley using an electron-microprobe analyser[39]. He found it not to be analogous to the pleochroic mineral in high alumina cement, although related to it structurally. The analysis of the magnesium compound was equivalent to $Ca_{35}Mg_4Al_{24}(Al_2O_7)_8\ (AlO_4)_{13}(SiO_4)_5$ while the formula of the mineral from high alumina cement was $Ca_{22}Fe_3Al_{14}(Al_2O_7)_8\ (AlO_4)_4(SiO_4)_2$. Further work on these compounds is required to resolve doubts about their compositions.

The high alumina part of the quaternary system, of interest in connection with refractory high alumina cement, has been studied by Rao[40]. He investigated the sub-system CA-CA_2-MA-C_2AS. Spinel and CA_2 form a simple binary system and MA-CA_2-C_2AS is a true ternary system. The pseudo-system MA-CA-C_2AS contains an area of CA_2 as primary phase. Figure 1 shows the four sides of the quaternary sub-system. The first liquid formation for any composition is at 1475°C, but since the invariant point lies outside the tetrahedron studied, the amount of liquid cannot be calculated. It was considered that the silica content of high alumina cement should be restricted to avoid excessive liquid development.

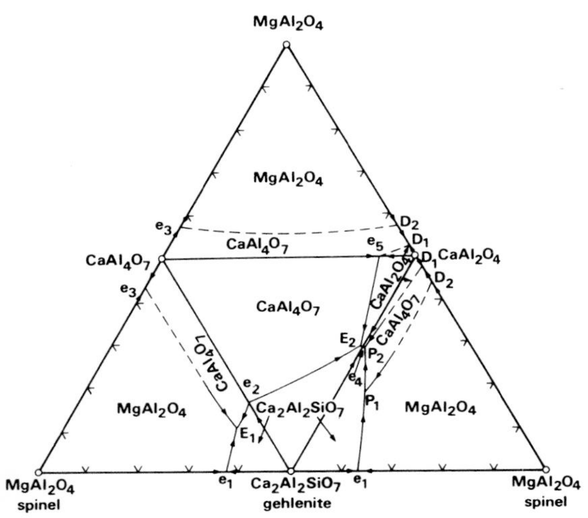

Figure 1 Composition tetrahedron of the quaternary subsystem $CaAl_2O_4$-$CaAl_4O_7$-$Ca_2Al_2SiO_7$-$MgAl_2O_4$ (after Rao[40])

THE SYSTEM CaO-Al_2O_3-Fe_2O_3

Dayal and Glasser[41] traced a series of isothermal planes in this system at 1140, 1175, 1300 and 1330°C. They delineated the liquidus diagram, with C_3A, $C_{12}A_7$, CA, CA_2, C_2F, CF and CF_2 as primary phases; a new ternary compound was also located with a composition on the join 'CF_3' -'CA_3' . The compound melts incongruently to give CA_6 and liquid. There is extensive solid solution in the system, Fe replacing Al and vice versa.

Because of their importance in the production of high alumina cement, Imlach and Glasser[42] studied the sub-solidus phase relations in the system CaO-Al_2O_3-Fe-Fe_2O_3. Extensive solid solution was evident, but no quaternary compounds were discovered. The authors point out that high alumina cement is fused under conditions of low partial pressure of oxygen and that on cooling the clinker is protected from oxidation by an impervious skin of glass and crystalline matter. Thus the ratio of ferrous to ferric iron is preserved and this determines the crystallisation paths.

SYSTEMS WITH FLUORIDES AND PHOSPHATES

Gutt and Osborne[6] studied the system CaO-$2CaO.SiO_2$-CaF_2; their diagram is reproduced in Figure 2. A field is shown for the incongruently melting compound $(3CaO.SiO_2)_3.CaF_2$ and the compound $(2CaO.SiO_2).CaF_2$ occurs at subsolidus temperatures. A very similar diagram was given by Tanaka, Sudoh and Akaiwa[43], but the incongruent compound was identified as $11CaO.4SiO_2.CaF_2$. The bearing of this system on the use of fluoride in cement manufacture

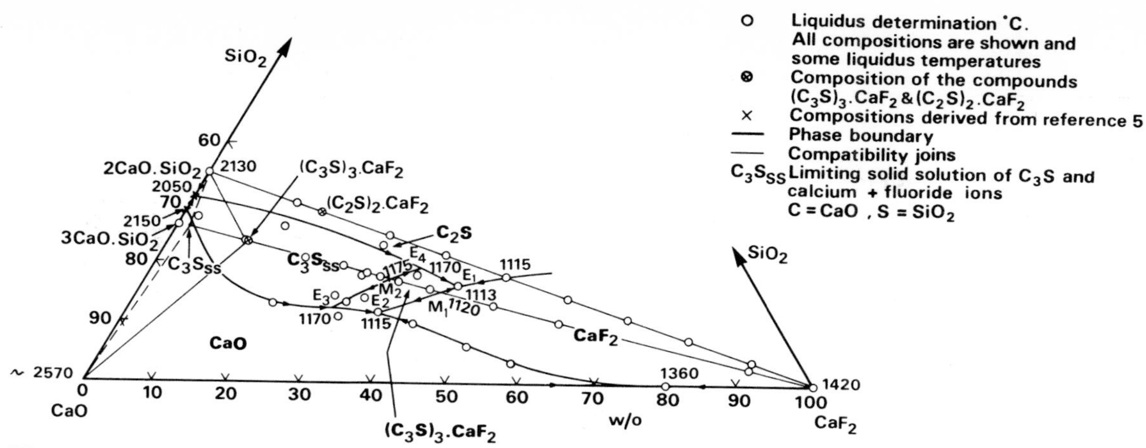

Figure 2 The phase diagram of the system CaO-2CaO. SiO_2 - CaF_2 (after Gutt and Osborne[6])

has been discussed by Gutt[44]. The C_3S forms as a solid solution containing fluoride ions of limiting composition at 1175°C of 98.5 per cent C_3S, 1.5 per cent CaF_2. After heating at 1250°C in steam, $(3CaO.SiO_2)_3.CaF_2$ decomposes into triclinic C_3S and CaO. When 5 per cent CaF_2 was added to a typical raw mix and fired at 1130°C $(3CaO.SiO_2)_3.CaF_2$ was formed, but it decomposed on heating to 1200°C, giving monoclinic C_3S containing some fluoride ions.

The compounds $12CaO.7Al_2O_3$ and $11CaO.7Al_2O_3.CaF_2$ form[45] a solid solution in which the O^{2-} ion is partially replaced by $2F^-$. Point B in Figure 3 containing 2.30 per cent fluorine co-exists[46] in the system $CaO-Al_2O_3-SiO_2-CaF_2$ as follows:

1 $C_3S-C_2S-11CaO.7Al_2O_3.CaF_2-CaF_2$

2 $CaO-C_3S-11CaO.7Al_2O_3.CaF_2-CaF_2$

3 $C_3S-C_2S-(12-x)CaO.7Al_2O_3.xCaF_2$ (x = 0·8 to 1·0)

4 $CaO-C_3S-(12-x)CaO.7Al_2O_3.xCaF_2$ (x = 0·8 to 1·0)

5 $C_3S-C_2S-C_3A-11.2CaO.7Al_2O_3.0{\cdot}8\ CaF_2$

6 $C_2S-C_3A-(12-x)CaO.7Al_2O_3.xCaF_2$ (x = 0 to 0·8)

It is seen that contrary to earlier findings, C_3A is a stable phase in the system.

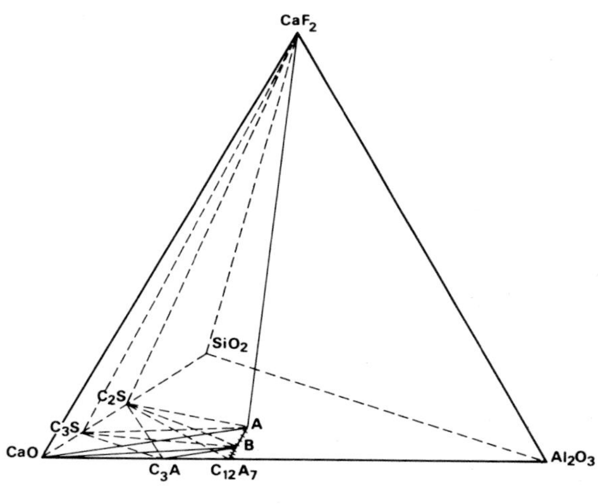

Figure 3 The system CaO-SiO_2 - Al_2O_3 - CaF_2 (after Massazza and Pezzuoli[46])

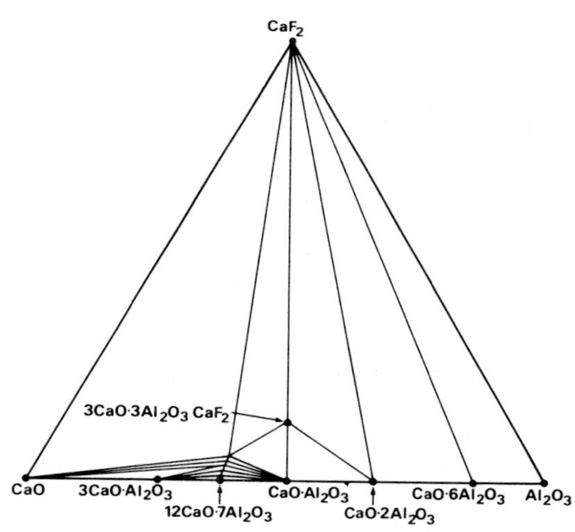

Figure 4 The system CaO-Al_2O_3 - CaF_2 (after Brisi and Rolando[47])

The region of the system richer in alumina has been investigated[47] and, as shown in Figure 4, the equilibria involve the ternary compound $3CaO.3Al_2O_3.CaF_2$. An analogous compound was not found in the system with $CaCl_2$. The join CA-CaF_2 was studied by Gutt, Chatterjee and Zhmoidin[48]; a wide zone of liquid immiscibility was found. The compound $3CaO.3Al_2O_3.CaF_2$ melted congruently at 1507°C and formed eutectics with CA and CaF_2.

Massazza and co-workers[49] extended their study to the system containing the ferrite phase, $CaO-Al_2O_3-Fe_2O_3-CaF_2$. No compounds formed in the sub-system $CaO-Fe_2O_3-CaF_2$ at 1000°C, tie lines being found between CaF_2 and CaO, C_2F, CF or Fe_2O_3. The lines in the lime-rich

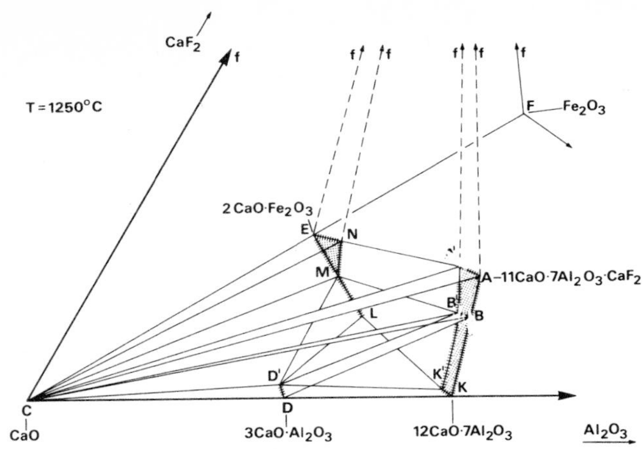

Figure 5 Sub-system CaO-2CaO. Fe_2O_3 - 12CaO. $7Al_2O_3$ - 11CaO. $7Al_2O_3$. CaF_2 (after Massazza et al[49])

portion of the system CaO-Al_2O_3-Fe_2O_3-CaF_2 are the ferrite solid solution series excepting those compositions close to C_2F (Figure 5).

Details of the synthesis of C_2S.NaF and the corresponding compounds with KF and LiF are given by Balmer and Silverman[50] with X-ray data in a separate publication[51]. Teoreanu and van Huynh[52] carried out microprobe analyses and found the compositions given in Table 2 for alites formed in the presence of the named mineralisers.

Table 2 Composition of alite in presence of mineralisers (per cent by weight)

Mineraliser	Temperature (°C)	CaO	SiO_2	Al_2O_3	Fe_2O_3	MgO	CaF_2
None	1450	72.75	24.60	1.85	0.75	-	-
0.3% Na_2SiF_6	1450	72.61	25.10	1.18	0.76	-	0.320
0.5% K_2SiF_6	1350	71.95	24.75	1.73	0.76	-	0.316
0.5% K_2SiF_6	1450	72.35	24.80	1.70	0.75	-	0.305
0.5% $MgSiF_6$	1400	71.88	24.75	1.76	0.75	0.09	0.305
0.3% $MgSiF_6$	1450	71.80	24.85	1.81	0.79	0.05	0.287
0.5% $MgSiF_6$	1450	71.90	24.80	1.76	0.76	0.09	0.306
1.0% $MgSiF_6$	1450	71.85	24.75	1.69	0.74	0.07	0.600
1.5% $MgSiF_6$	1450	71.75	24.70	1.61	0.75	0.10	0.871

CaO-MgO-CaF_2 is[53] a simple eutectic system with eutectic composition CaO 19, MgO 10 and CaF_2 71 per cent at 1343°C.

Relevant systems containing P_2O_5 or P_2O_5 and F were summarised in a review by Gutt at the Fifth Symposium[44]. New work will be found in the supplementary paper by Gutt and Smith.

Matsumo, Miyahoshi and Ando[54] studied the system $Ca_3(PO_4)_2$-$CaNaPO_4$ and found that at high temperatures a continuous series of solid solutions was formed. On cooling suitable compositions in the range 40-50 wt per cent $CaNaPO_4$, the compound $Ca_5Na_2(PO_4)_4$ formed.

The system CaO-P_2O_5-SiO_2-iron oxide has been investigated by Margot-Marette and Riboud[55]. The study was carried out with an atmosphere of CO_2 and H_2 in the ratio 9:1 and was limited to parts of the system $SiO_2 + P_2O_5 \leq 50$ per cent and FeO < 50 per cent. Liquidus diagrams for planes with 0, 7.5, 10, 15, 20, 25, 30, 40 and 50 per cent FeO, and 5 and 10 per cent SiO_2 are presented. Their diagrams for the lime-rich corner of the system CaO-P_2O_5-SiO_2-FeO with FeO = 0, 7.5 or 10 per cent do not include a primary field for tricalcium silicate although they do show C_3S fields in the planes with 15, 20, 25, 30 and 40 per cent FeO. No C_3S field is shown at 50 per cent FeO. In view of earlier work[7] on the system CaO-SiO_2-P_2O_5 in which a C_3S field extending to about 10 per cent P_2O_5 was established, the diagrams presented by Margot-Marette and Riboud for the lower levels of FeO (0, 7.5 and 10 per cent) must be in error.

SULPHATE IN CEMENT CLINKERS

Increasing attention is being paid to the occurrence of SO_3 in cement clinker, one reason being that when the SO_3 is high, it is difficult to add sufficient gypsum to reach the optimum value while still complying with standards. In normal clinker the SO_3 arises from sulphate or pyrites in the clay or by oxidation of sulphur compounds contained in the fuel. The effect of sulphate is of special importance in the combined process for production of cement and sulphuric acid. The role of calcium sulphate in the burning of clinker has been reviewed by Gutt and Smith[56], and the same authors have recently studied the reactions in the combined process[57].

Phase equilibrium studies[56] showed calcium sulphate to form a solid solution with C_3S, the limiting value for SO_3 being 2.9 per cent. This compares with 3.8 per cent SO_3 found by electron-probe analysis of C_3S formed in the system CaO-SiO_2-SO_3 at a relatively high overall SO_3 content (10 per cent) and values of 1.0 to 5.2 per cent in sulphate-rich cement clinkers[29]. The cementing value of C_3S containing sulphate ions, alone and in combination with other ions, is impaired[58] whereas small amounts of $CaSO_4$ in solid solution with C_2S lead to enhanced strengths with the β modification[59]. Studies[29] in which the extent of solid solution of sulphate in C_2S has been determined by electron-microprobe analysis have been referred to above.

The compound $(2CaO.SiO_2)_2.CaSO_4$ decomposes in the solid state at 1298°C into α $CaSO_4$ and $\alpha'C_2S$ solid solution. The unit cell dimensions of this compound have been established[57,60] and its structure described by analogy with apatite and silicocarnotite[61]. Initial liquid formation in the system CaO-C_2S-$CaSO_4$ occurs at about 1270°C (Gutt and Smith[56]).

The compound $3CaO.3Al_2O_3.CaSO_4$ is of importance as a constituent of expanding or shrinkage-compensated cements and has been studied[62] in the system CaO-Al_2O_3-$CaSO_4$ at 1350°C. It decomposes in the presence of alkalis[63], but is compatible at 1200°C with $(2CaO.SiO_2)_2.CaSO_4$ and C_4AF.

Both $2C_2S.CaSO_4$ and the aluminosulphate $3(CaO.Al_2O_3).CaSO_4$ are formed as intermediate compounds in the formation of cement minerals in the cement/sulphuric acid process and desulphation of the raw meal does not occur until temperatures significantly above those usually assumed for the process are reached[57].

In the pure system CaO-SiO_2-$CaSO_4$, C_3S is stable even at very high SO_3 contents[56]. This is not the case however in the system CaO-Al_2O_3-SiO_2-$CaSO_4$, or in some technical clinkers, in which a solid solution of C_2S with aluminium and sulphate ions is formed in preference to C_3S in the clinker; Fe_2O_3 does not have the same effect. It is a finding of some importance that the decomposition of C_3S by SO_3 and Al_2O_3 is offset by the presence of MgO, and the authors[56] conclude that MgO is beneficial in clinkers manufactured by the combined process and should be carefully controlled. Sodium sulphate may have an adverse effect on alite formation in cement clinker[29].

CALCULATION OF PHASE COMPOSITION OF CEMENT

Yamaguchi and Takagi[64] studied a number of Japanese clinkers by several methods. The original clinkers were analysed and a part was separated by heavy liquids to give an alite and a belite fraction. These fractions were subjected to chemical and X-ray analysis and further treated with salicylic acid/methanol solution to determine the amount of non-silicate impurity. From these data the chemical compositions of the alite and belite were calculated. The original clinker was also extracted with salicylic acid to give the amount and analysis of the total interstitial. This fraction was further extracted with acetic acid to give a ferrite sample which was subjected to both chemical and X-ray analysis. Thus estimates of the amount and composition of both the ferrite and aluminate part of the interstitial were obtained. The 'standard formulae' for the cement minerals deduced from the average of twelve results for alite and belite and four results for the interstitial phases are given in Table 3. These formulae, but ignoring the alkalis, were used for a modified Bogue analysis instead of the stoichiometric formulae of the pure compounds. The results obtained are given in Table 4.

Table 3 Standard compositions of cement minerals (per cent by weight)

	SiO_2	Al_2O_3	Fe_2O_3	CaO	MgO	Na_2O	K_2O
Alite	24.83	1.24	0.49	72.23	0.98	0.09	0.14
Belite	32.50	2.63	1.03	62.83	0.52	0.20	0.30
Ferrite	3.61	24.51	22.08	44.50	4.36	0.37	0.57
Aluminate	5.88	27.43	7.81	53.49	1.97	2.27	1.15

Table 4 Mineral composition determined by several methods [63]

Clinker	Method	Alite	Belite $\alpha+\beta$ (α)	Ferrite	Aluminate	Others
N_1	Bogue (B)	52	26	8	9	
	Microscopy (M)	52	30	10	9	
	Chemical (C)	51	32	8	7	2 { $K_3Na(SO_4)_2$ 1.4, CaO 0.6 MgO 0.03
	X-ray (X)	55	26(8)	9	9	
	New method (N)	54	28	13	5	
N_2	B	42	34	10	10	
	M	52	32	11	5	
	C	59	23	10	6	2 { $K_3Na(SO_4)_2$ 0.7, CaO 0.08 MgO 0.6
	X	55	26(6)	11	7	
	N	49	31	13	7	
N_3	B	57	24	10	7	
	M	65	20	7	8	
	C	61	22	10	6	1 { $K_3Na(SO_4)_2$ 0.2, MgO 0.7
	X	58	24(0)	11	7	
	N	61	23	13	2	
N_4	B	63	13	10	9	
	M	69	17	9	6	
	C	72	10			{ MgO 1.05, CaO 0.31
	X					
	N	65	13	14	7	
N_5	B	45	33	10	10	
	M	50	36	8	6	
	C	50	36			CaO 0.12
	X					
	N	52	30	12	7	
R	B	64	16	9	8	
	M	65	18	9	8	
	C	67	19	14		CaO 0.03
	X					
	N	73	11	8	8	
M_1	B	48	31	11	7	
	M	48	32	10	10	
	C	51	33	9	2	5 CaO 0.02
	X	49	33(0)	14	3	
	N	58	27	14	0	
M_2	B	49	30	11	7	
	M	55	26	10	9	
	C	56	27	17		CaO 0.03
	X					
	N	58	26	14	1	

It will be seen that estimates by the five methods agree fairly well as regards alite and belite. Rather larger discrepancies are apparent for the ferrite, and the results vary over a wide range for the aluminate. The 'new' (modified Bogue) method gives consistently lower results for the aluminate phase.

Hansen's method

Assuming that equilibrium has been established and that crystallisation is complete, Hansen[65] assumes that the ferrite of clinker is C_6A_2F and not C_4AF and that the aluminate is $C_3A_{0.9}F_{0.1}$. He gives detailed instructions for calculating the potential contents of these compounds, recommending that Al_2O_3 analyses should be corrected for TiO_2 and P_2O_5. Hansen carried out these calculations for 21 cements which had been given a rating for resistance to sulphate attack; cements of types II IV and V (ASTM) all gave zero calcium aluminate by the new calculation, although they had positive results by XRD. He explains this by incomplete attainment of equilibrium during cooling. For the type I cements the aluminate by the new calculation was closer to the X-ray figure than the Bogue value.

Midgley's method

Midgley[66,67] analysed a number of clinkers and then determined the composition of the individual phases by microprobe analysis. He used the results to establish linear relationships between the amount of a given minor oxide in a phase and the amount in the total analysis of the clinker. Using these relationships, the compositions of each of the four main phases of any cement are calculated in terms of the oxide constituents CaO, SiO_2, Al_2O_3, Fe_2O_3, MgO, Na_2O, K_2O, TiO_2 and P_2O_5. The calculation of the composition of the ferrite is more complicated because of the considerable variation in composition even without the minor constituents. In this case, the composition of the ferrite was assumed to be related to the Al_2O_3/Fe_2O_3 ratio.

Having obtained the composition of each phase, alite, belite, aluminate and ferrite, the quantities were obtained by setting up simultaneous equations as in the Bogue calculation. A computer program is available for this stage.

Twenty-eight cements were analysed and correlation coefficients were calculated between the calculated results for alite and belite and the values determined microscopically, and between the calculated results for aluminate and ferrite and the values determined by XRD. It was concluded that the proposed new method was no better than the simple Bogue calculation. The same conclusion was drawn from a comparison of Bogue with a calculation using a ferrite composition[68] derived from the ratio $Fe_2O_3/(Fe_2O_3 + Al_2O_3)$ (per cent by weight). Midgley[67] subsequently refined his method further.

Knofel and Sphohn's method[69]

These authors attribute the differences between the calculated Bogue composition and that obtained by microscopy to the influence of minor constituents, especially MgO. The microscopically determined alite exceeded the Bogue figure by about 30 per cent and the belite was less than the calculated figure by about 20 per cent. Aluminate and ferrite were nearer to the Bogue values, especially for sulphate-resisting cement.

CONSTITUTION OF PORTLAND CEMENT

It first has to be decided which criterion can be used to test a method for obtaining constitution. It is fairly widely accepted that C_3S and C_2S are best determined microscopically, by point-counting or by television microscope analysis; ferrite and aluminate are determined by XRD, the ferrite analysis giving also an estimate of ferrite composition. If this is accepted, more clinkers need to be examined, especially using those methods in which the composition of the ferrite is first determined. As at present, no method appears to be statistically superior to the Bogue calculation.

CONCLUSIONS

Phase studies related to the manufacture of Portland cement and related special cements remain a necessary and practically useful area of study. A more extensive knowledge of liquidus and sub-solidus relationships in phosphate, fluoride and sulphate-containing systems is required so that the mineralogical and other properties of cements containing these components can be more closely related to the chemical composition, and so that modified raw meal proportioning, lime-saturation and compound composition formulae can be developed.

A continuing major difficulty in assessing the value of attempts to improve upon the Bogue equations for calculating compound composition, is the need to improve upon the accepted physical methods for determining the constitution of clinkers, such as XRD, optical microscopy, etc, with which the comparisons must be made.

REFERENCES

1 **Guinier, A** and **Regourd, M.** Structure of Portland cement minerals. Proc 5th Int Symp Chem Cement, Tokyo, 1968, Vol 1, pp 1-32. Tokyo, Cement Assoc of Japan, 1969.

2 **Uchikawa, H** and **Tsukiyama, K.** Polymorphic transformations of tricalcium silicate. Yogyo-Kyokai-Shi, Vol 79, No 6 1971, pp 7-12.

3 **Trömel, G, Fix, W** and **Heinke, R.** High-temperature investigations up to 1900°C on calcium orthosilicate and tricalcium silicate. Tonind Ztg, Vol 93, No 1, 1969, pp 1-8.

4 **Welch, J H** and **Gutt, W.** Tricalcium silicate and its stability within the system CaO- SiO_2. J Am Ceram Soc, Vol 42, No 1, 1959, pp 11-15.

5 **Gutt, W.** High temperature phase equilibria in polycomponent silicate systems. PhD Thesis, London University, 1966.

6 **Gutt, W** and **Osborne, G J.** The system CaO-2CaO.SiO_2-CaF_2. Trans Br Ceram Soc, Vol 69, No 3, 1970, pp 125-130.

7 **Gutt, W.** High-temperature phase equilibria in the system 2CaO.SiO_2-3CaO.P_2O_5-CaO. Nature, Vol 197, No 4863, 1963, pp 142-143.

8 **Sycher, M M, Korneev, V I** and **Boigalina, L B.** Phase transitions in the highly basic region of the calcium oxide-silicon dioxide system. Izr Vyssh Ucheb Zaved, Khim Khim Teknol, Vol 11, No 12, 1968, pp 1370-1375.

9 **Butt, Yu M, Timashev, V** and **Kanshanskii, V E.** Stability of tricalcium silicate. Neorg Mater, Vol 4, No 3, 1968, pp 465-467.

10 **Woermann, E, Eysel, W** and **Hahn, Th.** Chemical and structural investigations of the formation of solid solutions of tricalcium silicate. Zement-Kalk-Gips, Vol 16, No 9, 1963, pp 370-375.

Woermann, E, Eysel, W and **Hahn, Th.** Chemical and structural investigations of the formation of solid solutions of tricalcium silicate. II Phase relations in the systems CaO-MgO-SiO_2 and CaO-Al_2O_3-SiO_2. Zement-Kalk-Gips, Vol 20, No 9, 1967, pp 385-391.

Woermann, E, Eysel, W and **Hahn, Th.** Chemical and structural investigations of the formation of solid solutions of tricalcium silicate. III Combined substitution of MgO and Al_2O_3 in tricalcium silicate. Zement-Kalk-Gips, Vol 21, No 6, 1968, pp 241-251.

Woermann, E, Eysel, W and **Hahn Th.** Chemical and structural investigations of the formation of solid solutions of tricalcium silicate. IV Combined substitution of Fe_2O_3, Al_2O_3 and MgO in Ca_3SiO_5. Zement-Kalk-Gips, Vol 22, No 5, 1969, pp 235-241.

Woermann, E, Eysel, W and **Hahn, Th.** Chemical and structural investigations of the formation of solid solutions of tricalcium silicate. V Alite phase in the CaO-MgO-Al_2O_3-Fe_2O_3-SiO_2 system. Zement-Kalk-Gips, Vol 22, No 9, 1969, pp 412-422.

11 **Fletcher, K E.** The effect of Fe^{3+} and Al^{3+} on the polymorphism of tricalcium silicate. Trans Br Ceram Soc, Vol 64, No 8, 1965, pp 377-385.

12 **Midgley, H G** and **Fletcher, K E.** The role of alumina and magnesia in the polymorphism of tricalcium silicate. Trans Br Ceram Soc, Vol 62, No 11, 1963, pp 917-937.

13 **Eysel, W** and **Hahn, Th.** Polymorphism and solid solution of Ca_3GeO_5 and Ca_3SiO_5. Zeit Kristallogr, Vol 131, 1970, pp 40-59.

14 **Shirvinskaya, A K, Grebenshchikov, R G** and **Toropov, N A.** The system calcium oxide-germanium oxide. Neorg Mat, Vol 2, 1966, pp 332-335.

15 **Butt, Yu M, Timashev, V V** and **Malozhon, L I.** Crystallisation of minerals in clinkers containing chromium oxide. Inorganic Materials, Vol 4, No 3, 1968, pp 431-435.

16 **Kondo, R** and **Yoshida, K.** Effect of the solubility of titanium or manganese in tricalcium silicate and alite on their hydration characteristics. Yogyo-Kyokai-Shi, Vol 77, No 881, 1969, pp 25-31.

17 **Kurdowski, W** and **Wollast, R.** Solid solutions of BaO in tricalcium silicate. Silicates Industriels, Vol 35, No 6, 1970, pp 153-159.

18 **Midgley, H G.** The composition of alite (tricalcium silicate) in a Portland cement clinker. Mag Concrete Research, Vol 20, No 62, 1968, pp 41-44.

19 **Midgley, H G.** The polymorphism of calcium orthosilicate. Sixth International Symposium

on Chemistry of Cement, Moscow, September 1974.

Midgley, H G, Bennett, M W, Retford, D and **Pettifer, K.** The mineralogy of dicalcium silicate. Private communication.

20 **Gutt, W.** A new calcium magnesiosilicate. Nature, Vol 190, No 4773, 1961, pp 339-340.

21 **Gutt, W.** The system dicalcium silicate-merwinite. Nature, Vol 207, No 4993, 1965, pp 184-185.

22 **Niesel, K** and **Thormann, P.** The stability fields of dicalcium silicate modifications. Tonind Ztg, Vol 91, No 9, 1967, pp 362-369.

23 **Lehmann, H, Niesel, K** and **Thormann, P.** The stability fields of the polymorphic forms of dicalcium silicate. Tonind Ztg, Vol 93, No 6, 1969, pp 197-208.

24 **Chromy, S.** The inversion of the $\beta - \gamma$ modifications of dicalcium silicate. Zement-Kalk-Gips, Vol 23, No 8, 1970, pp 382-389.

25 **Eremin, N I, Egereva, A I, Dmitrieva, A** and **Firfarova, I B.** Solid solutions of $2CaO.SiO_2$ with some metal oxides. Zh Prikl Khim (Leningrad), Vol 43, No 1, 1970, pp 18-24.

26 **Gutt, W** and **Osborne, G J.** The effect of potassium on the hydraulicity of dicalcium silicate. Cement Technology, Vol 1, No 4, 1970, pp 1-5.

27 **Fletcher, K E.** The analysis of belite in Portland cement clinker by means of an electron-probe. Mag Concrete Research, Vol 20, No 64, 1968, pp 167-170.

28 **Midgley, H G** and **Bennett, M.** A microprobe analysis of larnite and bredigite from Scawt Hill, Larne, Northern Ireland. Cement and Concrete Research, Vol 1, No 4, 1971, pp 413-418.

29 **Gutt, W** and **Smith, M A.** Studies of sulphates in Portland cement clinker. Cement Technology, Vol 2, No 5, 1971, pp 143-147.

30 **Aggarwal, P S, Gard, J A, Glasser, F P** and **Biggar, G M.** Synthesis and properties of dicalcium aluminate, $2CaO.Al_2O_3$. Cem Concr Res, Vol 2, No 3, 1972, pp 291-297.

31 **Imlach, J A, Dent-Glasser, L S** and **Glasser, F P.** Excess oxygen and the stability of $C_{12}A_7$. Cem Concr Res, Vol 1, 1971, pp 57-61.

32 **Nurse, R W.** Phase equilibria and formation of Portland cement minerals. Proc 5th Int Symp Cement, Tokyo, 1968, Vol 1, pp 77-89. Tokyo, Cement Assoc of Japan, 1969.

33 **Tarte, P.** Structural research into the constitution of cements. Part II. Isomorphic substitution phenomena in tricalcium aluminates. Silicates Industriels, Vol 33, No 11, 1968, pp 333-339.

34 **Biggar, G M** and **O'Hara, M J.** Melting of fosterite, monticellite, merwinite, spinel and periclase assemblages. J Am Ceram Soc, Vol 53, No 10, 1970, pp 534-537.

35 **O'Hara, M J** and **Biggar, G M.** Phase equilibria aspects of the performance of basic refractories. Trans Br Ceram Soc, Vol 69, No 6, 1970, pp 243-251.

36 **Biggar, G M.** Phase relationships of bredigite ($Ca_5MgSi_3O_{12}$) and of the quaternary compound ($Ca_6MgAl_8SiO_{21}$) in the system $CaO-MgO-Al_2O_3-SiO_2$. Cem Concr Res, Vol 1, No 5, 1971, pp 493-513.

37 **Biggar, G M** and **O'Hara, M J.** Melilite crystallisation in the system $CaO-MgO-Al_2O_3-SiO_2$. Min Mag, Vol 38,No 300, 1972, pp 918-925.

38 **Gutt, W.** Crystallisation of merwinite from melilite compositions. J Iron and Steel Inst, Vol 206, No 8, 1968, pp 840-841.

39 **Midgley, H G.** The composition and possible structure of the quaternary phase in high-alumina cement and its relation to other phases in the system $CaO-MgO-SiO_2$. Trans Br Ceram Soc, Vol 67, No 1, 1968, pp 1-14.

40 **Rao, M R.** Liquidus relations in the quaternary sub-system $CaAl_2O_4-CaAl_4O_7-Ca_2Al_2SiO_7-MgAl_2O_4$. J Am Ceram Soc, Vol 51, No 1, 1968, pp 50-54.

41 **Dayal, R R** and **Glasser, F P.** Phase relations in the system $CaO-Al_2O_3-Fe_2O_3$. Science of Ceramics, Vol 3, 1967, p 191.

42 **Imlach, J A** and **Glasser, F P.** Sub-solidus phase relations in the system $CaO-Al_2O_3-Fe_2O_3$. Trans Br Ceram Soc, Vol 70, No 6, 1971, pp 227-234.

43 **Tanaka, M, Sudoh, G** and **Akaiwa, S A.** New compound $Ca_{12}Si_4O_{19}F_2$ in the system $CaO-SiO_2-CaF_2$ and the role of CaF_2 in the burning of cement clinker. Proc 5th Int Symp

Chem Cement, Tokyo, 1968, Vol 1, pp 122-135. Tokyo, Cement Assoc of Japan, 1969.

44 **Gutt, W.** Manufacture of Portland cement from phosphatic raw materials. Proc 5th Int Symp Chem Cement, Tokyo, 1968, Vol 1, pp 93-105. Tokyo, Cement Assoc of Japan, 1969.

45 **Brisi, C** and **Rolando, P.** Mechanism of the decomposition of tricalcium aluminate in the presence of calcium fluoride and calcium chloride. Ann Chim, Vol 56, 1966, pp 224-231.

46 **Massazza, F** and **Pezzuoli, M.** Solid state equilibria in the quaternary system $CaO-Al_2O_3-SiO_2-CaF_2$. Revue des Matériaux, No 642, 1969, pp 81-86.

47 **Brisi, C** and **Rolando, P.** Decomposition of monocalcium aluminate in the presence of chlorides and solid state equilibria in $CaO-Al_2O_3-CaF_2$ system. Ann Chim, Vol 57, No 11, 1967, pp 1304-1315.

48 **Gutt, W, Chatterjee, A K** and **Zhmoidin, G I.** The join calcium mono-aluminate-calcium fluoride. J Mats Sci, Vol 5, No 11, 1970, pp 960-963.

49 **Massazza, F, Pezzuoli, M** and **Gilioli, C.** Solid state phase equilibria in the quaternary system $CaO-Al_2O_3-Fe_2O_3-CaF_2$. Revue des Matériaux, No 663, 1971, pp 357-363.

50 **Balmer, M K** and **Silverman, G M.** Preparation and phase transformations of dicalcium silicate-alkali fluoride complexes. J Am Ceram Soc, Vol 54, No 2, 1971, pp 98-101.

51 **Balmer, M K** and **Silverman, G M.** Cell constants for dicalcium silicate-alkali fluoride complexes. J Am Ceram Soc, Vol 54, No 2, 1971, pp 101-102.

52 **Teoreanu, I** and **van Huynh, T.** Effect of fluoride constituents on the formation of crystals and on the mineralogical constituents in Portland cement. Revue des Matériaux, No 666, 1971, pp 73-77.

53 **Schlegel, E.** $CaO-MgO-CaF_2$ system. Cercetari Metalurgice, Vol 9, 1967, pp 785-792.

54 **Matsumo, S, Miyahoshi, T** and **Ando, J.** Kogyo Kagaku Zasshi, Vol 70, No 10, 1967, pp 1638-1640.

55 **Margot-Marette, H** and **Riboud, P V.** Study of liquidus temperatures of the system $CaO-P_2O_5-SiO_2$-iron oxide. Mem Sci Rev Met, Vol 69, No 9, 1972, pp 593-604.

56 **Gutt, W** and **Smith, M A.** Studies of the role of calcium sulphate in the manufacture of Portland cement. Trans Br Ceram Soc, Vol 67, No 10, 1968, pp 487-509.

57 **Smith, M A** and **Gutt, W.** Studies of the mechanism of the cement/sulphuric acid process. Cement Technology, Vol 2, No 6, 1971, pp 163-167 and 177.

58 **Smith, M A** and **Gutt, W.** Effect of magnesium and sulphate ions on the hydraulicity of C_3S. Cement Technology, Vol 1, No 6, 1970, pp 187-191.

59 **Okorokov, S D, Golynko-Wolfson, S L** and **Korneev, F I.** Change in the strength of dicalcium silicate depending on the nature of the stabilising admixture. Tr Leningrad Technol. Inst Lensoveta, No 56, 1960, pp 93-98.

60 **Pryce, M W.** Calcium silicosulphate in lime-kiln wall coating. Min Mag, Vol 38, No 300, 1972, pp 968-971.

61 **Scholze, H** and **Hildebrandt, U.** Compounds containing carbonate and sulphate in cement manufacture. Zement-Kalk-Gips, Vol 23, No 12, 1970, pp 573-579.

62 **Turriziani, R** and **Massazza, F.** Solid state equilibria in the system $CaO-Al_2O_3-SO_3$. Ann Chim, Vol 56, No 10, 1966, pp 1172-1180.

63 **Budnikov, P P.** Effect of alkali and gypsum on the syntheses of calcium aluminate and Portland cement clinker. Cement Wapno Gips, Vol 23/34, No 9-10, 1967, pp 281-284.

64 **Yamaguchi, G** and **Takagi, S.** Analysis of Portland cement clinker. Proc 5th Int Symp Chem Cement, Tokyo, 1968, Vol 1, pp 181-218. Tokyo, Cement Assoc of Japan, 1969.

65 **Hansen, W C.** Potential compound composition of Portland cement clinker. J Mats, Vol 3, No 1, 1968, pp 100-106.

66 **Midgley, H G.** Compound calculation of the phases in Portland cement clinker. Cement Technology, Vol 1, No 3, 1970, pp 79-84.

67 **Midgley, H G.** Compound calculation of the phases in Portland cement clinker. Cement Technology, Vol 2, No 4, 1971, pp 113-116.

68 **Midgley, H G** and **Moore, A E.** The ferrite phase in Portland cement clinker. Cement Technology, Vol 1, No 5, 1970, pp 153-156.

69 **Knofel, D** and **Sphohn, E.** The quantitative phase content in Portland cement clinkers. Zement-Kalk-Gips, Vol 22, No 10, 1969, pp 471-476.

The use of by-products in concrete (CP 53/74)

W. Gutt

INTRODUCTION

The construction industry requires large quantities of low cost materials, for example 180 M tonnes of aggregates were used in Great Britain in 1968[1]. Industrial by-products contributed 13 M tonnes to this total but over 100 M tonnes[1,2] of by-products potentially usable in construction are produced currently per annum. If these by-products could be utilised more extensively, as substitutes, natural materials could be conserved. Simultaneously, the build-up of spoil tips and the consequent land dereliction and possible water pollution of rivers and coastline could be avoided. It should be made clear, however, at this stage, that any proposal to introduce a waste material substitute as a building material must be based on a planned appraisal of the new material. First, the by-product must fulfil the engineering requirements in terms of its physical properties and must not contain excessive amounts of deleterious components which may lead to corrosion or instability of the concrete. While these requirements apply equally to natural materials, they are more difficult to fulfil with wastes which tend to be inhomogeneous and may contain deleterious matter. It is essential therefore to establish standards and these cannot be written without prior research to define acceptance limits and to develop methods of test. Technical appraisal and standardisation should be accompanied by an economic analysis including not only manufacturing and transport costs, but also social benefits and energy savings which would arise if the by-products were used.

Evaluation of by-products for use in construction has been a part of the research programme of BRE for many years. Recently, a comprehensive survey for England and Wales of the current production, stockpile, location and present or possible utilisation of the major waste materials has been carried out[2] on behalf of the DOE Aggregates and Waste Materials Working Group. This survey formed a part of the report[1] which the Working Group presented to the Advisory Committee on Aggregates. A supplementary survey of waste materials in Scotland is at present in progress. In the present paper, the existing knowledge on the present utilisation of the major by-products as aggregates or cements in concrete will be described for each material separately. Future prospects will be discussed. For all materials the following stages constitute the technical assessment:

1 Determination of physical properties

2 Chemical analysis

3 Mineralogical examination by microscopy and X-ray analysis to identify compounds present

4 Stability tests

5 Use in concrete as aggregate or cement followed by tests of the concrete for strength and durability

6 Assessment of the life of steel reinforcement in the concrete containing the by-product

7 Determination whether the material can be used directly, or after heat treatment. Such treatment can often be selected by correlation of mineralogy with practical properties.

BLASTFURNACE SLAG

Blastfurnace slag is a by-product of iron manufacture and results from the fusion of limestone, with ash from coke, and the siliceous and aluminous residue remaining after the reduction and separation of the iron from the ore. The slag rises to the surface and is tapped off from the blastfurnace from time to time. The slag also has a chemical role in removing sulphur from the molten iron. Blastfurnace slag provides an example of an industrial by-product which is well established as a construction material and the 9 M tonnes produced in 1971 were all used. The applications of blastfurnace slag in cement manufacture and as aggregates have been reviewed in some detail recently[3,4] and since this by-product is well accepted, only a brief account will be given in the present paper. Typical elemental composition of blastfurnace slag is given in Table 1.

Table 1 Range of compositions of blastfurnace slag

Substance	Occurrence (per cent)
Calcium oxide CaO	30 - 50
Silicon dioxide SiO_2	28 - 38
Aluminium oxide Al_2O_3	8 - 24
Magnesium oxide MgO	1 - 18
Iron oxide Fe_2O_3	1 - 3
Sulphur S	1 - 2.5

By selective cooling three main types of slag are produced. If the slag is to be used as roadstone or dense aggregate for concrete it is slowly cooled in air to produce a crystalline product resembling igneous rock. Alternatively, the slag may be foamed with a limited amount of water to give a lightweight aggregate, or it can be cooled rapidly to produce a slag glass as granules. This quenched glass is used in the manufacture of slag cements[3]. In Table 2 the main British Standards for slag aggregates and cements are listed.

Table 2 British Standards for slag materials

BS 146 :1958	Portland blastfurnace cement	$<$ 65% bfs
BS 4246:1968	Low heat Portland blastfurnace cement	50–90% bfs
BS 4248:1968	Supersulphated cement	$\nless$ 75% bfs
BS 1047:1952	Air-cooled blastfurnace slag coarse aggregate for concrete	
BS 802 :1967	Tarmacadam with crushed rock or slag aggregate	
BS 1621:1961	Bitumen macadam with crushed rock and slag aggregate	
BS 877 :1973	Foamed or expanded blastfurnace slag lightweight aggregate for concrete	
BS 2028/1314:1968	Precast concrete blocks	

Air-cooled slag consists of a mixture of minerals of which melilites such as the solid solution series akermanite $2CaO.MgO.2SiO_2$-$2CaO.Al_2O_3.SiO_2$ are the main series. Metastable β-dicalcium silicate, which might undergo a structural transformation to the stable γ-form with a 10 per cent volume increase, can occur in blastfurnace slag. The existing BS 1047 therefore excludes from use slag containing $\beta 2CaO.SiO_2$. The stability criteria have been discussed in detail elsewhere [4,5]. Total sulphur is limited to 2.0 per cent and acid-soluble SO_3 to 0.7 per cent. The minimum bulk density is prescribed as 1250 kg/m^3.

The long term durability of plain and reinforced concrete containing recently produced blastfurnace slags as dense aggregate has been shown to be good[6,7]. Figure 1 shows the development of compressive strength of air and water-cured specimens up to two years as compared with that of control concretes and Figure 2 illustrates the use of slag concrete in the new steelmaking complex, the Anchor project, at Scunthorpe, where 0.8 M tonnes of slag were utilised. Foamed slag aggregate to BS 877:1973 is used for blockmaking, for roof and floor screeds and for structural reinforced concrete capable of yielding 28 day concrete cube strengths comparable to those obtained with other lightweight aggregates [8].

Slag cements, and granulated slag under the trade name 'Cemsave' for mixing on site, are available. Granulated slag does not set if it is mixed with water. It has to be activated by means of Portland cement as in Portland blastfurnace cements or by Portland cement and calcium sulphate as in supersulphated cements. It should also be explained that granulated slag is not a pozzolana and that its hydration reactions form specific hydrated compounds [9].

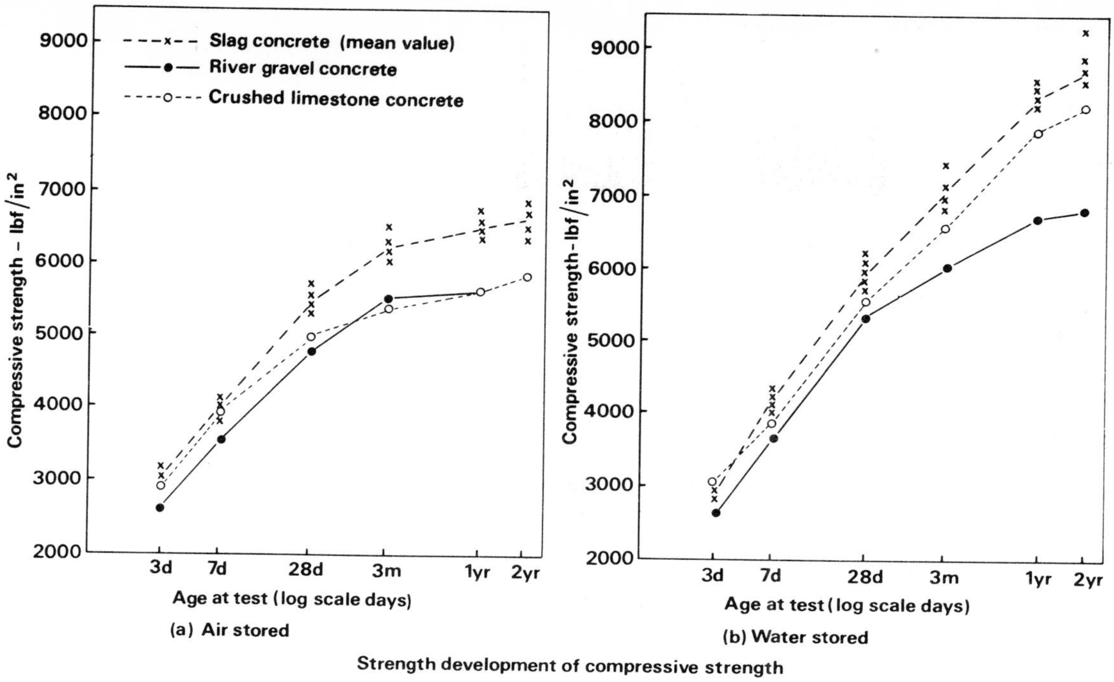

Figure 1 - Compressive strength of slag concrete

Potential activity of slag for cement manufacture is obtained from the modulus [10]

$$M = \frac{CaO + MgO + \frac{1}{3}Al_2O_3}{SiO_2 + \frac{1}{3}Al_2O_3} > 1.0$$

and 90 day strength of slag cements, S_{90}, from the equation: $(S_{90}-75) = 0.38(M-0.72)G$ where S_{90} = the 90 day strength of 1:2:4 concrete of w/c ratio 0.60 for a mixture of 65 per cent ordinary Portland cement/35 per cent blastfurnace slag and G = percentage of glass determined by a microscopic test.

Slag cements are suitable for all types of construction and have better sulphate resistance than ordinary Portland cement. They also provide low heat cements particularly suitable for use in massive concrete construction such as dams where excessive temperature rise during hydration may lead to cracking.

Figure 2 - Anchor project soaking pits during construction

The use of blastfurnace slag as aggregate for concrete is likely to increase and since in some ironworks the bulk density has fallen below the minimum specified in BS1047 (ie 1250 kg/m^3) a five-year study of such slags as dense aggregates has now been carried out[11]. Concrete strength, sulphate resistance and protection of steel reinforcement have proved to be satisfactory and these findings are likely to lead to a lowering of the bulk density minimum and wider use of slag in concrete. It should be remembered that changes in the type of iron ore may lead to changes in slag composition and this must be kept under surveillance to protect the user. In the meantime improvements in the understanding of slag stability may result from high-temperature phase equilibrium studies[12] of the system CaO-SiO_2-Al_2O_3-MgO, which represents 96 per cent of blastfurnace slag composition, and of related systems.

The prospects for utilisation of foamed slag aggregate would improve with wider acceptance of lightweight aggregate concrete while the use of slag cements, as part substitutes for Portland cements, could be beneficial in helping to conserve limestone resources and fuel, since no heat treatment is involved in manufacture of the granulate which is also cheaper than ordinary Portland cement[13]. It is noteworthy that most blastfurnace slags currently produced are suitable for granulation but only a very small proportion is processed in this way. While greater use of slag cements would divert some blastfurnace slags from use as roadstone, the latter might be replaced by other materials not capable of conversion into cement which is worth approximately 10 times as much as aggregate.

STEEL SLAG

The conversion of pig iron to steel involves lowering and controlled adjustment of the content of various impurities. Refining is achieved by fusion of the pig iron under oxidising conditions with a limestone flux when carbon, silica, manganese, phosphorus and sulphur pass into slag, the composition of which depends on the origin of the ore and the type of steelmaking process. The main processes now in use are the Basic Oxygen and Electric Arc processes. Phosphate rich basic slags are valuable as fertilisers while high calcium, low phosphate slags are used, after weathering, as roadstone. The elemental and compound composition of steel slags is very variable and fundamentally different from that of blastfurnace slag. Steel slag consists essentially of calcium silicates $3CaO.SiO_2$ (minor phase), $\beta 2CaO.SiO_2$, free lime, $(CaFeMnMg)O$, and the solid solution series $2CaO.Fe_2O_3$ - $2CaO.2Al_2O_3.Fe_2O_3$. Due to the presence of metastable compounds, steel slag is not suitable as aggregate for concrete and as new steelmaking plants are built it will be necessary to evaluate the slags produced.

COLLIERY SPOIL

Colliery spoil originates from the sedimentary rocks with which the seams of coal are associated. The introduction of mechanised mining techniques has increased the proportion of waste to one half of the material extracted and over 50 M tonnes are produced per annum. 3000 M tonnes are stockpiled. Since only 7 M to 8 M tonnes are used per annum, colliery spoil represents a large and increasing reservoir of material which could make a great contribution to materials supply if new uses for it could be developed.

In coal preparation the material is first crushed to a top size of between 200 mm and 25 mm in order to liberate any trapped coal and then a coal washer is used to separate the coal and dirt. Washing employs dense medium separation, jigging or froth flotation and the reject from washing is the biggest component of colliery spoil. Two other components of spoil are pit dirt resulting from underground excavation procedures and fine material produced at the face or by crushing or degradation prior to washing. This fine material is often processed to recover coal.

Colliery spoil is dumped on heaps and until recently these were loose tipped and therefore sometimes burnt since sufficient air penetrated to sustain the combustion, started either by spontaneous heating or by dumping of hot ashes, of any coal. Modern heaps are constructed in layers and are compacted by earthmoving vehicles. This increases stability of the heaps and prevents burning and the material contained in these unburnt heaps is known commercially as minestone. Small quantities of spoil are also dumped in old quarries and 10 per cent of total spoil production is tipped into the sea principally off the North East coast. The analyses of three burnt and three unburnt spoils are given in Table 3 [14].

Physical and chemical properties have been reported[15]. The minerals associated with coal are quartz, the clay minerals (illite, kaolinite and chlorite), pyrites and carbonates of calcium, mangesium and iron. Unburnt spoil is blue grey whereas burnt spoil is red brown, so visual distinction is simple. In addition to reducing the content of combustible matter, burning partly decomposes the clay minerals and carbonates and oxidises sulphides to soluble sulphates and some sulphur is lost through volatilisation. The nett effect of burning is greater stability except for the increase in undesirable sulphate. It was estimated [2] in 1966, that about 11 000 hectares of land were covered by 2000 M tonnes of spoil in NCB ownership. Since that time, the NCB stockpile has risen to 2500 M tonnes and there are 500 M tonnes in private ownership.

The vast majority of the 7 M to 8 M tonnes used is employed as fill in road embankments and building sites and burnt spoil has generally been preferred for these purposes, but the NCB has been trying successfully to increase the use of unburnt spoil. For instance, 1.25 M tonnes have been used in the Liverpool outer ring road and 1 M tonnes have been employed in the construction of the M62 motorway. This last use was assisted by DOE policy circular 47/72 which encouraged the use of spoil by a dual tendering procedure for natural and waste materials respectively. The local authority could then accept the waste material proposal even if it were more expensive, if sufficient social benefits were judged to ensue.

Table 3 Chemical analysis of burnt and unburnt colliery spoils

(Analysis of burnt spoil after Sherwood and Ryley [14]; analysis of unburnt made at BRS)

	Burnt spoils			Unburnt spoils		
SiO_2	57.6	56.2	60.2	51.9	46.97	42.99
Al_2O_3	31.3	31.1	21.2	19.4	19.80	21.45
Fe_2O_3	3.86	4.33	8.02	6.1	5.5	4.70
TiO_2	0.22	0.24	0.17	1.03	0.9	0.95
CaO	0.36	1.03	0.44	0.66	0.68	1.30
MgO	0.92	0.82	1.01	1.21	1.49	1.26
Na_2O	0.23	0.20	0.48	0.44	0.14	0.12
K_2O	2.50	2.06	3.30	3.0	0.79	0.90
SO_3	0.10	1.39	0.89	0.35	0.49	0.46
S	0.02	0.01	0.10	0.02	Trace	Trace
loss on ignition	1.9	2.2	3.8	16.13	21.59	24.12
TOTAL	99.0	99.6	99.6	100.4	98.4	98.3

Lightweight aggregate is made from unburnt spoil at Gartshore in Dunbartonshire (capacity 200 000 tonnes per annum) where a spoil heap at a disused anthracite mine serves as the raw material and at Denby, Derbyshire, spoil serves as a part of the feed to a similar plant of 500 000 tonnes per annum capacity. A new lightweight aggregate plant at the NCB Snowdown Colliery in Kent will produce 500 000 tonnes per annum, mainly from unburnt spoil. Most of the lightweight aggregate now produced from spoil is used in the manufacture of precast concrete blocks at plants on the same site and the application of such aggregate is illustrated by Figure 3.

Figure 3 - Lightweight aggregate blocks made from colliery spoil

Colliery spoil is converted into lightweight aggregate commercially[16] by firing on a sinter-strand. It is important to note that the fuel present in the spoil makes a useful combustion to the energy required. In the process the spoil bloats due to gas evolution as the gas is trapped within the softening pellet causing it to expand or, alternatively, sintering of particles occurs leaving voids between them. The product meets the requirements of BS 3797:1964, Lightweight aggregate for concrete.

Emission of tarry volatile matter presents a pollution problem particularly if bituminous coal wastes are processed. Currently, there are plans to increase the production of lightweight aggregate from spoil and this will make a useful contribution to aggregate supply. The contribution of spoil could, however, be greatly enhanced if a dense aggregate could be produced, especially since this, unlike the lightweight aggregate, would find application in road construction both in concrete road pavements, as a roadstone, and in bridges. Moreover, lightweight aggregates of all kinds are at present used principally in the manufacture of blocks for internal partitions and inner leaves of cavity walls, whereas their utilisation in structural concrete is small. By contrast, a dense aggregate like dense air-cooled blastfurnace slag, would be more readily accepted for use in structural concrete. Against this background, a process for manufacturing dense aggregate from spoil is being developed at BRE[17]. Here also the fuel present in the spoil would be utilised.

It should be noted that the direct use of unburnt colliery spoil as dense aggregate is prevented by the presence of coal, clay minerals and pyrites which incurs failure of BS 882[18]. Burnt spoil contains soluble sulphate, and is therefore similarly excluded. It is possible, however, that concrete adequate for undemanding applications could be made with 'lower grade' aggregates than are at present used and untreated colliery spoil is included in BRS research to explore such potential aggregates. Burnt or unburnt spoil can be used as raw material in cement manufacture by the main' limestone/clay process where it replaces clay. It has been similarly utilised in the cement/sulphuric acid process in which calcium sulphate and clay serve as the raw material[19].

The stockpiles of spoil are very large and landscaping clearly provides an important method for dealing with spoil heaps. Nevertheless, utilisation of spoil in concrete manufacture deserves further serious attention since it is the one waste material available in sufficient quantity to make a real impact on aggregate supply in the future.

CHINA CLAY WASTE

The excavation of china clay, in the St Austell-Bodmin and Plymouth areas for use as paper filler and as a raw material in the ceramic industry, is accompanied by the production of 9 tonnes of waste per tonne of china clay. This consists of 3.7 tonnes of coarse sand, 2 tonnes of waste rock, 2 tonnes of overburden and 0.9 tonnes of micaceous residue[2]. The sands are the most promising of these for use in concrete and their properties are therefore given in Table 4[2].

At present, the coarse sand, overburden and waste rock are stockpiled in the large white conical heaps conspicuous in Cornwall. An average conical heap contains 0.25 M tonnes, is 50 metres high and occupies 1.25 hectares of land[20]. Most of the micaceous residue is discharged in suspension into mica dams where the bulk of the solid matter settles out and the water, still carrying some fine residue, is disposed of in local rivers. In future, however, to avoid river pollution, the majority of the residue will be stored in disused pits.

English Clays Lovering and Pochin produce 2.5 M tonnes of china clay per annum, 80 per cent of the total production in the UK[2]. The stockpile of waste includes 125 M tonnes of coarse sand and the total is estimated at 280 M tonnes[2], the tips occupying approximately 800 hectares of land. The current production of coarse sand is 10 M tonnes per year, of micaceous residue 2 M tonnes per year and of waste rock and overburden about 10 M tonnes per year.

Present utilisation is small, 1 M tonnes of sand is used in the manufacture of building materials and as bulk fill in road construction. In the St Austell area[2], the waste is screened to produce 4 grades of sand: a $\frac{3}{8}$ in gravel, a Zone 2 BS 882 sand, fine sand to BS 1200 and a non-standard sand used locally. Concrete blocks and calcium silicate bricks are also made. Figure 4 illustrates one use of precast units from china clay waste.

In view of the quantities of china clay waste produced and the expected expansion of production it would clearly be desirable to find new uses not confined to the local market for these wastes. This is first of all a transport problem since the china clay industry is remote from the South-East and other centres where the demand for aggregates is great. The possibility of bulk transport to special depots has received attention and should not be ruled out. China clay

Table 4 Chemical and physical properties of china clay waste sands

	Coarse Sand
	Typical properties of sands from 6 pits
Mineralogical analysis	Percentage
Quartz	60 - 80
Feldspar	1 - 15
Tourmaline	2 - 10
Mica	0.5 - 15
Chemical analysis	Percentage by wt
SiO_2	75 - 90
Al_2O_3	5 - 15
Fe_2O_3	0.5 - 1.2
TiO_2	0.05 - 0.15
CaO	0.05 - 0.5
K_2O	1.0 - 7.5
Na_2O	0.02 - 0.75
MgO	0.05 - 0.5
% loss on ignition	1 - 2
Physical properties	
Specific gravity	2.6 - 2.65
Apparent specific gravity	2.63 - 2.7
Water absorption	0.5 - 1.0
Particle size	9 mm to 75 μm

sands provide white concrete and can be screened to meet the requirements of BS 882 Table 2 Zones 1, 2 and 3 and they have been used for the production of low to medium strength concrete up to high strength pre-stressed concrete [2]. The mica present has adverse effects on strength and workability however and to achieve comparable strengths more cement has to be used than with river sand. Manufacture of synthetic aggregates from the sands by heat treatment, and use of rock waste as coarse aggregate deserve attention. Equally, some low quality clays unusable directly may provide raw material for white cement manufacture, but they would have to meet stringent requirements for iron, chromium and manganese contents.

Figure 4 Tater-du Lighthouse, Cornwall. Aggregate from china-clay wastes used for precast horn panel units and blocks

SLATE WASTE

Only a small part of the slate produced at the quarries is converted into roofing slates and other marketable products while 90 per cent of the quarried output accumulates in mountains of loose tipped waste such as those at Blaenau Ffestiniog. Slate is a metamorphic rock, fine textured, and consisting of chlorite, sericite, quartz, haematite and rutile. The slate waste is inert but the slate tips are unattractive particularly when they intrude into National Parks. About 1.2 M tonnes of slate waste are now produced per annum from quarries in Wales, the Lake District, and Devon and Cornwall. A typical analysis of slate is given in Table 5.

Table 5 **Chemical composition of slate (average of 4 samples from different Scottish quarries)**

	wt %
SiO_2	54.7
Al_2O_3	18.1
FeO	5.4
Fe_2O_3	1.7
TiO_2	1.0
Mn_2O_3	0.1
CaO	1.7
MgO	3.9
Na_2O	1.5
K_2O	2.8
SO_3	0.4
F	0.02
P_2O_5	0.15
Sulphide S (other than pyritic)	0.13
Pyritic and Organic S	0.4
Loss on ignition	6.7
	98.7

The utilisation of slate waste is at present very small. Until recently, a lightweight aggregate named Solite was produced near Wrexham, in a plant of 150 000 tonnes per annum capacity while ground and sized waste is used as inert filler in bitumen road surfaces, rubber, paint, paper, linoleum and plastics and as granules to provide a surface on roofing felt.

To achieve wider application of slate waste in concrete, manufacture of more lightweight aggregate is the obvious step. As early as 1932, BRS surveyed the quality of Welsh [21,22] and Scottish slate [23] in this respect [22]. When heated rapidly to an optimum temperature, certain slates expand giving a porous product light enough to float on water. The expansion is due to the generation of gases from material decomposed within the slate. For successful expansion, giving 3 to 7 times the original volume, there must be sufficient gas evolution when the slate is in a plastic state and the plastic range must be fairly long, approximately 100°C, to allow for variation in the kiln temperature.

To produce lightweight aggregate, crushed slate is heated between 1100°C and 1300°C in a rotary kiln. Nodules of expanded slate are composed of small cells separated by glassy walls, the whole being covered by a vitrified skin impervious to water. The strength and durability and dimensional stability of concrete with such aggregate are satisfactory.

As with china clay waste, the obstacle to utilisation lies in the geographical location of the slate waste. However, if methods of transport, perhaps by sea, could be developed, slate waste would have good prospects as a lightweight aggregate particularly since the processing costs are lower than for any other synthetic aggregate and there should be no air pollution problems such as with colliery spoil.

WASTE FROM COAL BURNING POWER STATIONS

The generation of electricity consumes about 45 per cent of all the coal used in England and Wales. In most of the other main uses of coal, the residual ash is either consumed in the manufacturing process, for instance in iron and cement manufacture, or the residue is widely distributed in small quantities as the ashes from domestic fires. In 1970 [2], 136 out of a total 187 power stations were coal fired and 75 of these stations are fired with pulverised coal and produce pfa. The remainder use lump coal and produce fused ash or clinker. In 1970-71, 9.9 M tonnes of pfa and 2.3 M tonnes of clinker were produced. As the coal residue changed from the lump clinker of pre-war years to pfa, new uses were sought for this new by-product and CEGB set up its Ash Marketing Organisation to sell ash and reduce disposal costs. Pfa has now become accepted for various purposes and in 1971-72, 64 per cent of the 9.9 M tonnes were utilised. It should be noted that the CEGB charter does not allow the Board to operate any commercial enterprise other than the generation of electricity.

The principal waste from pf burning stations is a fine dust collected in cyclones and electrostatic precipitators. The particles of this ash are predominantly spherical and a small proportion consists of hollow spheres of very low density. The fineness resembles that of cement varying between 250 and 500 m^2/kg. The residual carbon content may vary from 3 per cent in modern stations to 10 per cent in the older stations, particularly those using anthracite coals. Chemical analysis of pfa is given in Table 6 [24].

Table 6 Principal chemical constituents of pfa

Chemical	Percentage		
	Max	Min	Typical
Silica (as SiO_2)	58	38	48
Alumina (as Al_2O_3)	40	20	26
Iron Oxides (as Fe_2O_3)	16	6	10
Calcium (as CaO)	10	2	4
Magnesium (as MgO)	3.5	1	2
Sulphate (as SO_3)	2.5	0.5	1.2
Alkalis (as Na_2O, K_2O)	5.5	2.0	4.5

The presence of about 50 per cent finely ground silica leads to useful pozzolanic properties and BS 3892:1965 covers superficially the properties of pfa for use in concrete but this standard does not include tests for pozzolanic activity.

Approximately 25 per cent of the coal ash in pf burning stations clinkers in the furnace to form conglomerates too heavy to be carried by the updraught. This waste collecting at the base of the furnace is furnace bottom ash which is a fused and porous material. After quenching, it is stored in sizes varying from coarse sand to 300 mm or more. BS 3797:1964 covers the use of furnace bottom ash in concrete.

Another fraction (0 to 5 per cent) which has special properties is cenospheres, the hollow spherical particles often known as floaters. They have a bulk density of 250 kg/m^3 as compared with 1000 kg/m^3 for pfa. These particles are 50 to 125 microns in size and their silica content is high being 55 to 60 per cent [2].

Furnace clinker from burning of lump coal on travelling grates or from retort stoker firing consists of glass with inclusions of refractory fragments and nodules of unburnt coal. Sulphur compounds present preclude the use of clinker in reinforced concrete. Residual fuel content can be high, leading to the limit of 10 to 25 per cent for loss on ignition in BS 1165:1966 which governs use of furnace clinker in concrete.

The disposal of wastes is considered when the sites for new pf burning power stations are selected. The storage capacity may be sufficient for the life of the station or may provide enough space to cover delays in ash removal. Disposal costs are borne by the CEGB either in terms of capital outlay on disposal sites or in transport costs and tipping rights.

Dry ash can be transported in road tanker wagons but it is readily air-borne and if aerated will flow like water. Accordingly, water addition is made in bulk disposal by discharging the ash as 40/60 ash/water slurry into large settling tanks or lagoons. The ash settles and the water flows away over a weir. The lagoons are visually unattractive but in time they are grassed over. Alternatively, ash can be discharged into a continuous mixer in which water is added to produce 'conditioned ash'. The aim is to produce a moist semi-dry material which will compact and is transportable in open top vehicles or rail wagons either for stockpiling or for use as a fill. Disposal in clay pits and sea dumping are the other methods of disposal. Figure 5 illustrates disposal of ash. Some furnace bottom ash is used in blockmaking and furnace clinker is all used. In Table 7 [2] the geographical distribution of power station waste is shown.

Table 7 Residues from pf burning power stations 1970-1971

pfa areas		Total residue tonnes	pfa tonnes	Furnace bottom ash (by difference) tonnes
1	Central England	6 780 000	5 090 000	1 690 000
2	London - SE England	1 350 000	1 020 000	330 000
3	NE England	905 000	675 000	230 000
4	S Wales and Bristol	830 000	620 000	210 000
	TOTAL	9 865 000	7 405 000	2 460 000

Table 8 Production of furnace clinker

		1970-71 tonnes	1971-72 tonnes
1	Central England	1 559 000	1 074 000
2	London and SE England	532 000	218 000
3	NE England	50 000	54 000
4	S Wales and Bristol	74 000	38 000
5	SW England	116 000	34 000
	TOTAL	2 331 000	1 418 000

Production of furnace clinker summarised in Table 8 shows the reduction in the supply of this waste between 1971 and 1972 but for detailed breakdown of uses the recent survey [2] should be consulted. It is important to note that in 1970-71, 73.4 per cent of pfa used was employed as fill, 16.9 per cent in concrete blocks and only 4.5 per cent as lightweight aggregate.

Figure 5 - PFA used for filling brick pits, Peterborough

Variability in the properties of pfa can present problems to the user and even when ash is used as a fill the 'conditioning' of ash to an optimum water content for the particular batch is vital to ensure maximum compaction.

Many other countries produce specified grades of pozzolanic Portland cements from mixtures of cement clinker and pfa and prospects for increased utilisation in the UK exist in cement manufacture. Pozament cement is of this type and has an Agrément Certificate. A company with experience of pfa in Australia has recently built a plant at Fiddlers Ferry Power Station to process and grade ash in order to sell it as a product of guaranteed pozzolanic quality and research is in progress at BRS to establish the engineering properties of concrete made with pfa currently available in the UK. Data will be obtained to enable revision of BS 3892, Pfa for use in concrete, especially by defining pozzolanic properties.

Substitution of pfa or ground granulated slag for Portland cement could lead to substantial direct energy savings in terms of fuel and grinding energy. Production of 1 tonne of Portland cement consumes about 6300 M Joules compared with about 50 M to 100 M Joules for graded and blended pfa and about 200 M to 300 M Joules for ground granulated blastfurnace slag. If 1 M tonnes of Portland cement are replaced by pfa, 6000 M Joules [2] or 0.21 M to 0.24 M tonnes of coal equivalent, costing £1.8M to £2.0M would be saved. The new pf burning power station to be commissioned at Didcot will produce 1 M tonnes of pfa per annum. This could make a useful contribution to material supply in the South-East.

It may be considered that pfa is really 'too good' to be used as fill. It is also often difficult to secure the large quantities needed over short periods. Now that colliery spoil is approved as fill, it would seem more worthwhile to upgrade the usage of pfa and at the same time save energy and limestone resources by using pfa for instance in cement manufacture.

BY-PRODUCT CALCIUM SULPHATE

Calcium sulphate is obtained as a by-product as anhydrite $CaSO_4$, hemihydrate $CaSO_4 \frac{1}{2} H_2O$, or as gypsum $CaSO_4 . 2H_2O$. The most important source is the manufacture of wet process phosphoric acid in which 5 tonnes of by-product gypsum are produced for each tonne of phosphoric acid leading to 2 M tonnes of the by-product per annum in the UK. Possible utilisation of the by-product must be viewed against the background of the usage of natural calcium sulphate.

Natural gypsum is used first of all in the manufacture of gypsum building plasters (35 per cent) and plasterboard (25 per cent) but 26 per cent is used as cement retarder. 84 per cent of the anhydrite mined is used in the Müller-Kühne process in which sulphuric acid and Portland cement are produced. This consists of roasting ground anhydrite, clay and coke (or pf) to 1200 to 1400°C when $CaSO_4$ is reduced to CaO which combines with the rest of the charge to form Portland cement clinker and gases containing 9 per cent SO_2. The SO_2 is converted into sulphuric acid.

Phosphogypsum is at present disposed of by pumping a slurry into the sea or is dry dumped [2], none is used. In considering possible uses for phosphogypsum it is necessary to take into account the phosphate and fluoride impurities which may cause difficulty and also small traces of radioactive matter. Nevertheless, phosphogypsum can replace natural gypsum and anhydrite in plaster and plasterboard manufacture and as retarder in cement or raw material in the cement/sulphuric acid process [2]. The application in plasters is outside the scope of this paper and it will only be mentioned that α-hemihydrate could be more economically produced from the by-product [2] and this could benefit the manufacture of glass reinforced gypsum (grg) [2,25]. The use of various by-products or chemical gypsums including phosphogypsum as set controllers in Portland cement has been studied by Murakami [2,26].

The phosphate and fluoride impurities present in phosphogypsum may delay setting excessively and reduce the rate of early strength development although strengths at later ages of test are not adversely affected. The adverse effects on setting and early strength are most pronounced with gypsum from the dihydrate process and can be avoided by suitable treatment of the phosphogypsum or by the use of gypsum obtained from hemihydrate processes. Use as retarder represents the largest potential outlet after use in plaster, since 740k tonnes of natural gypsum per annum are needed for this purpose.

Utilisation in the cement/sulphuric acid process is attractive since it solves a disposal problem in phosphoric acid manufacture and simultaneously leads to more sulphuric acid as shown in Figure 6.

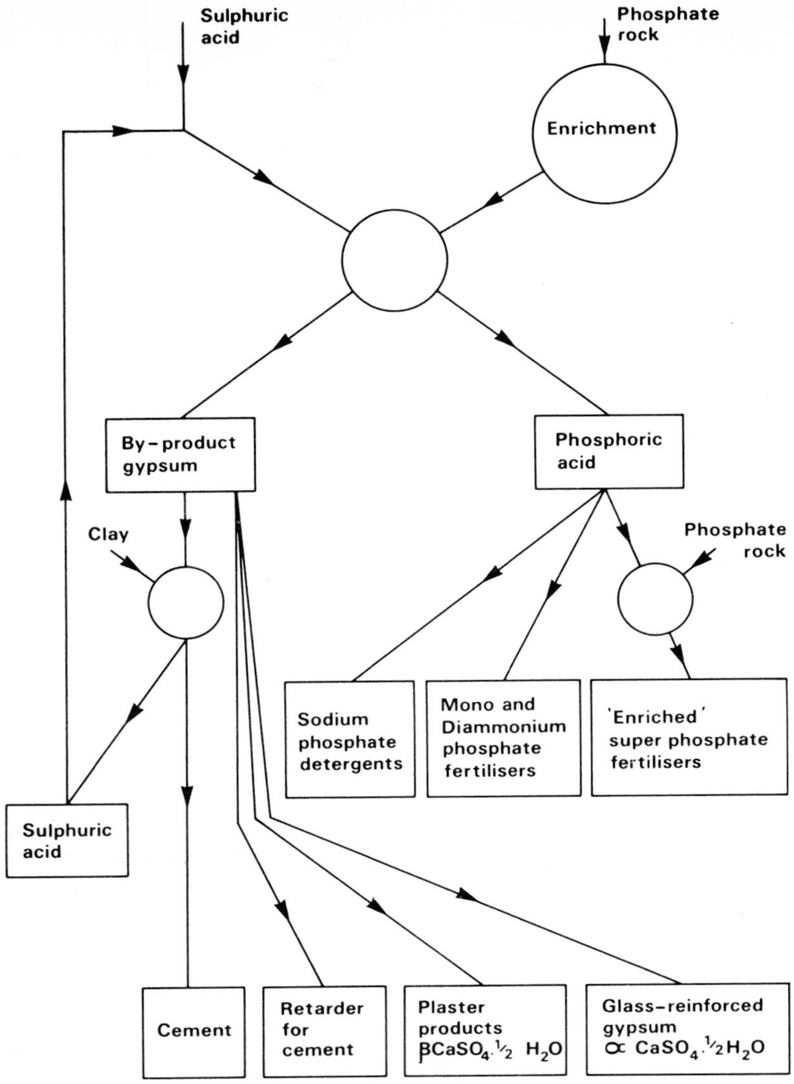

Figure 6 - Production and utilisation of phosphogypsum

Exploring this process, BRS showed in 1967 [27] that phosphogypsum from Whitehaven which had been reslurried and washed could lead to satisfactory cement clinker containing 1.2 per cent P_2O_5 and cements produced resembled those made from phosphatic/fluorinated raw materials using the clay/limestone process [28]. The presence of phosphate in the raw materials restricts the formation of alite, the main cementing phase present in cement clinker [28]. This effect is partly countered however by the presence of fluorides which on their own tend to be harmful but are beneficial in phosphatic cement. Since phosphatic cements are characterised by a slow rate of strength development at early ages, it is essential to limit the amount of phosphate in cement clinker to about 2.5 per cent P_2O_5. The present low price of imported sulphur, about £11 per tonne, makes use of phosphogypsum uneconomic but these circumstances could change. It is also noteworthy, that colliery spoil and phosphogypsum could replace clay and gypsum respectively in the cement/sulphuric acid process leading to the manufacture of cement and sulphuric acid entirely from waste materials [9,29].

The position on the radioactivity of phosphogypsum has recently been summarised [30,31].

Hydrofluoric acid is made by the reaction between fluorspar and sulphuric acid in heated kilns. The by-product of this reaction is anhydrite (β $CaSO_4$) 80k to 90k tonnes being produced per annum. A proportion of this is used to make 'Synthanite' floor screeds and the remainder is used as land fill or dumped [30].

In the USA where SO_2 is extracted from flue gases and oil there is a surplus of sulphur. This is not so in the United Kingdom where imports of sulphur cost £13 M per annum and where phosphogypsum represents a reservoir of sulphur the value of which could increase in the future.

RESIDUES FROM DIRECTLY INCINERATED REFUSE

40 direct incinerators have been planned and 18 are working[2]. These large installations, costing between £1M and £19M, take all the material directly into the incinerator. Even if alternative methods of refuse disposal prove more satisfactory, these installations are likely to remain in use for 10 to 30 years. Fundamental problems involved in this method include limitations placed on tall chimney effluents and loss of materials that might be recovered. Refuse disposal as a whole has been reviewed recently by a DOE Working Party[32] and Table 9[2] gives the principal methods of disposal and states their advantages and disadvantages. In 1970, total domestic and trade refuse collected by local authorities in England was 14 M tonnes[32] and this is expected to increase to 17 M tonnes by 1980. The collection, method of handling and transport of domestic refuse are being investigated at BRS and this could lead to a pulverised product reaching an incinerator by pipeline[33, 34].

Table 9 Methods used for refuse disposal

Method	Advantage	Disadvantage
Controlled tipping	Provides a means of land reclamation	May involve increased transportation
Pulverisation before controlled tipping	Higher density in tip. Less subsequent settlement. Less top covering required. More acceptable near housing. Recovery of metals possible.	Additional processing plant required.
Refuse baling before controlled tipping	As for pulverisation - maximum use of tipping space.	As for pulverisation.
Separation/incineration	Large reduction in volume. Efficient recovery of scrap for small installations.	High man-power requirement for unpleasant duties.
Direct incineration	Large reduction in volume. No fuel required. Ferrous metal extraction from clinker. Possible to further extract aluminium and copper. Possible to use waste heat	High installation and main-tenance costs. Pockets of unmixed trade wastes in the fire-bed disrupt burning control. Tall chimney effluent becoming more unpopular. Recovery limited to non-combustibles. Corrosion problems with waste heat utilisation.
Composting	Can be composted with sewage sludge. Attractive to areas where soil humus level is depleted.	Expensive process and still have to tip a proportion of the refuse
Pyrolysis	No air pollution; recycling of waste. Metal extraction processes applicable. Gaseous, liquid and solid by-product fuels produced in excess of those required in process. Other chemical by-products from condensate. Satisfactory outlet for by-products must be available for the advantages to be realised.	Only foreign pilot plant experience available at present. High installation cost.
Hydrolysis	Suitable for refuse with high paper content when it produces sugars, protein yeast etc.	Only theoretical exercises and small pilot projects on special trade wastes are at present available.

The aim of direct incineration is to accept large quantities of refuse continuously and to reduce the bulk by 90 per cent with the production of an inert residue. The majority of installations use the inclined hearth principle. The refuse is fed in at the top of the incline and the residual clinker is discharged at the bottom. Refuse delivery vehicles discharge their loads into pits and the refuse is extracted by overhead grab. No recovery from the refuse is attempted until the burnt out clinker has been water-cooled and is on a conveyor system to the final discharge hopper. Here, ferrous metal is extracted by an overhead magnetic band. Recovery of heat is practiced at Edmonton to generate electricity and the heat is also used for district heating at Nottingham and Newcastle. The total quantity of incinerated residue produced per annum is likely to reach 2 M tonnes in 1975. A typical analysis of the clinker is given in Table 10 and shows 23 per cent metal and 23 per cent glass to be present. Residual carbon in well burnt refuse amounts to 5 to 10 per cent. Extraction of ferrous metal is likely to reduce the metal content to about 12.5 per cent and increase the glass content to 25 per cent and the carbon plus ash content to 62.5 per cent.

Table 10 Composition of domestic refuse predicted for 1980

Constituent	% by weight of raw refuse	Estimated result of incineration	Proportions after reduction to 40%	% by weight of burnt refuse	% by weight after ferrous extraction
Metal	9	unchanged	9	23	12.5
Glass	9	unchanged	9	23	25
Paper	43	reduced to:	2.5	6	
Rag	3	reduced to:	0.5	1	
Vegetable and putrescible	17	reduced to:	4.0	10	62.5 mineral clinker and residual combustible (Paper to Unclassified debris)
Plastics	5	reduced to:	1.0	2	
Dust and cinder	12	unchanged	12	30	
Unclassified debris	2	unchanged	2	5	
	100		40	100	100

Warren Spring Laboratory have devised a method for extracting both ferrous and non-ferrous metal in which the clinker is broken down in hammer mills and crushing rolls and iron, aluminium and copper are separated by fluid bed methods. If the residual fines could be used[35], the economics of the process would be improved. The Warren Spring metal extraction scheme yields a residue which is within one of the Standard grading zones for sand (Zone 2) and could therefore serve as a fine aggregate for concrete but deleterious components are present and have to be considered. Ferrous metal present may cause unsightly staining of concrete, the glass present could lead to alkali-aggregate reaction and lead and zinc might interfere with the setting of Portland cement. The generation of hydrogen due to interaction of aluminium and hydrating cement could be troublesome. Nevertheless, it may be worth considering the limited use of the material as fines in the manufacture of clinker blocks provided that the sulphate and carbon contents remain within the accepted BS limits. One important advantage of this material is its availability in areas of high construction activity.

The clinker and ash from direct incineration, not treated by the above process, are too variable and contaminated for direct use in construction except as fill.

MISCELLANEOUS WASTES

The properties of lead-zinc blastfurnace slag have been described[2], the quantities produced are small, 60 k to 70 k tonnes per annum. 100 k tonnes of copper slag are also produced[2]. The tin slag production is 75 k tonnes per annum[2]. None of these slags constitute a significant environmental problem nor are they promising for use in concrete.

100 k tonnes of red mud are produced per annum. This is the waste from the production of alumina from bauxite by the Bayer process, as a stage in the manufacture of aluminium by electrolysis of the alumina in fused cryolite. The main constituent is iron oxide. In the UK red mud is of minor importance.

Tin mine tailings are at present discharged into tailing dams. If prospecting for metals in the UK were to increase, utilisation of mine tailings would deserve attention.

There are some 250 quarries in England and Wales and they also produce waste material. Glass waste is more likely to find re-use in glass and glass fibre manufacture than in concrete.

CONCLUSIONS

The main findings of the recent BRE survey[2] of the locations, disposal and prospective uses of the major industrial by-products and waste materials were summarised and are reproduced in the summary table at the end of this paper which will help to place the information in this paper and the conclusions in perspective.

China clay waste is the only major by-product, apart from blastfurnace slag which is already fully utilised, that in the present state of knowledge is suitable for use as an aggregate in concrete without processing. Although china clay sand is extensively used as a fine aggregate in concrete in the South West, the costs of transporting it to areas of high construction activity have restricted its use up to now. Other untreated waste materials may be usable as aggregate in concrete where the performance requirements are undemanding but further research is necessary to demonstrate this in specific cases. Such assessment of colliery spoil, for instance, is in progress at BRE.

Manufacture of lightweight aggregates from colliery spoil, pfa, blastfurnace slag and slate waste is well established but small, only 1 Mm^3 of such synthetic aggregate being produced per annum. Air-cooled blastfurnace slag which is used as a dense aggregate has become accepted more readily than other by-products and if a viable synthetic dense aggregate could be developed from other wastes it is likely to find a bigger market than the lightweight aggregates. This is especially valid in relation to concrete for roads and bridges, where lightweight aggregate concrete is not employed.

Colliery spoil and china clay waste are produced in sufficient quantities to make a large contribution to aggregate supply, and their utilisation could simultaneously help to solve considerable disposal problems. Since the stock piles of these materials are very large they would provide security of supply for any manufacturing process dependent on them. Production of lightweight aggregate from slate waste could also make a useful contribution in view of simple processing but transport problems need to be overcome.

The use of pulverised fuel ash in the manufacture of cement would upgrade the use of this by-product and would help to conserve limestone resources used in Portland cement manufacture and as aggregate. Moreover, partial substitution of pfa or ground granulated slag for Portland cement could lead to substantial and direct savings in fuel and grinding energy.

By-product calcium sulphate represents a reservoir of sulphate which could be brought into use either as cement retarder or as a raw material for cement manufacture. This would help to solve disposal problems and in the latter case achieve savings in sulphur imports.

ACKNOWLEDGEMENT

The preparation of this paper was suggested by the Materials Technology Committee of the Concrete Society.

REFERENCES

1 Report of aggregates and waste materials Working Group. BRE Current Paper CP 31/73.

2 **Gutt, W, Nixon, P, Smith, M A, Harrison, W H** and **Russell, A D**. A survey of the locations, disposal and prospective uses of the major industrial by-products and waste materials. BRE Current Paper CP 19/74.

Table 11 Summary table of the locations, quantities and disposal or use of the major waste materials

The main figures for production and use refer to 1970/71; where the figures for 1971/72 are known to be significantly different these are given in parentheses. The quantities refer to Great Britain except those marked* which refer to England and Wales.

Type	Main Locations	Origin	Production M tonnes/ Year	Stockpile M tonnes	Method of disposal	Present uses	Existing or new uses deserving research and development or increased application	
							Existing	New
Colliery spoil	Northumberland, Durham Yorkshire, Lancashire, Midlands, South Wales, Kent, Central Scotland	Mining coal	about 50	3000	Mainly tipping on land, some in sea. 7-8M tonnes/ year used	As fill and in manufacture of bricks, cement and lightweight aggregate	Synthetic light-weight aggregate	Synthetic dense aggregate
China clay waste – overburden – sand – micaceous residue	Cornwall and Devon	Quarrying of china clay	22	280	Mainly tipping and in lagoons. About 1 M tonnes/ year used	As fine aggregate in concrete, in manufacture of bricks and blocks and as fill	Fine aggregate in concrete	Synthetic aggregate
Slate waste	Wales, Lake District and Cornwall/Devon. Old tips but no production in Scotland	Mining and quarrying slate	1.2	Over 300	Mainly tipping, some back-filling of old workings. Minor usage	Inert filler, granules, expanded slate aggregate and filter medium, road building	Expanded slate aggregate	Slate-lime bricks
Pulverised fuel ash and furnace bottom ash	Countrywide	Waste from power stations burning pulverised fuel	9.9 (9.4)*	Not known	6.3 (5.4) M tonnes/year used. Rest in old workings or artificial lagoons	As fill and in manufacture of cement, concrete blocks, light-weight aggregate, bricks, etc.	Pozzolana in cement manufacture. Light-weight aggregate manufacture	
Furnace clinker	Countrywide	Waste from chain grate power stations	2.3 (1.4)*	–	All used	Concrete block making		
By-product calcium sulphate	Bristol area, Runcorn, Rotherham, Immingham, Billingham, Whitehaven, Leith, Aberdeen	Manufacture of phosphoric acid and of hydrofluoric acid	2.1	–	Mainly in sea. Some dumped on land. 6k tonnes/ year used	In manufacture of floor screeds		Production of hemi-hydrate plaster
Blastfurnace slag	South Wales, Corby, Sheffield, Scunthorpe, Consett, Teesside, Glasgow area	Smelting of iron	9	–	Nearly all used	As roadstone, railway ballast, filter medium, aggregate for concrete, fertiliser and in manufacture of cement	Aggregate in concrete, manufacture of cement	

Table 11 Summary table of the locations, quantities and disposal or use of the major waste materials (continued)

Type	Main Locations	Origin	Production M tonnes/ Year	Stockpile M tonnes	Method of disposal	Present uses	Existing or new uses deserving research and development or increased application: Existing	New
Steel-making slag	South Wales, Corby, Sheffield, Scunthorpe, Consett, Teesside, Glasgow area	Steel making	4	Not known	2 M tonnes returned to blastfurnaces, rest either dumped or used as fill near steel works or sold	As roadstone	Skid resistant roadstone	
Incinerator ash	Countrywide	Residue from the direct incineration of refuse	0.6 (0.7)*	Not known	Most is dumped. Minor usage	As fill and for covering refuse tips		Component in aggregate for blockmaking
Zinc-lead slags	Avonmouth	Smelting of zinc and lead	0.06-0.07	Not known	Stockpiled and used locally	Bulk fill and some in pavement asphalt	Grit blasting	Fine aggregate in concrete slag-lime bricks
Copper slag	Widnes and Walsall	Smelting of copper	0.1	–	Complete utilisation	Grit blasting	–	–
Tin slag	Hull and Liverpool	Smelting of tin	0.074	Not known	Major utilisation. Some in tips	Grit blasting and road building	–	–
Red mud	Burntisland (Fife)	Production of alumina	0.10	Not known	Minor utilisation rest in lagoons	Pigment in paints and plastics	–	Synthetic roadstone
Tin mine tailings	Cornwall	Tin mining	0.46	Not known	Minor utilisation. Tailing lagoons and discharge into sea	Aggregate for concrete	–	Manufacture of bricks and aerated concrete blocks
Fluorspar mine tailings	Derbyshire and Durham Pennines	Fluorspar mining	0.23	Not known	Minor utilisation. Tailing lagoons	Aggregate for roadmaking and concrete	Aggregate for concrete and roadmaking	
Quarry wastes	Countrywide	Quarrying	Not known	Not known	Some utilisation. Rest tipped	Roadmaking, brickmaking	Manufacture of bricks and as fine aggregate in concrete	
Waste glass	Countrywide	Waste glass within domestic refuse	about 2	Not known	No utilisation. In domestic refuse	–	Re-use in glass making and in manufacture of glass fibre	

3 **Gutt,** W. Manufacture of cement from industrial by-products. Chemistry and Industry, No 7, 1971, p 189. (Also BRS Current Paper CP 19/71.)

4 **Gutt,** W. Aggregates from waste materials. Chemistry and Industry, 1972, p 439. (Also BRS Current Paper CP 14/72.)

5 **Parker,** T W and **Ryder,** J F. J Iron and Steel Inst, Vol 146, Part II, 1942, p 211.

6 **Gutt,** W, **Kinniburgh,** W and **Newman,** A J. Mag Concr Res, Vol 19, 1967, p 71; Discussion, p 125.

7 **Everett,** L H and **Gutt,** W. Steel in concrete with blastfurnace slag aggregate. Mag Conc Res, Vol 19, 1967, p 83.

8 **Teychenné,** D C. Lightweight aggregates their properties and use in concrete in the UK. Proc Int Congress on lightweight aggregates, London 1968. (Also BRS Current Paper CP 73/68.)

9 **Nurse,** R W. The chemistry of cements. Edited by H F W Taylor, London, Academic Press, Vol 2, 1964, Chap 13.

10 **Parker,** T W and **Nurse,** R W. DSIR Building Research Technical Paper No 7, London, HMSO, 1949.

11 **Gutt,** W, **Teychenné,** D C and **Harrison,** W H. To be published.

12 **Gutt,** W. J Iron and Steel Inst, Vol 210, 1963, p532, Vol 202, 1963, p 770, Vol 205, 1968, p 840.

13 Cement Lime and Gravel, Vol 47, No 3, 1972, p 67.

14 **Sherwood,** P T and **Rigley,** M D. Road Research Laboratory Report LR 324, 1970.

15 **Fraser,** C K and **Luke,** J R. Road Research Laboratory Report LR 125, 1967.

16 **Cook,** P B. Int Symp on Lightweight expanded clay and sintered fly ash aggregate, 1969, Belgrade.

17 **Gutt,** W and **Russell,** A D. Patent application no 1979/74 UK 1974.

18 BS 882:1965. Specification of aggregates from natural sources for concrete.

19 **Gutt,** W and **Smith,** M A. Cement Technology, Vol 2, Part 2, 1971, p 41, and Vol 2, Part 3, 1971, p 91.

20 Derelict Land, Civic Trust. London.

21 **Colman,** E H and **Nixon,** P J. A survey of possible sources in Wales of raw materials for the manufacture of lightweight expanded slate aggregate. BRE Current Paper to be published.

22 **Colman,** E H. Concrete and Constructional Engineering, 1935.

23 BRS unpublished work.

24 PFA Data books. CEGB, London, 1967

25 **Ryder,** J F. Proceedings of the International Building Exhibition Conference 1971, HMSO, 1971, p 69.

26 **Murakami,** K. Proceedings of the 5th International Symposium on the Chemistry of Cement, Tokyo 1968. Tokyo, Cement Association of Japan, 1969, p 457.

27 **Gutt,** W and **Smith,** M A. Cem Technol, Vol 2, Part 2, 1971, p 41, and Vol 2, Part 3, 1971, p 91.

28 **Gutt,** W. Proceedings of the 5th International Symposium on the Chemistry of Cement, Tokyo 1968. Cement Association of Japan, Vol 1, 1969, p 93.

29 Smith, M A and Gutt, W. Cement Technology, Vol 4, Part 1, 1973, p 3.

30 Gutt, W and Smith, M A. Chemistry and Industry, July 1973.

31 O'Riordan, M G, Duggan, M J, Rose, W B and Bradford, G E. National Radiological Protection Board, Report No 7, 1972, London, HMSO.

32 Refuse Disposal. DOE 1971, Report of the DOE Working Party.

33 Courtney, R G and Sexton, D E. BRE Current Paper CP 4/73.

34 Sexton, D E and Smith, J T. BRE Current Paper CP 12/73.

35 Miles, J E and Douglas, E. Surveyor, 8.12.72, p 36.

DISCUSSION

Prof R W Cahn. What has become of the apparently promising initiative by BISRA (as was) in researching and marketing a new glass-ceramic made by crystallising blastfurnace slag? It was, I understand, a superb flooring material.

W Gutt. So far this sophisticated process has not fulfilled commercially its early promise but I may not be up to date on this point and BISRA should be consulted. Simpler applications of blastfurnace slag are being successful.

Dr R J Murray. APCM Ltd. We are carrying out experiments on the use of by-product calcium sulphate as a cement retarder but this poses problems. As far as the production of blended cements from pfa is concerned, the transport costs can be high, and we have doubts about the economics of this process. Will Dr Gutt please comment.

W Gutt. Pulverised fuel ash (pfa) is used abroad more than in UK as a partial substitute for cement. An Australian firm offers blended pfa in the UK for site addition to concrete. It is cheaper - pfa costs £4 a ton whereas cement costs £10 per ton.

M A Smith. About 70 per cent of the cement used in France contains either pfa or slag. In other European countries 20 to 25 per cent cement is made from slag. Very little pfa is used in cement manufacture in UK. A reason to expect higher usage in France in the future is a recent government directive to save 10 per cent of fuel used to make cement.

Mr J N Wanklyn. I should like to ask Prof Page what role he sees for modern science and technology in the development of building materials. Some years ago Buckmaster Fuller pointed out that up to about 1880 available materials were stronger in compression than in tension, so that the construction of buildings demanded the piling up of large weights of material. With newer materials, eg high strength alloys, this should no longer be necessary; and large savings of weight might be possible. These savings could outweigh the other costs involved in the use of 'high technology' in place of traditional technology. I should be interested to hear Prof Page's views on the balance between these two types of technology and the role that modern materials science has to play in the development of building methods.

W Gutt. The development of glassfibre reinforced cement composites is the subject of a research and development programme at the Building Research Establishment. This work has centred on the invention by Dr A J Majumdar of BRS of an alkali-resistant glass fibre. (Ordinary glass such as is used to reinforce plastics cannot be used to reinforce cement since the alkali liberated in the hydration of cement attacks the ordinary glass.) The BRE has patented the special alkali-resistant glass fibre via the NRDC, and Pilkingtons have an exclusive licence to manufacture it. It has the trade name of CEM-fil, and is at present available only to the 50 to 60 firms licenced by Pilkingtons. At present, and until durability studies are further advanced, grc is recommended only for nonstructural applications. Prospective users are welcome to consult BRE. Since the development of glass reinforced cement provides strong, thin, lightweight and fire-resistant sections which may replace heavier components in various applications, eg in cladding panels for buildings, this work is relevant to the conservation of materials. Application of grc is also relevant to replacement of asbestos which is associated with certain health hazards and is also an imported material. BRE has also developed glass reinforced gypsum (grg) for internal use which is exceptionally fire resistant and strong. Both grc and grg are made from indigenous materials.

Dr C V Phillips. I presume the BRE has made specification tests on aggregate and fine screened sand produced from china clay production. Such material is extensively used in the South West, and must contain some felspar which is still subject to kaolinisation with subsequent degradation. Does this process stop on cementitious use, or can the degradation lead to future weakness of structures?

W Gutt. Over the years BRS has made a number of studies on china clay sands. The kaolinisation process is hydrothermal and therefore is not a problem when the material is used in concrete; other degradation processes which can affect felspars are of no practical importance in relation to their use in Portland cement concrete.

The use of lighter-weight blastfurnace slag as dense coarse aggregate in concrete (CP 93/74)

W. Gutt, D.C. Teychenné and W.H. Harrison

SYNOPSIS

The use of imported iron ores has led, in recent years, to a lowering in the bulk density of some blastfurnace slags produced in the United Kingdom. To take account of this change, it was decided to examine the properties of this lighter-weight slag and its suitability as dense coarse aggregate for concrete. Accordingly, studies have been made of three slag aggregates having bulk densities below the minimum of 1250 kg/m³ specified in British Standard 1047:1952, which governs the use of dense coarse slag aggregates for concrete. This investigation, which included chemical and mineralogical examination of the aggregates and concrete durability studies up to five years from casting, has now been completed. It has been established that the strength and durability of concretes made with the lighter-weight slags did not differ significantly from those of control concretes made with flint gravel, crushed limestone and normal density slag, provided that adequate cement content and workability levels were ensured by appropriate mix design techniques. The investigations carried out included examination of reinforcing steel taken from the concrete. These findings could lead to a lowering of the bulk density minimum in BS 1047, and further encourage the utilization of blastfurnace slag as dense coarse aggregate for concrete.

Introduction

British Standard 1047:1952, *Air-cooled blastfurnace slag coarse aggregate for concrete,* requires that the bulk density of aggregate shall be not less than 78 lb/ft³ (1250 kg/m³), when tested in accordance with Appendix B. Slag fulfilling this requirement continues to be produced, but the trend towards the use of imported ores relatively rich in iron has resulted in the production of many slags with low bulk densities. In 1967, nine out of ten works surveyed by the Transport and Road Research Laboratory[1] produced slag having bulk densities between 1120 and 1265 kg/m³ and a considerable proportion of slag currently produced has a bulk density below 1250 kg/m³. In these circumstances, the British Slag Federation asked the Building Research Station to investigate the suitability of this 'lighter-weight slag', as it has come to be known, for use as coarse dense aggregate for concrete.

This work forms a logical extension of the preceding investigation at the BRS of slags passing the present minimum bulk density in BS 1047:1952. A comparative study of the properties of concretes made with such slag (of bulk density from 1250 to 1440 kg/m³) and with limestone or gravel as coarse aggregates was made over a period of six years. The results of this earlier study were presented in two papers at a short symposium held at the RIBA, London, in February 1967, and were published in the same year in the *Magazine of Concrete Research*[2,3] together with the discussion which took place[4].

In the present investigation, three 'lighter-weight slags' have been tested and one denser slag, a flint gravel (Thames Valley) and a crushed limestone (Cheddar) served as controls. The programme of work included a full examination of the slag composition by chemical analysis, microscopy and X-ray analysis. Concretes made with the aggregates under test as the coarse aggregate were examined for strength, volume stability, permeability and sulphate resistance. The behaviour of steel reinforcement in such concretes was also examined.

Mineralogical examination of aggregates

The bulk density and method of air-cooling of the four slags selected for investigation are listed in Table 1. Metallic iron was detectable in polished and trans-

TABLE 1: **Analysis of the four slags.**

	Slag			
	B419	B435	B428	B431 (control)
Method of air-cooling	—	pit	pit	ladle
Bulk density* (kg/m^3)	1185	1175	1240	1420
Free iron (%)	0·5	4·8	0·2	—
CHEMICAL ANALYSIS (%)				
SiO_2	32·16	31·51	33·95	33·81
Total Fe as FeO	1·10	0·90	0·42	0·74
TiO_2	0·88	0·53	0·44	0·70
Al_2O_3 by diff.	13·40	16·95	12·92	20·05
CaO	39·00	38·26	37·91	33·43
MgO	8·97	8·53	10·88	7·24
MnO	0·65	0·67	0·86	1·11
BaO	0·25	0·17	0·15	0·26
SrO	0·09	0·12	0·05	0·26
Na_2O	0·67	0·36	0·38	0·44
K_2O	0·80	0·75	0·53	1·13
S (sulphide)	1·48	0·76	0·89	0·98
Sulphate SO_3	0·39	0·34	0·54	0·17
Total S (calculated)	1·64	0·90	1·11	1·05
Total S (determined)	1·71	0·93	1·24	1·07

*measured on 19–4·8 mm fraction.

parent sections of the slags and it constituted a definite separate phase. Rounded inclusions of iron were seen and they were segregated from the melilite, the main phase in the slags. The percentages of free iron determined by chemical analysis of magnetic fractions washed free of slag are shown in Table 1; the full chemical analyses of the slags after removal of this iron are also shown. The sulphur contents are below the maximum in BS 1047.

The analytical results in Table 1 were used to investigate the composition of the slags for 'lime' unsoundness in accordance with equations A and B of BS 1047:1952. All the slags complied with the requirement of the British Standard in respect of both equations, as shown in Table 2.

The mineralogical composition of the slags was examined by microscopy and by X-ray analysis. Polished sections for examination by reflected light and transparent sections for examination by transmitted light were used for the microscopic work. The results are given in Table 3. For slags complying with the 'lime' unsoundness requirements of equations A and B in BS 1047, further compliance testing using the Microscope Test is not required. However, the BS 1047 Microscope Test was carried out in the course of this

TABLE 2: **Tests for 'lime' unsoundness (to BS 1047, equations A and B).**

Equation A: $CaO + 0{\cdot}8\ MgO \leqslant 1{\cdot}2\ SiO_2 + 0{\cdot}4\ Al_2O_3 + 1{\cdot}75\ S$

Equation B: $Cao \leqslant 0{\cdot}9\ SiO_2 + 0{\cdot}6\ Al_2O_3 + 1{\cdot}75\ S$

Slag	Analytical test						Microscope test		
	Equation A			Equation B					
	Left-hand side	Right-hand side	Result	Left-hand side	Right-hand side	Result	C_2S	Fe	CaS
B419	46·2	46·5	complies	39·0	39·6	complies	not detected	present	not detected
B435	45·1	45·9	complies	38·3	39·9	complies	detected in β and γ forms	high	present
B428	46·6	47·5	complies	37·9	39·9	complies	not detected	present	present
B431 (control)	39·2	50·3	complies	33·4	44·2	complies	not detected	present	present

TABLE 3: **Mineralogical analyses of slags.**

Slag	Examination of transparent sections by microscopy	Phases detected by X-ray analysis
B419	Melilite, C_3S_2, C_3MS_2 or CMS or CMS_2 Oldhamite* (single grains)	Mainly melilite, C_3S_2, CMS C_3S_2 possibly
B435	Melilite, β-C_2S, γ-C_2S, α-CS, CMS Oldhamite* (dendrites)	Mainly melilite, CMS, C_3MS_2 $(CaO)_{1.7}(MgO)_{0.3}SiO_2$† trace
B428	Melilite, C_3MS_2, CMS or CMS_2 Oldhamite* (aggregates and dendrites)	Mainly melilite, C_3MS_2, CMS
B431 (control)	Melilite, CMS_2 Oldhamite* (strings)	Mainly melilite, CMS_2

*Oldhamite is a calcium manganese iron sulphide which occurs in slags in the various forms indicated.

†In some lumps fluorescing under ultra-violet light, a substantial proportion of $(CaO)_{1.7}(MgO)_{0.3}SiO_2$ was found and the presence of γ-C_2S could not be excluded. Lumps fluorescing under ultra-violet light were found to have a skin containing α quartz.

investigation and the results are shown in Table 2 including a comment on the presence of Fe and CaS. It will be noted that, although the slags all comply with the stability requirement of the Standard, the presence of α-C_2S and γ-C_2S was detected in slag B435 and this apparently anomalous situation is discussed by Gutt and Russell in a paper to be published presently.

Slag B435 also revealed the presence of the calcium magnesio-silicate $(CaO)_{1\cdot7}(MgO)_{0\cdot3}SiO_2$ characterized by Gutt(5,6) and discussed recently.(7,8) The presence of skin containing α quartz was found on some lumps of slag and oldhamite was identified in all the slags. Microscopic examination provided also a visual comparison of the porosity of the slags of different bulk density which were tested. This and other features observed by microscopy are illustrated in Figures 1 and 2.

Physical properties of the aggregates

Representative samples of the four slags were taken and tests were carried out in accordance with BS 1047:1952 and BS 812:1960 as appropriate. The results of these tests are given in Table 4.

The slags were supplied in two nominal sizes, 19·0–9·5 mm ($\frac{3}{4}$–$\frac{3}{8}$ in.) and 9·5–4·8 mm ($\frac{3}{8}$–$\frac{3}{16}$ in.). The sieve analyses of these separate sizes and of combined gradings consisting of 2 parts by weight of 19–9·5 mm material and 1 part by weight of 9·5–4·8 mm material are given in Table 4. In all cases, the combined gradings complied with those specified in BS 882:1965 for combined 19–9·5 mm aggregate but one of the slags, B428, failed to comply with the requirements of BS 1047:1952 for this combined aggregate. The passing 4·8 mm fraction exceeded the 5% maximum by 3%. BS 1047:1952 specifies that the compacted bulk density of the slag shall be greater than 1250 kg/m³ (78 lb/ft³). Slag B431, which was used as a control slag, easily met this requirement; the three lighter-weight slags had bulk densities below this value, the lightest being B435 which had a bulk density of 1175 kg/m³.

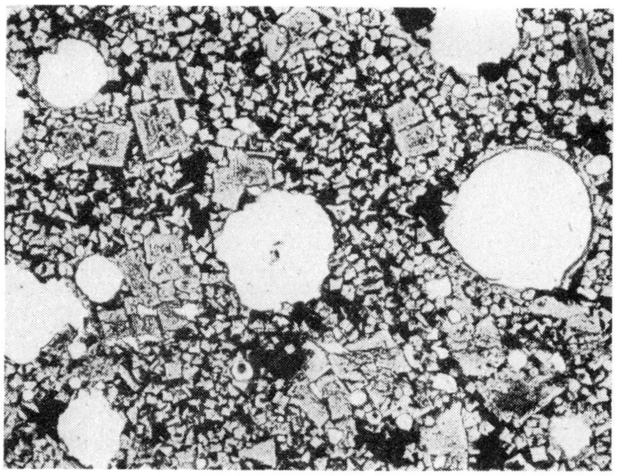

(a) Slag B431 (control): bulk density 1420 kg/m³

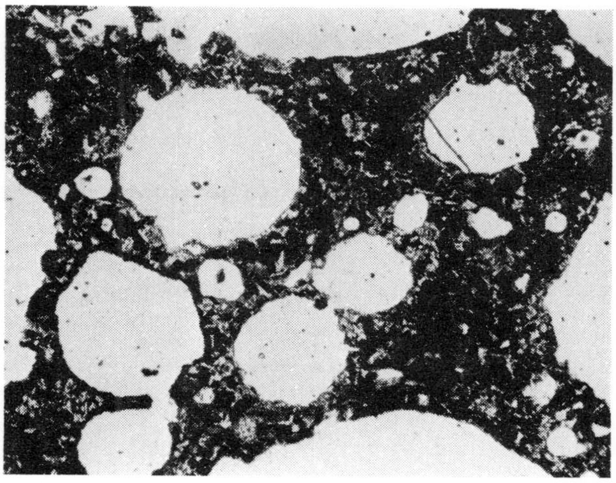

(b) Slag B419: bulk density 1185 kg/m³

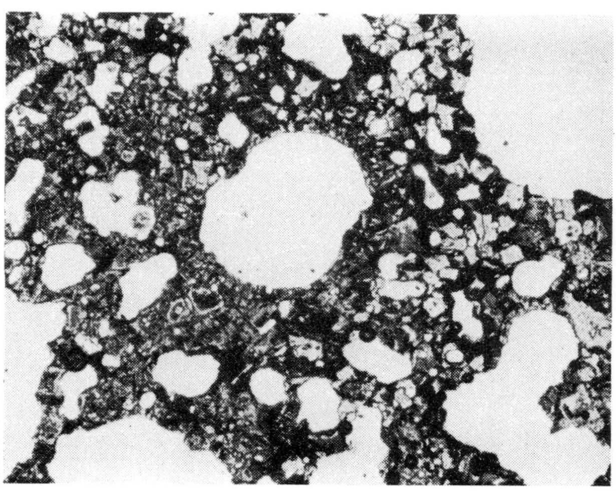

(c) Slag B435: bulk density 1170 kg/m³

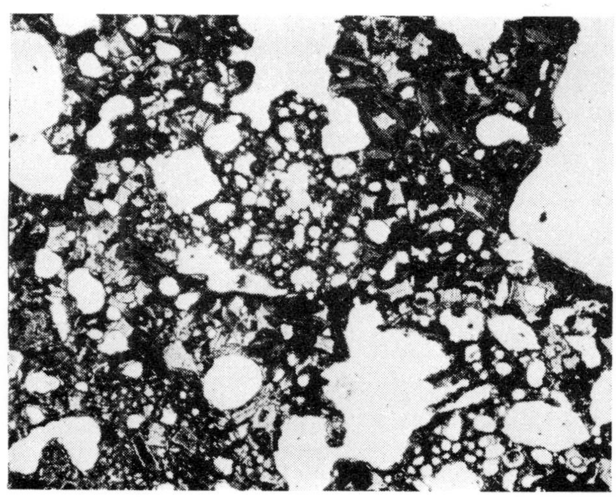

(d) Slag B428: bulk density 1240 kg/m³

Figure 1: Relative porosity and correlation with bulk density. Clear white areas are pores, transmitted light, × 30.

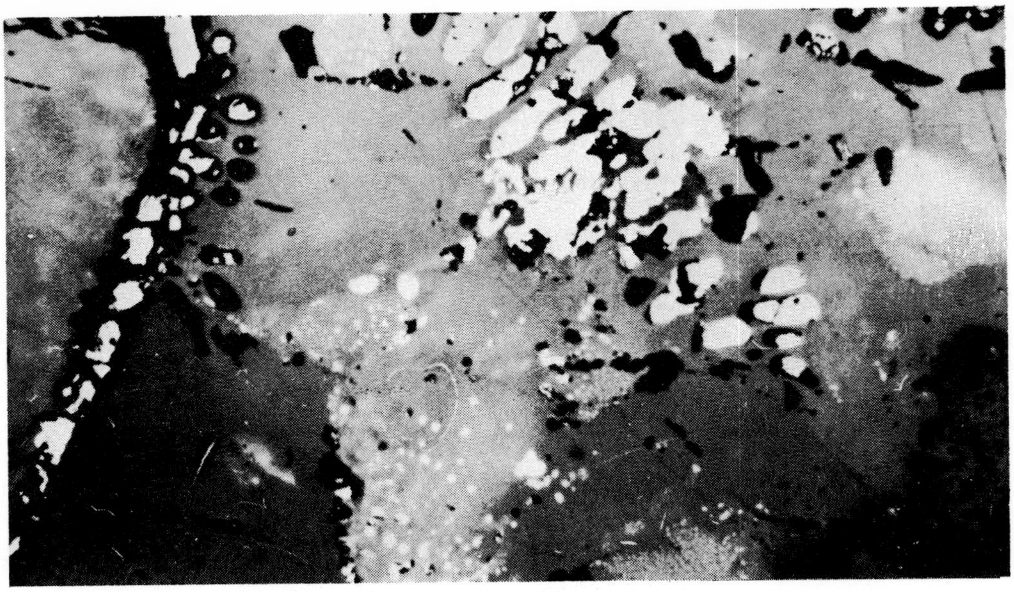

Figure 2: Oldhamite crystals in slag B428. The light areas show the white crystals of oldhamite in this unetched polished section in reflected light, × 300.

TABLE 4: **Physical properties of the coarse aggregates.**

	Slag B419			Slag B435			Slag B428			Slag B431 (control)		
	9·5–4·8 mm	19–9·5 mm	Combined 19–4·8 mm	9·5–4·8 mm	19–9·5 mm	Combined 19–4·8 mm	9·5–4·8 mm	19–9·5 mm	Combined 19–4·8 mm	9·5–4·8 mm	19–9·5 mm	Combined 19–4·8 mm
GRADING (% BY WEIGHT PASSING SIEVE SIZE STATED)												
19·0 mm	100	93	96	100	94	100	100	99	99	100	98	99
12·7 mm	98	14	42	100	53	69	100	57	71	100	46	64
9·5 mm	86	4	31	90	9	36	86	9	35	99	16	44
4·8 mm	6	2	3	6	2	3	24	0	8	6	2	3
2·4 mm	3	0	1	4	0	1	9	0	3	3	0	1
COMPLIANCE WITH BS GRADING LIMITS OF COMBINED 19–4·8 mm FRACTION												
BS 1047			complies			complies			fails			complies
BS 882			complies			complies			complies			complies
BULK DENSITY (kg/m³) TO BS 812												
Uncompacted	1105	1000	1090	1070	1030	1075	1190	1070	1140	1315	1250	1285
Compacted	1200	1110	1185	1175	1140	1175	1280	1170	1240	1465	1380	1420

	Slag B419	Slag B435	Slag B428	Slag B431 (control)
On 12·7–9·5 mm fraction				
Uncompacted	1090	1060	1100	1160
Compacted	1130	1100	1160	1190
OTHER PROPERTIES TESTED ON 12·7–9·5 mm FRACTION TO BS 812				
Specific gravity	2·27	2·22	2·24	2·55
Water absorption (%)	5·97	5·77	5·36	2·12
Angularity No.	16	13	12	11
10% fines value (tonnes)	8	10	8·5	15
Flakiness index	7	10	5	16

Unlike other specifications, BS 1047 does not specify the grading of the slag on which the density is measured although it specifies four different gradings, viz. 38–4·8 mm, 19·0–4·8 mm, 12·7–4·8 mm and a nominal 19·0 mm single-sized material. The density tests were carried out on four different fractions: 19–4·8 mm, the separate sizes of 19–9·5 mm and 9·5–4·8 mm and the 12·7–9·5 mm size used for BS 812 tests. Table 4 shows that the density of a particular slag varies according to the size of the material tested.

BS 802:1967, *Tarmacadam with crushed rock or slag aggregate,* requires a minimum compacted bulk density of 1089 kg/m³ when measured on the 12·7–9·5 mm fraction. Slags B419 and B435 had densities of 1130 kg/m³ and 1100 kg/m³ respectively when measured on this fraction, and thus this investigation covers slags with bulk densities similar to the minimum limit specified in BS 802.

As would be expected, the specific gravity of these slags is lower, the average being 2·24 compared with the 2·55 of the control slag, and the absorption is higher, the average being 5·7% compared with the 2·12% of the control slag. The more open texture of the lighter-weight slags is also shown by a lower 10% fines value, indicating the presence of weaker aggregate particles. The average 10% fines value of the lighter-weight slags is 8·8 tonnes compared with 15 tonnes for the control slag.

These lighter-weight slags contain a larger proportion of 'honeycombed' particles. Samples of all four slags were taken and the honeycombed and solid particles were sorted by hand on a visual basis. There is obviously a range in the degree of honeycombing, producing particles having a range of specific gravities and bulk densities, and thus a single split into honeycombed and dense particles relies on subjective judgement. The percentage of honeycombed slag contained in the four slags and the compacted and uncompacted bulk densities of the honeycombed and the solid particles of each slag are given in Table 5. Figure 3 shows the honeycombed and solid particles of the 19–9·5 mm fraction of a lighter-weight slag and the control slag.

It is seen from Table 5 that there is a greater difference between the bulk densities of honeycombed and solid slag in the 19–9·5 mm fraction than in the 9·5–4·8 mm fraction. Of the lighter-weight slags, B435 shows the greatest difference in uncompacted densities between the honeycombed and solid particles, but B419 contains the lightest honeycombed material, see Figure 3. The control slag B431 produces slag with the greatest difference in bulk density between honeycombed and solid material, and with the clearest distinction between these types of material, Figures 3c and 3d. The solid B431 material has the greatest density and contains more flaky particles and this is reflected in the flakiness index given in Table 4.

The cements used in the concrete mixes

Most of the concrete mixes were made with an ordinary Portland cement (BRS Batch No. 717). A sulphate-resisting Portland cement (BRS Batch No. 728) was used in some of the mixes tested for sulphate resistance. The cements complied with the relevant British Standards, as shown in Table 6.

The ordinary Portland cement had a calculated tricalcium aluminate content of 8·5% and this is within the range normally found in British cements although possibly a little below the average. The amount of this compound present has a considerable effect upon the sulphate resistance of Portland cements and in sulphate-resisting Portland cement it must not exceed 3·5%. With very low C_3A contents, it is recognized that the method of calculation given in BS 4027 can yield a negative value, as shown in Table 6 for cement 728.

Concrete mixes

Three different sets of mix proportions were used for the three series of tests as follows.

TABLE 5: **Proportions and bulk densities of honeycombed and solid slag.**

Slag	Size (mm)	Percentage of slag honeycombed, by weight	Bulk density (kg/m³) Uncompacted Honeycombed	Bulk density (kg/m³) Uncompacted Solid	Bulk density (kg/m³) Compacted Honeycombed	Bulk density (kg/m³) Compacted Solid
B419	19–9·5	46·0	895	1050	990	1165
	9·5–4·8	48·5	1070	1100	1125	1210
B435	19–9·5	47·5	990	1180	1100	1225
	9·5–4·8	46·5	1090	1105	1160	1255
B428	19–9·5	49·0	1060	1100	1135	1180
	9·5–4·8	50·0	1170	1205	1210	1230
B431 (control)	19–9·5	41·0	950	1355	1030	1485
	9·5–4·8	46·0	1245	1285	1320	1535

(*a*) *Slag B419: Honeycombed, uncompacted bulk density 895 kg/m*^{3}

(*b*) *Slag B419: Solid, uncompacted bulk density 1050 kg/m*^{3}

(*c*) *Slag B431* (*control*)*: Honeycombed, uncompacted bulk density 950 kg/m*^{3}

(*d*) *Slag B431* (*control*)*: Solid, uncompacted bulk density 1355 kg/m*^{3}

Figure 3: Honeycombed and solid particles from lighter-weight slag B419 and control slag B431, 19–9·5 mm fractions.

Series 1: Trial mixes, 1:2:4 by volume
Series 2: Main strength mixes, 1:2·4:3·6 by volume
Series 3: Durability mixes, cement content 330 kg/m^3

The original intention was to match this work on lighter-weight slags to the previous work on normal-weight slags[2] and hence trial mixes were made with cement:sand:coarse slag aggregate proportions of 1:2:4 by volume at different water/cement ratios (Series 1). However, these proportions produced rather harsh mixes and so, for the concrete tests to measure the strength characteristics of the slag concretes, mix proportions of 1:2·4:3·6 by volume were used (Series 2).

When it came to the mixes required for durability tests, the criterion of comparable cement content recommended by BRS in Digest 90[9] for sulphate resistance and since adopted for all durability requirements in CP 110[10] was used. A level of cement content of 330 kg/m^3 was chosen for the concrete used for the durability tests (Series 3). The cement contents of the strength mixes (Series 2) designed on the basis of fixed cement:aggregate ratios were slightly lower and varied between 295 and 320 kg/m^3.

MIX PROPORTIONS

In all the concrete mixes, the coarse aggregate, slag, flint gravel or crushed limestone was composed of the nominal 19–9·5 mm fraction and the 9·5–4·75 mm fraction combined in the proportions of 2:1 by weight. The fine aggregate was a river sand consisting of 40% by weight of the 4·8 mm–600 μm fraction + 60% by weight of the 600–150 μm fraction, producing a grading complying with grading zone 3 of BS 882:1965.

The nominal mix proportions of 1:2:4 or 1:2·4:3·6 by volume were converted to proportions by weight based on a bulk density of cement of 1440 kg/m^3 and the uncompacted bulk densities of the coarse aggregate and the river sand. The uncompacted bulk densities of the combined slag coarse aggregates are given in Table 4 and for the flint gravel, crushed limestone and river sand are 1420 kg/m^3, 1425 kg/m^3 and 1570 kg/m^3 respectively. For trial mixes, Series 1, i.e. those nominally 1:2:4 by volume, an uncompacted bulk density of the river sand of 1600 kg/m^3 was used to convert to mix proportions by weight. The actual mix proportions by weight are given in Tables 7 and 8.

For the mixes for durability studies (Series 3), a

TABLE 6: **British standard tests on the cements.**

	Ordinary Portland cement		Sulphate-resisting Portland cement	
	BS 12 requirement	Result on cement 717	BS 4027 requirement	Result on cement 728
PHYSICAL TESTS				
Fineness:				
Specific surface (cm^2/g)	>2250	3190	>2500	4060
Consistence:				
Water requirement (%)		27·0		26·8
Setting times:				
Initial (h min)	> 0 45	4 35	> 0 45	2 25
Final (h min)	<10 00	6 20	<10 00	4 25
Soundness:				
Expansion by Le Chatelier test (mm)	<10·0	2·0	<10·0	1·0
Compressive strength (N/mm^2)				
(1) Average strength of 1:3 vibrated mortar at				
1 day		9·0		12·5
3 days	>15·2	30·5	>15·2	31·0
7 days	>23·4	47·2	>23·4	49·9
28 days		64·6		58·4
(2) Average strength of concrete at				
3 days	> 8·3	15·1	> 8·3	16·5
7 days	>13·8	23·4	>13·8	24·6
28 days		38·3		37·2
CHEMICAL TESTS				
Analysis (%)				
SiO_2		21·17		20·10
Fe_2O_3		1·73		5·71
TiO_2		0·37		0·35
Al_2O_3		4·32		2·88
CaO		65·65		64·52
MgO	<4·0	1·31	<4·0	1·59
Mn_2O_3		0·14		0·05
Na_2O		0·13		0·13
K_2O		0·38		0·30
SO_3	<3·0	2·75	<2·5	2·17
P_2O_5				0·16
Loss on ignition (%)	<3·0	1·79	<3·0	1·63
Lime saturation factor	0·66–1·02	0·97	0·66–1·02	
Ratio of alumina to iron	>0·66	2·50		
Calculated C_3A content		8·5	<3·5	−ve

cement content of 330 kg/m^3 and a free water/cement ratio of 0·50 were chosen for all the aggregates both with ordinary Portland and with sulphate-resisting Portland cements. The mix proportions used and the cement contents achieved are given in Table 9.

In Series 1 which consisted of trial mixes only, three water/cement ratios were used with each aggregate to obtain three levels of workability (see Table 7). The same procedure was used in the trial mixes of Series 2 but only one water/cement ratio was used with each aggregate in the main Series 2 strength mixes. It was agreed that the mixes should be compared on an equal workability basis with the different aggregates, and this resulted in the use of different water/cement ratios.

It was agreed to use water/cement ratios midway between those used in trial mixes /5 and /6 (Table 8) which should give a compacting factor of 0·92 ± 0·02. These trial mixes are W419, W435, W428, W431, WHR and WL in Table 8 and the same mix proportions were used in the main strength tests.

MIXING PROCEDURE

The two sizes of coarse aggregate, two sizes of sand, the cement and the water were batched by weight and the concrete was mixed in an 80 litre capacity mixer of the rotating pan and paddle type. The aggregates were placed in the pan and mixed with some of the mixing water for 1 minute; the cement and the remainder of the mixing water were then added and the concrete mixed for a further 2 minutes.

WORKABILITY TESTS

Workability tests were carried out on Series 1, the

TABLE 7: **Series 1. Trial mixes, based on nominal 1:2:4 proportions.**

Slag batch (Rodded density kg/m³)	Mix proportions: 1:2:4 by volume by weight	Mix proportions: Water/cement ratio by weight	Mark	Workability tests: Slump (mm)	Workability tests: V.B. time (s)	Workability tests: C.F.	Workability tests: Comments	Wet density (kg/m³)	Cement content (kg/m³)	Crushing strength of water-stored cubes (N/mm²): 7 days	Crushing strength of water-stored cubes (N/mm²): 28 days
B419 (1190)	1:2·22:3·02	0·60 (0·39)	19/1	13	9·5	0·84	Large aggregate. Rather stony. More sand? Hand and Kango.	2280	333	43·0 (2340)	53·3 (2330)
		0·66 (0·45)	19/2	10	4·2	0·89	Reasonably good—still short of sand.	2280	330	33·4 (2320)	43·0 (2340)
		0·75 (0·54)	19/3	Coll.	2·0	0·94	Wet mix, near segregation, short of sand. Series rather harsh.	2270	324	25·4 (2320)	35·3 (2320)
B435 (1180)	1:2·22:2·98	0·60 (0·40)	35/1	2	10·0	0·83	Slightly harsh, better than B419; more sand? Hand.	2260	333	38·9 (2330)	51·3 (2330)
		0·66 (0·46)	35/2	20	7·5	0·91	Harsh? Quite good. Vibrates O.K.	2250	328	29·4 (2310)	41·3 (2310)
		0·75 (0·55)	35/3	Coll.	1·5	0·95	Good wet mix, still needs more sand.	2240	321	19·6 (2290)	27·4 (2290)
B428 (1240)	1:2·22:3·18	0·60 (0·40)	28/1	2	14·7	0·82	Not as stony as 19/1. O.K. into moulds.	2280	326	37·5 (2320)	52·2 (2330)
		0·66 (0·46)	28/2	12	6·0	0·89	Good fatty, workable mix.	2280	323	30·0 (2320)	42·0 (2320)
		0·75 (0·55)	28/3	Coll.	3·0	0·94	Wet mix, not too cohesive, rather short of fines.	2270	317	21·6 (2280)	31·9 (2280)
Control slag B431 (1420)	1:2·22:3·58	0·60 (0·49)	31/1	18	4·5	0·88	Most workable mix of the set, stony but more cohesive.	2380	321	31·9 (2410)	48·1 (2420)
		0·66 (0·55)	31/2	36	3·0	0·95	Very workable mix, stony and still short of sand.	2370	318	22·8 (2390)	37·0 (2400)
		0·75 (0·64)	31/3	Coll.	1·0	0·97	Very sloppy, near segregation bleeding.	2350	311	15·4 (2380)	25·8 (2380)
Flint gravel (1420)	1:2·22:3·93	0·75 (0·65)	HR 1	2	34·0	0·78	Dryish mix, sets quickly, compacts with Kango but not easily.	2350	311	47·7 (2400)	59·9 (2410)
		0·56 (0·45)	HR 2	14	7·5	0·86	Rather stony but works well, easily compacted.	2340	304	33·8 (2380)	47·2 (2390)
		0·65 (0·54)	HR 3	56	3·0	0·94	Wet mix, but cohesive and highly workable.	2340	300	25·9 (2340)	37·8 (2280)

Figures in parentheses beneath the water/cement ratios are approximate free water/cement ratios.
Figures in parentheses beneath the crushing strengths are the bulk densities in kg/m³.
Coll. = collapse. C.F. = compacting factor.

original trial mixes, the trial mixes of Series 2 and on the Series 3 mixes used to cast the specimens for the sulphate resistance tests. The workability of the concrete was measured by the slump, V.B. and compacting factor tests in accordance with the requirements of BS 1881:1952.[11] The results of these tests and observations on the working properties of the concrete are given in Tables 7 to 9.

RESULTS OF TRIAL MIXES

From Table 7 it may be seen that the 1:2:4 mix proportions used in Series 1 are unsatisfactory, in that these proportions produce harsh mixes. Increasing the water/cement ratio only results in increasing the risk of segregation. Better mixes are obtained by increasing the fines content and in Series 2 the proportions were changed to a nominal 1:2·4:3·6 mix by volume. These mixes with a fines content of 40% by volume were more workable and had higher compacting factors at the same total water/cement ratios.

The crushing strengths of the strength Series 1 and 2 trial mixes at 7 days and 28 days are given in Tables 7 and 8 and shown plotted against water/cement ratio in Figures 4 and 5. These show that the crushing strengths of concretes made with lighter-weight slag

TABLE 8: **Trial mixes of Series 2 based on nominal 1:2·4:3·6 proportions.**

Slag batch (Rodded density kg/m³)	Mix proportions: 1:2·4:3·6 by volume by weight	Mix proportions: Water/cement ratio by weight	Mark	Workability tests: Slump (mm)	Workability tests: V.B. time (s)	Workability tests: C.F.	Workability tests: Comments	Wet density (kg/m³)	Cement content (kg/m³)	Crushing strength of water-stored cubes (N/mm²): 7 days	Crushing strength of water-stored cubes (N/mm²): 28 days
B419 (1190)	1:2·62:2·72	0·59 (0·40)	19/4	5	10·4	0·84	More workable than 1:2:4 mix. Compacts by hand. Finishes easily.	2260	327	41·1 (2340)	51·6 (2345)
		0·66 (0·47)	19/5	18	5·5	0·91	Workable mix, easy to compact by hand and finish.	2250	320	32·4 (2295)	44·4 (2320)
		0·70 (0·50)	W419	20	5·8	0·94		2300	320	28·9 (2290)	39·2 (2310)
		0·75 (0·56)	19/6	75	2·2	0·98	Highly workable. No segregation, easy to finish.	2230	315	24·8 (2295)	32·8 (2295)
B435 (1180)	1:2·62:2·68	0·60 (0·41)	35/4	9	10·7	0·86	Not as cohesive as the same mix for slag B431. Easy to compact and finish.	2240	324	36·1 (2330)	48·0 (2330)
		0·66 (0·47)	35/5	12	9·0	0·92	Appears dry in mixer but works well into cubes and finishes easily.	2230	320	30·7 (2300)	41·8 (2315)
		0·70 (0·51)	W435	35	5·0	0·93		2250	320	28·2 (2285)	39·9 (2315)
		0·75 (0·56)	35/6	65	1·5	0·97	Considerably wetter than 35/4 or 35/5. Good fatty mix easily worked and finished.	2230	317	21·6 (2280)	32·3 (2300)
B428 (1240)	1:2·62:2·86	0·60 (0·42)	28/4	3	11·7	0·85	Looks drier in cubes. Not so easy to compact but finishes easily.	2250	318	32·7 (2335)	47·7 (2340)
		0·66 (0·48)	28/5	8	6·2	0·90	Rather dry looking, stiffens rapidly. Plenty of fines, finishes easily.	2230	313	29·6 (2300)	42·1 (2310)
		0·70 (0·51)	W428	25	6·2	0·90		2260	314	26·1 (2310)	35·8 (2310)
		0·75 (0·57)	28/6	38	2·5	0·97	Highly workable cohesive mix; easy to compact and finish.	2220	308	22·7 (2295)	33·9 (2300)
Control slag B431 (1420)	1:2·62:3·22	0·60 (0·50)	31/4	18	6·7	0·89	Fairly workable mix. No signs of harshness or segregation. Easy to compact and finish.	2330	314	34·6 (2390)	49·0 (2405)
		0·66 (0·56)	31/5	18	5·2	0·93	Good workable mix. Compacts and finishes easily.	2330	311	29·4 (2395)	43·9 (2405)
		0·67 (0·57)	W431	55	3·5	0·92	Very good workable mix. Easy to compact and finish.	2330	311	23·1 (2410)	36·6 (2405)
		0·75 (0·65)	31/6	Coll.	1·5	0·99	Wet sloppy mix but no segregation. Adequate fines for this workability.	2320	305	18·9 (2340)	30·5 (2355)
Flint gravel	1:2·62:3·54	0·52 (0·42)	HR 4	2	32·0	0·82	Sandy looking mix. Does not compact easily.	2320	302	37·9 (2395)	51·4 (2400)
		0·60 (0·50)	HR 5	9	7·7	0·90	Sandy. Compacts moderately well.	2300	295	29·2 (2355)	42·6 (2350)
		0·64 (0·54)	WHR	9	8·7	0·90	Good workable mix. Easy to compact and finish.	2310	295	27·2 (2385)	40·6 (2365)
		0·67 (0·57)	HR 6	30	3·7	0·97	Wet mix but cohesive and highly workable.	2280	292	23·7 (2350)	37·2 (2350)

continued on next page

TABLE 8 *continued*

Slag batch (Rodded density kg/m³)	Mix proportions 1:2·4:3·6 by volume by weight	Water/cement ratio by weight	Mark	Workability tests: Slump (mm)	V.B. time (s)	C.F.	Comments	Wet density (kg/m³)	Cement content (kg/m³)	Crushing strength of water-stored cubes (N/mm²): 7 days	28 days
Crushed limestone	1:2·62:3·56	0·52 (0·48)	LS 1	3	10·5	0·85	Dry, non-cohesive mix, difficult to vibrate and finish.	2380	308	42·1 (2420)	55·8 (2440)
		0·60 (0·56)	LS 2	8	9·0	0·92	Cohesive mix, easy to compact and finish.	2390	306	33·1 (2410)	46·3 (2420)
		0·64 (0·60)	WL	80	3·5	0·95	Very workable mix. Looks oversanded.	2380	304	28·3 (2400)	40·3 (2420)
		0·66 (0·62)	LS 3	38	2·7	0·95	Highly workable cohesive mix. Not too wet, easy to finish.	2370	302	26·2 (2410)	37·6 (2410)

Figures in parentheses beneath the water/cement ratios are approximate free water/cement ratios.
Figures in parentheses beneath the crushing strengths are the bulk densities in kg/m³.
Coll. = collapse. C.F. = compacting factor.

TABLE 9: **Details of concrete mixes for durability tests (Series 3).**

Coarse aggregate	Mix proportions Cement:Sand:Coarse Agg. By volume	By weight	Water/cement ratio Total	Free*	Type of cement	Workability: Compacting factor	Slump (mm)	V.B. time (s)	Cement content (kg/m³)	Crushing strength at 28 days on 101·6 mm cubes (N/mm²)	Drying shrinkage to BS 1881 (%)
Slag B419	1:2·32:3·48 (1:5·8)	1:2·53:2·63	0·68	0·50	OPC SRPC	0·96 0·99	55 61	4·1 2·4	327 327	38·1 37·6	0·03 0·03
Slag B435	1:2·32:3·48 (1:5·8)	1:2·53:2·58	0·68	0·50	OPC SRPC	0·94 0·98	58 65	3·2 2·1	331 328	36·8 35·6	0·03 0·03
Slag B428	1:2·24:3·36 (1:5·6)	1:2·44:2·67	0·68	0·50	OPC SRPC	0·95 0·98	65 82	4·2 2·0	329 330	39·0 39·0	0·04 0·04
Slag B431	1:2·32:3·48 (1:5·8)	1:2·53:3·11	0·60	0·50	OPC SRPC	0·94 0·96	35 69	5·4 4·8	324 325	44·7 44·7	0·02 0·03
Flint gravel	1:2·16:3·24 (1:5·4)	1:2·36:3·18	0·59	0·50	OPC SRPC	0·92 0·98	29 110	4·9 1·2	324 324	44·1 41·8	0·03 0·03
Crushed limestone	1:2·20:3·30 (1:5·5)	1:2·40:3·26	0·53	0·50	OPC SRPC	0·93 0·95	15 20	7·4 5·1	328 334	51·1 57·0	0·02 0·02

*estimated

aggregate follow water/cement ratio relationships similar to those of concretes made with other aggregates.

Figure 4 shows that, at a given age and total water/cement ratio, there is a considerable range in the crushing strengths for the different aggregates. Concretes made with the lighter-weight slags are generally stronger than the other concretes tested, but it must be realized that at the same total water/cement ratio the workability of these concretes is less and their cement content is higher. This is well illustrated by comparing mixes 19/2, 35/2, 28/2, 31/2 and HR3 in Table 7, which all have a total water/cement ratio of 0·66 or 0·65 This difference in workability is due to the difference in absorption of the coarse aggregates, and the increase in the cement content is due to the paste filling the larger interstices of the honeycombed slag particles. The absorption of the 13–10 mm fraction of the slags is given in Table 4 and the values for the flint gravel and crushed limestone are 2·0% and 0·18% respectively.

When, as in this case, a range of aggregates is used, the crushing strength is best related to the 'free' water/cement ratio, which is the total water/cement

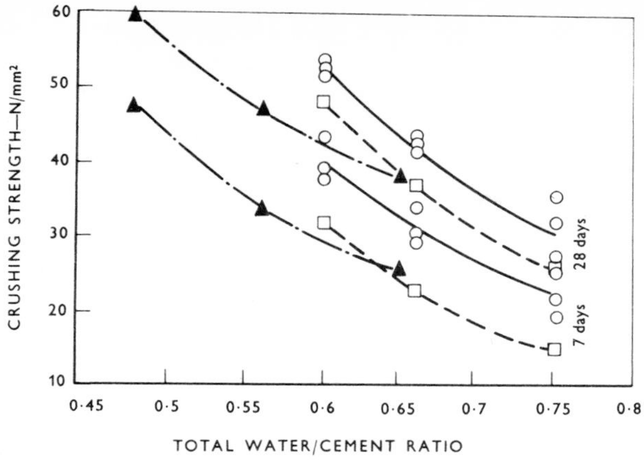

(*a*) *Series 1. 1:2:4 mix*

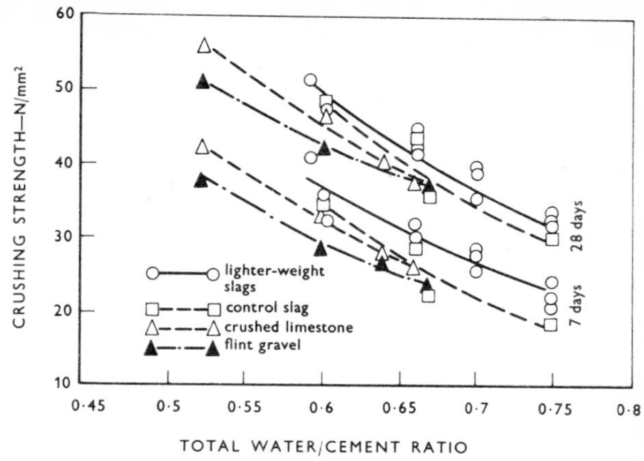

(*b*) *Series 2. 1:2·4:3·6 mix*

Figure 4: Relationship between crushing strength of trial mixes and total water/cement ratio.

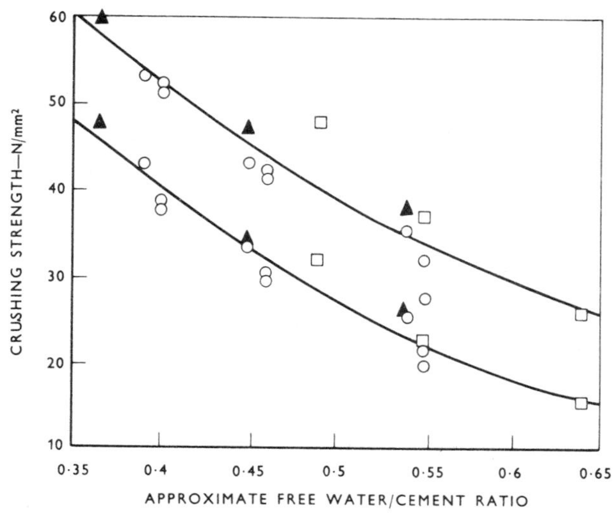

(*a*) *Series 1. 1:2:4 mix*

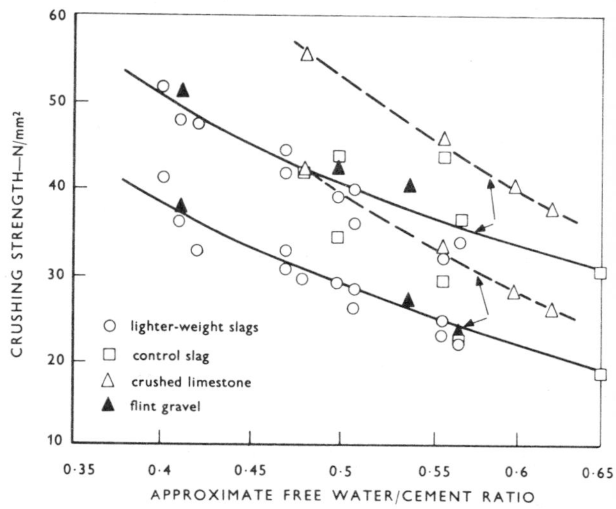

(*b*) *Series 2. 1:2·4:3·6 mix*

Figure 5: Relationship between crushing strength of trial mixes and 'free' water/cement ratio.

ratio minus the absorbed water/cement ratio. In these tests, the absorptions were not measured on the range of aggregate sizes used but only on a single size as described above; thus the 'free' water/cement ratios calculated on this basis are only approximate. Figure 5 shows that the crushing strength is better related to the 'free' water/cement ratio and there is a reduced scatter of results as compared with Figure 4 except for the crushed limestone mixes which were used in Series 2.

Concrete strength tests (Series 2 and Series 3)

From each mix, twelve 508 × 101·6 × 101·6 mm beams were cast. After being stored in moist air for 24 h, the beams were demoulded and six were stored in air at 18°C and 65% R.H. and six were stored in water at 18°C. One each of the beams from air and from water storage was tested at ages of 3, 7 and 28 days, 3 months, 1 year and 2 years.

Each beam was tested to determine the flexural strength in accordance with BS 1881:1952. One half of the broken beam was then crushed between 101·6 mm × 101·6 mm steel plates to determine the 'equivalent cube' crushing strength, and the other half was split along its longitudinal axis between half-round wooden strips to measure the indirect tensile strength. For each material, three identical mixes were made; thus the results given in Tables 10 and 11 are the mean values from three tests and these are shown plotted in Figures 6 to 8 as curves of strength development with age.

Two sets of strength specimens were prepared from the concrete mixes used in the Series 3 durability mixes. Some of the durability tests used both ordinary

TABLE 10: **Crushing, flexural and indirect tensile strengths of 1:2·4:3·6 concretes stored in air (Series 2 tests).**

Coarse aggregate and mix proportions by weight	Total water/cement ratio by weight	Cement content (kg/m³)	Equivalent cube crushing strength (N/mm²)						Flexural strength (N/mm²)						Indirect tensile strength (N/mm²)					
			3 days	7 days	28 days	3 months	1 year	2 years	3 days	7 days	28 days	3 months	1 year	2 years	3 days	7 days	28 days	3 months	1 year	2 years
Slag B419 1:2·62:2·72	0·70	320	21·3 (2210)	28·1 (2215)	35·5 (2195)	36·5 (2195)	45·0 (2180)	42·1 (2290)	2·77	2·98	3·48	4·78	4·76	4·49	2·23	2·53	3·21	2·99	3·35	3·65
Slag B435 1:2·62:2·68	0·70	320	16·8 (2200)	25·5 (2185)	34·8 (2190)	38·3 (2170)	41·7 (2170)	42·3 (2275)	2·49	3·11	3·72	4·73	4·76	4·80	1·93	2·54	3·12	3·53	3·41	3·36
Slag B428 1:2·62:2·86	0·70	314	17·6 (2215)	25·8 (2205)	36·7 (2190)	40·9 (2115)	43·1 (2165)	44·1 (2290)	2·62	2·92	3·69	4·47	4·96	4·60	1·74	2·28	3·22	3·43	3·66	3·74
Mean of lighter-weight slags			18·6	26·5	35·7	38·6	43·3	42·8	2·58	3·00	3·63	4·66	4·82	4·63	1·97	2·45	3·18	3·32	3·47	3·58
Slag B431 1:2·62:3·22	0·67	311	17·6 (2320)	24·1 (2290)	34·1 (2275)	39·8 (2285)	45·8 (2275)	37·5 (2390)	2·77	2·85	3·24	4·19	4·67	4·26	1·84	2·31	2·86	3·76	3·27	3·21
Mean of all slags			18·3	25·9	35·3	38·9	43·9	41·5	2·66	2·96	3·53	4·54	4·79	4·54	1·94	2·41	3·10	3·43	3·42	3·49
Flint gravel 1:2·62:3·54	0·64	295	16·6 (2280)	23·4 (2285)	31·5 (2280)	34·8 (2270)	38·5 (2265)	37·8 (2395)	1·97	2·21	2·86	3·67	3·62	4·23	1·54	1·92	2·56	2·98	2·81	3·05
Crushed limestone 1:2·62:3·56	0·64	304	15·5 (2470)	18·9 (2310)	28·5 (2345)	34·6 (2300)	35·3 (2300)	33·1 (2445)	2·55	2·77	3·42	4·17	4·41	4·35	1·67	2·24	2·91	3·01	2·76	2·95

The figures in parentheses beneath the crushing strength are the bulk densities in kg/m³.

TABLE 11: **Crushing, flexural and indirect tensile strengths of 1:2·4:3·6 concretes stored in water (Series 2 tests).**

Coarse aggregate and mix proportions by weight	Total water/cement ratio by weight	Cement content (kg/m³)	Equivalent cube crushing strength (N/mm²)						Flexural strength (N/mm²)						Indirect tensile strength (N/mm²)					
			3 days	7 days	28 days	3 months	1 year	2 years	3 days	7 days	28 days	3 months	1 year	2 years	3 days	7 days	28 days	3 months	1 year	2 years
Slag B419 1:2·62:2·72	0·70	320	22·1 (2270)	28·7 (2280)	37·0 (2280)	44·0 (2285)	47·7 (2290)	46·3 (2420)	2·81	3·43	3·96	4·78	4·91	4·77	2·14	2·61	3·21	4·10	3·59	4·18
Slag B435 1:2·62:2·68	0·70	320	17·2 (2250)	25·1 (2260)	34·9 (2270)	42·2 (2265)	47·1 (2285)	47·4 (2400)	2·42	3·24	4·39	4·86	5·18	5·30	1·73	2·54	3·59	3·64	3·22	3·48
Slag B428 1:2·62:2·86	0·70	314	18·2 (2265)	26·9 (2275)	38·3 (2280)	32·4 (2210)	50·6 (2280)	49·5 (2415)	2·51	3·23	4·20	4·69	5·38	5·30	1·78	2·72	3·36	3·66	3·31	4·09
Mean of lighter-weight slags			19·2	26·9	36·7	43·2	48·5	47·7	2·58	3·30	4·18	4·77	5·16	5·12	1·88	2·62	3·39	3·80	3·37	3·88
Slag B431 1:2·62:3·22	0·67	311	17·2 (2355)	25·4 (2360)	37·2 (2360)	44·8 (2370)	51·5 (2380)	48·6 (2495)	2·43	3·27	4·17	4·79	5·11	5·28	1·74	2·53	3·23	3·83	3·66	4·66
Mean of all slags			18·7	26·5	36·8	43·6	49·2	47·9	2·54	3·30	4·18	4·78	5·15	5·16	1·85	2·60	3·35	3·81	3·44	4·08
Flint gravel 1:2·62:3·54	0·64	295	13·9 (2345)	28·9 (2350)	40·7 (2355)	42·5 (2345)	52·5 (2370)	49·5 (2475)	2·58	2·88	3·79	3·99	4·39	4·39	1·83	2·43	3·16	3·25	3·76	3·71
Crushed limestone 1:2·62:3·56	0·64	304	16·7 (2375)	24·3 (2380)	38·2 (2380)	47·1 (2380)	52·6 (2400)	49·5 (2520)	3·19	3·52	4·56	5·12	5·44	5·59	2·01	2·76	3·93	4·01	3·42	4·24

The figures in parentheses beneath the crushing strength are the bulk densities in kg/m³.

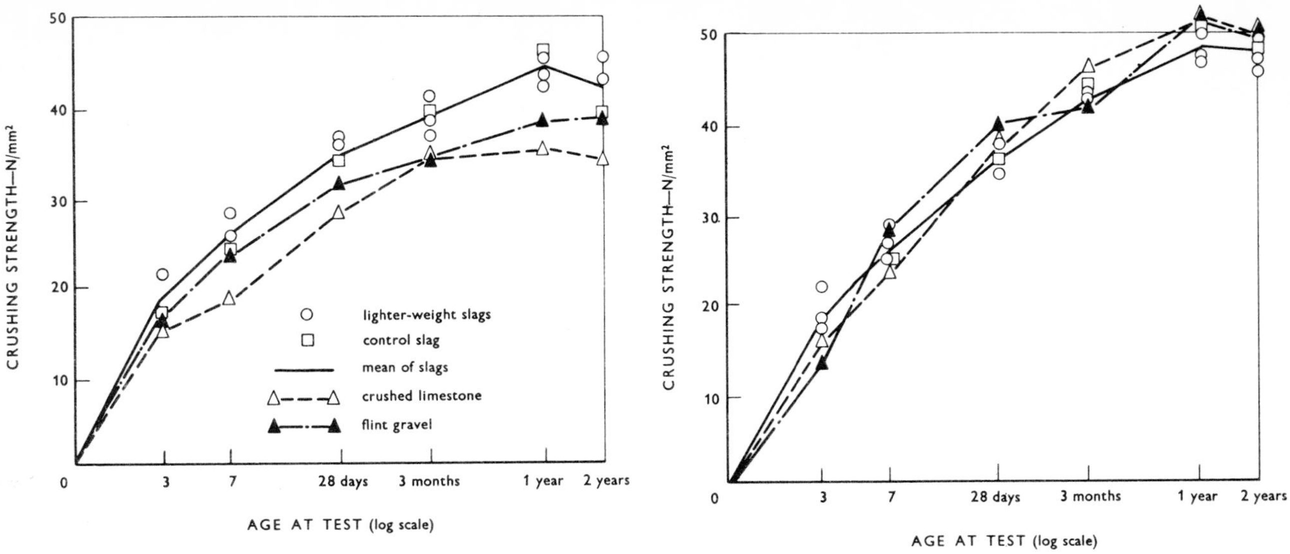

(*a*) *Stored in air at 18°C and 65% R.H.* (*b*) *Stored in water at 18°C*

Figure 6: Crushing strength development of slag and control concretes (Series 2 tests).

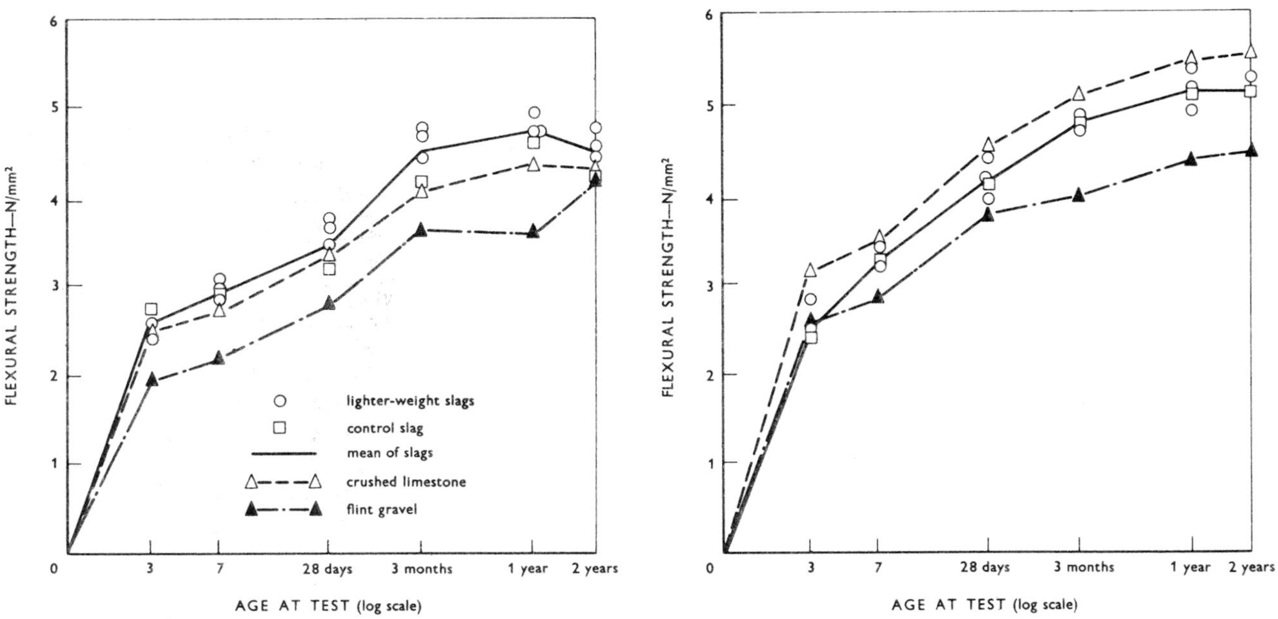

(*a*) *Stored in air at 18°C and 65% R.H.* (*b*) *Stored in water at 18°C*

Figure 7: Flexural strength development of slag and control concretes (Series 2 tests).

Portland cement and sulphate-resisting Portland cement and, from the mixes used to cast the sulphate-resistance specimens, three 101·6 mm cubes were cast for determining the 28 day water-stored crushing strengths. The results of these tests are given in Table 9.

In order to make a direct comparison with the main strength series results, a set of mixes was made with the durability mix proportions using ordinary Portland cement only. From these, twelve 508 × 101·6 × 101·6 mm beams were cast; six were stored in air at 18°C and 65% R.H. and six in water at 18°C. Three specimens from each set of storage conditions were tested at 28 days and at 2 years to give link points with the tests in the main strength series. The results of the flexural, indirect tensile and equivalent cube crushing strengths are given in Table 12.

When comparing the results obtained from the concretes in the strength Series 1 and 2 made with the three lighter-weight slags, the control slag B431 and the flint gravel and crushed limestone, it should be noted that the mixes used had the same volumetric proportions and workability. Owing to the character-

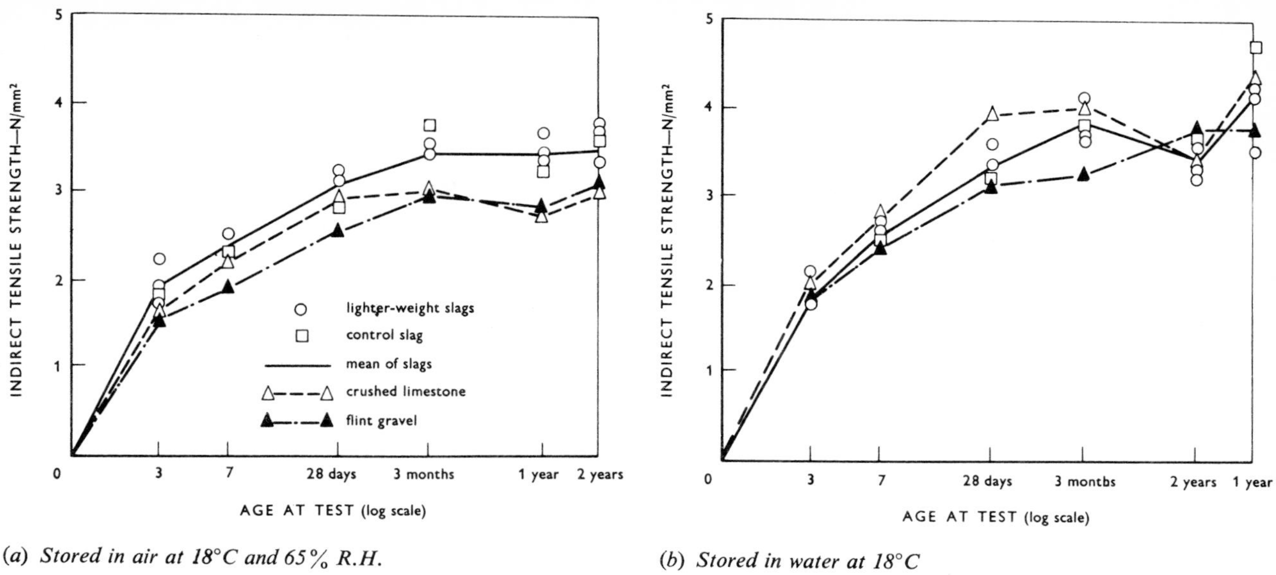

Figure 8: Indirect tensile strength development of slag and control concretes (Series 2 tests).

istics of the aggregates, different total water/cement ratios were required and there are significant differences in the cement content in kg/m³ of the concrete made with the different aggregates.

It is seen from Tables 10 and 11 that the bulk density of the concrete made with the lighter-weight slags is reduced by about 5% for both air and water storage, compared with the concrete made with the control slag.

The strengths obtained at 28 days and at 2 years given in Tables 10 to 12 have been expressed as a percentage of the strengths of the concrete made with flint gravel and are shown plotted in Figure 9 for Series 2 and 3.

Tables 10 and 11 show that generally there is no significant difference between the strengths of concrete made with the three lighter-weight slags, nor between the lighter-weight slags and the control slag. At ages up to 28 days there is a tendency for the lighter-weight slag concrete to be slightly stronger than concrete made with the control slag, but this tendency is reversed at later ages of test.

Owing to the similarity in the results from the slag concretes, the lines plotted in Figures 6 to 8 for the slag concretes are based on the mean value of all four slags. The lack of increase in strength beyond 1 year is probably a characteristic of the cement used. Figure 6 shows the usual increase in crushing strength with age, water-stored concrete being stronger than air-stored concrete, particularly at later ages. The difference in strength between air- and water-stored concrete is less marked with the slag aggregates than with the natural aggregates of flint gravel and crushed limestone. The air-stored slag-aggregate concretes are stronger than the flint-gravel and crushed-limestone concretes, but the water-stored concretes have similar crushing strengths irrespective of the type of aggregate used; however, the cement content of the slag-aggregate concretes in Series 2 was about 20 kg/m³ greater than the natural-aggregate concretes.

In Series 3, the mixes were designed on the basis of an equal cement content, and it is seen from Table 12 and Figure 9 that the crushing strength of the concrete made with the lighter-weight slags is about 5 to 10% lower than the strength of the control slag concrete. In this series, the crushing strength of the slag concrete is generally lower than that of the concrete made with natural aggregates.

Compared with the mixes used in the strength Series 2, the Series 3 durability mixes were slightly richer, particularly in the case of the natural aggregates where the cement content was about 25 kg/m³ greater. This resulted in the use of slightly lower water/cement ratios and this is reflected in slightly higher strengths as shown in Table 12 compared with Tables 10 and 11.

The flexural strength of concrete made with all four slag coarse aggregates is similar, and Figures 7 and 9 show that air-cooled-slag concrete is similar to crushed-limestone concrete in that its flexural strength is greater than that of flint-gravel concrete. The flexural strength of slag concrete is slightly higher than that made with crushed limestone when stored in air, and in Series 2 slightly lower when stored in water.

Table 12 shows that the flexural strength of the slag concrete is higher than the flint-gravel concrete but is not as high as the crushed-limestone concrete of equal cement content in Series 3.

The indirect tensile strength shown in Figure 8 of slag-aggregate concrete relative to the two natural-aggregate concretes is similar to the flexural strengths shown in Figure 7. Indirect tensile strengths are always lower than the corresponding flexural strengths; in this investigation, the indirect tensile strengths are about 80% of the flexural strengths. Figure 8 shows

TABLE 12: **Crushing, flexural and indirect tensile strengths of concretes used in the durability mixes (Series 3 tests).**

Coarse aggregate and mix proportions by weight	Total water/cement ratio	Cement content	Stored in air at 18°C and 65% R.H.						Stored in water at 18°C					
			Equivalent cube crushing strength (N/mm²)		Flexural strength		Indirect tensile strength (N/mm²)		Equivalent cube crushing strength (N/mm²)		Flexural strength		Indirect tensile strength (N/mm²)	
		(kg/m³)	28 days	2 years	28 days	2 years	28 days	2 years	28 days	2 years	28 days	2 years	28 days	2 years
Slag B419 1:2·53:2·63	0·68	327	35·4 (2200)	37·6 (2190)	3·78	4·47	3·26	3·71	38·0 (2310)	43·4 (2325)	4·40	4·95	3·21	3·66
Slag B435 1:2·53:2·58	0·68	331	33·2 (2165)	—	3·08	—	2·96	—	34·8 (2265)	—	4·43	—	3·57	—
Slag 1:2·44:2·67	0·68	329	37·2 (2195)	42·6 (2180)	3·85	4·66	3·19	3·58	38·7 (2275)	46·9 (2300)	4·50	5·34	3·47	4·14
Mean of lighter-weight slags			35·6	40·1	3·57	4·56	3·14	3·64	37·2	45·2	4·44	5·14	3·42	3·90
Slag B431 1:2·53:3·11	0·60	324	38·3 (2295)	41·2 (2295)	3·55	4·80	3·10	3·52	41·6 (2385)	48·5 (2410)	4·55	5·36	3·58	4·16
Mean of all slags			36·3	40·5	3·56	4·64	3·13	3·60	38·3	46·3	4·47	5·22	3·46	3·99
Flint gravel 1:2·36:3·18	0·59	324	38·4 (2290)	43·0 (2280)	2·99	4·14	2·96	3·30	41·2 (2380)	49·5 (2360)	4·10	4·51	3·32	3·87
Crushed limestone 1:2·4:3·26	0·53	328	43·2 (2360)	44·9 (2350)	4·05	5·38	3·33	3·59	48·7 (2420)	57·2 (2450)	5·23	5·91	4·07	4·78

The figures in parentheses beneath the crushing strengths are the bulk densities in kg/m³.

that there is little increase in the indirect tensile strength beyond 3 months.

Table 12 and Figure 9 show that, for the Series 3 mixes, although the indirect tensile strength of the slag concrete is higher than that of the flint-gravel concrete, the increase is less than for the Series 2 mixes.

Concrete durability tests

CONCRETE SPECIMENS FOR LONG-TERM TESTS

Specimens were cast from the mixes shown in Table 9 with a nominal cement content of 330 kg/m³ and put into storage for periods of up to 5 years for the following tests.

Sulphate resistance tests (OPC and SRPC)
Volume stability tests (OPC and SRPC)
Corrosion of reinforcement (OPC only)
Initial surface absorption tests (OPC only)

The cubes and prisms for the sulphate resistance and volume stability tests were compacted by hand-ramming and those for the corrosion of reinforcement and air permeability tests were compacted by vibration. The moulded specimens for all the tests were stored in moist air for 24 h, demoulded and then stored under water for a further 27 days. Prisms used for determination of drying shrinkage and for volume stability tests had stainless steel balls cemented into their end faces during the first 7 days' curing.

SULPHATE RESISTANCE TESTS

Concrete cubes (101·6 mm) were stored in the following six sulphate solutions and in tap water as a control.

Magnesium sulphate solutions containing 3·5, 0·75 and 0·35% SO_3
Sodium sulphate solutions containing 3·5 and 0·35% SO_3
Mixed salt solution containing 0·35% SO_3

(The mixed salt solution contained 0·375% magnesium sulphate, 0·177% sodium sulphate and 0·142% calcium chloride.)

Twelve cubes of each mix were stored in separate polythene tanks and the sulphate solutions were replaced with 36 litres of fresh solution every three months. Three cubes were removed after storage for 1 year, 2 years and 5 years for compressive strength tests. The mean percentage ratios of sulphate-stored to water-stored cube strengths at 5 years are recorded in Table 13 and shown in Figure 10 with the results obtained at 1 and 2 years.

Inspection of the results from the sulphate resistance tests reveals that there is sometimes considerable variation between values obtained from replicate cubes. This is usually found with this type of test where the normal within-batch variation of cast cubes is amplified by varying rates of attack by aggressive solutions on individual specimens. For this reason, little significance

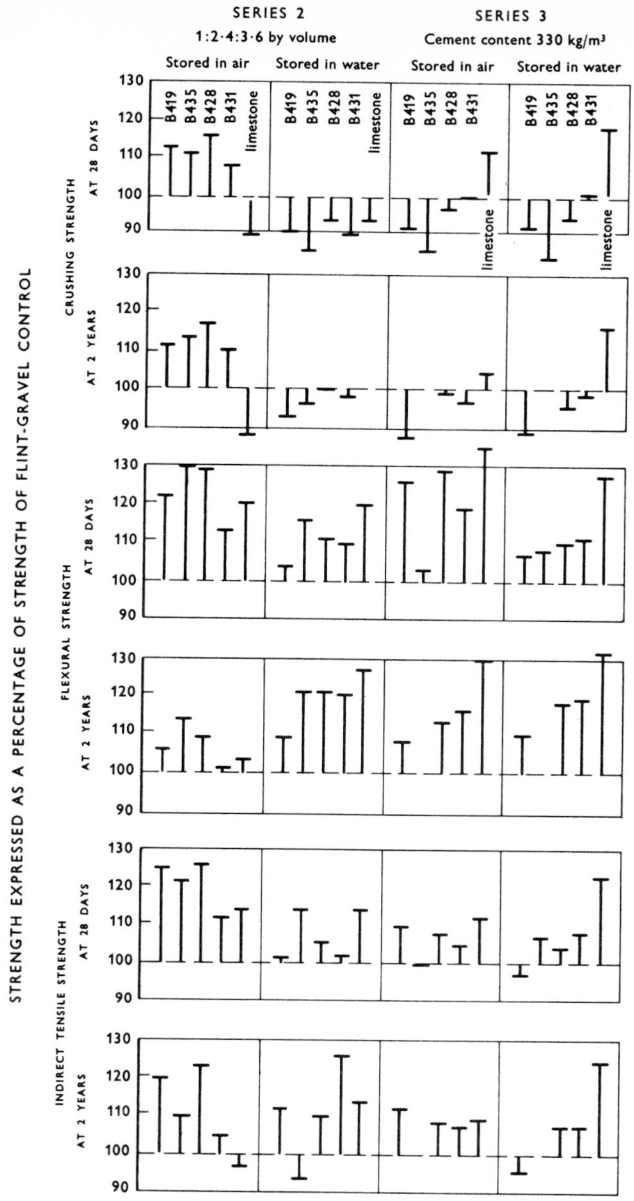

Figure 9: Strengths at 28 days and at 2 years of slag and crushed-limestone concretes compared with flint-gravel concrete.

should be attached to relatively small variations in the mean results derived from tests on three cubes. Nevertheless the general trend of performance of the concretes in the various solutions is demonstrated in Figure 10. Attack to varying degrees on all the OPC concretes is shown to have occurred in all the solutions except the weak sodium sulphate but the SRPC concretes have been attacked only in the strong magnesium sulphate solution. It is also demonstrated that, in general, there is no difference between the performance of concretes made with lighter-weight slag aggregates and the control concretes.

TABLE 13: **Percentage ratio of sulphate-stored to water-stored cube strengths after 5 years' immersion in sulphate solutions.**

	Sulphate solution					
	Magnesium sulphate		Sodium sulphate		Mixed salt	
	Type of cement					
	OPC	SRPC	OPC	SRPC	OPC	SRPC
STRONG SOLUTION ($3·5\%\ SO_3$)						
Slag B419	49	67	53	89		
Slag B435	43	74	54	106		
Slag B428	40	79	54	95		
Control slag B431	33	67	66	94		
Flint gravel	35	64	60	91		
Crushed limestone	44	76	46	93		
MEDIUM SOLUTION ($0·75\%\ SO_3$)						
Slag B419	57	88				
Slag B435	58	107				
Slag B428	48	94				
Control slag B431	60	89				
Flint gravel	76	103				
Crushed limestone	58	98				
WEAK SOLUTION ($0·35\%\ SO_3$)						
Slag B419	82	85	102	100	71	84
Slag B435	78	98	94	110	66	86
Slag B428	64	89	105	103	71	91
Control slag B431	71	96	99	100	77	87
Flint gravel	87	105	97	102	61	83
Crushed limestone	75	92	109	102	84	90

OPC = Ordinary Portland cement
SRPC = Sulphate-resisting Portland cement

VOLUME STABILITY TESTS

Nine 200 × 75 × 75 mm prisms from each of the twelve mixes were measured for their initial wet length after the curing period in water. The method of measurement, using a simple frame which incorporates a micrometer to measure between stainless-steel balls cemented into the centre of each end face, is described in BS 1881:Part 5:1970. Three prisms were returned to water storage and three were put onto open racks in a room controlled at 20°C and 65% R.H. The remaining three prisms of each batch were placed on a concrete slab at ground level on the exposure site at BRS. Periodic measurements were made on all the prisms, and the results after 1 year and 5 years are recorded in Table 14. Prisms from the exposure site were placed in a cabinet at 20°C for 4 to 5 h before measurements were made.

There were no indications of any significant differences between the movements recorded on the slag and on the control concretes stored under water on the exposure site. The shrinkage on storage at 65% R.H. was, on average, a little higher for the slag concretes

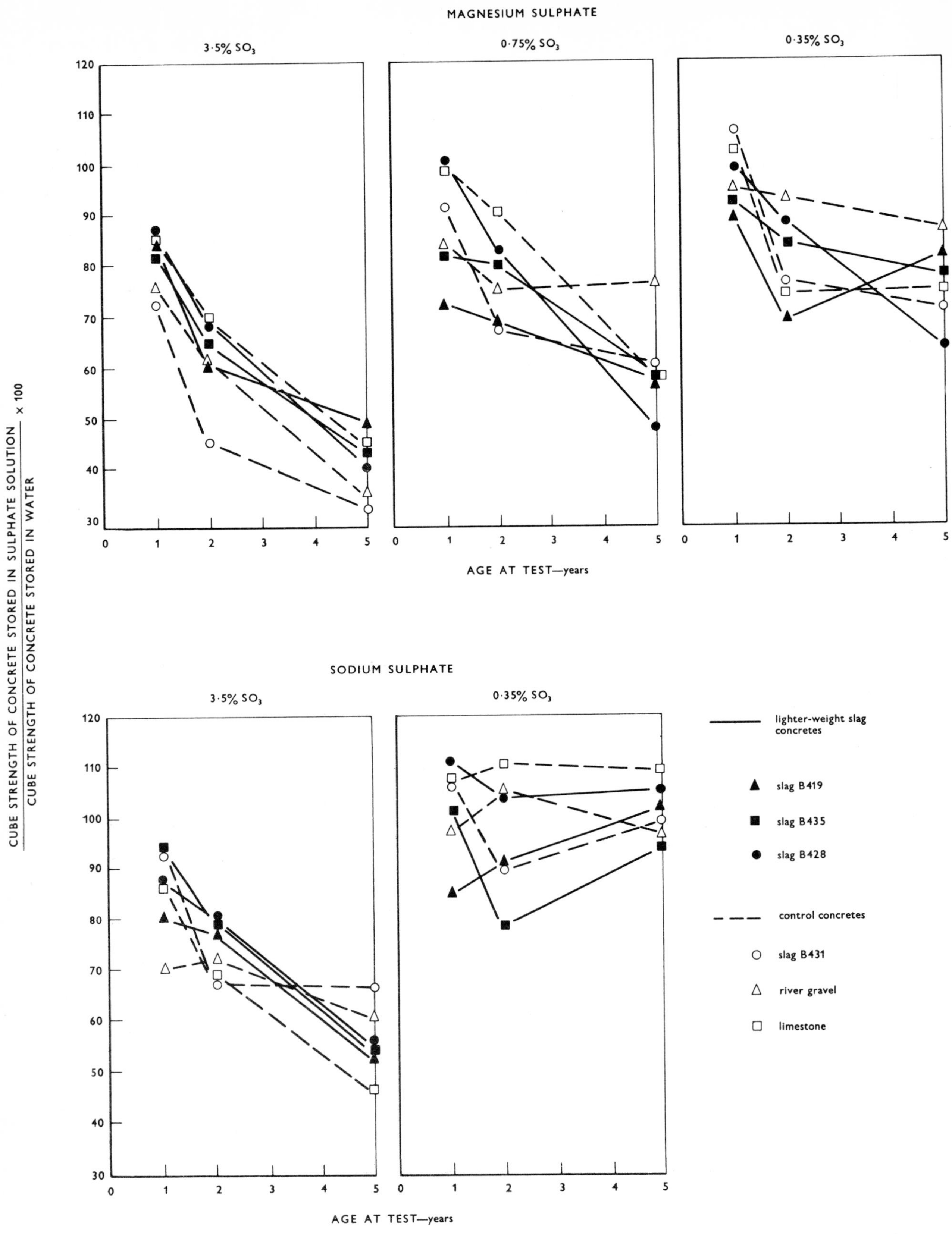

(a) *Ordinary Portland cement concretes*

Figure 10: Comparative tests over 5 years on concrete cubes in sulphate solutions.

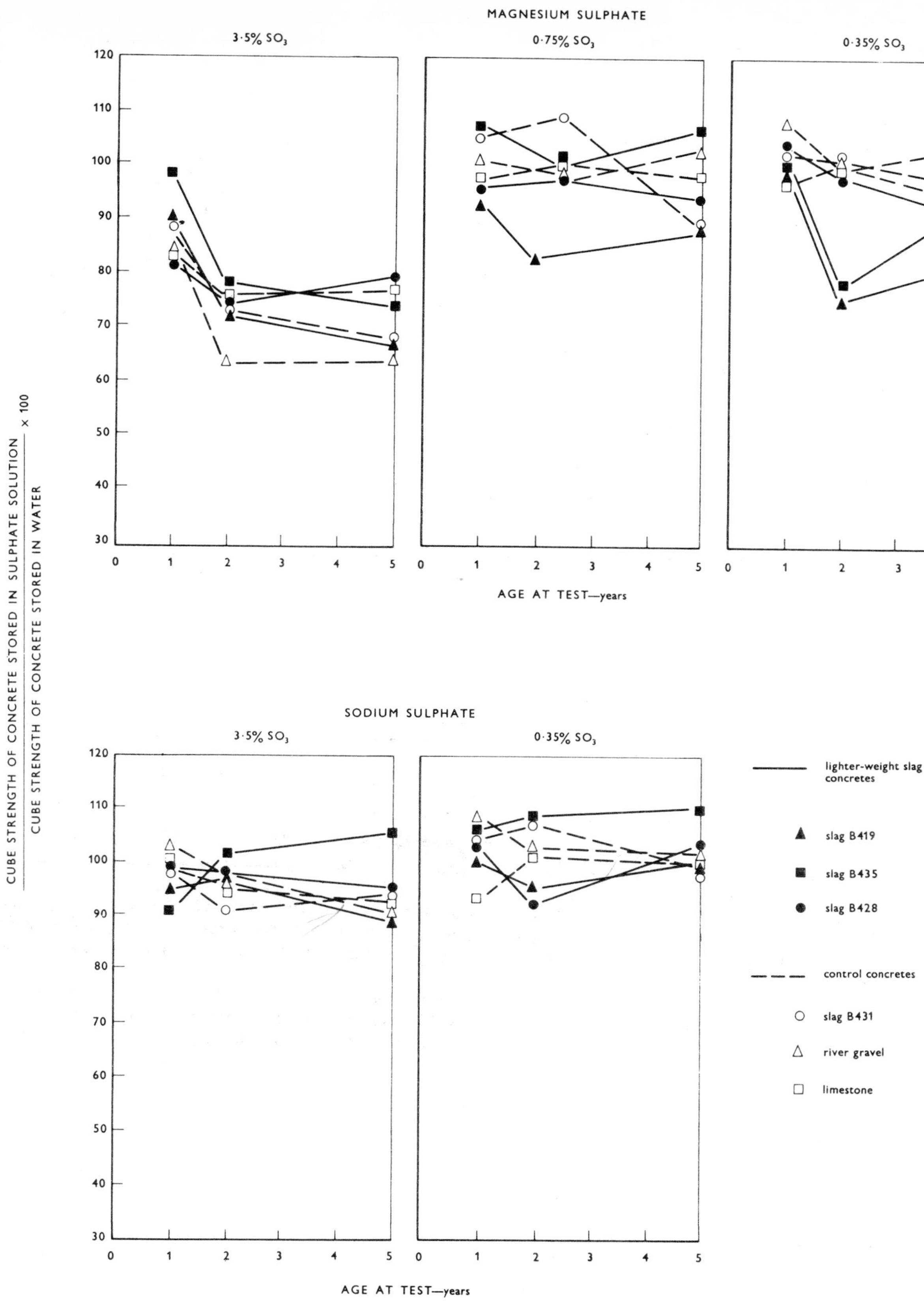

(*b*) *Sulphate-resisting Portland cement concretes*

Figure 10 continued.

than for the controls. Drying shrinkage tests were carried out in accordance with the requirements of BS 1881:1952 and the results are given in Table 9.

DEPTH OF CARBONATION AND DEGREE OF REINFORCEMENT CORROSION

The concrete prisms used for estimating the depth of carbonation of the concrete and degree of reinforcement corrosion were of dimensions 305 × 153 × 153 mm. Bright-steel reinforcing rods 12·7 mm diameter and cleaned of rust were held by means of a framework at distances of 13, 25, 38 and 51 mm from the face of the concrete. By adjusting the length of the bars, the depth of cover at the ends was made the same as along the length of the bar. The arms of the framework positioning the bars in the concrete were electrically insulated from the reinforcing steel by sheaths of polythene tubing. The specimens were placed upright on ceramic pots on an exposure site in an area of high industrial atmospheric pollution.

After 5 years, the specimens were all in good condition with no superficial cracking. At 3 years and 5 years, specimens were broken open and the degree of carbonation of the concrete was shown by moistening the broken face with phenol phthalein. The depth of carbonation after 5 years was found to be between 2 and 3 mm for the lighter-weight slag concretes and between 1 and 2 mm for the control concretes. After 3 years' exposure, little difference in the permeability to air was detected between the lighter-weight slag concretes and those made with the control slag and flint-gravel aggregates; that of the crushed limestone concrete was much lower than the rest. The reinforcing steel was examined visually. There was no evidence of corrosion in any of the concretes and the results were completely satisfactory.

INITIAL SURFACE ABSORPTION TEST

The Initial Surface Absorption Test measures the rate of absorption of water into the surface of the concrete and the conditions of test are laid down in BS 1881: Part 5: 1970. Initially the cubes were dried to a constant weight at 40°C in a reduced pressure oven. The results on 101·6 mm cubes stored for 3 years under water and on the exposure site are shown in Table 15 together with the results on specimens after storage

TABLE 14: **Percentage change in length from the initial wet length after curing under water of concrete prisms stored under different conditions.**

Coarse aggregate	Stored in air at 65% R.H.				Stored under water				Stored on exposure site			
	OPC		SRPC		OPC		SRPC		OPC		SRPC	
	1 year	5 years	1 year	5 years	1 year	5 years	1 year	5 years	1 year	5 years	1 year	5 years
Slag B419	−0·048	−0·056	−0·046	−0·058	+0·007	+0·009	+0·010	+0·012	−0·006	−0·014	−0·006	−0·002
Slag B435	−0·043	−0·051	−0·038	−0·045	+0·007	+0·011	+0·006	+0·011	+0·002	−0·009	−0·005	nil
Slag B428	−0·048	−0·055	−0·045	−0·049	+0·005	+0·004	+0·007	+0·015	−0·009	−0·016	−0·006	−0·016
Slag B431	−0·040	−0·048	−0·033	−0·041	+0·005	+0·009	+0·007	+0·009	−0·002	−0·010	−0·003	−0·001
Flint gravel	−0·035	−0·044	−0·037	−0·041	+0·005	+0·001	+0·004	+0·006	−0·005	−0·005	−0·004	−0·008
Crushed limestone	−0·037	−0·039	−0·036	−0·040	+0·005	+0·005	+0·009	+0·011	−0·005	−0·008	−0·010	−0·009

TABLE 15: **Initial surface absorption (ml/m² s) on lighter-weight slag and control concretes.**

Storage	Measured after*	Coarse aggregate					
		B419	B435	B428	B431 (control)	Flint gravel	Crushed limestone
Exposure site for 1 year	10 min	0·38	0·41	0·40	0·42	0·38	0·32
	30 min	0·24	0·26	0·27	0·27	0·25	0·21
	1 h	0·18	0·20	0·21	0·21	0·20	0·16
	2 h	0·14	0·16	0·16	0·16	0·15	0·13
Exposure site for 3 years	10 min	0·06	0·04	0·07	0·07	0·07	0·04
	30 min	0·03	0·02	0·04	0·04	0·04	0·02
	1 h	0·02	0·02	0·03	0·03	0·03	0·02
	2 h	0·02	0·01	0·02	0·02	0·02	0·01
Under water for 3 years	10 min	0·29	0·39	0·35	0·32	0·26	0·28
	30 min	0·16	0·21	0·18	0·16	0·16	0·15
	1 h	0·12	0·17	0·13	0·13	0·11	0·11
	2 h	0·09	0·14	0·10	0·11	0·08	0·09

*Specimens stored at 20°C for 4 to 5 h before test.

for 1 year on the exposure site. The rates of surface absorption are given in ml/m^2 s after measurement at 10 min, 30 min, 1 h and 2 h from the application of water to the surface. No significant differences in surface absorption were recorded between the lighter-weight slag concretes and the controls after storage on the exposure site for 1 and 3 years. It will be noted that all the concretes have shown a marked decrease in surface absorption between 1 and 3 years' exposure to air, whilst the results after 3 years' storage under water are nearer to the values for the cubes stored in air for 1 year.

Conclusions

(1) The compacted bulk density of slag aggregates, like that of other aggregates, depends upon the size fraction on which it is determined, larger sizes having lower densities. BS 1047:1952 requires a density greater than 1250 kg/m^3 when determined on any size of slag aggregate, whereas BS 802:1967 requires a density greater than 1089 kg/m^3 when determined on the 12·7–9·5 mm fraction of aggregate. In this investigation three lighter-weight slags, with densities lower than the BS 1047 requirement and approaching the BS 802 minimum limit have been used and their properties established.

(2) The lowest bulk density of the three lighter-weight slags was 1175 kg/m^3, which was 75 kg/m^3 lower than the minimum requirement in BS 1047:1952. The bulk density of the control slag sample complied with the density requirement and all four slags complied with the remaining requirements of BS 1047:1952, apart from slag B428 which failed the grading limit set for the amount passing the 4·8 mm ($\frac{3}{16}$ in.) sieve.

(3) These aggregates have been tested in concretes which were compared with concretes made with a normal-weight slag, a flint gravel and crushed limestone coarse aggregate. The bulk density of the lighter-weight slag concretes was about 5% less than the control slag concrete.

(4) For mixes having the same volumetric mix proportions and workability, the lighter-weight slag concretes required a slightly higher water/cement ratio and the cement content was about 20 kg/m^3 higher than the natural-aggregate concretes. On this basis, there was no significant difference between the strengths of the lighter-weight slag concrete and the normal-weight slag concrete. The strengths of the air-stored slag concretes were generally higher than those made with flint gravel and crushed limestone, and the difference between air and water storage was less marked.

When compared on the basis of equal cement contents, the compressive strengths of the lighter-weight slag concretes were slightly lower than those of the control-slag and natural-aggregate concretes, but the flexural and indirect tensile strengths were higher than those of the flint-gravel concrete but less than the crushed-limestone. However, the differences were small, and lighter-weight slags can be considered as being similar to normal slags and natural aggregates.

(5) Durability studies involving the examination of volume stability, initial surface absorption, resistance to sulphate attack up to 5 years, and protection to reinforcement against corrosion, have shown little difference in performance between the lighter-weight slag concretes and the control concretes having a cement content of 330 kg/m^3.

(6) On the basis of this investigation these lighter-weight slags are acceptable as coarse aggregate for concrete, provided that the mixes are properly designed. It would seem feasible that the minimum compacted bulk density specified in BS 1047:1952 could be reduced, but consideration should be given to linking this value to a specific size of aggregate.

ACKNOWLEDGEMENTS

The work described has been carried out as part of the research programme of the Building Research Establishment of the Department of the Environment and this paper is published by permission of the Director. The work received financial support from Slag Section of the British Quarrying and Slag Federation.

REFERENCES

1. LEE, A. R. Slag for roads, its production, properties and uses. *Journal of the Institution of Highway Engineers.* Vol. 16, No. 2. February 1969. pp. 11–29.
2. GUTT, W., KINNIBURGH, W. and NEWMAN, A. J. Blastfurnace slag as aggregate for concrete. *Magazine of Concrete Research.* Vol. 19, No. 59. June 1967. pp. 71–82.
3. EVERETT, L. H. and GUTT, W. Steel in concrete with blast-furnace slag aggregate. *Magazine of Concrete Research.* Vol. 19, No. 59. June 1967. pp. 83–94.
4. Slag research at the Building Research Station. Discussion at a symposium, 15 February 1967. *Magazine of Concrete Research.* Vol. 19, No. 59. pp. 121–126.
5. GUTT, W. A new calcium magnesiosilicate. *Nature.* Vol. 190, No. 4773. 22 April 1961. pp. 339–340.
6. GUTT, W. The system dicalcium silicate-merwinite. *Nature.* Vol. 207, No. 4993. 10 July 1965. pp. 184–185.
7. MIDGLEY, H. G. The polymorphism of calcium ortho silicate. Paper submitted to the Sixth Symposium on the Chemistry of Cement, Moscow, 1974.
8. GUTT, W. and NURSE, R. W. Phase equilibria composition of Portland cement clinker. Paper submitted to the Sixth Symposium on the Chemistry of Cement, Moscow, 1974.
9. BUILDING RESEARCH STATION. *Concrete in sulphate-bearing soils and ground waters.* London, H.M.S.O., July 1968. pp. 5. Digest 90 (second series).
10. BRITISH STANDARDS INSTITUTION. CP 110: Part 1: 1972. *The structural use of concrete. Part 1. Design, materials and workmanship.* London. pp. 154.
11. BRITISH STANDARDS INSTITUTION. BS 1881: Part 4: 1970. *Methods of testing concrete for strength.* London. pp. 28.

An investigation into the production of sintered pfa aggregate (CP 2/75)

W. H. Harrison and R.S. Munday

INTRODUCTION

Pulverised fuel ash (pfa) is the residue from the burning of powdered coal, and some 10 Mtonnes are produced annually from coal-fired electricity generating stations. These pfa stations have now largely replaced the chain-grate fired boilers from which came most of the clinker used in the concrete block industry. Building Research Station investigations on behalf of the Central Electricity Generating Board and extending over many years from about 1950 to the present time, have looked at a number of possible outlets for this ash in the building industry. This paper deals only with that part of the programme concerned with the production and testing of a sintered lightweight aggregate. After rudimentary feasibility studies at BRS had produced some encouraging results, further development at BRS continued whilst a commercial company worked along similar lines. A number of full-scale plants have now been built by this company and they produce an aggregate known as Lytag. Some of the information obtained from the work carried out at BRS has been summarised in CEGB publications [1,2], and during the years 1967-1969 a team comprising Messrs R S Munday, G J Seal and J G Whiteley from the South Eastern Region of CEGB worked on the project at BRS.

Much of the ash, especially that from older and less efficient power stations, contained varying amounts of residual fuel. In order to utilise this as part, or all, of the fuel required to sinter the ash into an aggregate it was necessary to produce a permeable fire-bed thus allowing access of oxygen. A process of pelleting the ash was evolved and a bed of pellets was then ignited and burned by a through current of air. This paper traces the development of the pelleting and sintering techniques to the pilot-scale stage. It also includes an appraisal of the aggregate produced.

1 PROPERTIES OF PFA RELEVANT TO ITS CONVERSION INTO AGGREGATE

The chemical and physical properties typical of pfa currently produced have been given by CEGB[1]. Those properties particularly relevant to aggregate production and therefore determined in many of the bulk samples received at BRS over the years of this investigation, are discussed below. Some of the properties of the bulk samples received between 1967 and 1969 are listed in Table 1 and compared with the typical values quoted by CEGB[1].

1.1 Fuel content

The most important single constituent variable in the ash is the carbon content. All pfa has some residual fuel. In practice the true carbon content is seldom determined because of the somewhat elaborate equipment involved: instead, loss of weight sustained by the ash when ignited to about 800°C in a muffle furnace for some pre-determined time is taken to represent the fuel content.

Bulk samples of ash were obtained from various power stations, selected principally on the basis that the station was considered to be a possible location for an aggregate plant or it used a particular source of coal (see Table 2). Variations in the fuel content of ash, particularly from solder stations, can arise for a number of reasons such as the efficiency of the coal grinding and the mode of operation of the boiler. Older stations often are only brought into full use at times of peak demand. The losses on ignition of the samples used in this investigation were mainly in the range 3-10 per cent. Values for 24 deliveries in bulk from one station over a 2-year period (shown in Figure 1) ranged from 3.5 to 10.5 per cent with a mean of 6.4 per cent and a standard deviation of 1.7 per cent. It should also be added that a few samples were rejected at BRS by visual inspection when the dark colour of the ash indicated an unacceptably high fuel content. The within-batch variation of samples with high fuel content can also be quite high and one batch received dry in 100 bags showed a range between bags for loss on ignition of 4.1-12.1 per cent with an average of 7.5 per cent and a standard deviation of 2.1 per cent.

Table 1 Some properties of bulk samples of pfa received at BRS from 1967-1969

Year	Power station	Carbon content (%)	Soluble salt content (%)	Specific surface area (m^2/kg)	Percentage, by weight, passing sieve of size		Relative density
					150μm	75μm	
1967	West Thurrock	2.9	2.9	535	98	93	2.25
	Northfleet	5.1	2.5	373	97	87	2.05
	Northfleet	4.1	1.9	322	98	90	2.09
	West Thurrock	4.7	2.3	438	98	91	2.15
	West Thurrock	7.4	2.5	336	93	78	2.02
	West Thurrock	5.5	2.7	445	96	87	2.15
1968	West Thurrock	6.9	3.4	288	93	80	1.99
	Bold 'B'	3.3	3.0	420	98	92	2.11
	High Marnham	1.5	4.5	294	96	85	2.08
	Aberthaw	11.8	1.7	405	98	90	2.09
1969	West Thurrock	1.6	2.6	357	97	89	2.14
	Northfleet	4.4	2.5	302	98	89	2.26
	Northfleet	3.4-7.1	2.1	153	89	63	2.00
	West Thurrock	3.4	4.0	279	95	79	2.04
	Ratcliffe	1.9	5.1	353	97	88	2.22
1972	Average values given by CEGB for precipitator ash		2.0-3.0	320-470		>97	1.9-2.3
	Base load stations	<3.0					
	Older stations	Up to 10% or more					

Table 2 Ashes used to represent coal supplies of the Board

Station	Source of coal
Bold 'B'	Lancashire
High Marnham Ratcliffe	East Midlands
West Thurrock Northfleet	Lothian and North Eastern
Aberthaw	South Wales

As might be expected the loss on ignition was consistently higher than the carbon content and a relationship was established as shown in Figure 2. The derived equation $y = 0.896x - 0.134$ shows that an approximate value for the carbon content may be obtained by multiplying the loss on ignition by 0.9. The loss on ignition was obtained by heating a sample of 1 g (after grinding to pass 150μm sieve) to a temperature of 750-850°C for 2 hours. The carbon content was determined by heating a similarly ground sample to 1300°C for 4 minutes in a stream of oxygen and weighing the carbon dioxide produced after absorption by soda-asbestos.

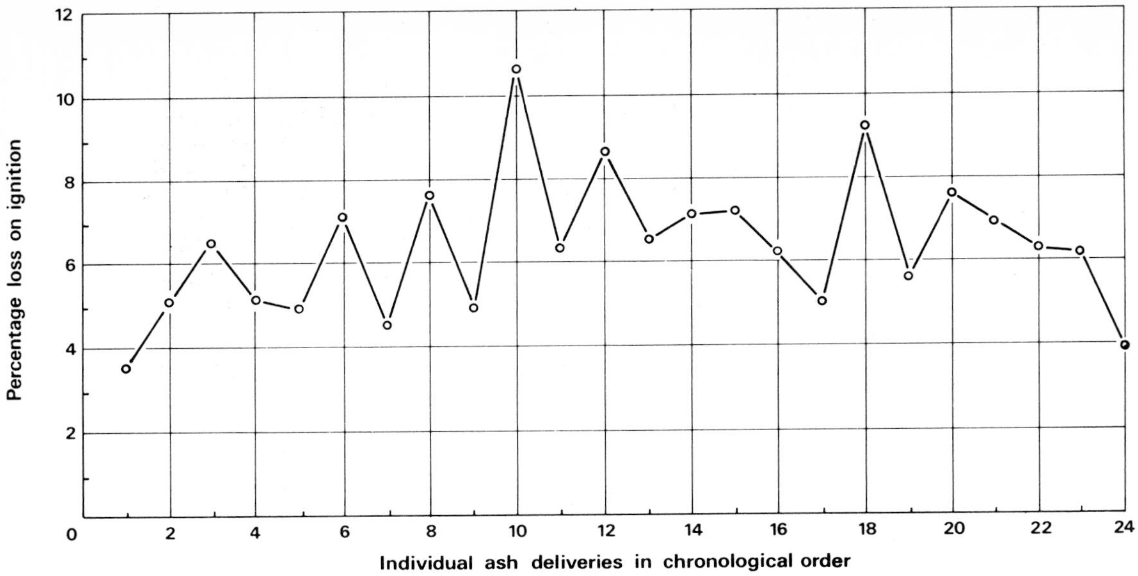

Figure 1 Fuel content expressed as loss on ignition in successive deliveries of pulverised fuel ash from one power station over 2 years

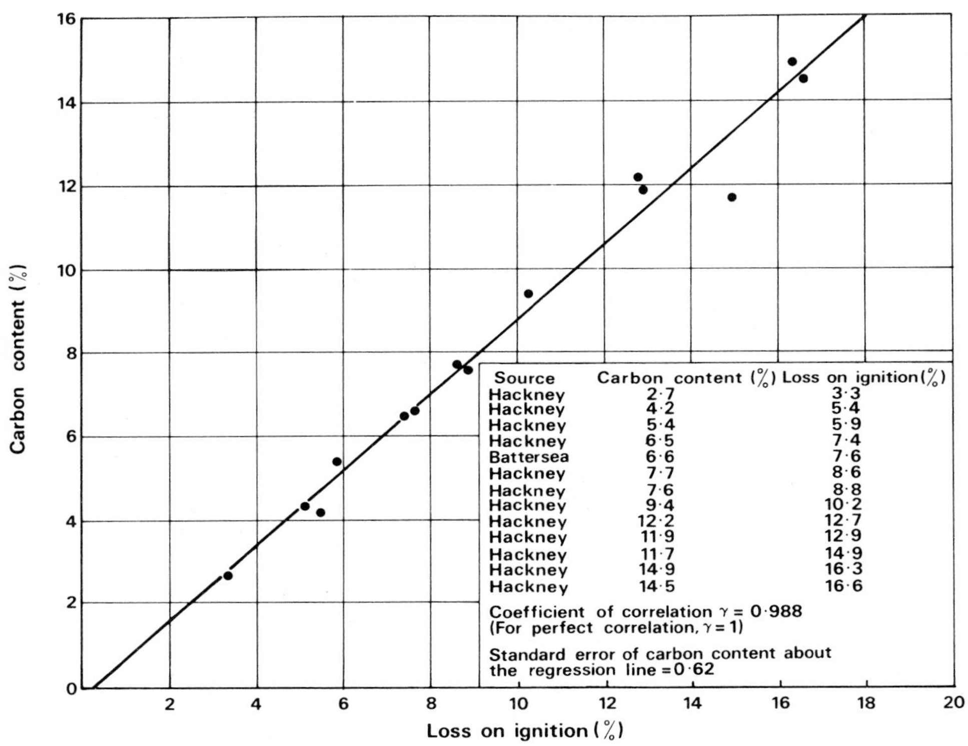

Figure 2 Carbon content/loss on ignition for a number of pfa samples

It will be seen subsequently that the level and uniformity of the fuel content is a major factor in deciding on the most suitable method of sintering, ie determining the rate of burning achieved, and it is reflected in the final properties of the aggregate produced.

1.2 Other chemical constituents

The influence of iron and alkali metal contents on the optimum sintering temperature was small in comparison with the fuel content so they were rarely determined. The sulphur appeared to be present mainly as calcium sulphate. Sulphur oxides were evolved on firing and several sulphurous compounds including hydrogen sulphide and carbon disulphide have also been identified in the evolved gases. The residual sulphur in the sintered product was negligible.

The level of soluble salts present in the ash was generally between 2 and 3 per cent and had some influence on the quality of pellet produced. Chemical examination of the water-soluble extract of pfa showed it to be principally calcium sulphate. The extract, when evaporated to dryness, gave a white powder which, when examined microscopically, showed needle-shaped crystals typical of $CaSO_4.2H_2O$ (with a small amount of $CaSO_4.\frac{1}{2}H_2O$ round the margin, presumably due to over-heating during the evaporation). X-ray diffraction confirmed the identification of gypsum ($CaSO_4.2H_2O$). It might be expected that the calcium sulphate would be anhydrous when the ash was formed and that subsequent wetting to form pellets would cause hydration. This seems to be bound up with the fact that pfa when once wetted is more difficult to pelletise than the material direct from the precipitators. Subsequently, X-ray analyses of the virgin ash showed lines for anhydrite but no lines for gypsum or hemi-hydrate. The importance of soluble salts is further discussed in Section 2.4 where the addition of small amounts of various salts was found to improve pellet strength. The pH of the water extract and its change with time was also found to be important in the pelleting process (see Section 2.4).

1.3 Physical properties

The principal physical property which affects the conversion of pfa into aggregate is the size distribution of the ash particles. An indirect comparison of the average fineness of different powders is given by the specific surface area (ie the surface area of all the particles in 1 g of a representative sample). The higher the surface area, the greater is the number of particles and hence the finer is the ash. In some cases a sieve analysis was made grading the particles between 63 and 500μm(0.063 and 0.5 mm).

The material retained on the 63μm sieve consisted chiefly of glassy spheres together with irregularly shaped particles formed by the agglomeration of very small spheres (probably each 1-10μm) and fragments of unburnt coal and coke. Both clear and opaque glassy spheres were present and the clear ones were often hollow, as shown by light reflected internally as well as by the shape of broken fragments. These are known as cenospheres and collect on the surface of ash disposal areas using water-borne ash transportation.

If the large particles had consisted only of large glass spheres it might be supposed that they had resulted from coarsely ground coal, but the presence of agglomerations of very small spheres indicates that these at any rate were derived from very fine coal particles, and it might well be that the larger glass spheres were also derived from small ones and represent a further stage of fusion. It is therefore suggested that the coarseness of the ash is not necessarily related only to the particle size of the powdered coal, but to combustion conditions in the furnace which are favourable to the formation of agglomerations. A full investigation into the formation and properties of cenospheres has been carried out by Raask[3] at the Central Electricity Research Laboratories.

The particle size analysis of pfa presents difficulties, since the ash is of a heterogeneous nature both chemically and physically. The Coulter counter to a large extent overcomes these problems by determination of the number and size of particles suspended in an electrically conductive liquid which are forced through a small immersed aperture having an electrode on each side. As each particle passes through the aperture, it replaces its own volume of electrolyte within the aperture, momentarily changing the resistance value between the electrodes. This produces a voltage pulse of short duration having a magnitude proportional to particle volume, and the resultant series of pulses is electronically amplified, scaled and counted. Some anomalies occur and the apparatus has to be calibrated with particles of known size. Results from five ashes are shown in Figure 3. Differences between ashes were demonstrated and quite good reproducibility was achieved.

In the more recent work a scanning electron microscope (stereoscan) was used to examine a number of ashes. Figures 4 and 5 illustrate the variability of the size distribution of the particles. The packing and compaction of the ash in the production of pellets is dependent, in part, on the size distribution of the ash particles. The pfa sample shown in Figure 4 exhibited a sufficient range in particle size to give a well compacted pellet, whereas the close size range of the sample shown in Figure 5 gave a less compact structure.

2 PELLETING OF PFA

As in the approach to the use of pfa for brickmaking, a prime consideration was the production of a building material solely from a single raw material. In the case of brickmaking this proved impracticable, and mixing the pfa with 15 per cent of clay was found necessary to produce a mix capable of moulding and subsequent handling before firing[2]. The dependence of a process on considerable additions of a second raw material limits its general application apart from adding to the cost. In the production of pfa pellets which could withstand handling and thermal shock, it

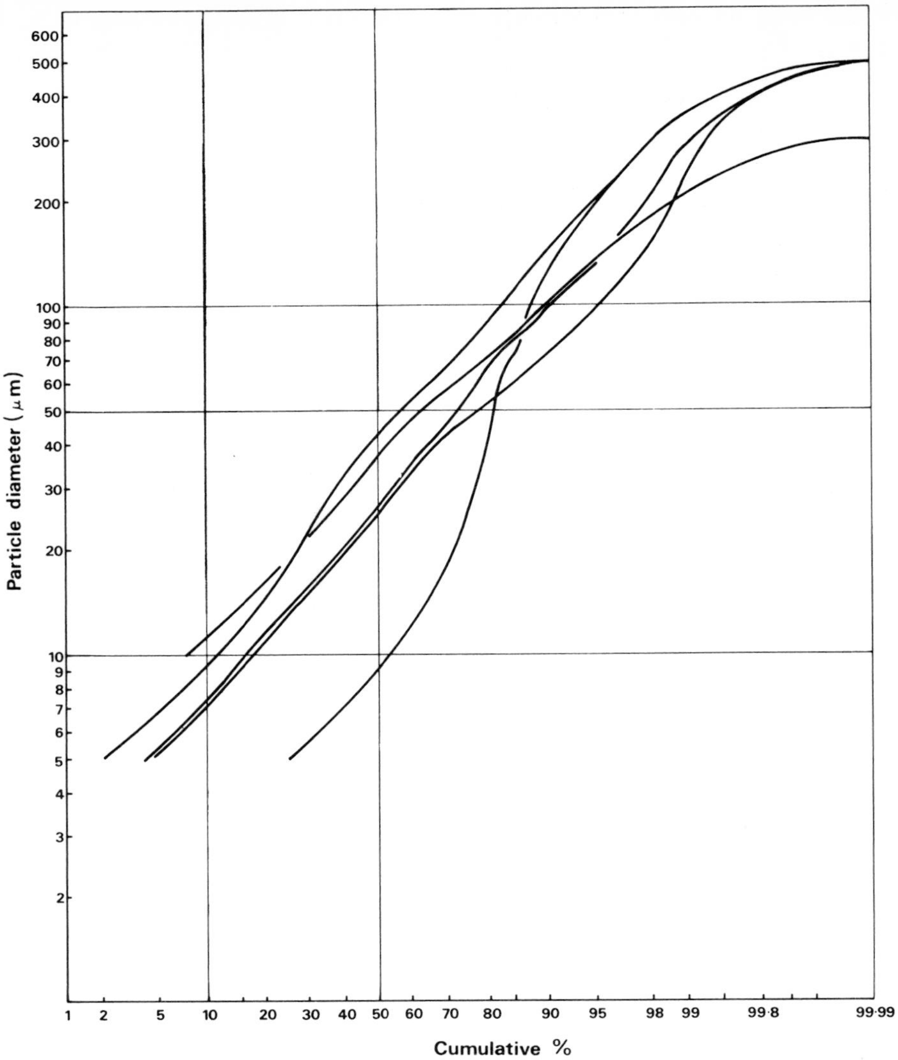

Figure 3 Particle size distribution of five pfa samples, measured by a Coulter counter

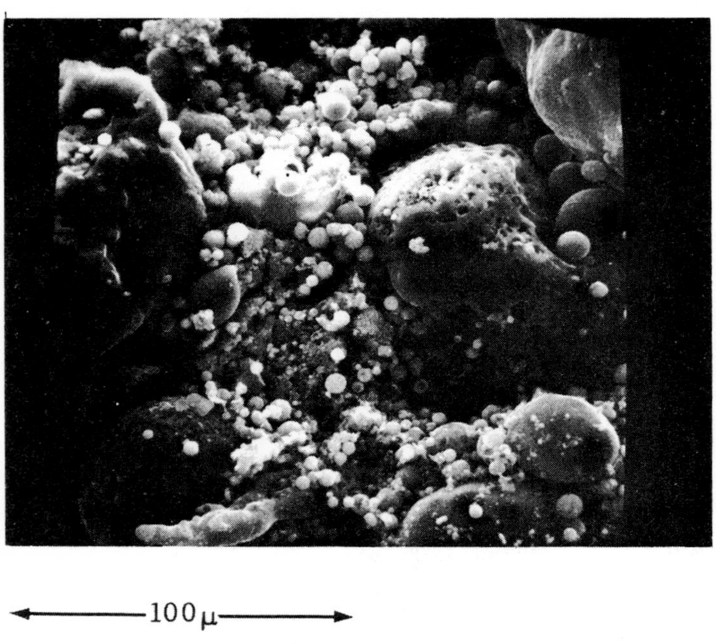

Figure 4 Stereoscan photograph of pfa with a wide range of particle sizes

← 100μ →

Figure 5 Stereoscan photograph of pfa with a narrow range of particle sizes

was found possible with many ashes to pelletise using water only. This section of the paper is primarily concerned with the optimisation of this process, a study of the mechanisms of pellet formation, and the addition, in very small amounts, of substances to extend the process to any type of ash. Additions of clays, shales or similar binders were not covered in this work since additions of 10-15 per cent would introduce limitations similar to those met in the brickmaking study. However, in certain locations, for example where pfa and colliery shale are available together, some advantages may be derived from their use in combination.

2.1 Development of the inclined pan process for pelleting

Water/pfa mixes have not sufficient plasticity for extrusion processes. Pressing techniques were considered but owing to the abrasive nature of the ash and the limited rates of output likely to be achieved, these were not pursued. Two methods of producing pellets by a tumbling action were investigated, one using a slightly inclined rotating cylinder and the other a pan rotating on a considerably inclined plane. Some trials with the rotating cylinder or drum pelleter are discussed later but the inclined pan process was found to be preferable for ash/water mixes. It is also used in many industrial processes for the pelleting of powders including cement and fertilizer manufacture. Early trials were made with a 2 m diameter pan replacing the tilting drum of a small concrete mixer. The only other essential requirement was a scraper to prevent build-up of ash and to maintain a smooth bed of ash over the base and rim of the pan. This apparatus was subsequently replaced by proprietary equipment fitted with a 5.6 kW motor driving a 1.5 m diameter pan with a rim height adjustable between 200 and 350 mm. The speed was variable between 12 and 18 rev/min and the angle of inclination to the horizontal was variable between 25° and 65°. As discussed later, the angle of inclination of the pan and the speed of rotation were not found to be the most important parameters in control of the process and angles of 46-48° and speeds of 14-15 rev/min were generally adopted. The ash was delivered dry by tanker and stored in silos of approximately 10 tonne capacity. An air-fluidised system transported the ash to a worm feeder which controlled the rate of feed of the ash to the pelleting pan. Thus a stage was reached where a continuous output of moist pellets was possible if stable conditions could be achieved.

2.2 Pellet production

Pellets of pfa bound only with water needed to be well consolidated in order to withstand handling in the moist state. The optimum water content varied with different ashes but was generally between 20 and 25 per cent of the dry weight. Excess water caused loss of capillary forces at the surface of the pellet followed by distortion from the spherical shape and coalescence of groups of pellets to form lumps. Too little water produced feathery, weak pellets. The process of formation required the continual generation of 'seed' pellets to provide a nucleus on which to build successive layers of ash as the pellets rotated in the pan. When dry ash was fed directly onto the pan it was found necessary to provide a continuous supply of seeds in order to control the size of the output pellets[4]. These were produced by breaking down a small proportion of the output pellets and re-cycling. Further development work showed that considerable improvement in the overall rate of pellet production and in the self-generation of seeds was afforded by mixing the dry ash with a proportion of the total water required for pelleting before feeding to the pan and by using a stepped pan.

It is generally accepted[5,6] that the formation of granules takes place with three distinct stages of liquid bonding: pendular, funicular and capillary (Figure 6). Wet particles are forced into intimate contact by the action of the rotating pan to form primary agglomerates of particles held by discrete lens-shaped liquid pendular bonds. Rolling of these primary agglomerates gives rise to the funicular state where the particles have been partially compacted with excess

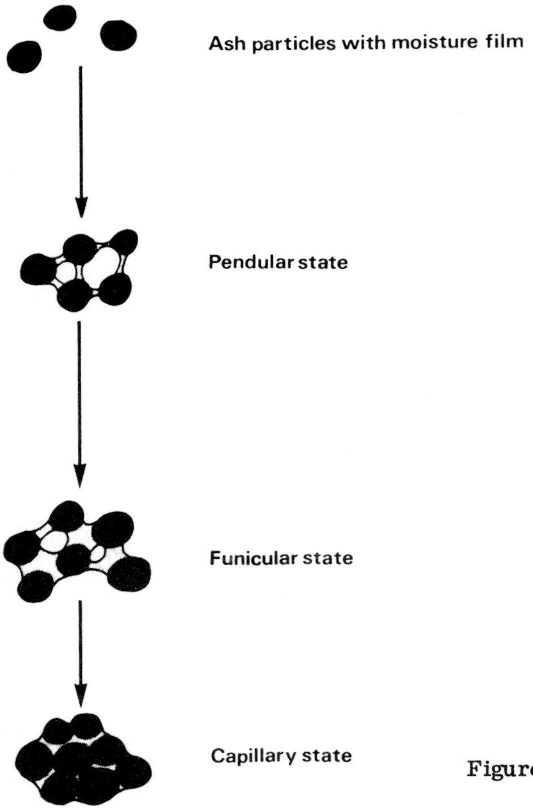

Figure 6 Diagrammatic representation of pellet formation

liquid squeezed to the surface. The pellet grows by absorbing dry or liquid-deficient particles onto the wet surface, further growth being promoted by spraying with liquid. In the absence of these particles two or more agglomerates in the funicular state will coalesce to give a large agglomerate, with excess moisture resulting in the formation of a mud. The final capillary state occurs in the fully compacted pellet with ideally 100 per cent liquid filling of the voids between particles. In general, liquid fillings of 80-95 per cent are cited in the literature compared with 65-85 per cent obtained with pfa. Theoretically the higher the liquid filling, the greater will be the capillary forces and suction potential and hence the green pellet strength. An apparent high liquid filling does not necessarily give high strength, as a wet pellet with excess surface moisture will give a high calculated liquid filling although it may not be fully compacted, some portion of the pellet remaining in the funicular state. An optimum liquid filling and compaction for maximum green pellet strength is therefore obtainable (Figure 7). As would be expected a relationship exists between liquid filling and porosity (Figure 8). It is of interest to note that in the case of trials with High Marnham ash, all of which gave satisfactory products, the liquid filling was generally less than the optimum as shown by Figure 7. Although the wet-pellet strengths generally obtained on the pilot plant were not at the optimum as shown by the graph, they were sufficient to allow the green pellets to be transferred to the sinterstrand and fired without break-down occurring.

It is apparent that the formation of seeds or nuclei is promoted by the presence of wet particles whereas the growth of the resultant seeds is dependent on the presence of dry or moisture-deficient particles. The build-up of a pellet by the bonding of particles to the surface is due to the surface tension of the binding liquid. The porosity and compaction of the final pellet is very largely dependent on the mechanical forces in the pelleting pan[7]. These forces derive from the fall of the pellet down the pan and are therefore dependent on the pan diameter and the angle of inclination. The drop height of the pellet decreases with decreasing inclination but the retention time increases. As these two effects act in direct opposition only very narrow limits of control of the porosity can be obtained by changes in the angle of inclination.

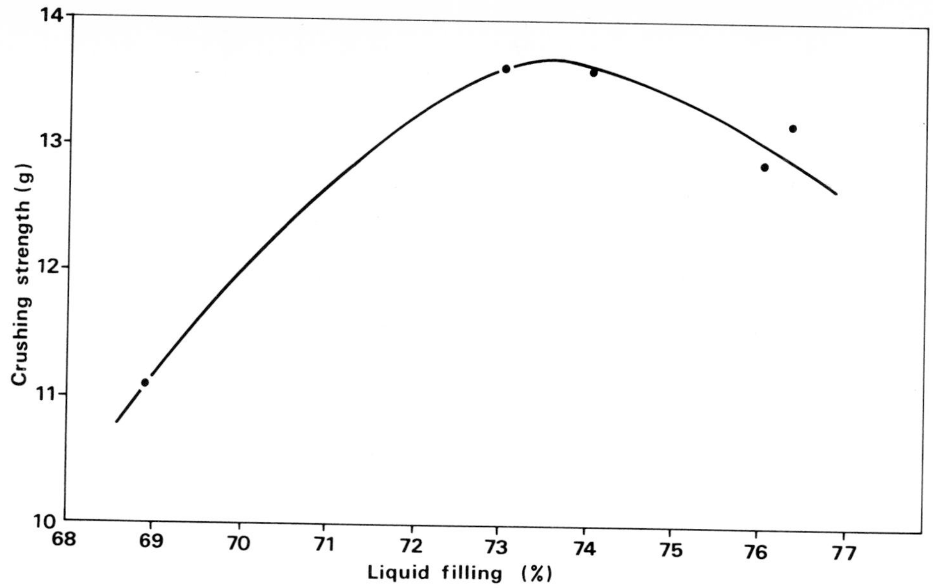

Figure 7 Crushing strength/liquid filling of green pellets

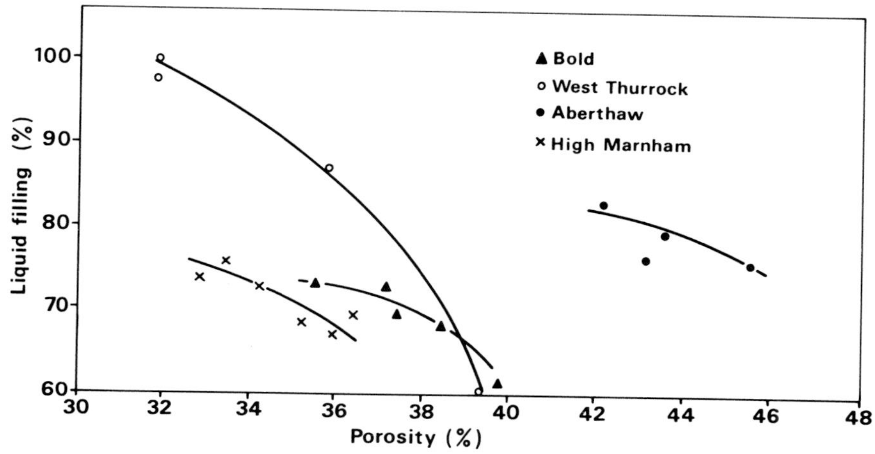

Figure 8 Liquid filling/porosity of green pellets

The pan diameter is therefore the prime factor controlling the compacting forces, large diameter pans increasing compaction and reducing porosity (Table 3). The rim height and throughput control the retention time and therefore the degree of compaction. Increased rim height also increases the compaction forces by increasing the weight of material in the pan but requires a higher motive power. The throughput controls the formation and build-up process of the pellet.

In the flat-bed pelleting pan seed/nuclei fall the full height of the pan (Figure 9(i)). Owing to the low strength of the pendular bond the rate of fracture rather than compaction will be high. The surviving seeds lying against the rim suffer impact from the larger pellets until they are covered and carried up round the pan for the process to be repeated. The capillary forces in the larger compacted pellets can be shown to be 3000–4000 times those existing in the pendular state. The seed/nuclei are therefore further depleted by the action of the larger pellets. The crushing action of the larger pellets is not confined to seeds in the pendular or funicular states. Capes and Dankwerts[8], using sands, concluded that the growth of the largest pellets occurred by crushing of the smallest size, with all the pellets at their optimum moisture content, ie the fully compacted capillary state. The wet strength of pellets, taken as the load required to crush the pellets when applied between two parallel plates, is shown later in Section 2.4 to give a plot of the form $S = KD$ where S = crushing strength and D = diameter of pellet.

Deviation occurs at the smaller sizes as would be expected with pellets partially in the funicular state. This is in contrast to the findings of Conway-Jones[9] who, using single-sized sands, obtained $S = KD^2$. This was, however, with a fixed charge of pellets in a sealed drum compared with the equilibrium conditions existing in the pilot-plant pelleting pan.

Table 3 **Pellet porosity as a function of pan diameter in cement manufacture**[7]

Pan diameter (m)	Porosity (%)
0.4	30–35
0.8	25–30
4.0	20

The force exerted by a pellet on impact with the rim of the pan is proportional to its mass:

$$F \propto M \propto D^3$$

where F = force of impact

M = mass of pellet

D = diameter.

While the strength of the pellet is proportional to its diameter, the compacting forces are proportional to the cube of the diameter of the pellet. Thus the wider the range of sizes present in the pelleting pan the greater will be the tendency for the larger sizes to grow at the expense of the smaller. To optimise the pelleting process, it is therefore essential to segregate the seeds/nuclei from the growing pellet and to reduce the compaction forces exerted on them by the pelleting pan.

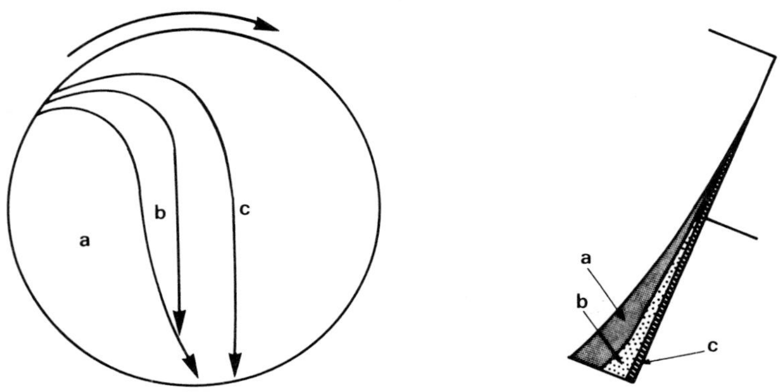

(i) Flat-bed pelleting pan

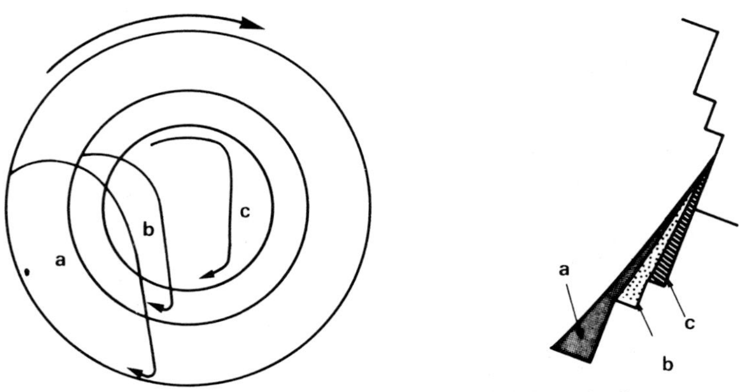

a Fully formed pellets
b Growing pellets and ash
c Seeds/nuclei and ash

(ii) Stepped pelleting pan

Figure 9 Distribution of pellets in pan

The use of a stepped pan as shown in Figure 9(ii), will give these conditions. In a pan at equilibrium the small seed pellets, travelling round the pan, descend into the central region where they are caught on the innermost step. Large pellets which break away early tend to fall to the outer rim. The cumulative effect is for the smaller seeds to gravitate to the bottom and inner step, and the large pellets to remain on the surface of the bed and in the outer rim, with intermediate pellets spaced between the two extremes. As can be seen from the cross-section of the bed, a large pellet will, when falling, skim over the smaller pellets to the outer rim with no collision occurring between widely differing sizes of pellet.

The flat-bed pan (Figure 10) was converted to multistep (Figure 11) and trials were carried out. It was immediately obvious that the potential output of the pan was much greater and that seed formation and pellet growth once established were more stable. Deliberate short-term variations of water and/or ash supplies failed to disturb the equilibrium within the pan. At no time during these trials, which were carried out over a period of several months using a variety of ashes, was any product produced which was not suitable for sintering. This is in contrast to the single-step pan which was difficult to control and often gave a poor product, ie too large, too small or insufficiently compacted. Construction details of the stepped pan and scrapers are shown in Figure 12.

Figure 10 Single-step or flat-bed pan

Figure 11 Multistep pan

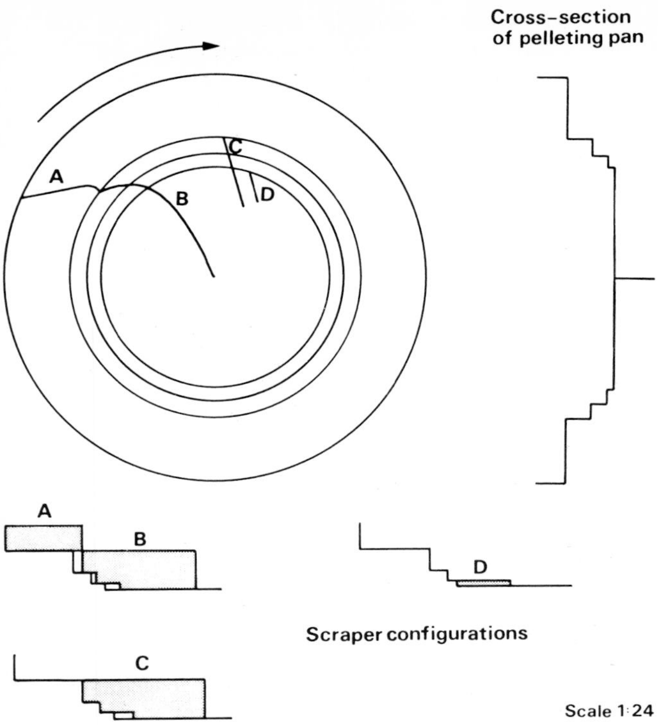

The flat-bed pelleting pan was converted to multistep by the addition of a rim to the existing pan and changing the design of the scraper blades. The modified scraper blades allowed ash to build up to form steps and were so arranged that the separation of sizes was enhanced and the maximum use made of the available bed area. From the scraper configurations shown, it can be seen that the smallest sizes are allowed furthest round the pan before being deflected.

Figure 12 Constructional details of pelleting pan

With the seeds and growing pellets in separate areas, the differing moisture demands could be achieved. Typical results obtained with the pan at equilibrium were:

Moisture content of seeds	23.4 per cent
Moisture content of pellets	22.3 per cent,

ie approximately 1 per cent excess moisture is required to promote seed formation. The separation of sizes into definite areas of the bed was clearly observed and the cross-section shown in Figure 9(ii) was obtained when the pan was stopped and the bed dug out.

To estimate the increase in output obtained with the multistep pan, allowance had to be made for its increased diameter. Klatt[10] states that as adjustment of pan angle, speed, etc, are very limited for optimum conditions, the output can be given by:

$$M = YD^2$$

where M = throughput

Y = constant

D = pan diameter

Other workers[11,12] have derived empirical and theoretical equations to predict output involving additional parameters. Pietsch[13] has shown that all the methods give approximately the same calculated dimensions for pelleting pans but notes the need for work on the actual size of the pan chosen. Using Klatt's formulae to obtain the output of a single-step pan of the same dimensions as the multistep pan used in this work, the latter was shown to have 3.5 times the output of the single-step unit.

2.3 Attempts at drum pelletising

Drum pelletisers are of large capacity and are therefore generally used by the high throughput industries, ie iron ore dressing. They do not classify the product so that a sieving operation with crushing of oversize and recycling is necessary. It was hoped, from the information gained by operating the stepped pan, that a drum could be designed which would eliminate the need for sieving and recycling. This would naturally give a higher output from a given size of drum. The suggested design consisted of a drum separated into four compartments (Figure 13(i) and (ii)). Various scraper shapes were tried in order to produce classification of the product by different bed profiles. Unfortunately, vibration set up by the eccentricity of the drum prevented deep steps being formed in the bed. It was however apparent that the approach was incorrect. The use of steps, although improving the classification, did not offer a complete answer as the pellets still moved in one direction only, down through the drum, compared with the continuous re-classification occurring in the pelleting pan.

Figure 13 Pelleting drum (i) General view (ii) Internal compartments

It was originally thought that the drum speed would be sufficient to cause cascading of the pellets. The use of deflectors could then force the smaller pellets back up the drum and improve the classification. The critical speed of a drum or pan, ie the speed at which the centrifugal forces overcome that due to gravity so that the pellet is carried right round the circumference of the pan or drum, is, according to Brooke[14], after conversion to metric units, given by:

$$\text{Critical speed} = \frac{138.5}{\sqrt{d}} \quad \text{rev/min}$$

where d = diameter of drum in metres,
and optimum pelleting occurs at half critical speed. The mass of the drum was however greater than expected and the speed was limited to 10 rev/min, the critical speed being 36 rev/min.

A further difficulty with the drum was control of moisture. The seeds once formed were compacted and the excess moisture squeezed to the surface. The addition of dry ash at this stage is essential to promote pellet growth; this proved difficult as each segment of the drum contained a complete range of pellet sizes with little or no classification along the length of the drum.

The shortcomings of the plant resulted in a product of poor quality. The pellets were generally weak and insufficiently compacted with a size range of 5-20 mm.

The work has highlighted the difficulties associated with the operation of a single-pass drum pelletiser. The need for accurate moisture addition and the controlled feed of ash at various points through the length of the drum would result in a both complex and expensive system of feeders. A high drum speed to give cascading which allows the use of deflectors for improvement of classification would require a drum manufactured to close tolerances to prevent heavy wear on the drive and roller systems. It is therefore to be expected that a once-through drum pelletiser would be complex and costly and not viable compared with the pelleting pan.

2.4 Benefits from minor quantities of additives

Despite the progress made in producing well compacted pellets at high output rates under controlled conditions on an inclined pan, some ashes were still difficult to pellet and tended to produce low strength pellets. Apart from green strength to resist damage during handling between pelletiser and kiln, a reasonably high dry strength was required to resist breakage under load during firing in all but the shallowest of fire-beds and also to resist the thermal shock associated with the fast drying and firing rates appropriate to aggregate production. Depending on the type of kiln, the pellets may also have to resist attrition.

During the development of the process, weakness of the dry pellets had been noted in ashes of low soluble salt content and additions of sodium or magnesium sulphate dissolved in the water used for pelleting were tried. These led to some improvement; for one ash it was shown that whilst a 5 per cent solution was sufficient to resist thermal shock, a 20 per cent solution was required to provide any significant improvement in the resistance to attrition. These observations were made during pilot-scale tests and verified from quantitative measurements of the amount of breakdown recorded on the final product after firing. Although it was shown that the residual sulphate content of well fired samples of aggregate was well within the maximum limit of 1.0 per cent SO_3 permitted in British Standard Specification 3797:1964, some other means of improving the dry pellet strength was preferable both on economic grounds and in consideration of the risk of excess sulphate remaining in any poorly sintered portions of the aggregate.

Consideration was given to other forms of additive, eg acids, alkalis and surface active agents, which it was thought might improve the pelleting process and other properties of the pellets produced. The only appeal of the surface active additive was its ability to increase the rate of wetting of the powder and distribution of the liquid phase. Since this was achieved by reducing the very forces of surface tension by which the pellets were formed and retained strength in the wet state, one test run produced sufficient evidence to confirm that such additives were not helpful. Sulphuric acid was tried but showed no advantages over sodium or magnesium sulphate. The cheapest available source of alkali was lime (see Table 4). Its low solubility precluded its addition in the water used for pelleting but no great difficulty was experienced in metering small quantities into the ash feed to the pre-mixer. The improvements it produced in processing and pellet strength combined with its advantage in price and availability encouraged further investigation into the mechanisms involved.

Table 4 Cost of additives (1969)

Material	Approximate cost/ton (£)	Cost/ton aggregate based on 1% (wt/wt) addition (pence)
Sodium sulphate	14	14
Magnesium sulphate	20	20
Water glass 79.5°Tw ($\triangleq$ 50% $Na_2S_4O_9$)	15	30
Lime	6	6
Sodium carbonate	15	15
Sodium hydroxide	80	80

The effect on the wet strength of pellets made with ash from Bold 'B', High Marnham and Aberthaw power stations was determined by crushing 10 mm diameter pellets between parallel anvils attached to a suitable balance. A straight line relationship $S = KD$ between crushing strength S and pellet diameter D was obtained (Figure 14) where K is nearly independent of the additive but dependent on the moisture content. This is in agreement with the strength of a fully formed green pellet being derived from capillary forces acting within the pores of the particulate assemblies[5]. The capillary forces are dependent on the surface tension of the liquid phase and it is this which is affected by the additives, and which affects K.

The dry strength will depend on the adhesion arising by crystallisation of the solution or physical and/or chemical bonding between particles. It is expected that these will be time and temperature dependent. The rate of drying in practice is rapid but for laboratory purposes drying at 110°C for 24 hours was adopted. The results obtained on four ash samples with various amounts of lime are shown in Table 5, each being the average obtained during a trial. Results of some tests with water glass dissolved, ie the water used for pelleting, are also included in Table 5. From Tables 4 and 5 it is seen that lime shows distinct advantages on the basis of both dry strength and cost. The use of lime also precludes any possibility that deleterious impurities may be introduced into the final lightweight aggregate by the additive.

As well as dry strength, thermal shock resistance is also important. The two parameters tend to go together, and with lime little or no breakdown occurred during firing. During drying,

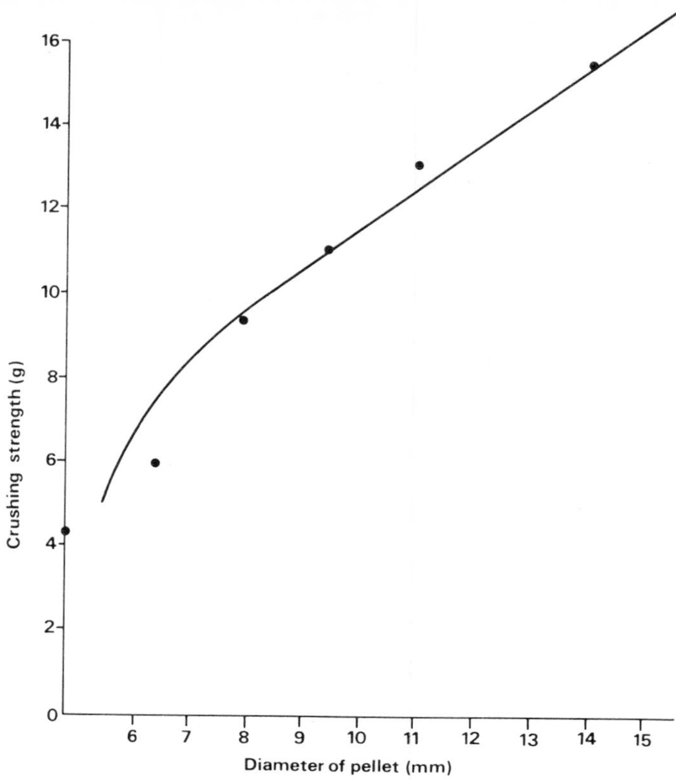

Figure 14 Strength of freshly prepared green pellets

Table 5 Dry strength of 10 mm diameter pellets (load in grams)

Additive	Per cent (wt/wt)	Bold	West Thurrock (1968)	High Marnham	Aberthaw
Lime	1.5	1082	412, 230		
	1.0	-		557	397, 346
	0.75	786, 645, 729, 418, 685			
	0.5	517			
Water glass	1.0	233	93		
No additive		50	35	<25	35

pellets made with lime additions also exhibited the property of being able to absorb or hold condensed moisture without collapse. This was beneficial in firing processes such as sinterstrand or shaft kiln where moisture is driven from a bed of pellets in advance of the fire and tends to condense on the cooler pellets.

The technical and economic advantages of lime (alkali) having been established, the availability in the ash and mode of action were investigated. Davies and Sherburn[15] have shown the variability of pH which can be obtained from an ash extract depending on its method of preparation. The ashes used by Davies and Sherburn were tested according to BS 1377: 1961, Method 9A, a complete plot of pH against time being obtained.

Figure 15 indicates the range of results obtained for 13 different ashes, each representing one generating station. These samples at the time of the test were over 2 years old and showed basically the same pattern. West Thurrock and Northfleet ashes in the current programme were of variable age but showed differing patterns when similarly tested (Figure 16). It should be noted that the West Thurrock ashes gave weaker green pellets with more breakdown and collapse during sintering than did the Northfleet ashes, indicating that the alkali of the Northfleet ash was more readily available and able to contribute to the process. A new sample of Bold 'B' ash was found to give a low rate of alkali release when compared with the 2-year-old sample from this power station[15]. This indicated that the age of the ash was a factor. The new sample of Bold 'B' ash was shown on storage in a sealed container to increase its rate of release of alkali

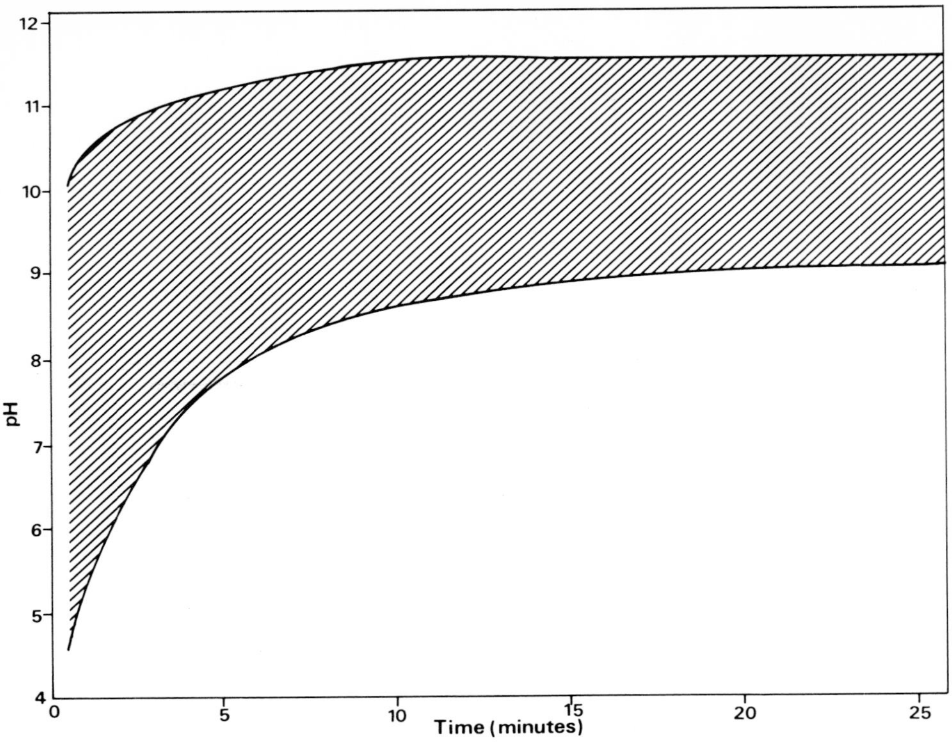

Figure 15 Extremities of pH/time relationship of pfa samples used by Davies and Sherburn[15]

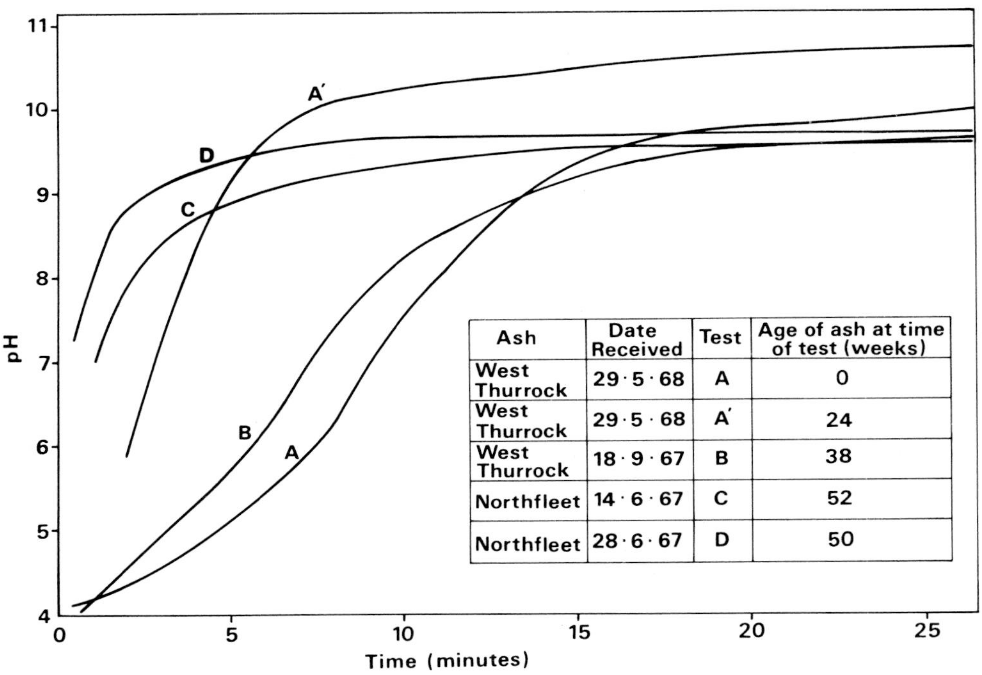

Ash	Date Received	Test	Age of ash at time of test (weeks)
West Thurrock	29·5·68	A	0
West Thurrock	29·5·68	A′	24
West Thurrock	18·9·67	B	38
Northfleet	14·6·67	C	52
Northfleet	28·6·67	D	50

Figure 16 pH/time relationship of ash-water mixtures

although maintaining a relatively constant final pH (Figure 17). Other ashes were subsequently shown to exhibit this effect, although those ashes giving an acidic reaction were found to be unaffected by time (Figure 18).

The natural availability of alkali in the pelleting process is therefore very variable and would account for the differing suitability of ashes for pelleting and sintering. The action of alkali in the pelleting process itself is not immediately obvious. Improved wetting with attack of the surface of the particles to give complex silicates and aluminates will occur. Pendular bonds will be more readily formed and packing and densification more easily accomplished. Brooke[14] has reported a drop in the rate of agglomeration of fertilizers when acid conditions are neutralised by addition of ammonia. Other workers have observed similar pH effects with the agglomeration of a variety of materials indicating that neutral conditions do not give good pelleting conditions. The marked increase in strength of the dried pellet is more readily

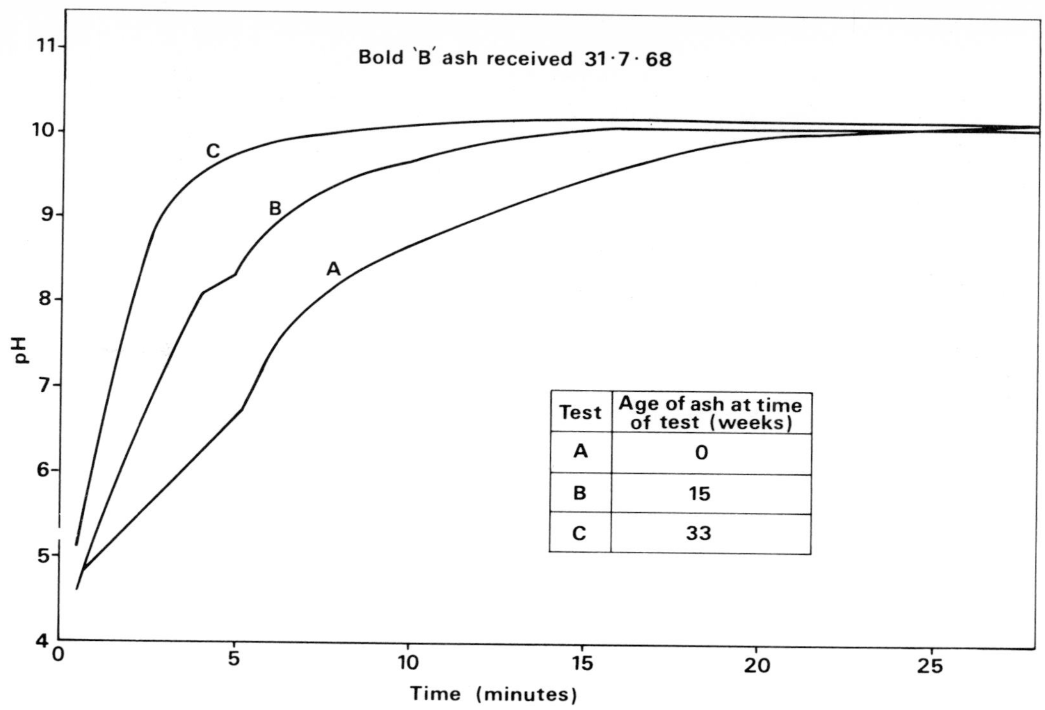

Figure 17 Effect of storage on the pH/time relationship

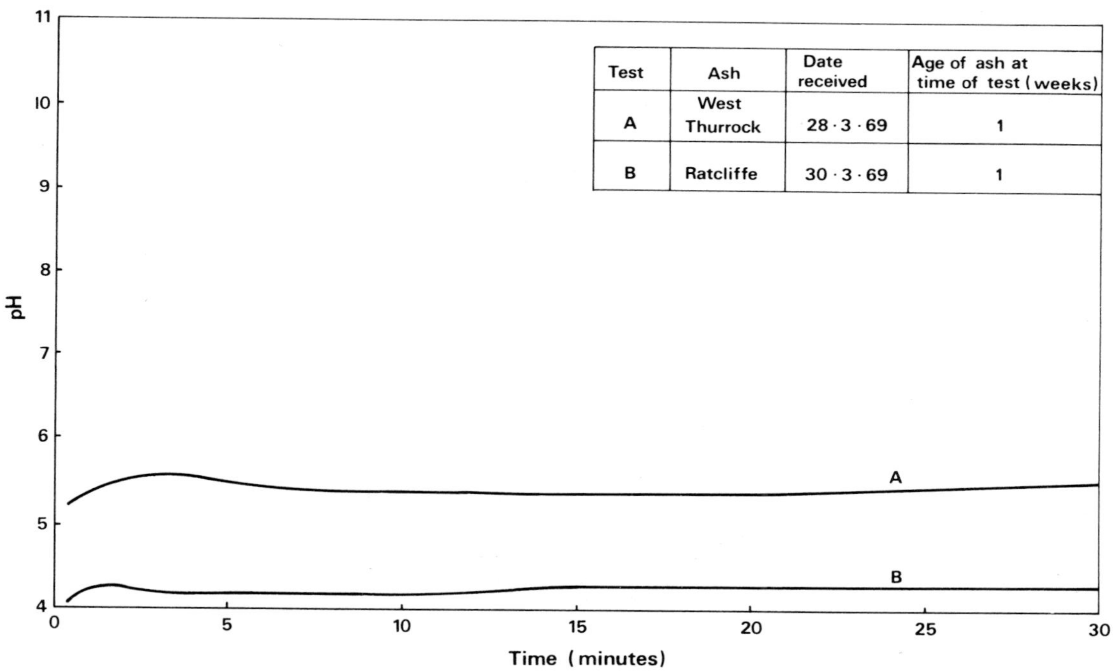

Figure 18 Acidic reaction of pfa (pH/time relationship)

understood; the complex silicates and aluminates formed by the alkaline attack of the ash produce excellent bonds between particles. The etching of the surfaces can be seen using a scanning electron microscope (Figure 19). The smooth glassy surface of the particles has been attacked, with the formation of aluminates and silicates. Whether this rapid bond formation between lime and pfa can be truly termed pozzolanic is open to discussion. Lea[16] defines a pozzolana as a material which reacts with lime in the presence of water at ordinary temperatures and forms a cementitious material, and he has shown, as have others[17,18], the marked temperature dependence of pozzolanic activity.

The increase in alkali availability with time indicates a change in the surface layers of the ash. Davies and Sherburn[15] have postulated that the ash has a layer of alkali-rich glass on which a skin of sulphate is formed while being carried through the later sections of the boiler, the skin of sulphates controlling the rate of alkali released to solution. Coward and Turner[19] showed that

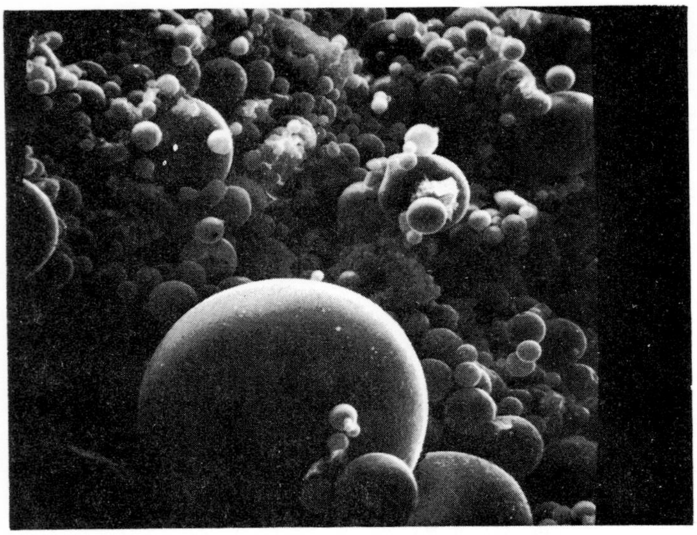

← 100μ →

Without additive

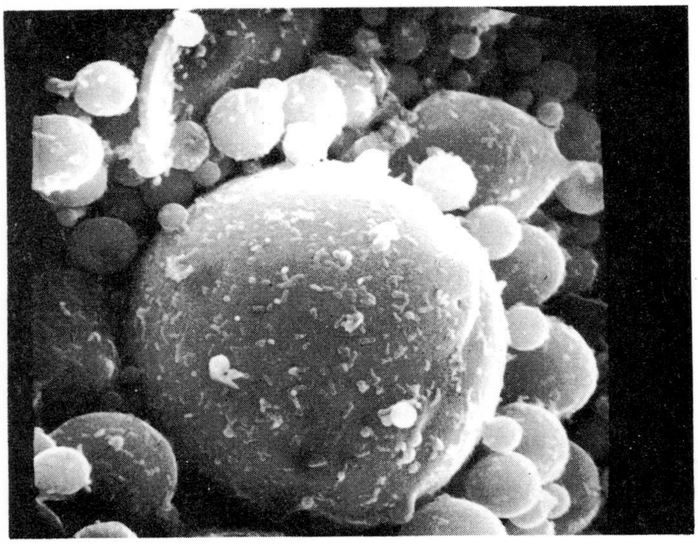

← 10μ →

With 1 per cent lime

Figure 19 Stereoscan photographs of interior of green pellet of West Thurrock ash

annealing of glass in sulphurous fumes reduced the water solubility of the unbroken surface and that the effect was about the same for a 1 per cent or 100 per cent atmosphere of sulphur dioxide.

The migration of alkali (Na_2O) to the surface under the influence of surface tension was suggested by Williams and Weyl[20], the surface energy of a glass being reduced by increasing the soda in the surface layers. The tendency for ion migration will be dependent on the surface tension and temperature. Although the particles are small, the rapid rate of cooling of ash in the flue gas would be expected to give rise to particles with a high surface tension. Douglas and Isard[21] have shown the temperature dependence of migration of ions and recorded ion migration at 20°C. The effect of temperature on the alkali availability of pfa is shown in Figure 20. A closed container enhances the effect at 110°C indicating that the presence of moisture is beneficial, as was shown by Douglas and Isard[21]. The closed container condition is similar to that of pfa held in store in a bulk silo.

3 SINTERING OF PFA

3.1 Vertical shaft kiln

The early sintering trials were made batchwise in small concrete sinter-box 150 mm square internally, mainly with the object of obtaining small quantities of aggregate for assessing its performance in concrete. A coal fire was established on the perforated base of the box which was then filled with wet pellets and covered with a perforated lid (Figure 21). Air was blown upwards through the packed pellets and the firing zone thus travelled to the top, first driving off moisture and then sintering the pellets at a temperature of between 1100°C and 1200°C. The sintered and partially caked pellets were then removed (Figure 22) and the process repeated.

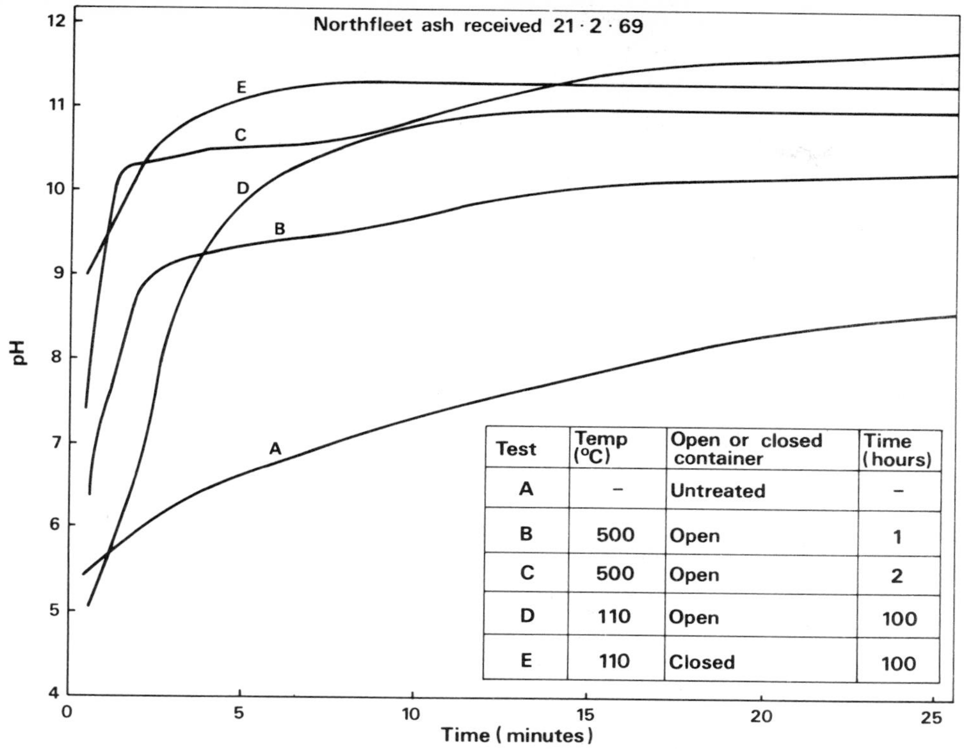

Test	Temp (°C)	Open or closed container	Time (hours)
A	–	Untreated	–
B	500	Open	1
C	500	Open	2
D	110	Open	100
E	110	Closed	100

Figure 20 Effect of pretreatment on the pH/time relationship

Figure 21 Sinter-box with air blower

Figure 22 Sinter-box discharged

This simple technique, modified to provide a semi-continuous process by having two interchangeable boxes fitting one above the other, was the forerunner of firstly a permanent brick kiln 300 mm square internally and lined with insulating refractory, and then with a battery of four such vertical shaft kilns 380 mm square internally and each working continuously but independently (Figure 23). Continuous operation was achieved by allowing the fire to approach the top of a cell and then discharging about half of the fired pellets by lowering the metal flaps at each side of the base. The cell was then re-charged with green pellets by means of the movable hopper shown in Figure 23. Ashes with a fairly wide range of fuel content (between 3.5 and 10.5 per cent loss-on-ignition) were successfully fired in this type of small cross-section kiln although, as the area of cross-section had progressively increased from the early sinter-box, the tendency towards slagging of the centre core of pellets with the higher fuel content ashes led to increasing problems of discharge. At the lower fuel contents, ideally between 4 and 5 per cent, lowering of the discharge flaps produced an immediate flow of well fired pellets. Ashes which proved difficult to form into pellets of good green and dried strength presented a problem. Breakage in handling or through thermal shock led to blinding of the bed and eventually extinguished the fire. When the kiln size was increased by the use of a steel cylinder 2.5 m high and 0.75 m internal diameter, a fuel content limit of 4.5 + 0.5 per cent was firmly established. The large kiln is

Figure 23 Battery of shaft kilns

Figure 24 Vertical shaft kiln in operation

shown in Figure 24 operating for a prolonged run; the fired material can be seen being ploughed from the slowly rotating pan base and deflected into a bin. The revolving pan rotated at about $\frac{1}{4}$ rev/min but this was adjustable. Superimposed on this was a central cone which penetrated into the cavity of the shell. This cone was not exactly central, but slightly eccentric, the amount of eccentricity being adjustable. This imparted a slight crushing action on the material in the annular space around the cone, thus loosening the agglomeration which frequently occurred. The two ploughs were fitted one on each side of the kiln, and were adjustable to provide added control on the rate of extraction.

The feed of green pellets to the upper part of the kiln was effected through a metal trunk leading from a hopper above the kiln, and terminating at a point over the centre of the fire-bed. The hopper and trunk were kept full of pellets so that as sintered product was ploughed out of the revolving pan at the bottom, a corresponding volume of green pellets was admitted at the top. Thus no interruption of the airflow occurred during charging, a constant bed level was maintained and the green pellets suffered little damage.

The controls of the kiln were twofold. Firstly, the rate of throughput of material could be varied by the rate of extraction, and secondly, the rate of airflow was adjustable by a butterfly valve in the ducting above the kiln. A number of inspection ports were placed round the body of the kiln as shown in Figure 25 but difficulty was experienced in exploring the fire zone within the body of the kiln. Thermocouples had been used with little success. Owing to the very viscous or solid conditions of the firing zone and the corrosive nature of the combustion gases, it had not been possible to find a suitable sheathing material with which to protect the thermocouple. Also, pushing, and at times forcing, the couple into the fire tended to disrupt the bed and often

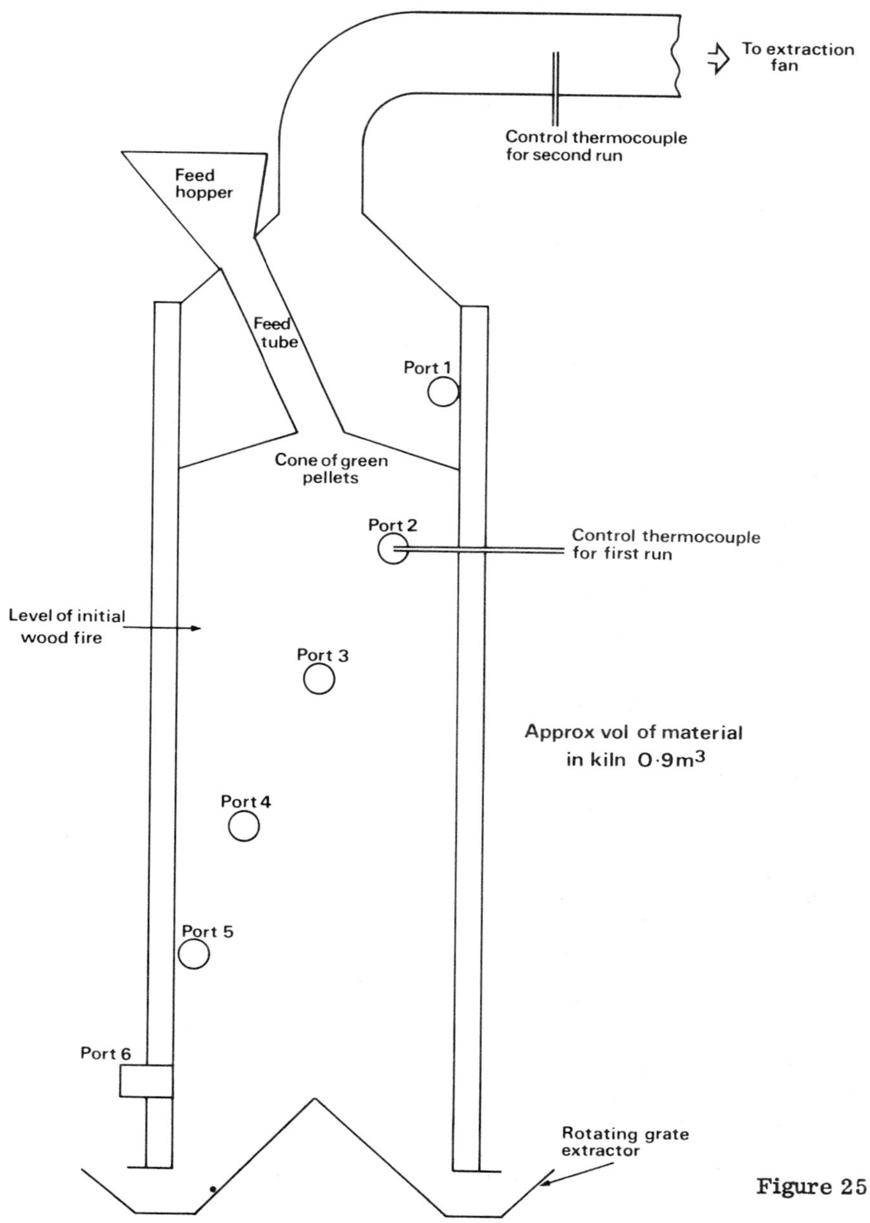

Figure 25 Diagram of vertical shaft kiln

increased locally the airflow past the couple. This altered the temperature just at the point where it was measured. Periodic examination of the state of the fire was possible, however, by forcing an iron bar horizontally through the kiln at various levels (ports 3 to 6 in Figure 25). After about 5 minutes the bar was quickly withdrawn and the parts glowing red were marked with chalk and subsequently measured. These inspections showed that the general pattern of the firing zone was an inverted cone extending to the wall only at the top.

To provide a measure of control of the feed and discharge relative to the rate of fire travel, a thermocouple was inserted through port number 2 (Figure 25) to a distance of one-quarter of the diameter of the kiln cross-section. Within limits the fire level was kept just below this couple, which was protected only by a thin metal tube in order to obtain quick response to temperature changes. As the fire approached the couple, the temperature rose and at 600°C the extraction mechanism was started and fresh green pellets automatically fed in at the top. The temperature continued to rise to 800-1000°C and then dropped, as the fire level receded due to extraction at the bottom. The temperature was allowed to drop to 300°C after which the extraction was stopped. Thereafter a further drop to 150-200°C was recorded before the temperature began to rise once more.

The thermocouple positions for the first and second runs shown in Figure 25 refer to two major runs made on the kiln after the rotating grate had been fitted. The first, of 19 hours duration, confirmed that the kiln could be run continuously if the carbon content of the ash was about 4.5 per cent. For convenience the ash was pelleted during the month preceding the test and had a moisture content of 24 per cent on leaving the pelleting pan. This had dropped to 19 per cent by the day of firing. During the first 8 hours of operation the airflow and the extraction rate were kept low in order to heat up the kiln. When fairly stable conditions had been established, the airflow was increased to 18.5 $m^3/min/m^2$ of kiln cross-section and the extraction rate was increased until the kiln was producing approximately 0.4 m^3 of good material per hour. This rate was maintained for the remaining 11 hours of the test and represented a displacement of 38 per cent of the total volume of material in the kiln each hour. Thus the material took $2\frac{1}{2}$ hours to pass through the kiln, and the rate of fire travel was 12 mm/min.

A measure of the rate of temperature rise and of the maximum temperature reached in the centre of the kiln was made at one stage of the test by a thermocouple inserted through port number 3 (Figure 25). The extraction was continued longer than usual until the fire-bed dropped well below this port and a recording was made of the rise in temperature as the fire regained its normal working level. It was found that over the range 250-1150°C the rate of rise of temperature was 130°C/min but that this rate was somewhat slower both below 250°C and above 1150°C as it approached a maximum of 1250°C. Some clinkering did occur but the extractor was able to deal with it and high quality aggregate was produced.

A second run was made in an attempt to fire an ash with a higher carbon content. The ash was from Battersea Power Station and had a carbon content of 8.8 per cent. Only sufficient material was pelleted in advance to provide an initial stock of 3 m^3 to heat up the kiln. For the remainder of the run freshly pelleted ash was used.

The kiln was lit near the top as in the first run, and at the level indicated in Figure 25. It was fed slowly at first and at a low airflow of 14.5 $m^3/min/m^2$. After 3 hours, and when 0.5 m^3 of the initial filling had been removed, the centre section of the fire-bed collapsed. The temperature had risen too high, causing 'bridging' of the fire-bed across the kiln. This had given way at the centre under the superimposed load of green pellets. Inspection showed that there was severe clinkering, and the kiln was completely emptied and refilled with loose fill to the level for relighting.

It was concluded that the fuel content of this ash was outside the range of the available controls on the kiln. Since these had been demonstrated to be capable of dealing with a lower fuel ash in a run of considerable duration an attempt was made to run the kiln using the high fuel pellets mixed with previously fired pellets in various ratios. Whilst it was recognised that the re-working of material was fundamentally uneconomical it was considered worthwhile to find whether, by this means, the kiln could be kept going until more suitable material became available. The kiln was re-lit and fed for the first $3\frac{1}{2}$ hours from stock material of 4.3 per cent carbon. The flow was 14.5 $m^3/min/m^2$ and the output over the period was approximately 0.2 m^3 per hour.

After this initial period a dilution of the high fuel pellets with 33 per cent re-work was used giving an overall carbon content of about 6 per cent. Nine hours from lighting up, the product of this 6 per cent mixture began to be extracted, and after a further $2\frac{1}{2}$ hours the symptoms of serious clinkering began to appear. The extractor was working continuously to little effect and the fire did not drop uniformly. Bridging had again occurred and the clinker took several hours to break and disperse, to form a fairly uniform red hot bed on which to continue the run.

The reduction of effective carbon from 8.8 per cent to 6 per cent was therefore still not sufficient to prevent serious clinkering, even at low airflow, and further reductions to $4\frac{1}{2}$ per cent and 3 per cent were made during the following 48 hours. No further clinkering was experienced but a considerable amount of dust was formed at times, which impeded the airflow and the action of the extractor. The formation of dust was attributed to low temperatures in the firing zone, producing under-fired pellets which were crushed by the extractor. With a low airflow of 14.3 $m^3/min/m^2$ both the $4\frac{1}{2}$ per cent and 3 per cent carbon mixtures produced large amounts of dust. On increasing the airflow the amount of dust diminished, and the rate of production of aggregate increased. The material from the 3 per cent effective carbon mixture (produced by use of 66 per cent re-work in the feed) was of fairly good quality when the airflow was 37 $m^3/min/m^2$ and it was extracted at 0.2 $m^2/hour$. This represented an overall output of only 0.07 $m^3/hour$, since two-thirds of the production was being used for re-work. These conditions were maintained for a period of 9 hours.

In the final period, the effective carbon was increased again to 4.5 per cent and at an airflow of 26 $m^3/min/m^2$ material of good quality and comparable to that of the first run was obtained. The rate of production was not determined owing to a failure of the main fan and the necessity of shutting down the plant, so leaving this superior material still inside the kiln. It was clear that a good rate of production would have been achieved had the operation been continued.

The position of the control couple had been moved for this run from its position in port number 2 to the centre of the 9 inch trunking where it leaves the top of the kiln. The sheath was removed to increase the sensitivity. This change was made because, owing to alterations in the feed, it was often necessary to allow the fire to burn through to the top, allowing visual inspection of the outer ring. As soon as the fire broke through to the surface the temperature of the extracted air rose sharply and this was used as a signal to start the extracting mechanism.

Subsequent tests on this batch of ash in the laboratory muffle furnaces showed it to have a poor resistance to thermal shock. Pellets were put into the furnaces held at temperatures of 300°C, 600°C and 900°C. At 300°C there was no trouble, but at 600°C and 900°C the moisture content had to be reduced below 1 per cent before the pellets remained unbroken. This factor no doubt contributed to the dust produced during the operation of the kiln.

Thus it was shown that by dilution of a high carbon ash feed to about $4\frac{1}{2}$ per cent by means of burnt pellets, the kiln could be kept running. For an ash containing twice this amount of fuel, dilution would naturally halve the overall output. Unless careful control is kept on the kiln and a high airflow maintained, the output figures might drop by 75 per cent or more.

A thermal balance worked out for the kiln showed that the minimum carbon content at which it could run under ideal conditions was about $2\frac{1}{2}$ per cent. This low theoretical carbon content gives a clear indication of the hazards likely to attend the use of ashes of much higher carbon contents on a commercial scale. In fact, the only commercial plant in Britain to attempt this form of firing failed through lack of control over the fuel content and quality of its input pellets.

3.2 Sinterstrand

As the shaft kiln work progressed it became apparent that the sinterstrand or travelling grate technique, widely used in ironworks for the sintering of fines before feeding to the blastfurnace, would have considerable advantages.

A sinterstrand generally consists of a chain-grate made up of cast-iron links and side plates passing over air-boxes exhausted by fans. An essential feature of this type of machine is that continuous ignition of the top surface is required whereas in the shaft kiln no fuel in addition to that contained in the raw feed is needed. Whilst the expense of the ignition is not negligible the higher probability of uninterrupted production is more than adequate compensation. The condition of the fire-bed at all stages can be easily judged by visual inspection from above the bed and below the grate bars and by thermocouples placed in the wind-boxes beneath the grate. The rate of travel of the grate and the airflow through each box can be varied to correct for changes in the fuel content of the ash. In the absence of analytical data a change of fuel content is soon detected and the appropriate alterations made to combat it. Moreover, in the unlikely event of an interruption, the plant can be started up again at the point where it stopped without wasteful and time-consuming removal of large quantities of material.

A pilot-scale machine was purchased (Figure 26) which provided a 450 mm wide chain-grate made up of cast-iron links spaced 9 mm apart. The maximum depth of bed was 150 mm to the top of the side plates and the length of horizontal travel was 3.5 m. The principal of operation

Figure 26 View of pilot sinterstrand

was to provide a feed of moist pellets directly from the pelleting machine and to form these into a bed of the required depth by means of a gated hopper at one end of the moving grate. This bed was then drawn beneath a drying hood where hot air was drawn down through the pellets into a draught-box beneath the bed. This warmed and dried the pellets in the upper layers of the bed before they passed beneath a multi-jet gas burner to ignite the residual (or added) fuel in the pellets. The fire thus established in these upper layers of pellets was drawn down through the bed by the downdraught air before the fired pellets were discharged at the end of the horizontal travel of the chain-grate. The product was often in the form of a lightly fused cake of pellets which easily broke down again into pellets during handling. Very high carbon ashes produced pronounced slagging and the coarse aggregate from this contained fragmental particles from slag that had required crushing. The first 1.0 m of horizontal travel was occupied by the feed hopper, drying hood and ignition zone. The exhaust gases from the draught-boxes beneath the first 2.5 m of travel were discharged to the atmosphere since they had a high water content, but the final draught-box air was re-cycled to provide the hot air for the drying hood. The rate of travel of the grate could be varied between 50 mm and 250 mm per minute. The normal working speeds varied between 70 and 120 mm per minute which corresponded to an output of approximately 0.35 and 0.50 m^3/hour respectively. The effective depth of bed of the sintered pellets was usually some 10 mm less than that formed from the raw pellets.

Ashes varying from 4.3 per cent to 14.9 per cent carbon have been sintered on the strand, indicating the flexibility of the equipment, but the preferred range was between 7 per cent and 9 per cent. The heat requirement for ignition was found to increase as the fuel content in the ash decreased. For an ash with a carbon content of 5.8 per cent the ignition heat required was found to be 372 MJ (89 000 kcal) per m^3 of aggregate for a final bed depth of 125 mm. Clearly a greater bed depth would reduce this ignition heat requirement per unit volume of aggregate. The characteristics of the shaft kiln and the sinterstrand can be summarised as follows:

	Shaft kiln	Sinterstrand
1	Bed of pellets moved vertically - some disturbance during firing	Bed of pellets moved horizontally - no disturbance during firing
2	Deep bed (2 m or more)	Shallow bed (normally 150-350 mm)
3	Thermally efficient 4-5 per cent fuel	Thermally inefficient 7-9 per cent fuel
4	Slagging or pellet collapse stops production	Slagging or pellet collapse only reduces output
5	Fuel content critical within narrow limits	Fuel content not critical
6	Few moving components	Moving bed with generally higher power requirements
7	Excellent product	Good product

If it were possible to combine the rapid start up of the sinterstrand and its ability to cope with defects in the sinter-bed with the thermal efficiency and excellent product of the shaft kiln the ideal plant would be produced. As a step towards this it was decided to deepen the bed of the sinterstrand and thus approach the deep bed of pellets present in the shaft kiln. From practical considerations the maximum depth of bed was considered to be approximately 550 mm or nearly four times the depth of the original strand. An end view of the modified strand is shown in Figure 27. It will be seen that the extended side plates were made of metal plates faced with insulating firebrick. As the fired material was discharged downwards, the extended side plates were fed forwards onto the channels from which they were taken by hand and returned to the feed end. This adaption worked well enough to show that, with well made pellets, a good quality product at increased rates of output could be made. The output of a given sinterstrand is not, however, dependent on the depth of the sinterbed but is controlled by the firing rate, ie the rate at which the flame front passes down through the bed. The heat released by the oxidation of carbon etc at the flame front is also carried down through the bed as a hot zone. Voice and Wild[22] have shown that for any given airflow the rate of advance of the hot zone through the bed is not dependent on that of the flame front. Thus the heat generated near the top of a sinter-bed may arrive at a lower level before, during or after combustion. Only when both rates are the same will maximum temperatures and firing rates be obtained. If too high an airflow is used, the hot zone advances in front of the flame front resulting in a cooling of the bed and spreading of the hot zone. This is useful if too much fuel is present, but otherwise results in an under-fired product. Too low an airflow (ie small or weak pellets) allows build-up of heat in the firing zone with a reduced flame front speed and possible slagging (Figure 28).

Figure 27 End view of deepened bed of sinterstrand

Thermocouples placed within the bed revealed the progress of the flame front. Figure 29 indicates the type of curve obtained. The initial rise in temperature of about 20°C is characteristic of all the curves and is probably associated with the drying of the pellets before ignition. The rapid rate of temperature rise (1100°C in 4 minutes) is clearly shown. For a pfa with a carbon content of 4.1 per cent it was found possible to run the strand at 90 mm/min which, at the greater bed depth, was equivalent to an aggregate output of 0.95 m^3/h. This was approximately twice the output recorded in earlier trials at normal bed depth with ash of a similar low carbon content. A strict comparison however is not valid owing to a lack of records of the other factors known to affect the rate of burning of the pellets, eg airflow, pellet size, moisture content and temperature in the drying hood. With a deeper bed a longer horizontal travel of the chain-grate would be required to investigate the full potential of the improvement in output quantity.

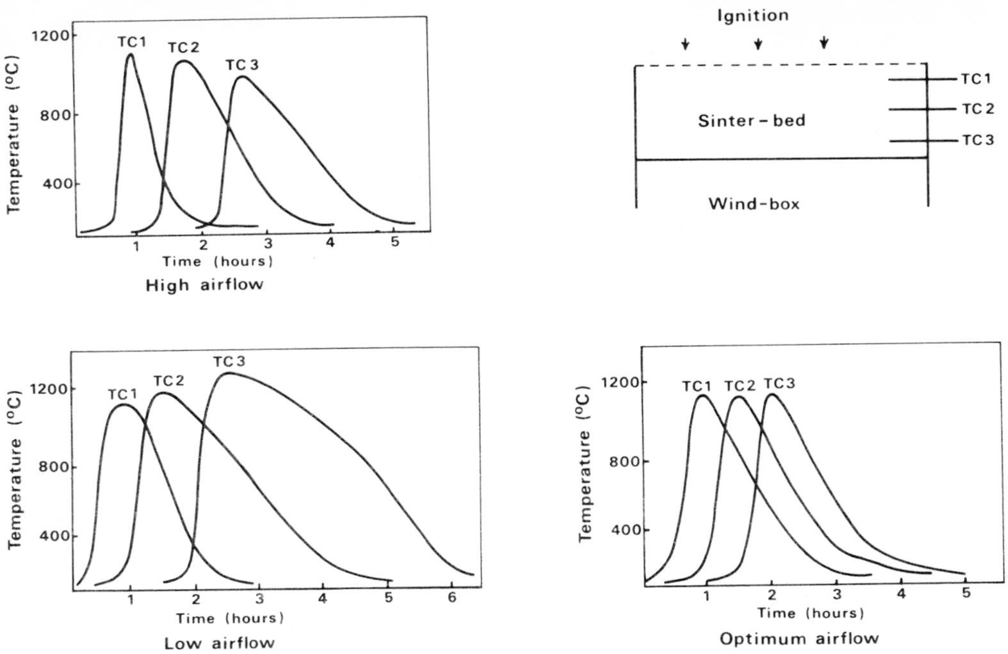

Figure 28 Effect of air flow on temperature rise on sinter-bed

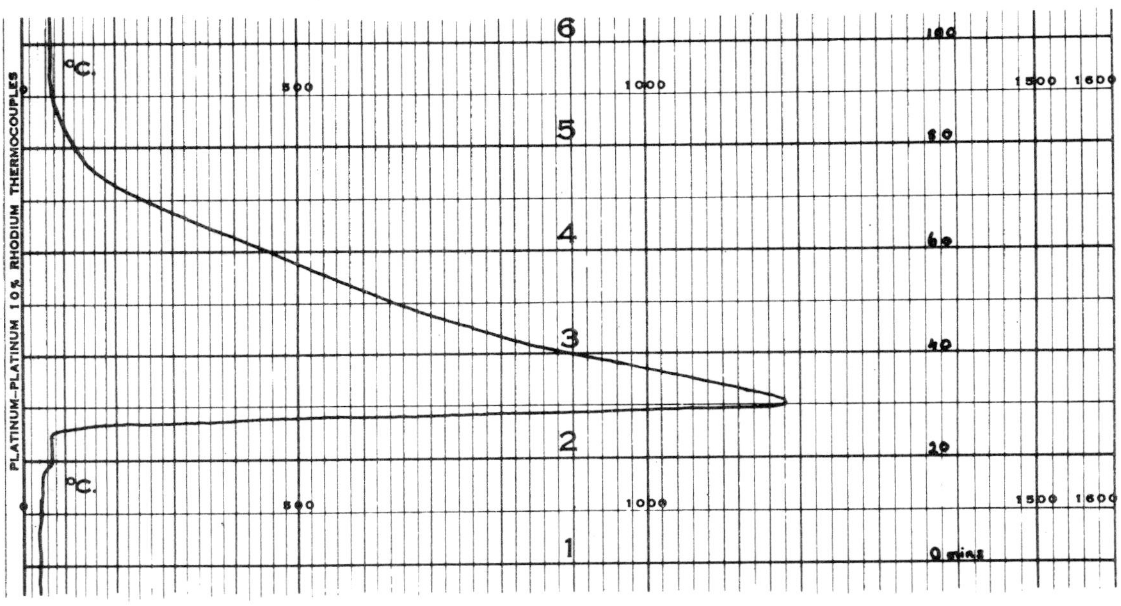

Figure 29 Rate of rise of temperature in bed of sinterstrand

The number of points of contact of uniformly sized pellets in a unit volume of the sinter-bed is dependent on the size of the pellets so that less inter-pellet contact and sintering will occur with larger pellets. In a bed of pellets the ratio of the number of points of contact per unit volume for 5 mm and 12.5 mm diameter pellets would be 15:1. The degree of sintering of the mass is therefore considerably reduced by increasing the pellet size thus producing a sinter-cake which breaks easily into individual pellets.

3.3 Rotary kiln

Inclined rotary kilns are normally used for firing materials which contain no fuel. The feed is introduced at the elevated end and moves down towards the burner at the exit end. The material should remain sufficiently cohesive even in the dry state to resist attrition by the motion of the kiln since very fine particles are easily carried away by the hot gases flowing counter to the inclination of the kiln. Thus it will be seen that this type of kiln is not likely to be successful with pfa unless ash with some form of binder is used. However, with care and some minor loss of material a small amount of good aggregate was made from one pfa using a 3 m long by 150 mm diameter pilot-scale kiln. At a later date a few test runs were made using a somewhat larger diameter kiln. These showed that pellet breakdown and consequent control difficulties were only improved by the use of considerable quantities of binder additives. In view of the improved output from the sinterstrand, no further work on rotary kilns was carried out.

3.4 Sintering mechanisms

Sintered pellets consist of a black core and a brown outer layer with a fairly sharp demarcation between the two areas. These colours are due to the oxidation state of the iron (ferrous or ferric) and the carbon content (Table 6). This effect has been shown to be due to the availability of oxygen during sintering which in turn is controlled by the time and temperature of sintering.

Table 6 Chemical analysis of a fired pellet

	C (%)	Fe^{2+} (%)	Fe^{3+} (%)
Outer brown layer	0.35	0.56	2.93
Inner black core	2.3	3.31	0.08

The structure and form of the sintered pellet will affect its strength. As the strength of the fired product is a prime factor some experiments were made to determine the effect of time and (sintering) temperature on the strength of the pellet.

Dry 10 mm diameter pellets containing 1 per cent added lime were fired in a laboratory muffle furnace for varying periods of time at different temperatures. The fired pellets were crushed individually using an Instrom machine. The pallets of the Instrom were set to crush the pellets at 2 mm/min, the load applied being plotted on a continuous strip recorder. The maximum load recorded was taken as the strength of the pellet. Fracture normally occurred as a vertical split completely through the pellet and in some cases spalling occurred before the main fracture. Ten determinations were made for each condition. The standard deviations obtained were large, as might be expected from the nature of the pellet. Despite this variability, similar curves were obtained at each temperature. These showed a fall in strength after the initial sintering which then recovered, to be followed in most examples by a further slow fall (Figure 30).

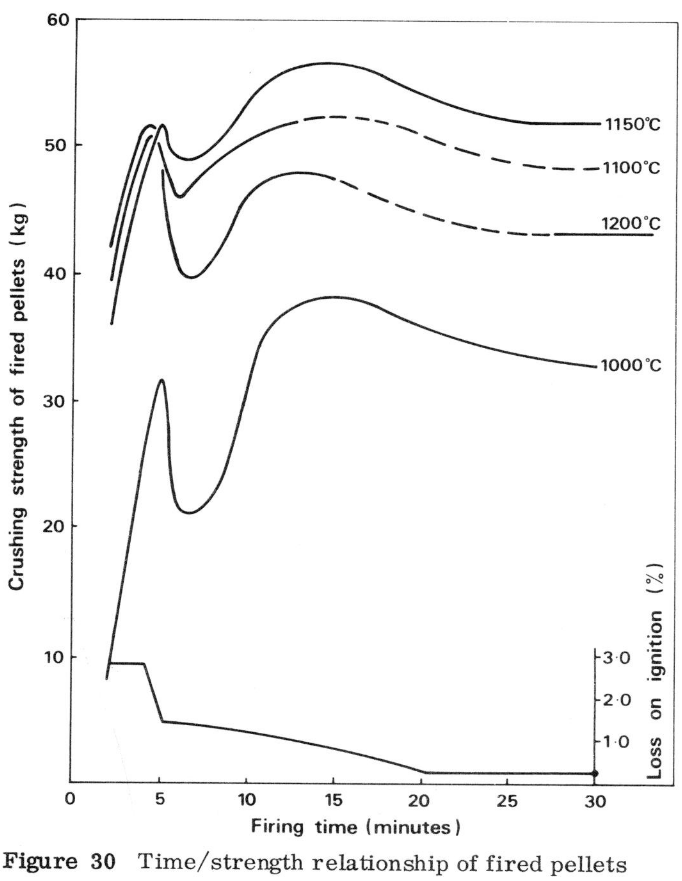

Figure 30 Time/strength relationship of fired pellets

Sintering occurs initially by surface diffusion of ions between particles at points of contact to give a chemical bond. At slightly higher temperatures surface fusion will occur with viscous flow at points of contact due to the surface tension[23]. With prolonged heating or even high temperatures, the fused material becomes a significant proportion of the whole and under the continued influence of surface tension rearrangement of material occurs to give a more cellular structure. Some bloating may also occur at this stage. This can be seen with a microscope and would account for the observed reduction in strength of the pellets with prolonged heating.

The loss of strength in the 5-10 minute period shown in Figure 30 is probably due to loss of combined water and is also shown to be associated with combustion of the carbon. Oxygen for combustion is obtained at least in part by reduction of the iron oxides. It is also known that the state of oxidation of the iron affects the sintering process. From the phase diagram shown in Figure 31, the reduction of iron to the ferrous state will give a system of lower melting point, in the Fayalite area. A higher proportion of material will therefore be in the fluid state with an increased rate of sintering.

No account has been made in the above of the inherent chemical inhomogeneity of the ash when examined particle by particle. The viscosity of a melt is dependent on its constituents. Fusion of the lower melting fraction will give a high viscosity medium, the viscosity of which will fall as successive fractions fuse. When completely molten, the viscosity will vary solely as a function of temperature. This effect can be seen on the viscosity curves of many ashes[26] (Figure 32). This indicates that controlled sintering occurs below 1100-1200°C; above this the viscosity falls rapidly giving conditions for rapid material transfer and re-arrangement. Reference to Figure 30 shows a maximum strength at 1150°C and pilot-plant trials have generally shown collapse of the pellet bed, with heavy sintering, when the temperature rises above 1200°C.

The results of the strength trials can be used to plot 'iso-strengths' against time and temperature from which the area of time and temperature available to give a certain minimum strength of product can be obtained (Figure 33).

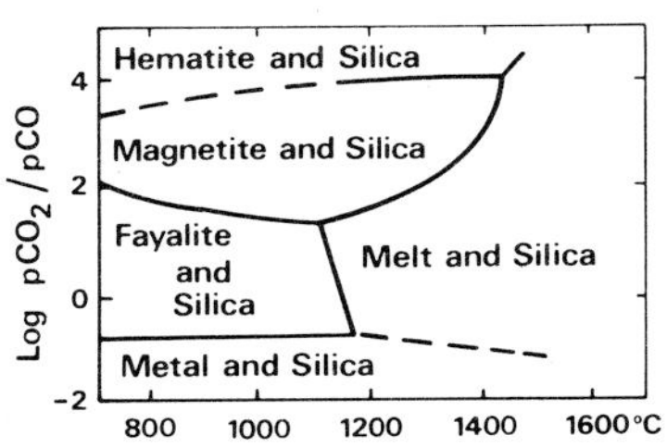

Stable phases in system Fe-Si-O in presence of solid silica[24]

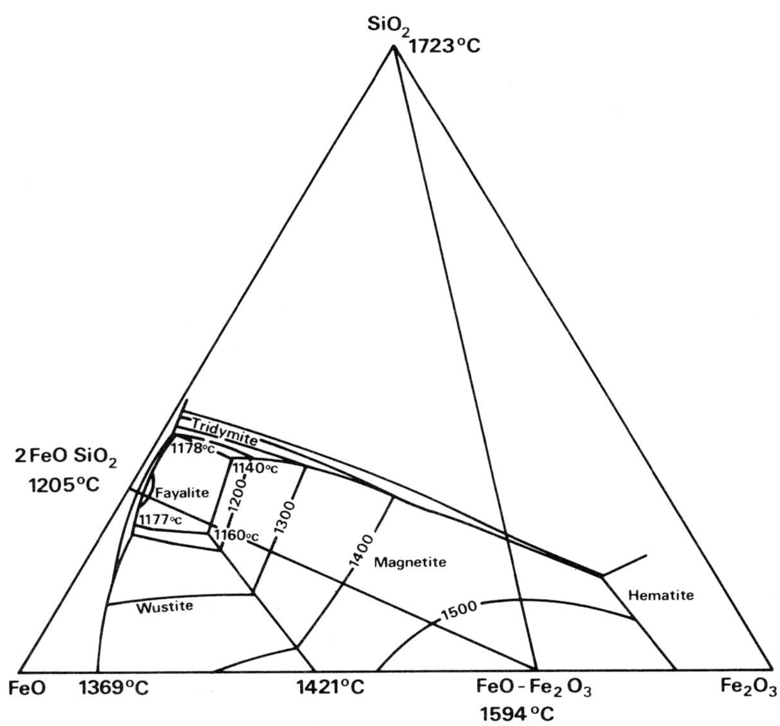

Figure 31 Phase diagram[25] of system FeO-Fe_2O_3-SiO_2

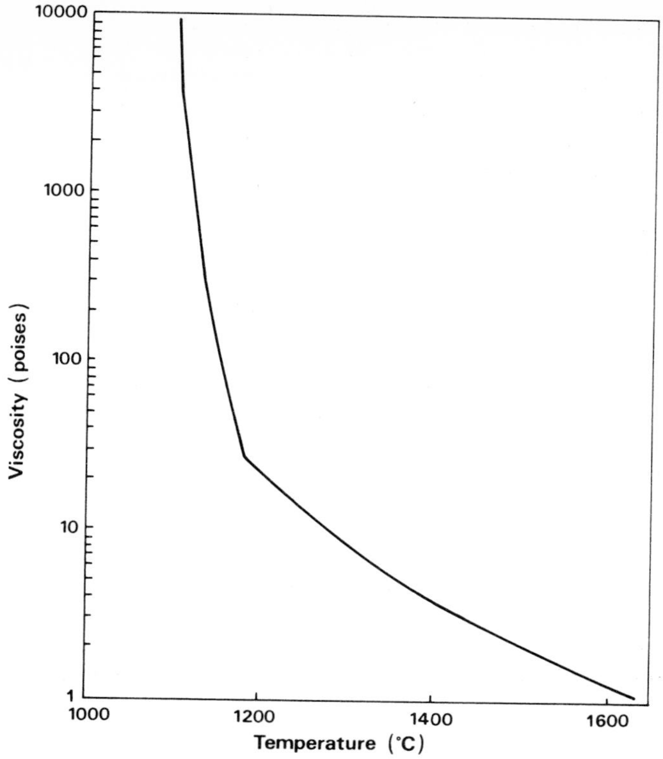

Figure 32 Viscometric data of a pulverised fuel ash (from the British Coal Utilisation Research Association[26])

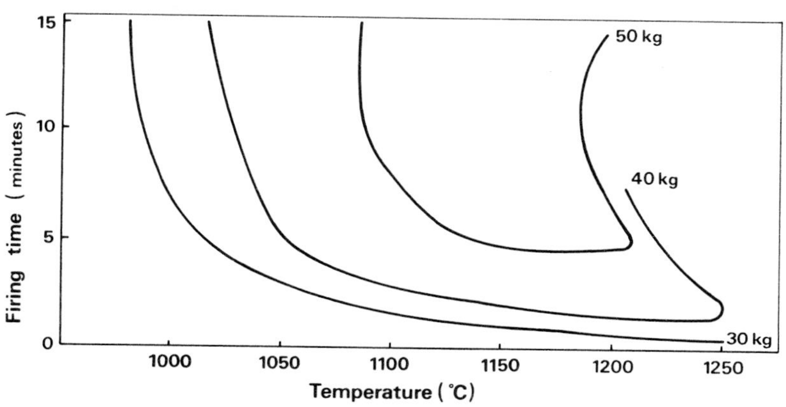

Figure 33 'Iso-strengths' of fired pellets

4 SINTERED PFA AGGREGATE

The product from any of the sintering processes described will, after light crushing and before grading, contain the following:

- Smooth-surfaced spherical particles
- Spherical particles with adhering rough sinter
- Irregular-shaped fused sinter
- Broken spheres
- Fine dust

The proportions of each of these will depend on the type of process, the size of feed pellets and the efficiency of combustion.

The use for which the aggregate is required will also influence the manufacturer in the control of the process. Most lightweight aggregate at present is used in blockmaking where a semi-dry concrete mix with good cohesion is required. This type of mix, on machine pressing, produces some interlocking of particles to provide a sufficient green strength to retain its moulded shape.

For blockmaking it is generally accepted that a high proportion of spherical particles is a disadvantage. Experience with a small semi-manual press at BRS indicated that mixes containing over 30 per cent of wholly spherical particles in the 10-12 mm diameter range showed a marked tendency to segregate. In filling the mould the spherical particles rolled free and accumulated to produce weak areas in the block. The mix was also difficult to compress. These difficulties were reduced with smaller diameter spheres or with broken spheres. In structural lightweight concrete mixes the spherical particles help to provide good workability which leads to better compaction. Also, with smooth-surfaced aggregate, there is little loss of cement into the surface pores during mixing.

Crushing of the product to break caked material and to produce a graded fines is required and a manufacturer can choose to operate between the extremes of making large-diameter pellets which will permit deeper bed operation with little clinker formation and making small-diameter pellets which will restrict the airflow limiting the depth of bed and tending to produce a clinkered cake. Typical values for the aggregate crushing value[27] for the material produced by the different processes are shown below:

Process	Carbon content of pfa (%)	Aggregate crushing value (%)
Small shaft kiln	5-7	48-55
Rotary kiln	5-7	36-40
Sinterstrand	5-11	55-65

A high crushing value indicates a low strength in the aggregate since it is related to the percentage of a sample that is reduced in size after a compression test. This indicates that the aggregate produced in the small shaft kiln and the rotary kiln was superior to that from the sinterstrand but it became clear that the quality was related primarily to the fuel content of the pfa used. This is shown in Table 7 for ashes with a range of fuel contents fired on the sinterstrand. The range of bulk density values is also given. The aggregate made in all the principal production runs complied with British Standard[28] requirements. Precast concrete blocks, complying with the requirements for Type B blocks in the British Standard[29], were made and tested from many of the batches. As already stated, the composition of a block-making mix has to be suited to the type of aggregate and also to the type of block-making machine in order to obtain the optimum results. Hence there is little value in quoting the compressive strength and drying shrinkage values on specific test runs.

Table 7 Comparison of aggregate quality and carbon content

Fuel content of raw ash (% of dry ash)	Aggregate crushing value. BSS 812: 1967 (%)	Bulk density of coarse aggregate (10-12.5 mm) (kg/m^3)
5.1	55	750
5.9	55	720
6.0	54	800
6.7	52	765
7.7	54	700
8.3	57	720
11.4	65	640
12.7	82	690
16.6	84	575

No full-scale programme of tests was carried out on fully compacted concrete mixes of the type used in structural concrete with the aggregate made at BRS. The relatively few small test specimens made showed that concrete with a compressive strength of at least 20 N/mm^2 was attainable. A comprehensive research programme comparing the properties of the various commercial lightweight aggregates was made and the results published[30]. In this programme the commercial sintered pfa Lytag was used. The process favoured by this company was to use a sinterstrand fed with fairly small diameter pellets. Some fusing and caking of parts of the bed resulted and the product was composed principally of graded spheres with a maximum size of approximately 10 mm, crushed sinter and a relatively high proportion of fine dust. After grading, the fines fraction (ie $<$5 mm) therefore tended to fall into grading zone L2 of BS 3797[28] for which an upper limit of 35 per cent passing a 150 μ m sieve is allowed. The research results showed that, in company with some of the other synthetic lightweight aggregates, concrete with a compressive strength of at least 50 N/mm^2 can be attained, albeit at a higher cement content than required with gravel aggregate. The pozzolanic properties of the fine dust would tend to increase the long-term strength and durability of sintered pfa aggregate concrete.

CONCLUSIONS

The variation in carbon content of the raw ash was shown to present the principal problem in control of any process for the conversion of pfa into a sintered aggregate and to have a marked effect on the quality of the product.

The preferred method of manufacture of sintered pfa aggregate is that using a sinterstrand. This permits considerable flexibility in its mode of operation to deal with variations in the properties of the raw material and in the type of product required. It was found possible to operate with a fairly deep bed providing that good quality feed pellets were made.

A shaft kiln was shown to be capable of producing high quality aggregate when the carbon content was fairly low and closely controlled.

An inclined disc pelleter with a stepped base was favoured to pellet the dampened ash and so provide a fairly permeable fire-bed. No additional binder was required although with many ashes higher output and improved pellet quality were obtained by the addition of 1 per cent of hydrated lime.

ACKNOWLEDGEMENTS

The work described was sponsored by the Central Electricity Generating Board. The earlier work described in this paper was carried out under the supervision of Mr W Kinniburgh of BRS and the completion of the later studies was to a large extent due to the considerable support and co-operation of the Scientific Services Department of the South Eastern Region of CEGB.

REFERENCES

1 **Central Electricity Generating Board.** PFA Utilisation 1972. London, CEGB.

2 **Central Electricity Generating Board.** PFA Data Book. London, GEGB.

3 **Raask, E.** Cenospheres in pulverised fuel ash. Journal of the Institute of Fuel, September 1968.

4 **Kinniburgh, W, Harrison, W H** and **Akam, E H.** British Patent No 1031352:1966.

5 **Newitt, D M** and **Papadopulus, A L.** The mechanism of granule formation. The Fertiliser Society, Proceedings No 55, 1959.

6 **Rumpf, H.** Fundamentals and methods of pelletising. Chemie-Ingenieur-Technik, Vol 30, No 3, 1958, pp 144-158.

7 **Papadakis, M** and **Bombled, J P.** The pelletisation of raw cement mixes. Revue des Matériaux de Construction et des Travaux Publiques, Vol 549, 1961, pp 289-299.

8 **Capes, C E** and **Dankwerts, P V.** Granule formation by agglomeration of fine powders. Transactions of the Institution of Chemical Engineers, Vol 43, 1965, pp 116-123.

9 **Conway-Jones, J M.** A contribution to the theory and practice of granulation. PhD Thesis, University of London, 1957.

10 **Klatt, H.** The adjustment of pelleting pans during operation. Zement Kalk Gips, Vol 11, No 4, 1958, pp 144-154.

11 **Kono, H.** The practical use of the dish noduliser with shaft kilns. Zement Kalk Gips, Vol 12, No 12, 1959, pp 549-554.

12 **Macavei, G H.** A general relationship for scaling up granulating pans. British Chemical Engineer, Vol 10, No a, 1965, pp 610-614.

13 **Pietsch, W.** Possibilities of influencing pelletising pan operation and their effects on the properties of the pelleted material. Aufberietungs-Technik, Vol 4, 1966, pp 177-191.

14 **Brooke, A T.** Development in granulation techniques. The Fertiliser Society, Proceedings No 47, 1957.

15 **Davies, I** and **Sherburn, F H.** National survey of pfa. Preliminary results. CEGB. RD/SE/RN7/67.

16 **Lea, F M.** The chemistry of cement and concrete. Arnold, 1970.

17 **Blanks, R.** Proceedings of the American Concrete Institute, Vol 46, 1950, p89.

18 **Sherwood, A T** and **Ryley, M D.** The use of stabilised pfa in road construction. Transport and Road Research Laboratory Report No 49.

19 **Coward, J N** and **Turner, W E S.** Transactions of the society of Glass Technology. Vol 22, 1938, p309.

20 **Williams, H S** and **Weyl, W A.** Glass Industry, Vol 26, 1945, p275.

21 **Douglas, R W** and **Isard, J O.** The action of water and sulphur dioxide on glass surfaces. Journal of the Society of Glass Technology, Vol 23, October 1949.

22 **Voice, E W** and **Wild, R.** Importance of heat transfer and combustion in sintering. Iron and Coal Trades Review, Vol 10, No 175, 1967, pp 841-850.

23 **Herring, C.** Physics of powder metallurgy. Ed W Kingston. New York, McGraw Hill.

24 **Darken, L S.** Journal of the American Chemical Society, Vol 70, 1948, p2051.

25 **American Ceramics Society** and **Edward Orton Jnr.** Phase equilibrium diagrams of oxide systems. Ceram Foundation, 1960.

26 **British Coal Utilisation Research Association.** The chemical composition and viscometric properties of the slags formed from the ashes of British coals. 1963.

27 **British Standards Institution** BS812:1967. Methods for sampling and testing of mineral aggregates, sands and fillers. London, BSI, 1967.

28 **British Standards Institution** BS3797;1964. Specification for lightweight aggregates for concrete. London, BSI, 1964.

29 **British Standards Institution** BS2028:1968. Specification for precast concrete blocks. London, BSI, 1968.

30 **Teychenné, D C.** Lightweight aggregates: their properties and use in concrete in the UK. Building Research Station Current Paper CP73/68.

Review of standard specifications for fly ash for use in concrete (CP 8/75)

M.A. Smith

INTRODUCTION

Pulverised fuel ash (pfa) or fly ash is the by-product of burning pulverised coal in power station boilers. In 1971-72 about 10 M tonnes of pfa were produced in the UK by the Central Electricity Generating Board. The production, quantities and present disposal and utilisation have been surveyed recently[1] and its properties in relation to its various uses reviewed by Barber et al[2]. The pozzolanic properties of pfa, which allow it to react with lime at ordinary temperatures in the presence of water to give a composition with cementing properties, were first reported in 1937 by Davis[3] and his co-workers; its use in concrete started to develop in the USA in the following decade, particularly for mass concrete for dams. Later its use became wider and spread to many other countries. In 1971-72 in the UK the greatest utilisation of pfa was as bulk fill (67 per cent of total usage) and in concrete blocks (22 per cent). The use in ready-mixed concrete, dams and other on-site applications was very small.

Since 1937 a considerable amount of research has been carried out on pfa and its use in concrete and a comprehensive review was presented by Kokubu[4] in 1968 at the 5th International Symposium on the Chemistry of Cement held in Tokyo that year.

Standard specifications and codes of practice relating specifically to the use of pfa in concrete have been produced in several countries and in many others its use is governed by standards for pozzolanas and pozzolanic cements. The Standard Specifications dealing with fly ash are of three main types:

1 permitting small additions (not more than 20 per cent) of pfa to Portland cement during manufacture, the blended product still being termed 'Portland cement'

2 for Portland/fly-ash cements (usually containing not more than 40 per cent pfa)

3 for fly-ash for use in concrete 'at the mixer'.

The only British Standard Specification on pfa (BS 3892:1964, 'Pulverised-fuel ash for use in concrete') is of the last type and specifies limits for loss on ignition (this is essentially carbon content), SO_3, moisture content and surface area (ie fineness).

A comparison is made in the main body of the present paper between the requirements of this British Standard and the corresponding Australian, Austrian, Indian, Japanese, Turkish, American (ASTM) and Russian standards which are listed in Table 1. This comparison is followed by suggestions of some of the ways in which the BS standard might be revised in order to make it a more reliable guide of fly-ash quality. The chemical requirements of the standards are summarised in Table 2 and the more important physical requirements in Table 3. The main provisions of each standard are summarised and discussed in a standard format in the Appendix. The information on the Russian standard is taken from Kokubu[4].

For the USA, only the relevant ASTM specification (C618-72) has been considered. Mielenz[5] has recently compared and contrasted the requirements of this specification with those laid down by the US Army Corps of Engineers[6], the Federal Specification (SS-P-570B)[7] and other American authorities. It is believed that the most recent version of each standard has been obtained, but with the constant updating of national standards, it is possible that this is not entirely the case.

Standards for Portland/fly-ash cements exist in Australia, France and Japan (Table 4). In the USA and a number of other countries fly ash cements are covered by specifications for Portland-pozzolanic cements. These countries include[4] Bulgaria, China, Czechoslovakia, West Germany, Greece, Hungary, Italy, Mexico, Netherlands, Portugal, Romania, Spain, USSR, and Yugoslavia.

Table 1 Standard specifications for fly ash for use in concrete

Specifications

Country	Designation of standard		Year
Australia	AS 1129	Fly ash for use in concrete[a]	1971
	AS 1130	Code of practice for use of fly ash in concrete	1971
Austria	ÖNORM B 3319	Fly ash as hydraulic powdered admixture component for cement manufacture	1962
India	IS 3812		
	Part I	Fly ash for use as pozzolana	1966
	Part II	Fly ash for use as admixture in concrete	1966
	Part III	Fly ash for use as fine aggregate for mortar and concrete	1966
Turkey	Fly ashes for use with Portland cement clinker and Portland cement concrete (TS 639)		1968
Japan	JIS A6201	Fly ash[a]	1958 reaffirmed 1967
United Kingdom	BS 3892	Pulverised-fuel ash for use in concrete[a]	1965
USA	ASTM C618-	Fly ash and raw or calcined natural pozzolans for use in Portland cement concrete[b]	1971
USSR	GOST 6269 - 63	Binder active mineral additives	1963

Notes: a Methods of sampling and testing are included

b Methods of sampling and testing are determined in accordance with ASTM C311

Table 2 Standard specifications for fly ash for use in concrete

Chemical requirements

	Australia AS 1129	Austria ÖNORM B3319	India IS 3812 Part I	India IS 3812 Part II	India IS 3812 Part III	Turkey TS 639	Japan JIS A6201	UK BS 3892	USA ASTM C618	USSR GOST 6269
SiO_2 min per cent	-	-	-	-	-	-	45	-	-	40
(SiO_2 + Al_2O_3 + Fe_2O_3) min per cent	-	-	70	70	-	70	-	-	70	-
MgO max per cent	-	-	5.0	5.0	5.0	5.0	-	4.0	-	-
SO_3 max per cent	2.5[a]	3.5	3.0	3.0	5.0	5.0	-	2.5[a]	5.0	3.0
CaO max per cent	-	-	-	-	-	6.0	-	-	-	-
Loss-on-ignition max per cent	8.0[b]	7.0	12.0	12.0	12.0	10.0	5.0	7.0[b]	12.0	10.0
Available alkalis as Na_2O max per cent	-	-	1.5[c]	1.5[c]	1.5[c]	-	-	-	1.5[c]	-
Moisture content max per cent	1.5	-	-	-	-	3.0	1.0	1.5	3.0	-

Notes:

a where ratio of cement to pfa is less than 1.0 (by mass) the SO_3 content of the fly ash should not exceed 1.5 per cent.

b unless otherwise agreed between purchaser and vendor.

c applicable only where required for use in concrete containing reactive aggregate and cement required to meet a limitation on content of alkalis.

Table 3 Standard specifications for fly ash for use in concrete

Physical requirements

	Test	Australia AS 1129	Austria ÖNORM B3319	India IS 3812 Pt I	India IS 3812 Pt II	Japan JIS 6201	Turkey TS 639	UK BS 3892	USA ASTM-C618	USSR GOST 6269
Fineness	Amount retained when sieved on:-		Not relevant							
	200 μm sieve max per cent	-		-	-	-	0.3[d]	-	-	-
	150 μm sieve max per cent	10		-	-	-	-	-	-	-
	87 μm sieve max per cent	-		-	-	-	8.0[d]	-	-	-
	44 or 45 μm sieve max per cent	50		-	-	25	-	-	20	-
	Specific surface m^2/kg (Blaine or Lea and Nurse)	-	Not relevant	320	280	270	-	Zones[a] A125-275 B275-425 C >425	325[c]	-
Compressive strength of mortar	Percentage of control at 7 days, min	-	-	-	-	-	100	-	100	-
	Percentage of control at 28 days, min	-	-	80	-	63	100[e]	-	100	-
	Percentage of control at 91 days, min	-	-	-	-	80	70[f]	-	-	-
Soundness	Autoclave expansion of mortar bars	-	-	0.8	0.8	-	-	-	0.5	-
Drying shrinkage	Increase of mortar bars at 28 days	-	-	0.15	0.10	-	-	-	0.03	-
Water requirement	Percentage of control, max.	-	-	-	-	100	-	-	105	-
Pozzolanic activity	With Portland cement, 28 days, min % control	-	-	-	-	-	-	-	85	-
	With lime, normal cure, 7 days MN/m^2	-	-	4.0	4.0	-	-	-	-	-
	With lime, accelerated, 7 days MN/m^2	-	-	-	-	-	-	-	5.5	-
Reactivity with alkalis	Mortar expansion, 14 days, max %	-	-	-	-	-	-	-	0.20[b]	-

Notes:
- a By agreement between purchaser and vendor any other range of specific surface may be supplied
- b Optional
- c Limit is 650 m^2/m^3. If SG = 2000 kg/m^3 this is equivalent to 325 m^2/kg
- d Fineness limits defined by residues on 950 and 4700 apertures/cm^2 sieves
- e 25 wt per cent sand replacement
- f 35 vol per cent cement replacement. This is called test for 'pozzolanic activity' in Turkish standard

Table 4 Standard specifications for fly ash cements

Specifications

Country	Designation of standard		Year
Australia	AS A 181	Blended cements	1971
France	NF 15-302	Portland cement with secondary constituent	1964
Japan	JIS R5213	Fly ash cement	1964
USA	C 595	Blended hydraulic cements	1971
Other	a		

Note: a In a number of other countries fly ash cements are covered by specifications for pozzolanic cement. These include Bulgaria, China, Czechoslovakia, West Germany, Greece, Hungary, Italy, Mexico, Netherlands, Portugal, Romania, Spain, USSR and Yugoslavia.

2 COMPARISON OF NATIONAL STANDARDS

The standards range in complexity from the very simple Russian standard (GOST 6269), which apparently sets no physical requirements, to the very detailed ASTM standard (C618). With the exception of the British Standard (BS 3892) they all cover the use of fly ash as a pozzolanic material. The British Standard is exceptional in covering the use of fly ash only as an inert sand replacement and avoiding reference to its pozzolanic properties. In general there are few major differences in the provisions of the various standards and in some cases they are obviously derivative from one another, for instance the British and Australian standards. A brief indication of the need for each type of test is given, but the reader should consult Kokubu[4] or Abdun-Nur[8] for a more complete explanation.

The Austrian standard, which covers fly ash for grinding with cement clinker during the manufacture of blended cements, has been included in the comparison since the provisions on chemical composition and pozzolanic activity are relevant.

2.1 Chemical requirements

Loss-on-ignition

All the standards set limits on the loss-on-ignition, as an approximation of the carbon content, which has been shown in a number of studies to be related to the suitability of the ash for use in concrete, although the reasons why this is so are not fully known[8]. The limits set range from the 5 per cent in the Japanese standard (also the advisory limit for reinforced concrete in the Indian standard) to 12.0 per cent in the ASTM and Indian Standards. Britain at 7.0 per cent with provision for a waiver if purchaser and vendor agree, is comfortably in the middle of the range. The carbon content will affect the colour of the ash, and for this, if for no other reason, low carbon contents are preferable.

Sulphate

All the standards, apart from the Japanese, include a maximum limit on the sulphate content of the fly ash as do standards for Portland cements, and for the same reasons, ie that too much sulphate can lead to volume instability of concrete due to formation of ettringite. The limits range from 2.5 per cent SO_3 in the British and Australian standards (1.5 per cent SO_3 is the weight if the fly ash in the concrete mix is equal to or greater than the weight of the Portland cement) to 5.0 per cent in the Turkish and ASTM standards (also the lowest grading in the Indian standards). The variation in limits reflects difference between countries in the limits they set for Portland cement. The BS limit for certain Portland cements is likely to be revised upwards in the near future.

Major oxide analysis

The ASTM, Indian and Turkish standards each specify that $Al_2O_3 + SiO_2 + Fe_2O_3$ shall not be less than 70 per cent. The Japanese standards require a minimum SiO_2 content of 45 per cent and the Russian standard a minimum SiO_2 content of 40 per cent. These limits in each case are meant to be indicators of potential pozzolanic activity. The general applicability of such general criteria is not without controversy. The Turkish standard is unusual in specifying a minimum CaO content (6 per cent). The Australian, British and Austrian standards do not include any provisions for the major oxides.

Magnesia

The British Standard sets a limit of 4 per cent, the Indian and Turkish standards a limit of 5 per cent. The other standards do not set limits for magnesia although some of the US ones other than the ASTM do. The limit was included in BS 3892 to parallel the requirement in BS 12 for Portland cement. Limits on magnesia are included in this, and in most standards for Portland cement, since if too much is present, free periclase (MgO) may appear in the clinker and this can lead to long-term volume instability of concrete[13]. The important factor is the quantity of free MgO present and not the total MgO. It is doubtful[13] whether the test methods applied to Portland cements to determine volume stability due to presence of periclase are applicable to cements containing fly-ash.

Available alkalis as Na_2O

The ASTM and Indian standards both set a limit of 1.5 per cent available alkalis as Na_2O where required for use in concrete containing a reactive aggregate. These requirements are similar to those in the corresponding standards for Portland cement and are needed with certain aggregates which react with alkalis, leading to long-term disruption of concrete. There are no similar BS requirements for Portland cements.

Moisture content on delivery

The limits on moisture content range from 1.0 per cent (Japan) to 3.0 per cent (ASTM). The Austrian, Indian and Russian standards do not include provisions on moisture content. A low moisture content provides some indication that the ash has been handled correctly since leaving the power-station. There is also some evidence that too high a moisture content is associated with loss of activity.

2.2 Physical requirements

Fineness

Fineness is one of the principal variables affecting the suitability of a fly-ash in concrete. All the standards, with the exception of the Russian and Austrian, include provisions on fineness. Such a provision would not be relevant to the Austrian standard since it covers fly ash for grinding with cement clinker during the manufacture of blended cements. The Australian and Turkish standards set limits in terms of maximum residue on specified sieves; the Indian and British standards set limits in terms of minimum surface area as determined by an air permeability method; and the ASTM set limits on both residues and minimum surface area; the Japanese set residues and surface area as alternatives.

Apart from the British Standard which uses a specially developed method (Rigden cell with Lea and Nurse apparatus) the surface areas are to be determined by the Blaine method. The results obtained with the two methods are likely to be broadly comparable. The limits set range from 270 m^2/kg (Japan) to 425 m^2/kg for Zone C in the British Standard. Zone B in the British Standard, 275-425 m^2/kg, compares with minima in the other standards for pozzolanic materials of 270 (Japan), 320 (India), 325 (ASTM) and for admixture 280 (India). The different methods of determination must however be noted. The results of surface area measurements will be affected by the presence of carbon, and ashes with high carbon contents will give misleadingly high surface area values, although in practice high carbon tends to be associated with coarse ashes. It is for this reason that sieve residues are preferred as a measure of fineness or made an additional requirement by some authorities.

Limits on the residue on a 45 μm sieve are set by ASTM (20 per cent), Japanese (25 per cent) and Australian (50 per cent) standards. The Australian couples this much higher limit with a limit of 10 per cent residue on a 150 μm sieve. The Turkish standard couples residue on a 87 μm sieve (8.0 per cent max) with a residue on a 200 μm sieve (0.2 per cent). The relationship between surface area, particle size distribution, carbon content and performance in concrete is being investigated for a series of British fly ashes at BRS.

Compressive strength tests

Compressive strength tests are specified in the ASTM, Indian, Turkish and Japanese standards. In each case they are made on mortar specimens, and the water content of the fly ash mixes is adjusted to give a similar consistency or flow to the controls made with Portland cement alone. The ages of test are: India 28 days; ASTM and Turkish, 7 and 28 days; Japanese 28 and 91 days. In each case the test serves to demonstrate that the use of the fly ash will not adversely affect the properties of the Portland cement.

In the ASTM and Turkish tests, the fly ash is used to replace an amount of fine aggregate equivalent to 25 wt per cent of the Portland cement content, ie the Portland cement content of the mix remains essentially constant (since the fly ash mix will have a lower density the amount of cement per cube will be slightly reduced - about 3-4 per cent). In both cases the requirement is that the strength of the fly ash mix at 28 days shall not be less than that of the control.

The Japanese test is made using a replacement level of 25 wt per cent of the cement by fly ash and requires that the fly ash mix gives 63 per cent of the strength of the control at 28 days and 80 per cent at 91 days. Thus a small negative effect is allowed at 28 days and a small positive effect is considered a possibility at 91 days.

The Indian standard uses a replacement level of 20 wt per cent and the requirement at 28 days is a strength of at least 80 per cent of the control.

The Turkish standard includes a test for pozzolanic activity resembling that in the ASTM standard for pozzolanic activity with Portland cement (see below) but unlike the ASTM standard no accelerated curing appears to be involved and it therefore falls more appropriately in the category of a compressive strength test. In this test a 35 vol per cent replacement is used which is approximately equal to a 26 wt per cent replacement (assuming densities of 3100 and 2000 kg/m^3 for the cement and fly ash respectively). The requirement is for 70 per cent of the control strength at 28 days, that is, no positive contribution to strength is expected.

Pozzolanic activity

Tests for pozzolanic activity are included in the ASTM, Indian and Turkish standards. The Turkish test has been discussed above as it seems more appropriate to term it a compressive strength test. Pozzolanic activity relates to the long-term performance of the potential pozzolana in concrete rather than to shorter-term properties. These are governed to a greater extent by physical parameters such as fineness and water requirement.

The ASTM includes two tests: one with Portland cement and one with lime. The test with Portland cement compares the strength of a control mortar with one in which 35 vol per cent (approximately 26 wt per cent) of the Portland cement has been replaced by fly ash. The cubes are tested at 28 days after curing for most of this time at 38°C. This elevated temperature is used to accelerate the pozzolanic reaction. The fly ash mix is required to give 85 per cent of the strength of the control.

Both the ASTM test and the Indian test with lime set simple strength minima. The mix method of cure and form of specimen are all different and a direct comparison of the two tests is not possible.

Water requirement

The Japanese and ASTM standards include water requirements based upon the relative water content of unit volumes of a fly ash mix and a control mix without fly ash. The Japanese standard states that the water content of a unit volume of the fly ash mix (25 per cent replacement of cement) should not exceed 100 per cent of the water content of the same volume of the control. The ASTM standard employs 35 vol per cent replacement (approximately 26 per cent by wt) and sets a limit of 105 per cent.

Other requirements

The ASTM and Indian standards set limits on increase in drying shrinkage and on soundness by the autoclave expansion test. It is doubtful[9] whether this latter test is applicable to cements containing fly ash. The Turkish standard specifies soundness to be determined by the Le Chatelier test.

3 POSSIBLE CHANGES IN BRITISH STANDARD BS 3892

The purpose of this section of the paper is to identify, by comparison with the other national standards, the main points which should be taken into consideration if BS 3892 is to be revised so that it covers use of fly ash as a pozzolan.

3.1 Chemical requirements

There seems to be no pressing reason to amend the chemical requirements of the standard although some throught could be given to the need to retain the limit on MgO content and to a small increase in the sulphate limit in line with the tendency for limits in Portland cement to rise.

3.2 Physical requirements

Fineness

As discussed above there is no general agreement on whether the fineness of fly ash is best determined by surface area measurement or by sieve residues. Japan offers the two methods as alternatives and the ASTM standard specifies both. Fly ashes of similar surface area can vary markedly in their particle size distribution, much more so, for example, than Portland cements.

Air permeability methods can be difficult to apply to fly ashes, hence the special method in the present British Standard, and can be adversely affected by the carbon content of the fly ash which tends to give misleadingly high readings. Similar sieve residues can, however, also be given by ashes of markedly different particle size distribution.

Although the ASTM and Japanese standards are similar in specifying a maximum residue on a 45 μm sieve of 20 and 25 wt per cent respectively, the Australian standard allows a 50 wt per cent residue. The limits on surface area are close, ranging from 270 m^2/kg (Japan) to approximately 325 m^2/kg (ASTM). The Indians consider it possible to distinguish between pozzolanic fly ash (320 m^2/kg) and fly ash for use as admixture (280 m^2/kg) with a very small difference in surface area.

The need in revising BS 3892 will be to decide whether to retain the air permeability method or to adopt the attractively simple Australian approach of two sieve residue limits. It can be said at this stage that if this latter approach is adopted something closer to the Japanese and ASTM limits on a 45 μm sieve, than to the Australian limit, is likely to be required.

Compressive strength

The compressive strength tests on mortar specimens included in the Indian, Japanese, Turkish and American standards serve as a check that the fly ash has no adverse effect on the hydraulic properties of the cement, and that it is generally suitable for use in mortar or concrete. Should it be considered desirable to include such a test in the revised British Standard it might be sensible to base it on ISO type 40 x 40 x 160 mm mortar specimens since these will probably eventually replace the BS Vibrated 5000 mm^2 mortar cubes. Such a test could be based upon concrete cubes but the problems of selecting mix proportions that would give a fair result on all fly ashes would be difficult. A choice would have to be made between sand replacement on the Turkish and ASTM pattern, or of cement replacement following India and Japan. The latter seems preferable.

Pozzolanic activity

The Australian, Japanese and Turkish standards (if the authors' interpretation of the provisions of the last are correct) do not contain any test for pozzolanic activity. It would be desirable if the revised British Standard could also avoid the use of a test for pozzolanic activity, particularly since there is no general agreement on the way in which pozzolanic activity should be assessed. Chemical tests of the type detailed in ISO Recommendation No 863 (1968) and now incorporated in BS 4550 Part 2:1970, which can only decide between pozzolanic and non-pozzolanic materials, but which give no information on the degree of activity, are not suitable. Neither the tests with lime included in the Indian and ASTM standards, nor the test with Portland cement in the ASTM standard, seem particularly good.

Should it be felt necessary to include a test for pozzolanic activity, one modelled on the accelerated curing test devised by F M Lea[10] seems a possibility. In this test a comparison is made between the compressive strength of mortars cured at 18°C and 50°C. The pozzolanic reaction is accelerated more by the higher temperature than is the normal hydration reaction of Portland cement and provided the period of curing at the higher temperature is not too long, the increase of strength over that of identical specimens cured at ordinary temperatures is largely due to the pozzolana.

In the test six 5000 mm^2 1:3 mortar cubes are prepared from standard sand and either a 20/80 wt per cent or a 40/60 wt per cent mixture of pozzolana and Portland cement. The latter is more usual. Three of the cubes are stored for 1 day moist air, 6 days water 18°C and 3 cubes for 1 day moist air 18°C, 4 days water 18°C, 46 hours water 50°C, 2 hours water 18°C. Lea, who did his work in the 1930s, used hand compacted mortar cubes. More recently vibrated mortar cubes have been used at BRS and these will tend to give somewhat higher strengths so that the limits suggested by Lea may not be completely appropriate, although the differences are likely to be small. The cubes may be made either with a fixed water/cement ratio (0.4) or fixed consistency. For fly ash the latter is probably preferable. Lea found, for a given Portland cement, good correlation between the results of this test and the compressive strengths of both mortar and concrete at 180 days (Figures 1 and 2). This contribution to long-term strength is an essential feature of pozzolanic materials and the accelerated curing test will not normally

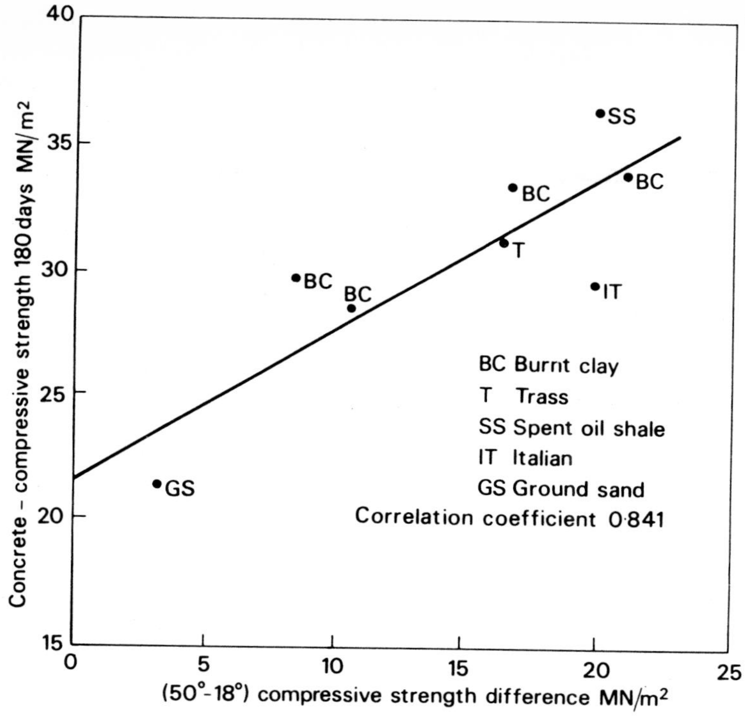

Figure 1 Concrete tested at 180 days (after Lea[10])

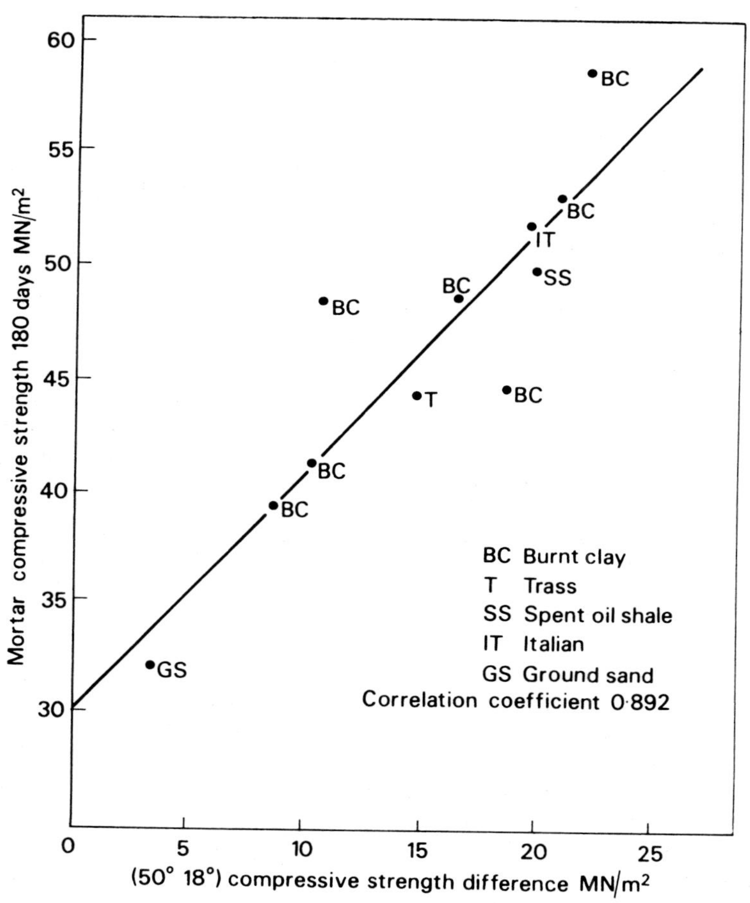

Figure 2 Mortar tested at 180 days (after Lea[10])

provide correlations with results obtained under normal conditions at say 28 days. Lea also found that while the 50°/18° strength difference varies somewhat with the Portland cement used as a base, it is within a reasonably close range and characteristic of the pozzolana.

Before incorporation into a standard a number of problems need to be settled:

1 The mortar mix to be employed.

2 The form of specimen (use of 40 x 40 x 160 mm prisms in preference to vibrated mortar cubes should be considered).

3 Very little information is available on the performance of fly ashes in the test. Sufficient samples have to be tested and the results correlated with long-term strength behaviour to enable statistically valid and realistic limits to be established.

4 The form of mix (mortar, concrete, mix design), with which the Accelerated Curing Test (ACT) results are to be correlated, has to be decided. Lea in his work used both the same mortar mix as for the ACT specimens and 4-inch cubes of 1:2:4 ($\frac{3}{4}$ in. gravel) concrete with water/cement ratio 0.6. The former would be more straightforward and likely to give a better correlation but the latter is more likely to satisfy the engineer.

The applicability of the test to fly-ashes is currently being investigated at BRS.

Water requirement

The inclusion of a water requirement test on the lines of the Japanese and ASTM tests should present no difficulties, although the need for such a test has to be established. There is a body of opinion however which believes that the water requirement of an ash is a good guide to its general suitability for use in concrete.

While it is probably true that the workability of concrete made with fly ash is related to the 'water requirement' of the ash, there is no agreement as to the best method of assessing the workability of the concrete in a practically useful way. This possible correlation of 'water requirement' and concrete workability does not seem to justify of itself inclusion of a water requirement test in a standard designed to measure the suitability of fly ash for use as a pozzolana. It should only be included if it can be shown that correlation exists between water requirement and either pozzolanic activity of some adverse effect on strength or durability if the water requirement is increased beyond a certain limit, for example 100 per cent of the control.

Other physical requirements

There is no evidence at present that there is any need to include either drying shrinkage or soundness tests.

4 RELATION OF BRITISH STANDARD FOR FLY ASH TO OTHER BRITISH STANDARDS AND TO CODES OF PRACTICE

The major problem relating to other standards, eg draft BS for specification of concrete, and Codes of Practice, eg CP 110, is the definition of 'cement' to be used in the various concrete mixes. At present this term applies only to a specified list of cement types and does not include pozzolanas within the definition of cement. The Department of the Environment has dealt with a similar problem relating to the use of ground granulated blastfurnace slag in concrete by issuing an amendment[11] to its Specification for Road and Bridge Works, allowing direct additions of the ground slag to concrete to be treated as though a Portland blastfurnace cement to BS 146[12] or BS 4246[13] was being used.

REFERENCES

1 **Gutt, W, Nixon, P J, Smith, M A, Harrison, W H,** and **Russell, A D.** A survey of the locations, disposal and prospective uses of the major industrial by-products and waste materials. BRS Current Paper CP 19/74.

2 **Barber, E G, Jones, G T, Knight, P G K** and **Miles, M H.** PFA Utilisation. Central Electricity Generating Board, London, 1972.

3 **David, R E, Carlson, R W, Kelly, J W** and **Davis, H E.** Properties of cements and concretes containing fly ash. Proc Amer Concrete Inst, Vol 33, 1937, pp 577-612.

4 **Kokubu, M.** Fly ash and fly ash cement. Proc 5th International Symposium on the Chemistry of Cement, Vol IV, pp 75-113. Cement Association Japan, Tokyo, 1969.

5 **Mielenz, R C.** Specifications and methods of using fly ash in Portland-cement concrete.

Proc Third International Ash Utilisation Symposium, Pittsburgh 1973.

6 Specifications for Pozzolan for use in Portland-cement Concrete, Corps of Engineers, US Army CRD-C 262-63, 1963.

7 US General Services Administration. Pozzolan (for use in Portland-cement Concrete): Federal Specification SS-P-570B, 1969.

8 **Abdun-Nur, E A.** Fly-ash in concrete - an evaluation. Highway Research Board, Bulletin 284, Washington DC, 1961.

9 **Gaze, M E** and **Smith, M A.** High-magnesia cements I: curing at 50°C as a measure of volume stability. Cement Technology, Vol 4, No 6, pp 224-6 (BRE Current Paper CP 27/74).

Gaze, M E and **Smith, M A.** High-magnesia cements II: the effect of hydraulic and non-hydraulic admixtures on expansion, Cement Technology 1974, Vol 5, No 2, pp 291-5 (BRE Current Paper CP 60/74).

10 **Lea, F M.** The testing of pozzolanic cements. Cement Technology, Vol 4, No 1, 1973, pp 21-25.

11 **Department of the Environment.** Specification for Roads and Bridge Works (1969). Technical Memorandum (Bridges) No BE9/72 (which amends clause 26095 of the main specification). HMSO.

12 **British Standards Institution.** BS 146: 1958 Portland-blastfurnace cement.

13 **British Standards Institution.** BS 4246: 1968 Low heat Portland-blastfurnace cement.

APPENDIX

SUMMARY OF MAJOR PROVISIONS OF NATIONAL STANDARD SPECIFICATIONS FOR FLY ASH FOR USE IN CONCRETE

The main provisions of the British, Australian, Austrian, Indian, Japanese, Turkish, American and Russian standards are listed below. The descriptions follow a generally similar format and the provisions of each standard have been converted as far as possible into SI units. The descriptions do not reproduce the detailed wording of the standards.

United Kingdom (BS 3892:1965)

The foreword to BS 3892, which covers pulverised-fuel ash for use in concrete states that: 'Some pulverised-fuel ashes may have pozzolanic properties, but there is at the present time no generally accepted method for their assessment'. Thus, although most applications of pfa in concrete seek to make use of its pozzolanic properties, the standard only really deals with its use as an inert admixture to concrete. The standard sets general requirements for moisture content (1.5 wt % maximum), loss-on-ignition (7.0 wt % maximum), sulphate (2.5 SO_3 wt % maximum) and magnesia (4.0 wt % maximum). The limits on moisture content and loss-on-ignition may be waived by agreement between the purchaser and vendor. The sulphate limit is reduced to 1.5% SO_3 if the weight of pfa in the mix is equal to or greater than the weight of cement.

Physical requirements are restricted to measurements of the surface area by an air permeability method. Three gradings or zones are recognised: Zone A is the range 125-275 m^2/kg; Zone B > 275-425 m^2/kg; Zone C, >425 m^2/kg subject to a range of not greater than 150 m^2/kg. In addition, by agreement between the purchaser and vendor, any other range of specific surfaces may be specified, eg 200-400 m^2/kg.

Sampling is on the basis of at least one composite sample of not less than 4.54 kg per 101.6 tonnes (100 tons) of pfa, each composite sample to consist of at least 40 equal sub-samples, ie not less than 1 per $2\frac{1}{2}$ tonnes taken from places evenly spaced throughout the consignment.

Australia (AS 1129 and AS 1130)

AS 1129 covers fly ash for use in concrete, and sets limits on moisture content (1.5% maximum), loss-on-ignition (8.0% maximum) and sulphate (2.5% SO_3 maximum).

The physical requirements are limited to a sieve analysis. The amount of fly ash retained when wet sieved on a 150 μm sieve should not exceed 10% and the amount retained on a 45 μm sieve should not exceed 50%.

In notes to the Standard it is stated that:

1 Loss-on-ignition may be taken as an accurate indication of carbon content. The deleterious effects attributable to carbon in fly ash are of two kinds:

(a) loss of strength due to physically weak particles

(b) a high total surface area of the particles leading to an excessive absorption capacity for air-entraining admixtures.

2 Fly ash with residue retained on a 45 μm sieve exceeding 50% may be used if specific performance tests have established its suitability to the satisfaction of the purchaser.

3 Generally it is seen that the lower the 45 μm sieve residue, the more will be the pozzolanic activity of the fly ash.

Sampling is on the basis of one 2 kg grab sample per 20 tonne container or part thereof. Of these grab samples one in five is randomly selected for testing in accordance with the Standard.

A standard Code of Practice (AS 1130) for the use of fly ash in concrete was issued jointly with the Standard Specification (AS 1129). This code recognises that there are difficulties in assessing fly ashes by the standard methods for pozzolanas; that there is a need to carry out preliminary trial mixes with the other materials (aggregate, Portland cement) to be used in the concrete, that for optimum results the mix design should allow for a proportion of the sand to be replaced by fly ash as well as part of the cement, and that all preliminary tests should be carried out under conditions which simulate curing conditions likely to be encountered on the job. This latter recommendation is related to the sensitivity of the pozzolanic reaction to the temperature of curing and because of the adverse effects of adverse drying at an early stage of the hydration process.

If concrete is to be specified by an engineer or architect, the engineer or architect should state in the specification for the concrete whether or not fly ash is permitted and if so any pertinent details, such as maximum cement replacement. Where the concrete is not specified by an engineer or architect it is permissible to use fly ash at the discretion of the supplier of the concrete. In this case the supplier must guarantee the compliance of the concrete with specified requirements. Other provisions of the Code cover storage (provision and records), accuracy of measuring equipment and identification of mixes containing fly ash in batching plant records and delivery dockets.

Austria (Ö NORM B3319 (1962))

The Austrian standard B3319 covers fly ash for use as hydraulic powdered admixture component for cement manufacture, that is it does not specifically cover fly ash for use directly in concrete in batching plant or on site. The standard sets limits for loss-on-ignition (7.0% maximum) and sulphate (3.5% SO_3 maximum). Physical requirements are limited to a test for hydraulic activity.

Before testing for hydraulic activity the fly ash is ground in a laboratory mill as necessary until a maximum residue of 7 wt% on a 63 μm sieve (DIN 4118) is achieved. Powdered quartz ground to the same fineness as the fly ash is also used in the test. Compressive strength tests are made at 28 days on 40 x 40 x 160 mm prisms made from 1 cement: 1 fine sand: 2 coarse sand mortars with a water : cement ratio of 0.60. The tests are made with 'cements' composed of: (1) 100% Portland cement; (2) 15 wt% fly ash : 85 wt% Portland cement and (3) 15 wt% ground quartz : 85 wt% Portland cement.

A hydraulic factor K is calculated based on the formula:

$$K = 100 \frac{(b - c)}{(a - c)}$$

where a = compressive strength at 28 days of control (Portland cement)
b = compressive strength at 28 days of control (fly ash mix)
c = compressive strength at 28 days of control (ground quartz mix).

For the fly ash to be acceptable, the hydraulic factor K should not be less than 30.

Sampling is not dealt with in the standard.

India (IS 3812 (Parts I-III - 1966)

The Indian standard for fly ash is divided into three parts: IS 3812 (part I) covers its use as a pozzolana; IS 3812 (part II) covers use as an admixture in concrete; IS 3812 (part III) covers use as a fine aggregate for mortar and concrete.

Parts I and II of the standard, which are those directly relevant, both set limits for loss-on-ignition (12.0% maximum), sulphate (3.0% SO_3 maximum), major oxides $SiO_2 + Al_2O_3 + Fe_2O_3 =$

70% minimum), silica (35% minimum) and magnesium oxide (5.0% maximum). No limit is placed on moisture content. The foreword advises that for reinforced concrete it is preferable to limit the loss-on-ignition to 5.0%.

The physical requirements of Part 1 (Pozzolana) are for surface area by the Blaine air permeability method (320 m^2/kg), lime reactivity, compressive strength in comparison with standard mortar control, drying shrinkage (0.15%), and soundness by autoclave expansion. Part 2 (admixture) sets a lower limit for surface area (280 m^2/kg), the same lime reactivity and autoclave expansion but lower drying shrinkage. It does not include a test for compressive strength development.

The lime reactivity test is made using 70.6 mm cubes of a (1 fly ash : 2 hydrated lime) : 3 sand mortar.

The cubes are tested at 7 days and should give a strength of not less than 4 MN/m^2 (40 kg/cm^2).

The compressive strength tests are made on 2.78 inch 1:3 vibrated mortar cubes in accordance with IS 269 for Portland cement. For the fly ash mix, 20 wt% of the cement is replaced by fly ash. In each case the water content is (0.25 P + 3)% where P is the percentage of water for normal consistency. The fly ash mix should give not less than 80% of the compressive strength of the control when both are tested at 28 days.

Shrinkage and autoclave expansion tests are both made using a 20 wt% fly ash : 80 wt% Portland cement mix.

The sampling requirements for all three parts are the same and basically one sample of at least 2 kg or more should be taken, by either grab or composite methods, for each 100 tonnes of fly ash. Individual samples on which all tests are to be conducted should weigh at least 4 kg.

Japan (JIS A6201)

JIS A6201-1958, which covers fly ash for use as an admixture in Portland cement mortar and concrete, set limits on moisture content (1% maximum), loss-on-ignition (5% maximum) and silicon dioxide (minimum 45%). The physical requirements comprise specific gravity (1950 kg/m^3 minimum), surface area by Blaine air permeability method (270 m^2/kg minimum), residue on 45 μm sieve (25% maximum), a water requirement test in comparison with standard mortar, and a comparison of compressive strength of mortars made with, and without, the fly ash.

For the water requirement test a standard 1 : 2 Portland cement: sand mortar is used with a water : cement ratio of 0.65. The flow and the weight of a unit volume of the standard mix is determined. For the fly-ash-containing mix, 25% of the cement is replaced by fly ash, the water content is adjusted to give a similar flow to that of the standard mix, and the weight of a unit volume of the mix determined. A comparison is then made between the weights of water in each unit volume of mortar: that in the fly-ash-containing mix should not exceed 100% of that in the standard mix. It is therefore assumed that fly ash will generally reduce the water content per unit volume of mortar.

The compressive strength tests are made on the same mortar as the water requirement test using 40 x 40 x 160 mm prisms. The strength of the fly-ash mixes should not be less than 63% of the control at 28 days, and 80% at 91 days.

Turkey (TS 639-1968)

TS 639 (1968) covers fly ashes for mixing with Portland cement clinker and Portland cement concrete. It defines two classes of fly ash: (1) clinker fly ash which is a coarse fly ash for grinding in with Portland cement clinker during the manufacturing process; (2) Portland cement concrete fly ash which is a fine fly ash for use on site.

For both classes the standard sets limits for moisture content (3 wt% maximum), loss-on-ignition (10 wt% maximum), sulphate (5.0 wt% SO_3 maximum), major oxides (SiO_2 + Al_2O_3 + + Fe_2O_3 = 70% minimum), and magnesium oxide (5 wt% maximum).

The relevant physical requirements are those for fineness, volume stability (Le Chatelier) compressive strength in comparison with standard mortar, and pozzolanic activity. The fly ash should leave not more than 0.3 wt% residue on a ~ 200 μm sieve (950 apertures /cm^2) and not more than 8% on a ~ 87 μm sieve (4700 apertures/cm^2).

Compressive strength tests are made on 70 mm 1:3 mortar cubes. For the tests with fly ash the cement content is held constant and the sand reduced by an amount of fly ash equivalent to 25%

of the cement content, that is 1/12 of the original sand content. The water content is determined in accordance with TS 24 using the Vicat gauge (ie a standard consistency is used). The compressive strength of the cubes is determined at 7 and 28 days and the strength of the fly ash mix expressed as a percentage of the control mix: it should not be less than 100%.

Pozzolanic activity tests are also made on 70 mm 1:3 mortar cubes but in this case the fly ash mix is prepared by replacing 35 vol% of the cement by fly ash. (Approximately 26% weight assuming densities of 3100 and 2000 kg/m^3 for cement and fly ash respectively). The tests are made at standard consistency, the water requirement being determined in accordance with TS 24. The cubes are cured for 28 days and then tested for compressive strength. The pozzolanic activity, which should not be less than 70%, is defined as the strength of the fly ash mix as a percentage of the control. No accelerated curing is employed.

The Le Chatelier test for unsoundness is made on a mixture of 20 wt% fly ash: 80 wt% Portland cement. Expansion should not exceed 10 mm.

Sampling should be in accordance with TS 23.

United States of America (ASTM C618-71)

ASTM C618-71 covers the use of fly ash and raw or calcined natural pozzolanas for use as an admixture in Portland cement concrete where pozzolanic action is desired, where a suitable fine material may be required to promote workability and plasticity or where both effects are to be achieved. It is used in conjunction with ASTM C311-68, which covers the sampling and testing of fly ash for use in Portland cement.

C618-71 sets limits on the moisture content (3.0% maximum) loss-on-ignition (12.0% maximum), sulphate (5.0% SO_3 maximum) and major oxides (SiO_2 + Al_2O_3 + Fe_2O_3, 70% minimum). The physical requirements include surface area by Blaine air permeability apparatus (650 m^2/m^3 = 325 m^2/kg for SG = 2000 kg/m^3) residue on 45 μm sieve (20% maximum), water requirement test, compressive strength in comparison with standard mortar control pozzolanic activity, increase of drying shrinkage, soundness by autoclave expansion.

The compressive strength tests are made on 2 inch 1:2.75 mortar prisms made and tested in accordance with ASTM C109. For the fly ash mix, the fly ash is used to replace a portion of the sand equivalent to 25% of the cement content (1/11 of the original sand content). Sufficient water is used in each mix to give a standard flow. The compressive strength of the fly ash mix should be not less than that of the control at both 7 and 28 days.

The pozzolanic activity is determined with Portland cement using 2 inch, 1:2.75 standard mortar cubes. For the fly-ash mixes, 35% of the absolute volume of the cement is replaced by fly ash. (Approximately 26$\frac{1}{2}$% by weight assuming densities of 3160 and 2000 kg/m^3 for cement and fly ash respectively). Sufficient water is used in each mix to give a standard flow. The prepared cubes are first stored for 20-24 hours in their moulds at 23°C (73.4°F) at high relative humidity, they are then demoulded and stored in airtight containers for 27 days at 38°C (100°F), followed by cooling to 23°C for testing. The pozzolanic activity of the fly ash is taken as the strength of the fly ash specimens as a percentage of the strength of the controls and should not be less than 85%.

The water requirement of the fly ash is the weight of water used in the fly ash mix used in the pozzolanic activity test with Portland cement, expressed as a percentage of water content of the control mix, and should not exceed 105%.

The pozzolanic activity is also determined with lime in accordance with ASTM C595-71 for 'Blended Hydraulic Cements'. The tests are made using 2 in. diameter cylindrical specimens 4 in. high made from a mortar consisting of 1 part hydrated lime by weight: 9 parts by weight standard sand: fly-ash-equavalent to twice the weight of the lime multiplied by a factor obtained by dividing the specific gravity of the fly ash by the specific gravity of the lime. The overall proportions are thus approximately 1 (lime + fly-ash) : 3 sand. The specimens are stored for 24 hours at 23°C(73.4°F) and then at 55°C(131°F) for six days, cooled to 23°C over four hours and the compressive strength determined. The specimens are kept in their moulds until 1$\frac{1}{2}$ hours before testing. The compressive strength of the lime mortar should not be less than 5.5 MN/m^2 (800 psi).

Tests for soundness are made in accordance with ASTM C151 for autoclave expansion of Portland cement using a paste composed of 20 wt% fly ash: 80 wt% Portland cement.

Tests for drying shrinkage are made using the same mortar mixes as for compressive strength ie 1:2.75 with 1/11 of the sand replaced by fly ash in the test mix, on bars 25 x 25 x 285 mm with an effective gauge length of 250 mm.

Samples for testing should weigh at least 1.81 kg (3lb) if they are to be composited and individual test samples on which all the tests are to be performed should weigh at least 3.63 kg (8 lb). Test samples, either individual or composite, should not usually represent more than 102 tonnes (100 tons). Composited samples should usually be taken on the basis of one sub-sample for each 2 tonnes (2 tons) of fly ash.

USSR (GOST 6269-63)

No information on this standard is available other than data due to Kokubu[4]. It sets limits for loss-on-ignition (10.0% maximum), sulphate (3.0% SO_3 maximum) and silica (40% minimum). There appear to be no physical requirements.

Properties of concrete

Recommendations for the treatment of the variations of concrete strength in codes of practice (CP 6/74)

D.C. Teychenné

Some time ago (Materials and Structures n⁰ 24, November-December 1971), Mr. Petersons published a paper referring to the work of Working Group III of the Joint CEB/CIB/FIP/RILEM Committee on the Statistical Control of Concrete. In the following, Mr. Teychenné gives a personal view of the ideas actually being discussed by Working Group I. The Joint Committee will at its next session discuss these proposals and hopes to finalize its work in the near future by publishing draft Recommendations on the Statistical Quality Control of Concrete.

This paper reports the work of one of three Working Groups set up by a joint CEB/CIB/FIP/RILEM Committee to study the variation in the strength of concrete produced in the factory and on the site. The Working Group consisted of members from five European Countries, Canada, Israel and the USA.

The Working Group recommends that the variation in concrete strength should be expressed as a standard deviation. It gives relationships between the standard deviation and the characteristic strength and proposes minimum values for specifying the target mean strength under various conditions for use in Codes of Practice.

1. INTRODUCTION

In 1966 the four International Organisations, CEB, CIB, FIP and RILEM approved the formation of a Joint Committee to consider the statistical control of concrete quality. At a meeting held in Paris three working Groups were formed to consider various aspects of the overall problem. The terms of reference of Working Group 1 were :

1. To examine the variation of concrete strength on the site and in the factory as measured by test specimens. To consider how this is affected by :
   (a) the size of the job and the length of the contract,
   (b) the supervision, workmanship and plant used,
   (c) the making, curing and testing of the specimens,
   (d) the variation in successive batches,
   (e) the variation in the constituent materials.
2. To consider methods of continuous control of the quality of concrete.

Initially, the Working Group consisted of members from Denmark, France, Netherlands and the United Kingdom, but subsequently members from Canada, the German Federal Republic, Israel and the USA attended meeting of the Working Group. In addition there were corresponding members in Canada and the USSR.

Four meetings of the Working Group have been held, i.e. London 1968, Amsterdam 1969, Prague 1970 and Copenhagen 1971. A draft report was discussed at the Copenhagen meeting and this report is developed from a revised draft that was circulated to all members of the Joint Committee for comment in November 1972.

The Working Group concentrated its activities on the first item in the terms of reference since it considered that the relative magnitude of the various factors should first be determined as this could influence the methods required for the continuous control of the quality of concrete.

2. SOURCES OF INFORMATION

In 1968 the British Building Research Station compiled a Bibliography No 215 " Quality control of concrete " [1] to which an Addendum was added in 1970. Although the search was made under key phrases such as " site control of concrete ", " variability of concrete, cement and aggregates ", and " statistical analysis of concrete strength " as suggested by the terms of reference, related items on subjects such as accelerated curing and analysis of fresh concrete were found and were included in the bibliography. The various topics under the general heading of quality control were divided into ten groups listed below with the number of references found in that group.

Group	Subject	No of References
I	Mixing procedures	8
II	Curing conditions	4
III	Shape and size of specimens and test methods	47
IV	Batching procedures	8
V	Material variations	21
VI	Site control	76
VII	Statistical analysis	55
VIII	Accelerated test methods	8
IX	Wet analysis of concrete mixes	6
X	Requirements of Standards and Codes	7

This Bibliography adequately covers the topics in Groups IV to VII, but is by no means a complete list of references for the subjects covered by the other Groups. Even so, the papers referred to in the other Groups include useful background information. For example the papers in Group I show how alterations in the mixing time or the order of mixing the mix constituents affect the strength of concrete. The papers in Groups II and III indicate the importance of standardising in the methods of making, curing and testing concrete specimens and in the size and shape of the test specimen. Many countries have their own National Standards for testing concrete and ISO Recommendation R 1920 [2] is a welcome first step in International standardisation. There is a number of papers which discuss the effect of the shape and size of the test specimen on the level of the strength obtained, but there is little information on the way in which specimen shape and size affect the variability of concrete strengths.

The Bibliography shows that there were peaks of interest in site control during the period 1925-1930 mainly in terms of methods of batching concrete on the site, and again during 1951-55 when statistical theory was beginning to be applied to the variability of concrete strengths. In the 24 years from 1936-60 there are only 16 references in Group VII on the statistical analysis of strength results, but these increased to 16 for the period 1961-65 and to 18 for the period 1966-69. More recent papers are included in the references given at the end of this report.

A total of 21 papers was presented to the Working Group, either by members themselves or by other authors from their own country, these are all included in the list of references given at the end of this report. From an examination of the many papers published on the variability of concrete strengths there are very few that have attempted to break down the overall variation into component factors under typical site production conditions. There is considerable information on the variation due to sampling and testing concrete and of the variation of the strength of cement tested under laboratory conditions. This information is discussed in section 6 of this report where data from Erntroy [3] Rusch [4] and others are reported as well as work by Rackwitz [5] to develop a theoretical study of how various factors influence the overall variation.

3. THE MEASUREMENT OF VARIABILITY : STANDARD DEVIATION OR COEFFICIENT OF VARIATION

There are many ways in which the variability of a set of numbers may be expressed and there have been many studies on which statistical parameter is the most appropriate one to express the variability of concrete strengths. This is of more than purely academic interest, since the margin between the specified strength and the target mean strength is based on one or other of these factors, and this may also affect the conditions for compliance.

Basically, the variation in concrete strengths may be expressed either in terms of a standard deviation or of a coefficient of variation. The standard deviation S is calculated from the equation :

$$S = \sqrt{\frac{\Sigma (x - \bar{x})^2}{n - 1}} \qquad (1)$$

where x = the individual result,
$\bar{x}$ = the mean of all the results,
n = the number of results obtained.

It is sometimes more convenient to use the equation :

$$S = \sqrt{\frac{\Sigma (x^2) - \frac{(\Sigma x)^2}{n}}{n - 1}} \qquad (2)$$

The coefficient of variation V is calculated from the equation :

$$V = \frac{S}{\bar{x}} \qquad (3)$$

Thus S is measured in the same units as the strength of the concrete whereas V is a ratio, or if multiplied by 100 it is expressed as a percentage.

Most Codes and Standards require that the concrete should be designed to have a target mean strength greater than the specified strength. However, the manner in which this " margin " or " over-design factor " is derived differs for different Codes. Some Codes require that the target mean strength should be the specified strength multiplied by a factor, and some that the target mean strength should be the specified strength plus a factor. In statistical terms the first method implies that the coefficient of variation remains constant irrespective of the strength level, while the second method implies that the standard deviation remains constant irrespective of the strength level. The practical implication of this is that on the constant coefficient of variation basis, the margin increases as the specified strength increases, but on the constant standard deviation basis the margin remains constant for the same control of the production process irrespective of the specified strength.

If a Standard or Code covers only a small range of specified strengths, e.g. in the early version of the British CP 114 [6] where the specified concrete strength only ranged from 21 N/mm² to 31 N/mm², then there is little practical difference whichever parameter is used. However, over the range of strengths likely to be used in modern Codes, say from 10 N/mm² to 60 N/mm² the use of margins derived from these two parameters has a significant effect on the target mean strength and thus the economy of mix design. This is illustrated in the following example.

If the specified strength x_s is a characteristic strength as in the CEB/FIP Recommendations [7] this is related to the target mean strength x_t by the equation :

$$x_t = x_s + kS \quad (4)$$

where k is a constant depending on the proportion of defectives permitted, at the 5 % level given in the CEB/FIP Recommendations, $k = 1.64$.

From eq. (3) $S = Vx_t$

$$\ldots x_t = x_s + k \cdot Vx_t$$

and hence

$$x_t = x_s \left(\frac{1}{1 - kV}\right) \quad (5)$$

Eq. (4) shows the " plus factor " based on the standard deviation and eq. (5) the " multiplying factor " based on the coefficient of variation.

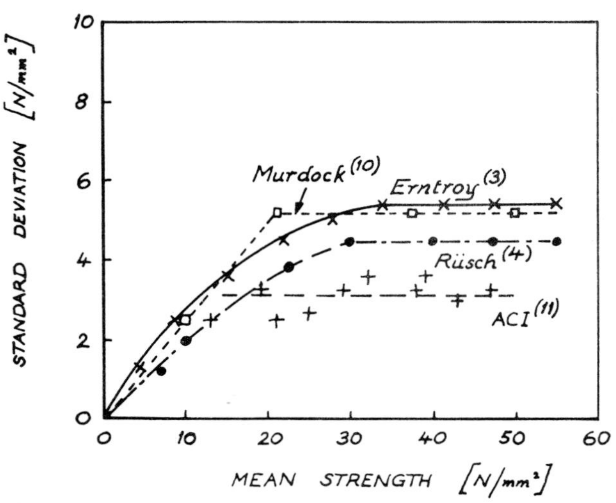

Fig. 1. — Relationship between standard deviation and mean strength as shown by Newlon [11].

If the specified strength is 20 N/mm² and the standard deviation is 6.1 N/mm² then $x_t = 20 + 1.64 \times 6.1$

$$x_t = 20 + 10 = 30 \text{ N/mm}^2$$

In this instance, the coefficient of variation $= \frac{6.1}{30} = 0.203$ or 20.3 %, and the multiplying factor $\frac{1}{1 - kV} = \frac{1}{1 - 1.64 \times 0.203} = 1.5$ thus $x_t = 20 \times 1.5 = 30$ N/mm²

It is now required to produce concrete under the same site conditions but with a specified characteristic strength of 40 N/mm².

If the standard deviation remains constant at 6.1 N/mm², the margin remains constant at 10 N/mm² and $x_t = 40 + 10 = 50$ N/mm².

If the coefficient of variation remains constant at 0.203, the multiplying factor remains at 1.5 and $x_t = 40 \times 1.5 = 60$ N/mm².

It is obvious that in these circumstances the mix based on the coefficient of variation requires more cement than that based on the standard deviation.

If the parameter chosen is to be a measure of the physical control of the production process, then it should remain constant irrespective of the mean level of production. Thus for concrete production using the same materials, batching and mixing plants and the same supervisory staff then either the standard deviation or the coefficient of variation should remain constant over a range of different strength levels. Unfortunately, there are little data covering this situation under practical site conditions. Laboratory work has been done by Neville [8] and Teychenné [9] which shows that the coefficient of variation is independent of the strength while the standard deviation increases with the strength. However, these tests were made under laboratory conditions where the variation in the raw materials and in batching was negligible and the tests were therefore confined mainly to the measure of the variation in sampling and testing at different strength levels. Since these conditions are not obtained on normal sites it would be incorrect to apply these conclusions to the wider variation obtained under site conditions.

It has been suggested that the standard deviation is independent of the mean strength of concrete, but this cannot be true at very low strengths since at zero strength the standard deviation must also be zero. In 1953 Murdock [10] suggested that the standard deviation increased linearly with the strength from the origin to a limiting value of strength and then remained constant irrespective of the strength. From a large survey of many jobs Erntroy [3] concluded that the relationship between the standard deviation and strength can best be represented by a smooth curve through the origin tending to become a horizontal straight line for high values of mean strength. This type of relationship has also been established by Rusch [4] and Teychenné [9] and the results of large scale surveys made in Europe and the USA are reviewed in a paper by Newlon [11]. Figure 1 which is reproduced from Newlon's paper shows the mean relationships obtained from various surveys.

One disadvantage of such surveys is that they tend to be " historical " in that they assemble data

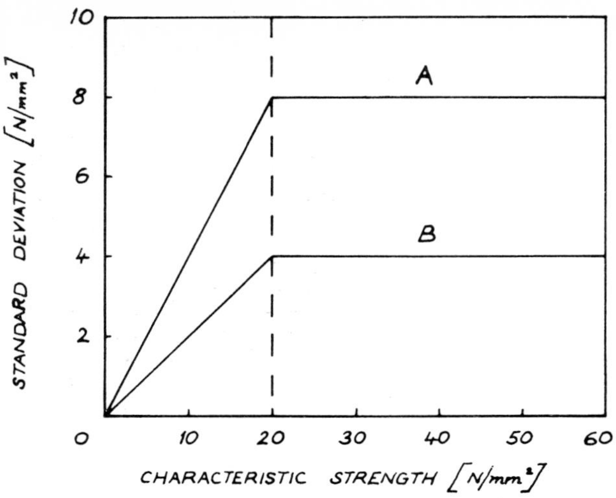

Fig. 2. — Relationship between the standard deviation and the characteristic strength.

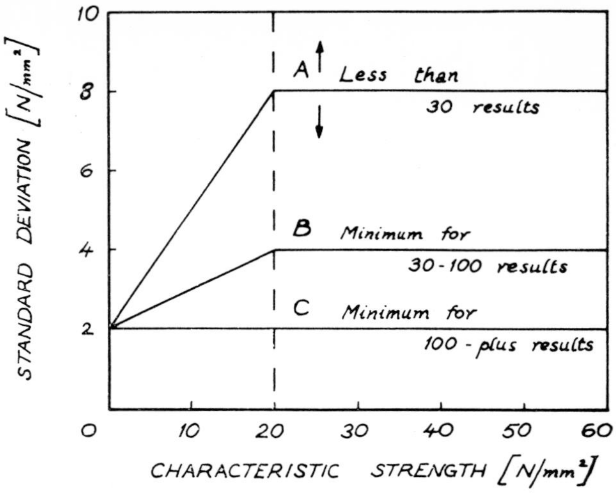

Fig. 3. — Proposed relationship between the standard deviation and the characteristic strength for use in codes.

after the work has been completed without full knowledge of the working conditions. However, these surveys have provided much useful information both on the magnitude of the standard deviation and its relationship to the strength of concrete such that the Working Group can make the following recommendation :

> When produced over a considerable period of time, the variability of the strength of concrete should be expressed as the standard deviation. The relationship between the standard deviation and the characteristic strength is as shown in figure 2, i.e. linearly variable from 0 to a specified strength (20 N/mm²) and constant for characteristic strengths above 20 N/mm²."

It should be noted that in figure 1 the standard deviation is plotted against the mean strength and that the various authors show that it reaches a constant value at a mean strength of between 20 N/mm² and 35 N/mm². In figures 2 and 3 the standard deviation is plotted against the characteristic strength and it is proposed that the strength above which the standard deviation remains constant is 20 N/mm². The magnitude of the constant standard deviation for characteristic strengths greater than 20 N/mm² depends upon the quality control applicable at the site, thus line *A* in figure 2 applies to a poorly controlled site and line *B* to a well controlled site. Most sites have standard deviations between these extremes as discussed in the next section.

4. THE MAGNITUDE OF THE STANDARD DEVIATION

Most structural concretes have a characteristic strength greater than 20 N/mm² and there are many data on the overall site variation which is expressed by some authors as a coefficient of variation rather than a standard deviation. A summary of the range of standard deviations quoted by various authors is given in Table I.

There is considerable scatter in the range of standard deviations quoted by each author and in the mean levels between different authors. It should be noted that lower standard deviations are quoted by USA and Canadian authors. It is not known whether this is due to the fact that the tests were made on cylinders instead of cubes, as is the European practice, or that better control is in fact obtained.

Metcalf [12] has analysed Erntroy's [3] data and the results of a more recent survey to show the distribution of jobs having different levels of standard deviations and Table II is derived from his work. There is little doubt that under most practical conditions the standard deviations obtained are considerably higher than the minimum requirements given in the CEB-FIP International Recommendations of June 1970 [7]. The replies to a questionnaire sent to the members of the Working Group indicated that they considered that the values quoted, 2 N/mm² for factory cast concrete and 3 N/mm² for in situ concrete were too low, and it was further thought that the different values quoted for different types of work could not, in general, be justified.

Many attempts have been made to correlate the variability (expressed either as the standard deviation or the coefficient of variation) with the type of work, the method of concrete production etc., so that an estimate could be made of the standard deviation to be expected on any new site. An interesting analysis of site variability has been made by Rusch, Sell and Rackwitz [13] which showed that one group of types of work had a mean standard deviation of about 5 N/mm² while another group had a lower mean value of about 4 N/mm². However, there was considerable variation of the standard deviation from job to job, even of the same type as is shown in the last column of Table I, and there is thus little reliance in predicting the standard deviation for a particular type of work.

The Working Group considers that it is not practicable to specify different standard deviation for different types of work, or for different types of production such as in situ, precast and ready mixed concrete.

However, full use should be made of existing data obtained from a particular concrete production plant in order to estimate the possible standard deviation to be obtained for future concrete production. If insufficient data are available, a conservative value for the standard deviation should be specified.

The Working Group recommends that for the purpose of fixing the margin for the concrete mix design and for establishing the rules for compliance in Codes of Practice, "the value of the standard deviation for normal concrete production should be derived from figure 3.

(a) If there are no previous data, or if there are less than 30 results obtained using the same plant, source of materials and supervision, the standard deviation shall be taken from curve A. Thus for characteristic strengths equal to or greater than 20 N/mm² the standard deviation is 8 N/mm² (see note).

(b) If there are between 30 and 100 results, the standard deviation shall be the value obtained but not less than that given by curve B, i.e. for characteristic strengths equal to or greater than 20 N/mm², the minimum standard deviation is 4 N/mm².

(c) If there are more than 100 results the standard deviation shall be the value obtained but not less than 2 N/mm² i.e. curve C ".

Note : the value of 8 N/mm² may be varied from 6 N/mm² to 10 N/mm² to meet National conditions.

Table I. — Standard deviations obtained under various conditions.

Author (shape of specimens)	Type of job or level of control	No. of jobs	Mean or range of crushing strengths at 28 days N/mm²	Standard deviation	
				Mean N/mm²	Range N/mm²
Dreux (cylinders)	Laboratory	–	9 – 52	–	0.5 – 4.6
	Ready mixed concreete	10	35 – 54	–	3.4 – 6.3
	Site mixed	15	36 – 47	–	3.1 – 7.1
Erntroy (cubes)	Laboratory a	13	29 – 63	2.1	1.4 – 2.8
	Laboratory b	9	25 – 50	3.9	2.4 – 4.8
	Laboratory c	10	12 – 55	4.1	2.8 – 5.5
	Control Aa	42	17 – 45	4.6	2.8 – 7.4
	Control Ab	104	10 – 60	5.5	2.6 – 10.0
	Control B	182	17 – 60	5.5	2.9 – 10.0
	Control C	127	29 – 57	6.0	2.8 – 9.8
	Control Da	6	29 – 47	6.5	3.8 – 7.5
	Control Db	20	21 – 39	4.9	3.2 – 7.4
Newlon (cylinders)	ACI 214	265	14 – 48	–	2.6 – 3.6
	Pavement	4	25 – 33	–	3.2 – 4.2
	Pavement	5	23 – 26	–	2.8 – 3.8
	Pavement	30	24 – 38	–	2.4 – 5.6
	Structural	4	30 – 33	–	3.0 – 5.3
	Prestressed	4	41 – 47	–	2.2 – 3.6
Philleo (cylinders)	Ready mixed concrete	2	14 – 17	2.6	–
		2	17 – 21	3.2	–
		19	21 – 24	2.7	–
		54	24 – 28	2.8	–
		89	28 – 31	3.2	–
		58	31 – 34	3.6	–
		21	34 – 38	3.2	–
		16	38 – 41	3.6	–
		3	41 – 45	3.0	–
Plewes (cylinders)	Consultants	35	28	3.0	–
	Ready mixed concrete	23	28	2.7	–
	Contractors	15	31	3.3	–
	Others	20	39	3.0	–
Rusch (cubes and cylinders)	Mass concrete	63	32	4.3	2.3 – 7.2
	Roads and Runways	39	44	5.2	3.9 – 7.6
	Bridge supports	56	45	5.2	2.6 – 9.3
	Normal structures	47	27	5.1	1.1 – 10.2
	Multi-storey flats	87	26	4.4	1.5 – 7.9
	Halls and factories	63	32	4.9	1.6 – 8.3
	Tunnels	25	33	4.0	2.3 – 6.2
	Precast	48	54	4.8	1.9 – 9.0
	Ready mixed concrete	71	33	3.8	1.1 – 6.3
Soroka (cubes)	Ready mixed concrete	4	20 – 65	–	2.5 – 8.5
	Site mixed	4	20 – 65	–	2.5 – 8.5
Teychenné (cubes)	Laboratory	3	16 – 42	–	0.6 – 1.6
	Building sites	29	29 – 43	–	3.5 – 10.0

Figure 3 differs from figure 2 in that the Working Group considered that an overall minimum standard deviation of 2 N/mm² should apply, and thus the lines for the standard deviation of concrete of strength less than 20 N/mm² originate from a standard deviation of 2 N/mm² rather than from the origin.

The three lines *A*, *B* and *C* are given to allow for differences in the amount of previous test data available. The standard deviation calculated from these data is only an estimate of the true standard deviation and is thus subject to error depending on the number of results from which it is calculated. Thus for less than 30 results it is advisable to assume a relatively high standard deviation (line *A*) of about 8 N/mm², note the high values shown in Table I, at the start of a new job until sufficient data have been obtained from the actual job on which to calculate a standard deviation. If the standard deviation is calculated from 30-100 results the value can be used provided that it is not less than that indicated by line *B*. If the standard deviation is calculated from 100 results there is only a 5 % probability that it differs from the true standard deviation by more than about 14 %. Since greater reliability can be placed on the value calculated from more than 100 results the minimum value to be used is allowed to fall to line *C*, i.e. 2 N/mm².

5. THE PROPORTION OF DEFECTIVE CONCRETE

Until recently, many Standards and Codes have been based on a minimum strength principle, with the implication that no concrete should be produced below this strength and that no test results should be obtained below this strength.

A study made by Metcalf [12] of about 500 jobs in the United Kingdom showed that 10 % of the jobs contained 10 % or more of defective concrete, and a survey by Plewes [14] of about 90 jobs in Canada showed that 18 % contained more than 10 % defective concrete. The level of the defective concrete that is permitted must be specified so that the target mean strength for mix design can be calculated from the equation :

$$f_m = f_c + kS$$

where f_m and f_c are the target mean strength and the characteristic strength respectively,

k is a constant depending on the proportion of defectives permitted.

S is the standard deviation.

Modern Codes follow the characteristic strength philosophy which permits a proportion of the concrete to have a strength below the specified characteristic strength. In this case it is inevitable that strength test results less than the specified value, often termed " defectives " will occur. These individual results should not be regarded as " failures " and the concrete in the structure that they represent should not be considered for rejection. However, new requirements to check the compliance of the concrete with the specified characteristic strength have to be established.

The characteristic strength can be defined to allow any agreed proportion of defectives sometimes called

" the defective fractile ", and various Codes and Specifications have been written with the defective rate varying from 1 % to 20 %. The new CEB-FIP Recommendations [7] specify the characteristic strength as having a 5 % defective rate and this value is now being incorporated in other European Codes of Practice.

6. THE DISTRIBUTION OF THE TOTAL VARIABILITY INTO CONSTITUENT COMPONENTS

Item 1 of the terms of reference listed five groups of factors for consideration; these can be regrouped as follows :

(1) Length and size of job (a).
(2) Materials variations (e).
(3) Batching variations (b and d).
(4) Sampling and testing (c).

One of the best analyses concerning item 1, was made by Rusch et al. [13].

Soroka [15] has examined the differences between daily and long term variation, and Metcalf [12] has estimated that from UK data the average within day standard deviation is 2.6 N/mm² while the day-to-day and longer term average standard deviation is 4.1 N/mm². However, it is not possible to make recommendations that generally apply.

So far as the other three groups of factors are concerned, i.e. materials, batching and testing, there are many data available but generally these are not in a form in which they can be applied to practical concrete production. For example, it is known that the strength of cement at 28 days varies considerably, but the figures usually relate to laboratory tests which are often made on mortar specimens such as the RILEM-Cembureau test rather than on concrete specimens as may be used in the British Standard for cement. Thus Bloem [16] has shown that cement from one works in the USA can have a yearly standard deviation of from 1.0 N/mm² to 3.5 N/mm², similar values have been quoted by Philleo who also reports standard deviations between different works of from 1.6 N/mm² to 7.2 N/mm². Petersons [16] has shown that Swedish cement can have a yearly standard deviation of from about 2 N/mm² to 6 N/mm² according to the works. Information from Israel shows that the standard deviation of the cement strength in concrete tests is about 2.5 N/mm², while in the UK the standard deviation from one works is about 3 N/mm² and between works about 5 N/mm².

There is also considerable evidence on the way that aggregates affect the strength of concrete. Newman and Teychenné [17] have shown how changes in the grading of sand can alter the strength of concrete by up to 9 N/mm² and other tests have shown that changing the type of natural aggregate can alter the strength of concrete by about 10 N/mm². However, these investigations are carried out in laboratories and there is little evidence of how changes in the grading, type or shape of aggregates affect the strength of concrete under practical site conditions.

Many Codes and Standards include workmanship Clauses dealing with the production of concrete and these often state requirements for the accuracy of the batching plant. These requirements generally call for the batching plant to have accuracies of the order of ± 1 or 2 % and these can readily be checked, however, it is more difficult to specify the accuracy of the operator and even more difficult to measure the combined plant and operator accuracy. There are indications that the variability from this source can be reduced by the use of automatic batching plants. As a result of the analysis of fresh concrete there is evidence that there are considerable variations in the actual proportions used in repeated batches.

Table II. — Distribution of standard deviations (derived from Metcalf [12]).

Standard Deviation N/mm²	Proportion of jobs with a standard deviation less than the stated value. Per cent			
	Metcalf's Survey	Erntroy's Survey	Highway Structures	
			Site mixed	Ready mixed
3	3	1	3	0
4	18	9	11	2
5	50	30	37	14
6	77	65	75	63
7	88	86	90	83
8	96	98	98	91

If the overall site variability is to be reduced, it would be helpful to know which group of factors contributes most to the total standard deviation. It is apparent from the literature survey and from the experience of the members of the Working Group that there are at present no site data on which to base any firm conclusions. However, the members of the Working Group were asked to estimate the distribution of the total variability, at three different levels of variability, 2.5, 5 and 7.5 N/mm² into the three groups of factors :

1. Materials (Quality of cement, gradings of aggregates, etc.).
2. Batching (Type of plant, workmanship and supervision).
3. Testing (Sampling, making, curing and testing concrete specimens).

These estimates are given in Table III, and it should be remembered that the total standard deviation S is not the sum of the component standard deviations but that,

$$S^2 = S_m^2 + S_b^2 + S_t^2$$

where S_m, S_b and S_t are the standard deviations due to materials, batching and testing respectively.

Some of the members of the Working Group commented that it was difficult to obtain a total standard deviation as low as 2.5 N/mm² and this is confirmed by the values given in Tables I and II. Estimator No 3 noted that with poor sampling and testing procedures it would be impossible to obtain a total standard deviation as low as 2.5 N/mm² and that this greatly alters the relative significance of the various factors at higher levels of variability as shown in Table III for estimator 3 A and 3 B.

Generally, there is considerable information concerning the contribution due to sampling and testing

Table III. — Estimated distribution of total standard deviation into components.

Estimator	Component	Standard Deviation N/mm²		
		2.5	5.0	7.5
1	Materials Batching Testing	1.0 2.0 1.5	2.0 4.5 1.5	3.1 6.5 2.0
2	Materials Batching Testing	0.7 1.8 1.6	3.9 2.7 1.6	6.5 3.5 1.6
3A	Materials Batching Testing	2.0 0.9 1.2	2.5–4.5 4.0–2.0 1.4	2.5–5.5 7.0–5.0 1.4
3B	Materials Batching Testing	- - -	2.3 2.0 4.0	3.5 4.5 4.0
4	Materials Batching Testing	1.5 1.3 1.5	3.4 3.4 1.5	4.0 6.0 2.0
5	Materials Batching Testing	1.0 2.0 1.1	1.4 4.7 1.1	1.7 7.2 1.1
6 Concrete 20	Materials Batching Testing	1.2 1.7 1.4	1.6 4.5 1.3	2.4 6.9 1.7
6 Concrete 40	Materials Batching Testing	1.1 1.7 1.4	1.6 4.5 1.5	2.7 6.7 2.0
7	Materials Batching Testing	1.1 1.8 1.4	2.2 3.6 2.7	3.3 5.3 4.1

when this is properly carried out. The standard deviation is generally in the order of 1.5 N/mm² - 2.0 N/mm² and increases slightly as the total variation increases but plays a small part in the total variation. Estimator No 7 considers that poor testing makes a major contribution to the high overall variation similar to 3 B.

At the higher levels of variability there is a majority opinion that more of the variation is due to batching errors than to variation in the quality of the materials, and this view is supported by a theoretical analysis by Rackwitz [5]. However, estimator No 2 considered that material variation contributed more to the high overall variation than did batching errors and estimator No 3 A allowed for a variety of circumstances. It is possible that the contribution due to batching has been overestimated, and that due to materials under-estimated. It is unlikely that the standard deviation due to variations in the cement quality will be less than 2 N/mm² and values much higher than this are reported at the beginning of this section.

7. METHODS OF CONTINUOUS CONTROL

The Working Group has not considered this topic in detail but it would appear from the opinion expressed during its meetings, and from the estimated distribution of the total variation that the major factor giving rise to high variability is batching errors. Thus if concrete is to be better controlled, improvements are required in the batching process, either to the plant used or to the supervision of the concrete production. It is interesting to note that in 1960 Erntroy [3] stated, " It is concluded that the degree of supervision on the site has a more important effect than the type of equipment employed ".

A clear distinction should be made between tests for control and tests for compliance. Unfortunately the terms " quality control " or " statistical control " have for some time been applied to the analysis of the results of cube tests which are usually carried out at an age of 28 days. In reality such testing is to see whether the concrete complies with the specified value. These tests may result in changes being made to the mix proportions because of failure to meet the specified requirements or because the concrete is too strong, but since such changes will be made at least 28 days after the concrete was produced they can hardly be said to control the concrete quality.

In order to control the concrete production effectively measures should be specified such that an early feed-back can be made to the actual production of the concrete. Such measures can include rigorous workmanship requirements in Codes dealing with such matters as the storage and handling of aggregates, regular frequent checks on the batching plant and the need to have trained operatives and supervisors. Any testing that is done should be such that the tests can have a very early effect on the mix proportions. These include test data from the cement manufacturers on the quality of the cement, data from the aggregate producers or on-site tests to measure the aggregate grading and its moisture content, workability tests on the concrete, and rapid analysis of fresh concrete. If strength data are required these can be obtained by accelerated curing techniques but generally it is only possible to apply the results to the concrete production about 1 day later.

8. CONCLUSIONS AND RECOMMENDATIONS

The Working Group recommends that :

1. The variability of the strength of concrete should be expressed as the standard deviation.
2. The relationship between the standard deviation and the characteristic strength is as shown in figure 2, i.e. linearly variable from zero to a specified strength (20 N/mm²) and constant for characteristic strengths above 20 N/mm², the value depending on the production control.
3. The minimum standard deviations quoted in the CEB/FIP Recommendations of 2 N/mm² and 3 N/mm² are too low for general application, and there is no justification for the different values quoted for different types of work in practice.
4. For the purpose of fixing the margin for the concrete mix design and for establishing the rules for compliance in Codes of Practice, the value of the standard deviation for normal concrete production should be derived from figure 3.

   (a) If there are no previous data, or if there are less than 30 results obtained using the same plant, source of materials, and supervision, the standard deviation shall be taken from curve *A*. Thus for characteristic strengths equal to or greater than 20 N/mm² the standard deviation is 8 N/mm² (see note).

(b) If there are between 30 and 100 results, the standard deviation shall be the value obtained but not less than that given by curve *B*, i.e. for characteristic strengths equal to or greater than 20 N/mm² the minimum standard deviation is 4 N/mm².

(c) If there are more than 100 results, the standard deviation shall be the value obtained but not less than 2 N/mm², i.e. curve *C*.

Note : The value of 8 N/mm² may be varied from 6 N/mm² to 10 N/mm² to meet National conditions.

5. At present it is not possible to establish reliable figures for the contribution that the components of (a) materials, (b) batching and (c) sampling and testing make towards the total standard deviation. When sampling and testing are not carried out in accordance with a strict specification, this factor can make a major contribution to the total variation. It is therefore essential that the sampling, making, curing and testing of specimens should be carried out in accordance with well-defined standards. The development of Internationally agreed test methods is thus most important.

At high levels of standard deviation it is probable that the major source of the variation is due to batching errors, but the variation in the materials, particularly the cement, could be of considerable significance.

9. ACKNOWLEDGMENT

The author acknowledges the assistance given by other members of the Working Group in the production of this report. The work described has been carried out as part of the research programme of the Building Research Establishment of the Department of the Environment and this paper is published by permission of the Director.

REFERENCES

[1] Building Research Station, Library Bibliography 215. — *Quality control of concrete* 1969 *with addendum* 1970.

[2] ISO Recommendation R 1920. Concrete Tests. — *Dimensions, tolerances and applicability of test pieces.* 1st Edition 1971.

[3] Erntroy H.C. — *The variation of works test cubes.* Research Report No. 10 Cement and Concrete Association. London 1960.

[4] Rusch H.H. — *Zur statistichen Qualitatskontrolle des Betons.* Materialprufung, November 1964.

[5] Rackwitz R. — *Zur Streung der Betondruckfestigkeit von Wurfelproben.* Beton 1971 (2).

[6] C P 114 : 1957. — *The structural use of reinforced concrete in building.* British Standards Institution. London.

[7] CEB-FIP. — *International recommendations for the design and construction of concrete structures.* Prague 1970.

[8] Neville A.M. — *The relation between standard deviation and the mean strength of concrete test cubes.* Magazine of Concrete Research Vol. 11, No. 32, July 1959.

[9] Teychenné D.C. — *The variability of the strength of concrete and its treatment in Codes of Practice.* Structural Concrete. Vol. 3, No. 1, January/February 1966.

[10] Murdock L.J. — *The control of concrete quality.* Proceedings of the Institution of Civil Engineers. Part 1, Vol. 2, No. 4, July 1953.

[11] Newlon H.H. — *Magnitude of concrete variability.* US-Japan Joint Seminar on Research on Basic Properties of Various Concretes, Tokyo, Japan, January 1968.

[12] Metcalf J.B. — *The specification of concrete strength,* Part 2. *The distribution of strength of concrete for structures in current practice.* Road Research Laboratory Report LR 300. Crowthorne 1970.

[13] Rusch H., Sell R. and Rackwitz R. — *Statistische Analyse der Betonfestigkeit.* Deutscher Ausschub fur Stahlbeton, H 206, Berlin 1969.

[14] Plewes W.G. — *A Canada wide survey of concrete cylinder test results.* National Research Council, Division of Building Research 1968.

[15] Soroka I. — *On compressive strength variation in concrete.* RILEM Bulletin Materials and Structures, Vol. 4, No. 21, May/June 1971.

[16] Bloem D.L. — *Quality control of concrete.* US-Japan Joint Seminar on Research on Basic Properties of Various Concretes. Tokyo, Japan, January 1968.

[17] Petersons N., Hellstrom B. and Hard R. — *Erfarenheter fran bestibtning av auktoriserade betonfabriker.* Cement-och Betonginstitutet, Stockholm. Report No. 6828, 1968.

[18] Newman A.J., Teychenné D.C. — *A classification of natural sands and its use in concrete mix design.* Proceedings of a symposium on Mix Design and Quality Control of Concrete. Cement and Concrete Association. London, May 1954.

OTHER PAPERS PRESENTED

SOROKA I. — *An application of statistical procedures to quality control of concrete.* Matériaux et Constructions, Vol. 5, No. 1, January/February 1968.

DREUX G. — *Contribution to the study of the statistical control of concrete.* BRS Library Archive No. 494.

DREUX G., GORISSE F. — *Vibration, ségrégation et ségrégabilité des bétons.* Annales de l'Institut Technique du Bâtiment et des Travaux Publics, janvier 1970, n° 265.

PLUM N.M. — *Quality control of concrete, its rational basis and economic aspects.* Proc. Institution of Civil Engineers, Vol. 2, 1953.

TSO W.K., ZELMAN I.M. — *Concrete strength variations in actual structures.* Canada Emergency Measures Organisation, Ottawa, October 1968.

DESOV A.C., MALINOVSKII A.G. — *Concrete strength variability control.* USSR, BRS Translation. Library Communication 1581, 1970.

HODE KEYSER J. — *Provisional standard for quality control.* Beton du Quebec, February 1967.

BLAUT M. — *Recommendations for quality control by the manufacturer.* BRS Library translation. Library communication 1635.

METCALF J.B. — *The specification of concrete strength, Part 1, The statistical implications of some current specifications and Codes of Practice.* RRL Report LR 299, Crowthorne 1970.

MATTHEWS D.H., METCALF J.B. — *The specification of concrete strength, Part 3. The design of acceptance criteria for the strength of concrete.* RRL Report LR 301. Crowthorne 1970.

DREUX G., GORISSE F. — *Contribution à l'étude statistique des contrôles du béton - mise au point d'un test accéléré.* ITBTP Conference, 9 Mai 1972.

BLAUT M. — *Statistiche Verfahren für die Gutesicherung von Beton.* Bauverlag GMBH Weisbäden und Berlin 1968.

SOROKA I. — *Length of concreting period and compressive strength variation in concrete.* Building Research Lecture, Israel Institute of Technology, Haïfa 1972.

The effect of rate of loading on plain concrete (CP 23/73)

P.R. Sparks and **J.B. Menzies**

SYNOPSIS

The paper describes an experimental investigation into the effect of the rate of application of steadily increasing loads upon the static strength and of fluctuating loads upon the fatigue strength of plain concretes in compression. Concretes made with gravel, limestone or Lytag aggregate were tested. Increasing the rate of application of load was found, in general, to enhance the static strength and fatigue strength of all three types of concrete. The increase in static strength was found to be greatest in concrete with a relatively weak and soft aggregate such as Lytag and least in a concrete with relatively strong and stiff aggregate such as limestone. In the fatigue tests, a strong correlation was found between the rate of strain increase per cycle during most of the life of a specimen and the endurance of the specimen, irrespective of the rate of application of the load.

Introduction

Engineers designing concrete structures in the past have not considered explicitly the time-dependent properties of concrete except in special circumstances such as in calculations of loss of prestress. From the engineering standpoint, it was expedient to obtain values for static strength and stiffness from standard tests on small specimens made at convenient rates of loading. The properties of the concrete in real structures under live loads were assumed to be the same as those measured in the standard tests on small specimens. The time-dependent creep properties associated with permanent loads were allowed for in design by simple techniques such as the use of a reduced stiffness for the concrete. Design values for fatigue strength were required for certain special structures and it was found necessary to obtain these by accelerated tests because tests at realistic rates would take too long to complete.

Essentially the procedure of basing design properties of concrete upon simple tests on small specimens coupled with simple techniques to allow for time-dependent effects is still in use. However, more realistic methods of design, incorporating the time-dependent properties of concrete, have gradually been introduced. It is becoming increasingly important, therefore, for the structural engineer to understand the influence of time-dependent effects upon the strength and behaviour of concrete in relation not only to complete structures but also to the tests on small specimens which are used to obtain design information.

Many investigators have reported that the static strength of concrete in compression when tested monotonically to failure is sensitive to the rate of application of the load over a wide range of rates[1]. Evans[2], however, when testing at rates in the range 10^{-1} to 10^{2} N/mm^2 s found no significant change in strength; at higher rates, he did detect an increase in strength with an increase in the rate of loading. It is not possible to determine from the literature the nature of the concretes tested, although Evans did report that 'weak' concrete appeared to be more sensitive to the rate of application of load than 'strong' concrete. Recently Spooner[3] reported tests in which concrete was loaded at one of two predetermined strain rates. An average decrease of 4% was observed for dry gravel concrete when the time to failure was increased from 5 min to between 15 and 20 h. The nature of the results, however, led Spooner to conclude that this difference was not statistically significant. When the concrete was water-cured and tested wet, this difference was found to increase to about 12%.

A logical deduction from the fact that the static strength is sensitive to rate of loading is to conclude that the fatigue strength of concrete should also be sensitive to the rate of application of a fluctuating load. This hypothesis, however, has rarely been tested. For expediency, tests to determine the fatigue strength of concrete are normally carried out at as high a rate as possible or, in the case of fixed-speed testing machines, at a rate dictated by the machine. These rates are normally well in excess of those experienced by civil engineering structures. Murdock[4], in a review paper in 1965, concluded that there was no significant rate effect for the fatigue strength of plain concrete in compression. This conclusion was drawn from work by Gray, McLaughlin and Antrim[5], who carried out tests at 8·3 and 16·6 Hz. It is unlikely that a mere doubling of the loading rate would have produced a significant difference in the fatigue life of the material, especially when the loading took the form of a sinusoidal wave in which the actual rate of loading was continuously varying throughout each cycle.

In the present investigation, static and fatigue tests in uniaxial compression have been carried out on concrete prisms made with gravel, limestone or Lytag aggregate. The prisms were loaded to failure in the static tests at rates between about 10^{-3} and 10 N/mm^2 s. A triangular wave-form was employed in the fatigue tests so that, within each cycle, the loading and unloading rates remained constant. These tests were performed at a number of different maximum-load levels, and the frequency of the loading was adjusted in each case so that the rate of loading was either 0·5 or 50 N/mm^2 s.

Test specimens and loading apparatus

The specimens used in the investigation were 102 × 102 × 203 mm rectangular concrete prisms cast in steel moulds. The specimens were cast generally in batches of 24 prisms together with 27 control cubes of 100 mm side. An electrical-resistance strain gauge was cast into each prism. In the case of the prisms of Lytag concrete required for the fatigue tests, some difficulty was experienced in preventing the strain gauges from sinking in the wet concrete. Measures taken to prevent this resulted in a slower casting procedure and it was necessary to cast the required number of specimens in two batches of 12 prisms with 15 control cubes with each batch. A total of 48 prisms of each type of concrete were cast, half for the static tests and half for the fatigue tests. After casting, each batch of prisms and cubes was stored for 24 h under damp sacking. They were then demoulded and stored under water at 18·3°C for 7 days. At the end of that time all the prisms and all but three of the cubes were removed to the test laboratory. The remaining cubes were kept under water until tested, usually at an age of 28 days. Details of the aggregates and concrete mixes are given in Table 1. The 28 day cube strengths were of the order of 30 N/mm^2 for the gravel and limestone concretes and about 20 N/mm^2 for the Lytag concrete.

The longitudinal axial strain in each concrete prism was measured by means of the electrical-resistance strain gauge embedded in the centre of it. The strain gauges were of 60 mm gauge length and consisted of a straining wire embedded in a polyester resin of stiffness similar to that of concrete. The surface of the resin was roughened to ensure good bond between the gauge and the concrete. A temperature-compensating gauge was embedded similarly in a concrete prism. Some small variations of ambient temperature in the laboratory were inevitable, but it was considered that these changes would not significantly influence the results of the tests.

One channel of a servo-hydraulic testing facility was used to load the prisms. A 450 kN hydraulic actuator reacting against a prestressed steel and concrete frame applied the load to the prisms through steel brush platens (Figure 1). These platens reduced the lateral restraint on the ends of the prisms and produced a near-uniform uniaxial stress in the concrete. The actuator was controlled by the closed-loop servo system. The loading rate in the static tests was determined by using a ramp function generator for all but the slowest rates of loading. At the slowest rates, a drum programmer generated the control signal for the

TABLE 1: **Details of aggregates and concrete mixes.**

Type of coarse aggregate	Concrete mix proportions by weight				Effective water/cement ratio*
	Ordinary Portland cement	Fines		Coarse aggregate	
		Sand	Lytag		
Ham River gravel	1	4·0†	—	4·0	0·70
Cheddar limestone	1	2·4	—	3·6	0·72
Lytag	1	—	2·4	3·6	0·52‡ 0·66‡

*The aggregates used in these mixes were laboratory-dried aggregates and, as such, absorbed a proportion of the mixing water. In determining the effective water/cement ratio, it was assumed that the amount of water absorbed by the aggregates was 1·3% by weight for gravel, 0·5% for limestone and 13% for Lytag.

†To meet the strength requirement, a higher fines content was necessary in this mix.

‡The batch of specimens used for the static tests was made with an effective water/cement ratio of 0·52; for the fatigue specimens, the ratio was 0·66.

Figure 1: Loading apparatus.

loading. In the fatigue tests, the triangular wave-form of the loading signal was generated by the function generator.

Experimental technique

For the static tests, the prisms were loaded to failure at rates in the range 10^{-3} to 10 N/mm² s. This represented times to failure from about 1 s to 3 h. Six or more rates of loading were chosen within the range and generally three prisms from each batch were tested at each rate. The active strain gauge in each prism and the temperature-compensating gauge were connected to a bridge amplifier. The output from this and from a load-cell mounted on the hydraulic actuator (Figure 1) were fed into an ultra-violet galvanometer recorder to give a plot of the relationship between the stress and strain in the prism as the load was increased at the predetermined rate. The output from the load-cell was also recorded, either on an X-Y plotter or on a sampling digital voltmeter, to enable the maximum load to be determined. At the same time as the prisms were being loaded, the control cubes which had been cast and cured under conditions similar to those for the prisms under test were loaded to failure at the standard rate of loading specified in the British Standard BS 1881[6], i.e. 0·25 N/mm² s.

Before the fatigue tests on each batch of prisms were carried out, six prisms were loaded statically to failure, two at a rate of 0·5 N/mm² s, two at 5 N/mm² s and two at 50 N/mm² s. These tests were performed to ensure compatibility with the results of the static tests on the other batch of prisms made with the same aggregate. The fatigue tests were carried out at loading and unloading rates of either 0·5 or 50 N/mm² s. For the purpose of these tests, the static strength (S_m) of the concrete was taken as that of a prism loaded at the logarithmic mean rate, i.e. 5 N/mm² s. Fatigue tests were then performed with maximum load levels between 90% S_m and 70% S_m. The minimum load level remained constant throughout the tests on each batch at one-third of the estimated static strength of the specimens at the rate of loading specified in BS 1881[6].

The frequency of the triangular-wave loading applied to the prisms varied from prism to prism, depending upon the level of the maximum load in the cycle. The frequencies were adjusted to obtain the required rates of loading and unloading of either 0·5 or 50 N/mm² s. Tests were continued until failure occurred or a time limit, normally two weeks, was reached. If the prism had not failed by this time, the test was stopped and usually the prism was made to fail under incremental static loading. Since the prisms had been stored in the laboratory for at least three months before they were tested, no special provision was made to inhibit the small changes in moisture content likely to occur during the fatigue tests. The instrumentation used in these tests was similar to that used in the static tests. An ultra-violet trace record was kept of the output of the embedded strain gauge during each test. For the short-lived prisms this was a continuous record, but for the longer-lived prisms a sampling procedure was adopted.

To monitor any changes in strength due to ageing and variations in compaction and curing within the batches of prisms for fatigue testing, a standard crushing test was carried out on a control cube at the time of each fatigue test.

Results

The prisms tested statically and those tested in fatigue failed similarly by longitudinal splitting along their whole length. This mode of failure, a typical example of which is shown in Figure 2, is characteristic of failures expected in a uniaxial stress field and indicated the effectiveness of the brush platens.

The results of the static tests are presented in Figure 3 in the form of a relationship between the strength of the prism, expressed as a percentage of the strength of the control cubes stored with the prism, and the rate of loading on a logarithmic scale. Inspection of the plotted information suggested that the data for each type of concrete lay strictly on a shallow curve. However, in all but one case, a straight line with a positive slope proved to be a satisfactory approximation. The one exception was the data for the Lytag concrete at rates of loading below about 3×10^{-2} N/mm^2 s, which was best represented by a line of negative slope. Regression analyses of static strength (F), expressed as a percentage of the appropriate cube strength, on log rate of loading ($\log_{10}R$), were made for the remaining

Figure 2: Prism after failure between brush platens.

data for the Lytag concrete and for all the data for the other types of concrete. The regression analyses yielded the relationships

$F = 67{\cdot}89 + 3{\cdot}96 \log_{10} R$ for gravel concrete

$F = 73{\cdot}17 + 1{\cdot}90 \log_{10} R$ for limestone concrete

and

$F = 87{\cdot}58 + 7{\cdot}67 \log_{10} R$ for Lytag concrete

which are drawn through the plotted points on Figure 3. Statistically the gradients of these lines are all positive and significantly different at the 95% confidence level.

Typical examples of the relationships between stress and strain in the prisms tested statically are given in Figure 4. At a given rate of loading, the limestone concrete is the stiffest, the Lytag concrete is the least stiff and the stiffness of the gravel concrete is between the two extremes. All three types of concrete have a greater stiffness at the fastest rate of loading than at the slowest rate.

The fatigue lives of the prisms are shown in Figures 5, 6 and 7 plotted against the maximum load level expressed as a percentage of their static strength. These static strengths were assumed to be the mean strengths at a loading rate of 5 N/mm^2 s as described above. Some scatter of the fatigue results is shown in Figures 5, 6 and 7, as would be expected from experience. The experimental information is not sufficient to allow the mean curve for the population to be estimated with confidence. However, if the mean is taken of the log (number of cycles) for each group of specimens tested at the same load, there is a clear indication for all three types of concrete that the fatigue life was enhanced at the fast rate of loading and unloading.

During the fatigue tests a record of the strain life of

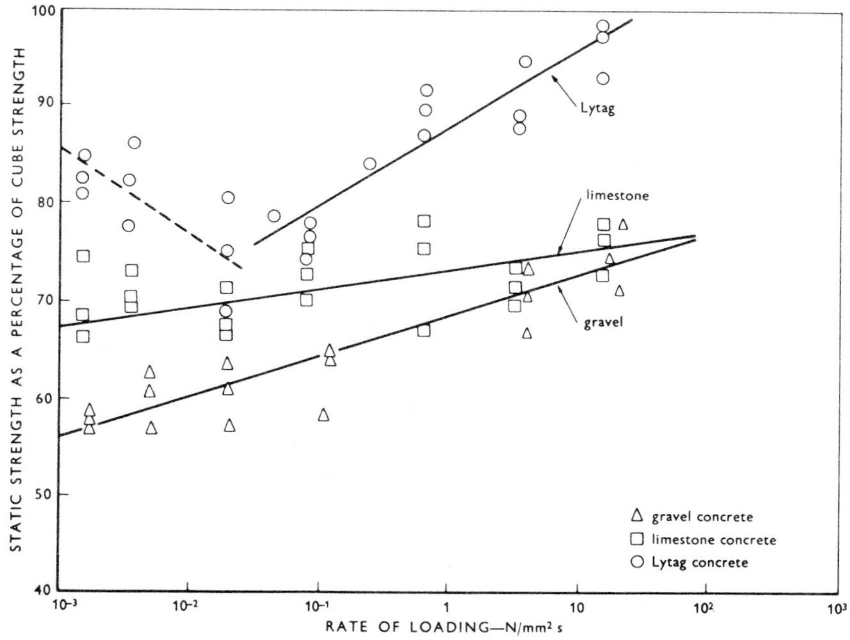

Figure 3: Results of the static tests.

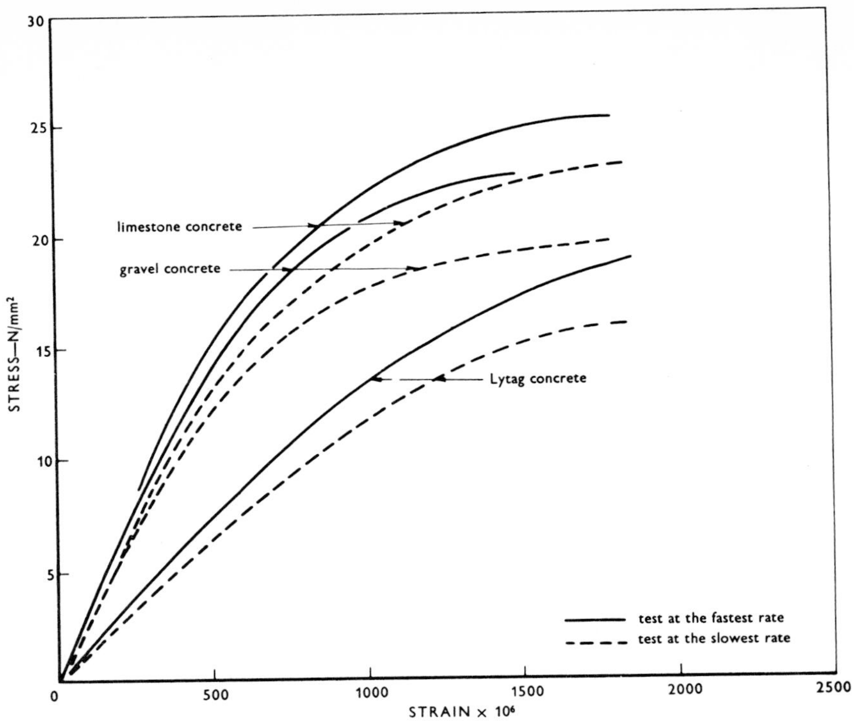

Figure 4: Typical relationships between stress and strain in the static tests.

most prisms was obtained from which plots of maximum strain per cycle against number of cycles were made. A typical example is shown in Figure 8. This curve has a pronounced three-part form which is more clearly illustrated in Figure 9. Here the rate of increase of maximum strain per cycle is plotted against the number of cycles of loading. Examination of the strain rates for the second part of all such curves has revealed a strong correlation between the 'secondary' strain rate and the endurance of the prism. Relationships were obtained between endurance (N cycles) and 'secondary' strain rate ($\partial\varepsilon/\partial n$ per cycle) as follows:

$$N = 2{\cdot}13 \times 10^{-3}\left(\frac{\partial\varepsilon}{\partial n}\right)^{-0{\cdot}89} \text{ for gravel concrete}$$

$$N = 1{\cdot}12 \times 10^{-3}\left(\frac{\partial\varepsilon}{\partial n}\right)^{-0{\cdot}98} \text{ for limestone concrete}$$

$$N = 1{\cdot}64 \times 10^{-4}\left(\frac{\partial\varepsilon}{\partial n}\right)^{-1{\cdot}06} \text{ for Lytag concrete}$$

These relationships appear to be independent of the rate of loading although they are dependent upon the type of aggregate (Figure 10).

Discussion

STATIC AND FATIGUE STRENGTHS

The results of the tests show that, in general, both the static and the fatigue strengths of concrete are increased at higher rates of application of load (Figures 3, 5, 6 and 7). In the static tests to failure, the findings of other research workers[1], except for Evans[2] and, to a certain extent, Spooner[3], are confirmed, certainly for rates of application of load above about 3×10^{-2} N/mm² s. Evans[2] may have used concrete with characteristics similar to those of the limestone concrete used in this investigation; if this assumption is correct, it would be possible to conclude that there was a negligible rate effect over this range of rates of loading after allowing for experimental scatter. Spooner's tests were carried out at constant straining rates and the nature of concrete is such that to maintain a constant rate of straining it is necessary to reduce the rate of loading as the strain increases and close to failure even to reverse the direction of loading. These conditions, particularly near failure, are quite different from those which exist in constant-loading-rate tests and thus Spooner's results are not strictly comparable with the results of this investigation or those referred to above.

At rates of loading below about 3×10^{-2} N/mm² s there is evidence of a reversal of the trend shown at the higher rates of loading. This is particularly marked for the Lytag concrete and is just detectable for the limestone concrete. It is possible that the limestone concrete and also the gravel concrete would show a more pronounced reversal similar to that of the Lytag concrete at rates of loading below 10^{-3} N/mm² s, the lower limit of the rates used in the present investigation.

The sensitivity of the static strength to the rate of loading appears to be related to the stiffness of the aggregate (Figure 3). Limestone, the stiffest of the

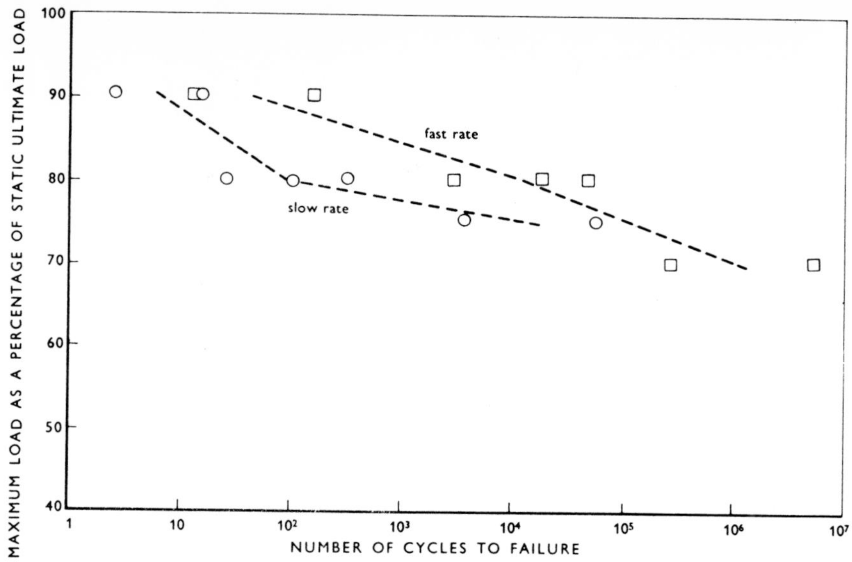

Figure 5: S-N curves for gravel concrete.

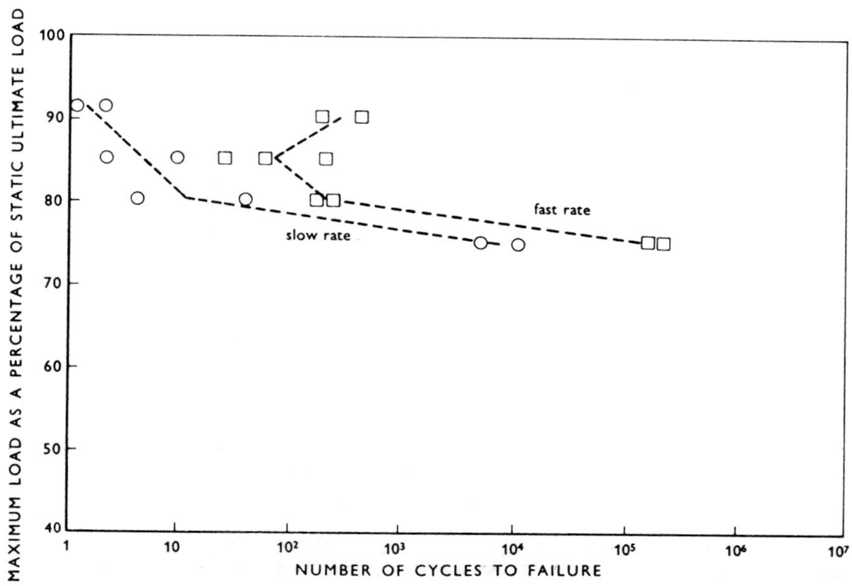

Figure 6: S-N curves for limestone concrete.

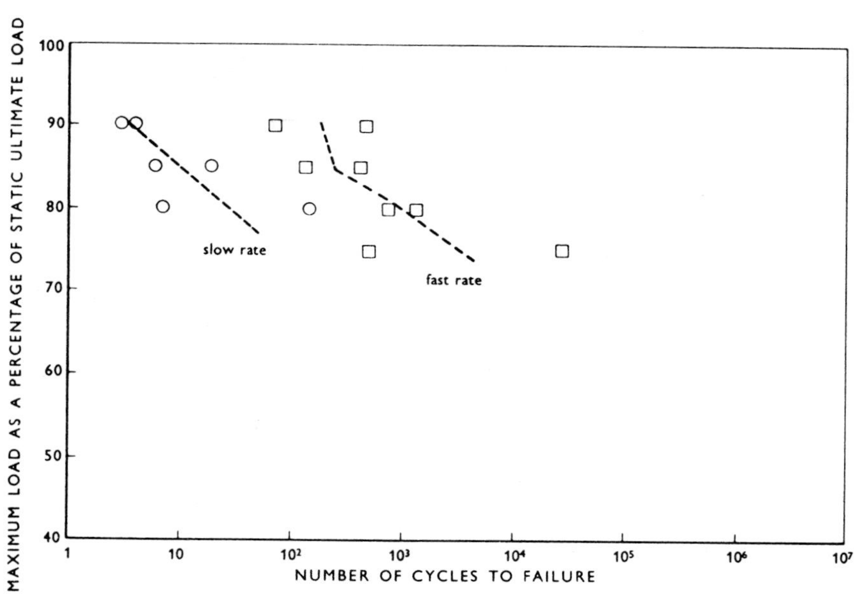

Figure 7: S-N curves for Lytag concrete.

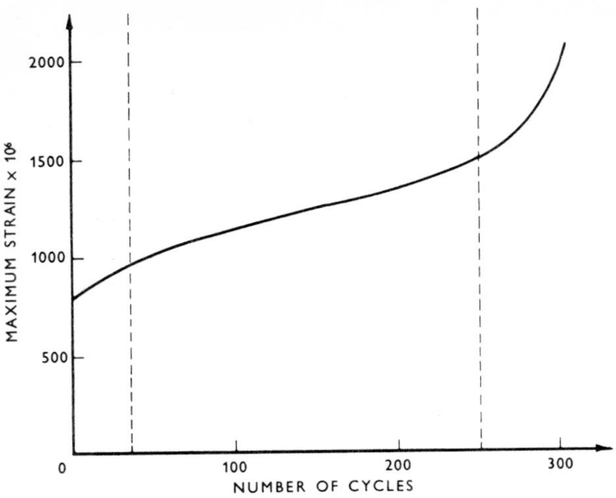

Figure 8: Typical relationship between maximum strain and number of cycles.

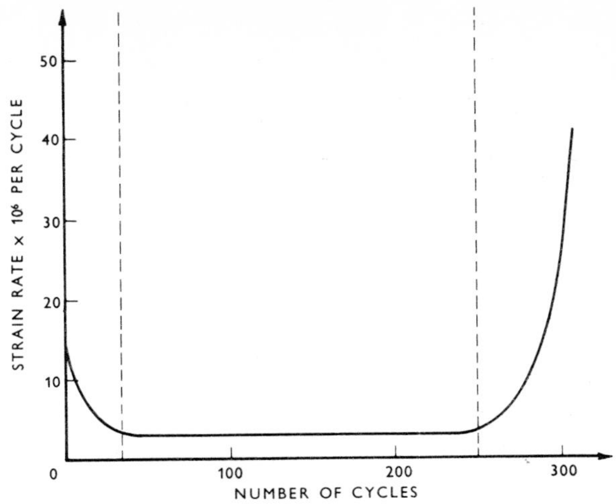

Figure 9: Typical relationship between strain rate and number of cycles.

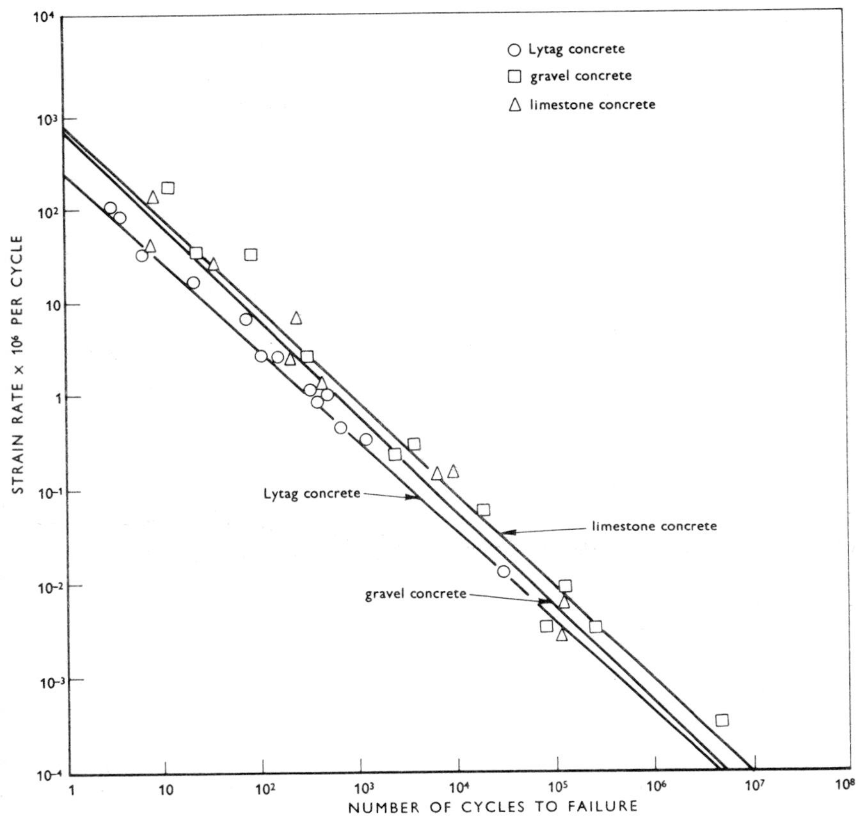

Figure 10: Relationship between secondary creep rate and fatigue life.

aggregates, produced concrete of which the static strength increased only 4% with a hundredfold increase in rate of loading. Gravel, a less stiff aggregate, produced concrete with an 8% improvement in static strength with the same increase in rate, whilst similarly Lytag, the least stiff of the aggregates, produced concrete with a 16% improvement in static strength (for rates above 3×10^{-2} N/mm² s).

Although it might be argued that the observed influence of rate of loading upon static strength is small and can be ignored in design, it should be taken into account in determining design strengths from test information.

The present fatigue tests do not confirm the conclusion of Murdock[4] that there is no significant rate effect for the fatigue strength of plain concrete in compression (Figures 5, 6 and 7). The degree to which the fatigue life was enhanced in the tests by the more rapid application of loading was dependent upon the level of the maximum load. Although it is not possible to put an exact figure on the improvement of life, it is generally of a lower order than the increase in the rate of loading. Thus, typically, a hundredfold increase in the rate of loading would produce an improvement in fatigue life of the order of tenfold. The specimens loaded at the slower rate therefore always lasted longer in terms of time before failure.

Most fatigue tests on concrete or concrete structures have been made at rates faster than those which obtain under service conditions. The results of the present investigation suggest that such tests may have overestimated the fatigue life of the concrete in service and the margin against fatigue failure may not be as great as assumed.

STIFFNESS AND STRAIN RATE

The stiffness of the aggregates was reflected in the stiffnesses of the concretes tested statically (Figure 4), the stiffest aggregate producing the stiffest concrete. For each of the concretes at a given stress, except when close to failure, the strain in the concrete loaded at the slowest rate was about 25% greater than that loaded at the fastest rate, i.e. there was an increase in strain of 25% for a decrease in rate of 10,000 times. This compares favourably with Spooner's work in which an increase of about 20% was observed for a decrease in rate of about 200 times[(3)].

The strain lives of the prisms (Figure 8) show development of maximum strain under fatigue loading which is similar to that generally observed under a constant 'creep' load. The increase of strain may be thought of as occurring in three stages: primary, secondary and tertiary. In the primary stage, the rate of increase of strain decreases until the secondary stage is reached when the rate is constant. At the tertiary stage, the rate gradually increases until failure occurs. During a large proportion of the life, the concrete is in the secondary stage of strain development. The independence of the secondary 'creep' rate of the rate of loading (Figure 10) suggests that a useful empirical guide to the fatigue life of concrete could perhaps be developed to form the basis for a fatigue-damage monitoring device. A method whereby the fatigue life of a structure might be predicted from deformations early in the life of the structure would be very valuable. However, it may well transpire that the secondary 'creep' rate is so dependent upon other factors, such as age of concrete and stress distribution, that a method based on it would be impracticable.

Conclusions

The main conclusions resulting from this investigation are summarized below.

(1) The stiffness and static strength in compression of each of the three types of concrete tested are, in general, enhanced by increases in the rate of application of load.

(2) The sensitivity of the static strength to the rate of loading appears to be related to the stiffness of the aggregate relative to that of the matrix. Limestone, the stiffest of the aggregates, produced concrete with an improvement of only 4% in static strength with a hundredfold increase in rate of loading. Lytag, the least stiff of the aggregates, produced concrete with a strength improvement of 16% (for rates above 3×10^{-2} N/mm^2 s) for a similar increase in rate of loading.

(3) The fatigue strength in compression of each of the three types of concrete is enhanced by increases in the rate of loading. Typically, a hundredfold increase in the rate of loading produces a tenfold improvement in fatigue life. This conclusion is of importance because it means that accelerated fatigue testing of concrete structures may produce an overestimate of their true fatigue life.

(4) For each of the three concretes, a linear relationship was found between the rate of secondary strain increase per cycle during the life of a prism and the endurance, irrespective of the rate of application of the load.

ACKNOWLEDGEMENTS

The authors wish to thank their colleague Mr D. Redfearn for assistance with the experimental work. The investigation was made as part of the research programme of the Building Research Establishment and this paper is published by permission of the Director.

REFERENCES

1. MCHENRY, D. and SHIDELER, J. J. Review of data on the effect of speed in mechanical testing of concrete. *Symposium on speed of testing non-metallic materials.* Philadelphia, American Society for Testing and Materials, 1956. Special Technical Publication No. 185. pp. 72–82.
2. EVANS, R. H. Effect of rate of loading on some mechanical properties of concrete. *Mechanical properties of non-metallic brittle materials. Proceedings of a Conference, London 1958.* Editor: W. H. WALTON. London, Butterworths Scientific Publications, 1958. pp. 175–192.
3. SPOONER, D. C. Stress-strain-time relationships for concrete. *Magazine of Concrete Research.* Vol. 23, No. 75–76. June–September 1971. pp. 127–131.
4. MURDOCK, J. W. *A critical review of research on fatigue of plain concrete.* Urbana, University of Illinois, 1965. pp. 25. Engineering Experiment Station Bulletin No. 475.
5. GRAY, W. H., MCLAUGHLIN, J. F. and ANTRIM, J. D. Fatigue properties of lightweight aggregate concrete. *Journal of the American Concrete Institute. Proceedings* Vol. 58, No. 2. August 1961. pp. 149–162.
6. BRITISH STANDARDS INSTITUTION. BS 1881. *Methods of testing concrete. Part 4. Methods of testing concrete for strength.* London. 1970. pp. 25.

Steel fibre reinforced concrete (CP 69/74)

J. Edgington, D.J. Hannant and R.I.T. Williams

Definitions of terms relating to fibre reinforced concrete

The composite The total constituents, ie cement paste, air void, aggregate and fibres.
Matrix That part of the composite which is not occupied by the fibres.
Fibre aspect ratio Fibre length/fibre diameter.
Fibre volume fraction $\frac{\text{volume of fibres}}{\text{volume of the composite}}$.

1 INTRODUCTION

At various intervals since the turn of the century short pieces of steel have been included within concrete in an attempt to endow the material with greater tensile strength and ductility. It was not, however, until 1963 when Romualdi and Batson[1] published the results of an investigation carried out in the USA on steel fibre reinforced concretes that any substantial interest was shown either by research organisations or by the construction industry. The claims made by Romualdi and Batson and subsequently by the Battelle Development Corporation[2], who filed a patent for the material later known as Wirand, were far reaching. In the development of the theory, it was assumed that concrete was a notch sensitive medium in which one could calculate the critical flaw size. It was claimed that the addition of short randomly distributed fibres to concrete would elevate the tensile cracking to at least 6.9 MN/m^2 (1000 lbf/in^2) when the average spacing of the included fibres was less than 7.6 mm. These claims aroused interest since, if the composite properties were as claimed, steel fibre reinforcement would provide a solution to the problem of tensile cracking that had for so long been an intrinsic deficiency of concrete.

It was with the intention of assessing both the validity of the claims and the viability of steel fibre reinforced concrete for use by the construction industry in this country that the Department of the Environment provided finance for this investigation. The main objectives were to assess the characteristics of the material during production and in the hardened state when incorporating various types and concentrations of steel fibres within cement pastes, mortars and concretes. A typical sample of steel fibre reinforced concrete is shown in Figure 1.

The detailed findings have been submitted to the Department of the Environment in a research report[3] and the aim of this paper is to present a summary of the work so that the knowledge

Figure 1 Typical sample of steel fibre reinforced concrete

gained regarding steel fibre reinforced concrete mixes is made available to the construction industry, particular emphasis being placed on mix design and on identifying the merits and limitations of the material.

2 PRODUCTION

The various fibre reinforced mixes were produced using a laboratory power driven pan and paddle mixer. The concrete constituents were mixed for two minutes after which a predetermined quantity of fibres was progressively added via a reciprocating fibre dispenser attached to the mixer. The dispenser ensured that the fibres entered the concrete matrix individually, a requirement which had previously been found to be important if uniform fibre distribution was to be achieved and essential when adding fibres having an aspect ratio greater than 100.

3 WORKABILITY

The workability of conventional structural concretes is normally chosen to suit compaction by vibration. When such mixes are discharged into formwork containing congested reinforcement, they still respond to vibration although the response may be slower than desired. In the case of concretes containing a high concentration of steel fibres, however, it may require careful mix design to achieve sufficient workability to permit compaction by vibration, since badly designed composites may not respond when vibrated.

3.1 Techniques

Whilst the slump test[4] is commonly used to assess the workability of conventional concretes, it is not generally suitable for fibre reinforced concretes because many fibrous mixes respond satisfactorily to vibration even though they have zero slump. As a result, the V–B consistometer test[4] was used for all composites and, in addition, the compacting factor test[4] was compared with the V–B test for mortars. The merit of the V–B test is that it simulates, at least in some respects, the compaction of concrete by vibration in practice.

3.2 Results

A wide range of steel fibres was incorporated at various concentrations into cement paste, mortar, 10 mm concrete and 20 mm concrete mixes. Details of the mixes are given in Table 1, and fibre types are shown in Table 2 and Figure 2.

Table 1 Mix proportions

Matrix type	Weight of mix constituents								
	Cement OPC	Aggregate (oven dry)*						Free water	Total water
		20–10 mm	10–5 mm	5–2.4 mm	2.4 mm–600 μm	Pass. 600 μm	Total		
Paste	1	–	–	–	–	–	0	0.26	0.26
Mortar	1	–	–	0.34	0.74	1.32	2.40	0.40	0.43
10 mm concrete	1	–	1.36	0.44	0.82	0.78	3.40	0.40	0.48
20 mm concrete	1	1.60	0.88	0.28	0.56	0.68	4.00	0.40	0.47

* The aggregate was an uncrushed irregular Thames Valley gravel which was presoaked prior to use to allow for absorption.

Typical relationships between workability and fibre content for fibre reinforced mortars containing different types of fibres are shown on Figure 3*, for the V–B test, and Figure 4 for the compacting factor test. From these results it is apparent that the workability of any mix decreases with increase in fibre concentration and fibre aspect ratio.

---

* In this and in subsequent figures, experimental points are omitted for clarity.

Table 2 Fibre types used, also shown in Figure 2

Fibre diameter (mm)	Fibre length (mm)	Surface characteristics	Fibre manufacturer
0.15	5	Brass coated	Steel Cords Ltd
0.15	10	Brass coated	Steel Cords Ltd
0.15	25	Plain round	National Standard Co Ltd
0.15	38	Plain round	National Standard Co Ltd
0.18	5	Brass coated	Steel Cords Ltd
0.18	10	Brass coated	Steel Cords Ltd
0.20	25	Brass coated	Steel Cords Ltd
0.25	20	Plain round	Johnson & Nephew Ltd
0.25	25	Plain round	National Standard Co Ltd
0.25	25	Duoform*	National Standard Co Ltd
0.25	25	Brass coated	N V Bekaert Ltd
0.25	25	Plain round	Tinsley Wire Industries Ltd
0.25	25	Brass coated	Steel Cords Ltd
0.25	30	Brass coated	Steel Cords Ltd
0.25	38	Plain round	National Standard Co Ltd
0.25	38	Plain round	Tinsley Wire Industries Ltd
0.25	50	Plain round	Johnson & Nephew Ltd
0.35	30	Plain round	N V Bekaert Ltd
0.38	12	Duoform*	National Standard Co Ltd
0.38	20	Brass coated	National Standard Co Ltd
0.38	25	Plain round	National Standard Co Ltd
0.38	25	Rusted	National Standard Co Ltd
0.38	25	Duoform*	National Standard Co Ltd
0.38	38	Plain round	National Standard Co Ltd
0.38	38	Duoform*	National Standard Co Ltd
0.50	38	Plain round	National Standard Co Ltd
0.50	38	Brass coated	National Standard Co Ltd
0.50	38	Duoform*	National Standard Co Ltd
0.50	50	Crimped	GKN Ltd

*Duoform is a patented fibre shape.

Although not apparent from Figures 3 and 4, it is worth emphasising that it is not the actual length or the diameter of the included fibres which is important but the ratio of the fibre length/diameter.

It may therefore be concluded for a given workability that (i) a higher volume fraction of low aspect ratio fibres may be incorporated into a mix, or conversely that (ii) the higher the aspect ratio of the included fibres, the lower will be the volume concentration than can be incorporated in the mix.

It can be seen from Figure 3 that the V-B test identifies a critical fibre content for each fibre aspect ratio beyond which the response to vibration rapidly decreases. This is not so for the compacting factor results shown in Figure 4 and this is regarded as a limitation of the test which was in any case suspect since the mixes all required rodding through the hoppers.

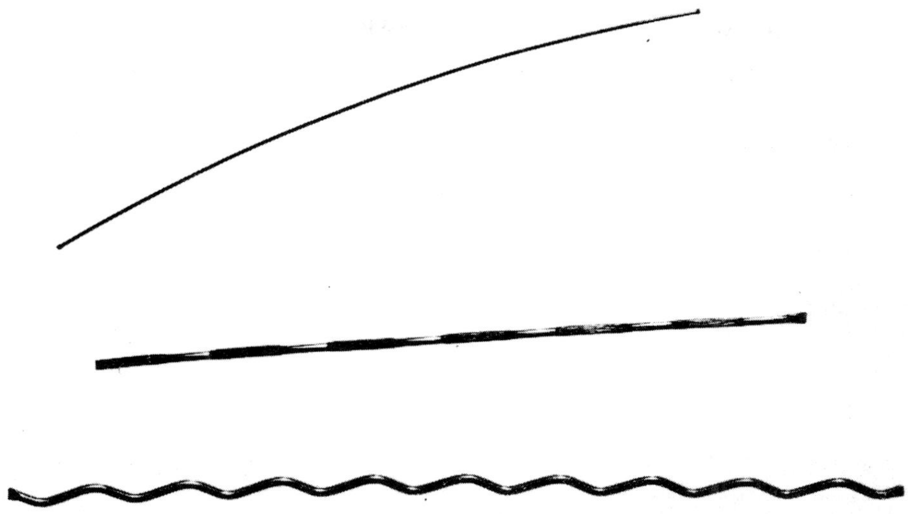

Figure 2 Fibre types used in the investigation

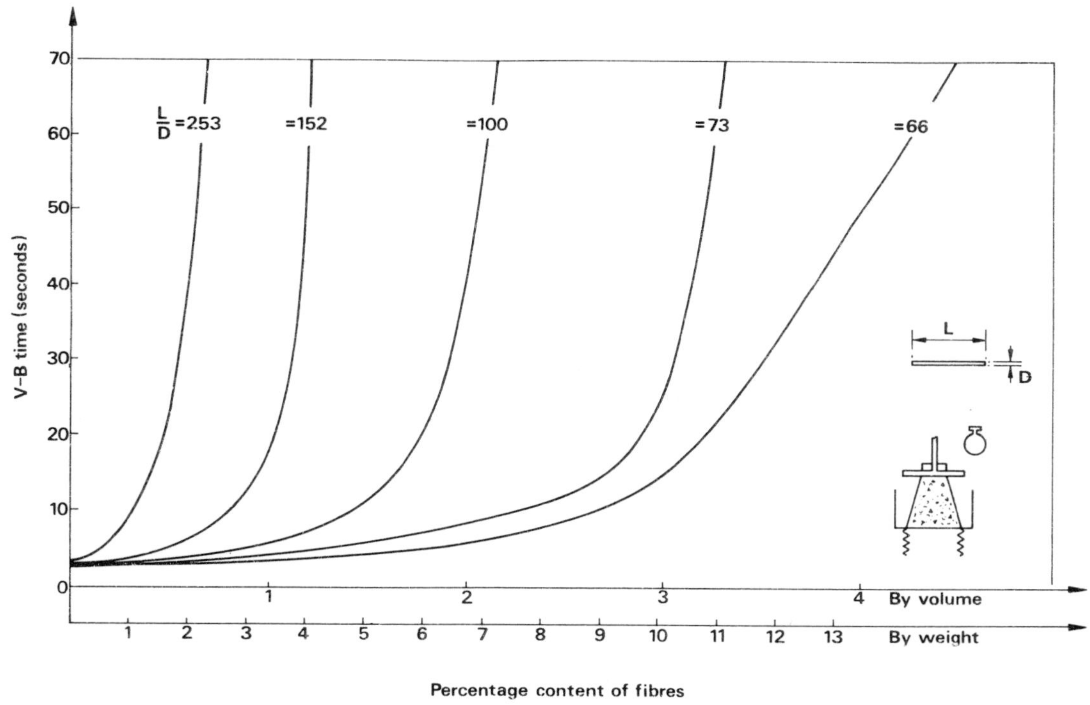

Figure 3 Effect of fibre aspect ratio on V-B time of fibre reinforced mortar

However, as a preliminary guide for work on steel fibre reinforced mortars, a comparison of V-B and compaction factor values is given in Figure 5, further work being necessary to establish whether or not this is applicable to concretes.

Figure 6 shows the relationship between fibre content and V-B time for fibres with an aspect ratio of 100 in matrices ranging from cement paste through to concretes having 20 mm maximum sized aggregate. From this it can be seen that the exponential type relationship established between fibre content and V-B time for mortar (Figure 3) exists for other matrices. Also the fibre content, beyond which the workability of the composite rapidly decreases, becomes less as the size of the coarse aggregate increases. It is also apparent from Figure 6 that the workability characteristics of fibre reinforced paste and mortar are broadly similar, thus indicating that the presence of aggregate particles up to 5 mm size has little influence on the compaction characteristics of fibre reinforced cement paste. Hence it would appear that for a given fibre type and orientation the workability of a mix decreases as the size and quantity of aggregate particles greater than 5 mm increases.

It has been found that a reasonable estimate of the fibre content required to make the concrete effectively unworkable can be obtained from the following equation:

$$PWc_{crit} = 75 \cdot \frac{\pi . SG_f}{SG_c} \cdot \frac{d}{L} \cdot K$$

where PWc_{crit} = critical percentage of fibres, by weight of concrete matrix

SG_f = specific gravity of fibres

SG_c = specific gravity of concrete matrix

$\frac{d}{L}$ = inverse of fibre aspect ratio

and $K = \frac{W_m}{W_m + W_a}$ in which

W_m = weight of the mortar fraction, ie that part of matrix whose particle size is less than 5 mm,

W_a = weight of the aggregate fraction whose particle size is greater than 5 mm.

In order to ensure, however, that the workability of a composite is sufficient to allow compaction by substantial external vibration it is recommended that the fibre content should not exceed $\frac{3}{4}$ PWc_{crit}.

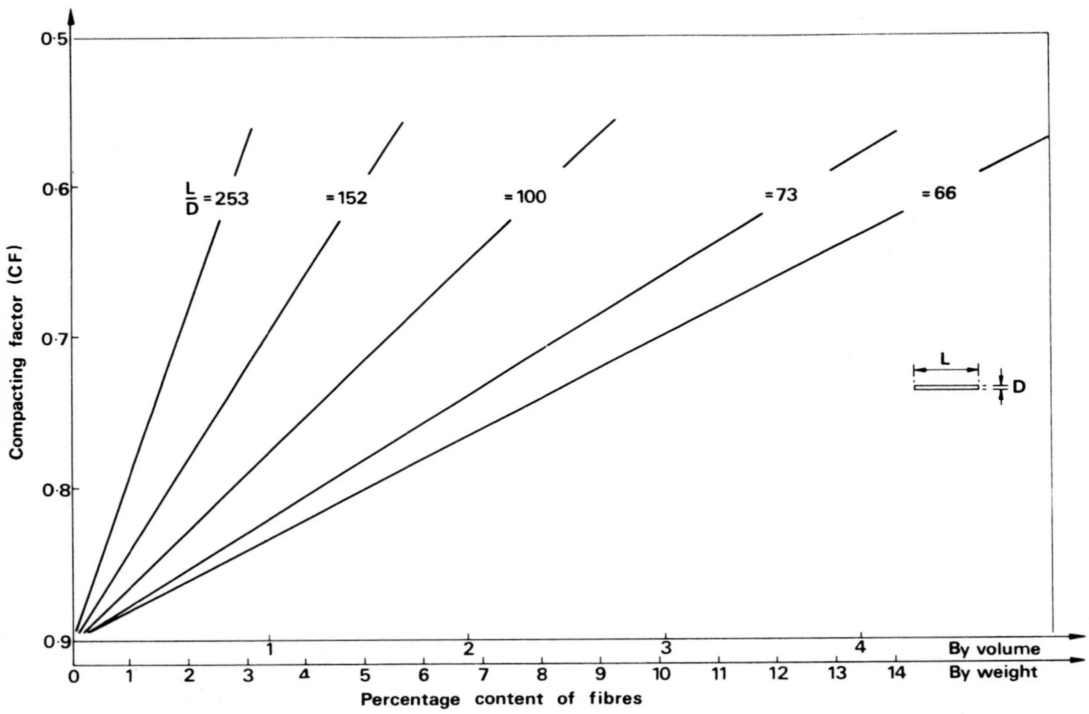

Figure 4 Effect of fibre aspect ratio on compacting factor of fibre reinforced mortars

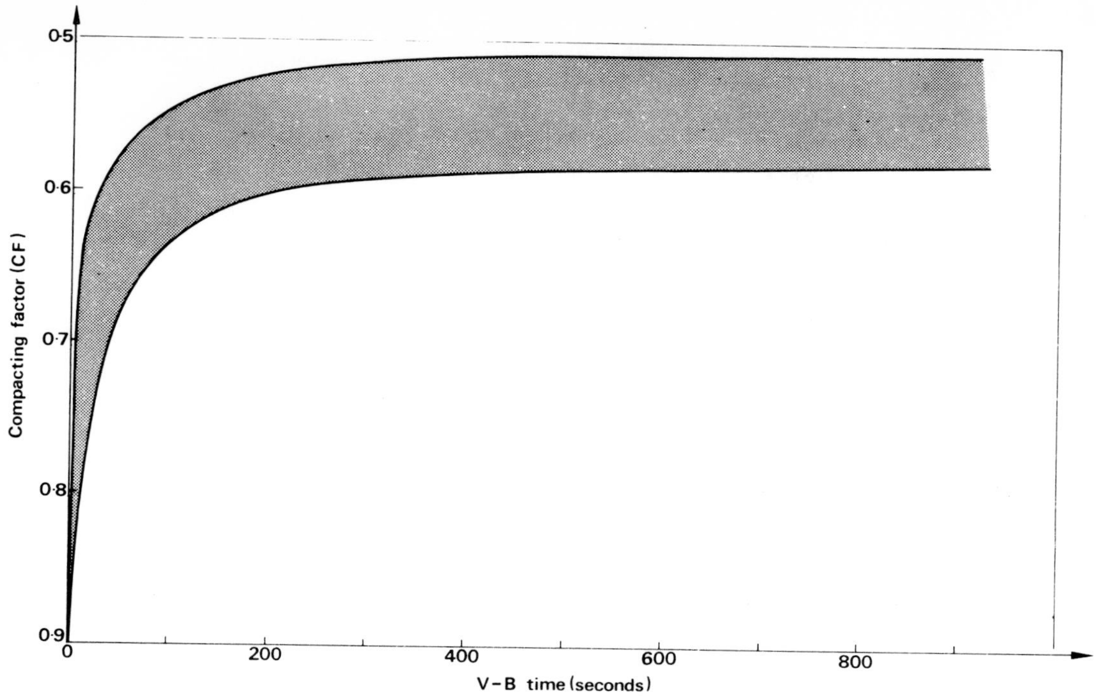

Figure 5 Relationship between compacting factor and V-B time for fibre reinforced mortars

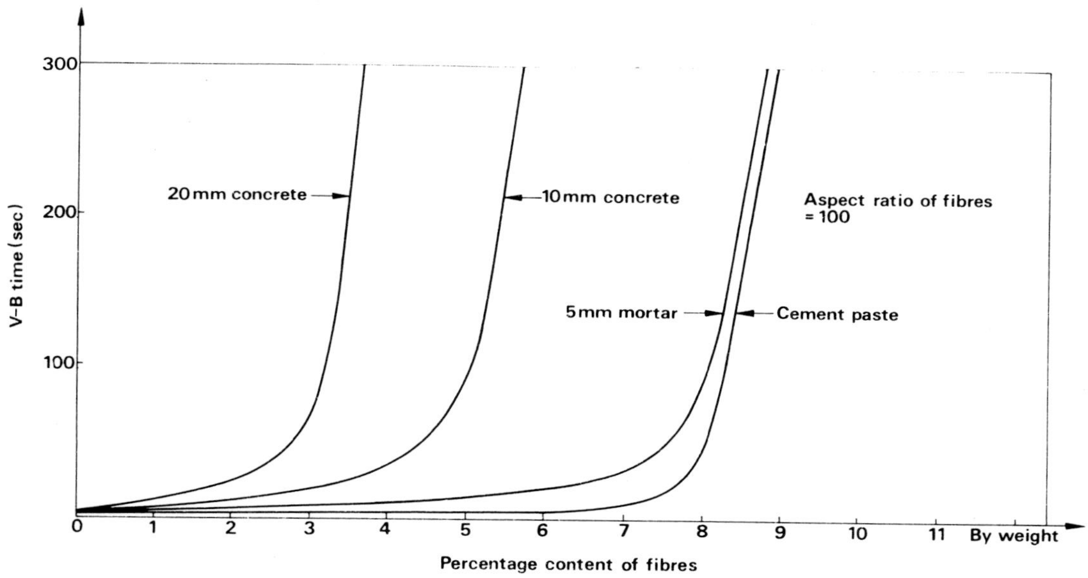

Figure 6 Workability against fibre content for matrices with different maximum aggregate size

4 COMPACTION

In the case of glass fibre reinforced cements, it has been reported [5] that the addition of glass fibres results in considerable air entrapment and, since air reduces the strength of concrete, a study was made of the air content of compacted steel fibre reinforced specimens.

The results of the study revealed that, provided the composite was capable of compaction on a vibrating table, the air content of the matrix within fibre reinforced concretes is no greater than that of the matrix without fibres. It was also found that in the case of fibre reinforced mortars there exists a trend of decreasing air content with increasing fibre content, and this may be due to damping effects of the fibres.

5 MODULUS OF ELASTICITY

From the results of uniaxial tensile stress-strain measurements on 100 mm x 100 mm x 500 mm plain and fibre reinforced specimens, it has been established that including fibres into the various cementitious matrices, shown in Table 1, only marginally increases the elastic modulus of the composite relative to that of the matrix (Table 3). This finding is consistent with the order of increase predicted by applying the laws of mixtures, the volume fraction of steel fibres being insufficient to change greatly the elastic deformation of the matrix.

Table 3 Effect of fibres on tensile modulus of elasticity

Matrix	Volume of fibre reinforcement (%)	Tensile modulus of elasticity (average of 3 specimens) GN/m^2
Cement paste	0	26.4
	2.70	28.4
Mortar	0	33.9
	2.34	34.8
10 mm concrete	0	39.7
	1.47	40.9

Compressive stress-strain measurements on plain and fibre reinforced 10 mm concrete showed similar small increases in modulus, the values in tension and compression being essentially equal.

6 CRACKING AND DUCTILITY

An investigation was undertaken to examine in some detail the cracking and failure behaviour of the material, the main objectives being to determine the effect of steel fibres on the cracking strain of the matrix and on the tensile cracking stress in relation to the 6.9 MN/m^2 claimed for the material by the patent [2].

6.1 Techniques

The onset and progression of cracking was determined on 100 mm x 100 mm x 500 mm specimens tested in flexure [4] and in uniaxial tension. The flexure specimens were instrumented so that continuous recordings could be made at increasing loads of the following parameters:

(i) Ultrasonic pulse time along the longitudinal axis of the tension zone of specimens. Changes in pulse time were considered to be indicative of cracking

(ii) Central deflection

(iii) Tensile strain near the bottom surface

(iv) Neutral axis position.

The deflection and strain measurements were made using linear variable differential transformers (lvdt) as shown with the ultransonic transducers on a test specimen in Figure 7.

In the case of the direct tension tests, longitudinal strains and pulse times were measured using similar techniques to those used in flexure.

6.2 Results

The results obtained in flexure tests on a typical plain and a typical fibre reinforced specimen are shown in Figure 8, from which the following observations are made:

(a) A change in pulse time occurs at a lower load than changes in slope of either the load-deflection or load-strain curves. Also cracking, as interpreted from the pulse time data, initiates at approximately the same load in both specimens and this was consistently the case throughout the test programme. It is therefore concluded that the presence of fibres does not have a significant influence in elevating the strain at which micro-cracking initiates in the matrix.

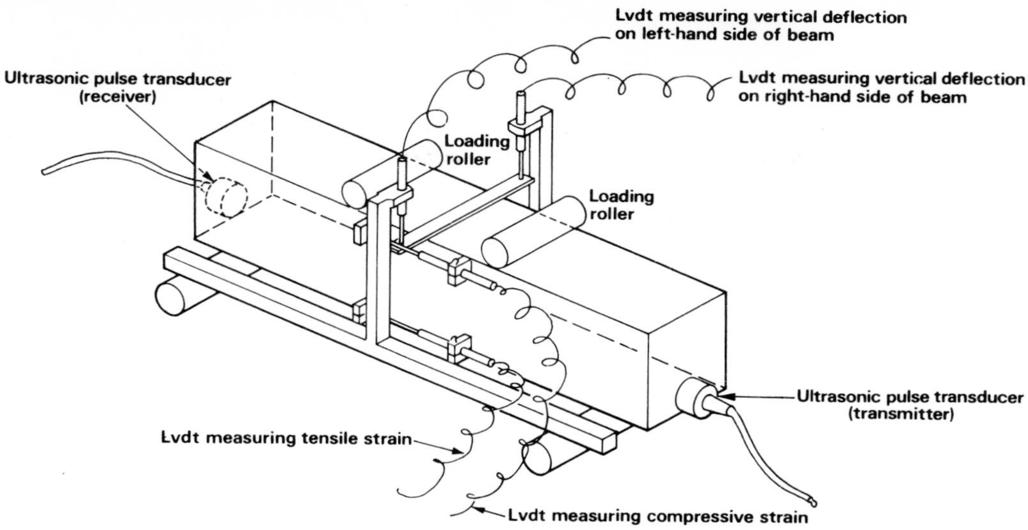

Figure 7 Equipment for detecting cracking under flexural loading

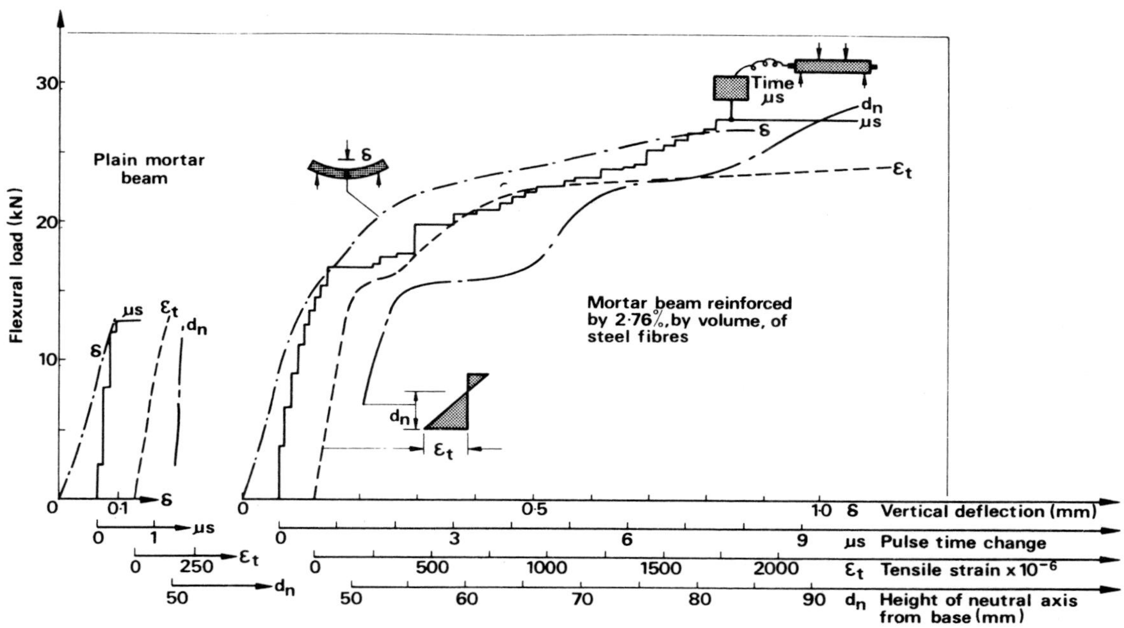

Figure 8 Cracking behaviour under flexural loading

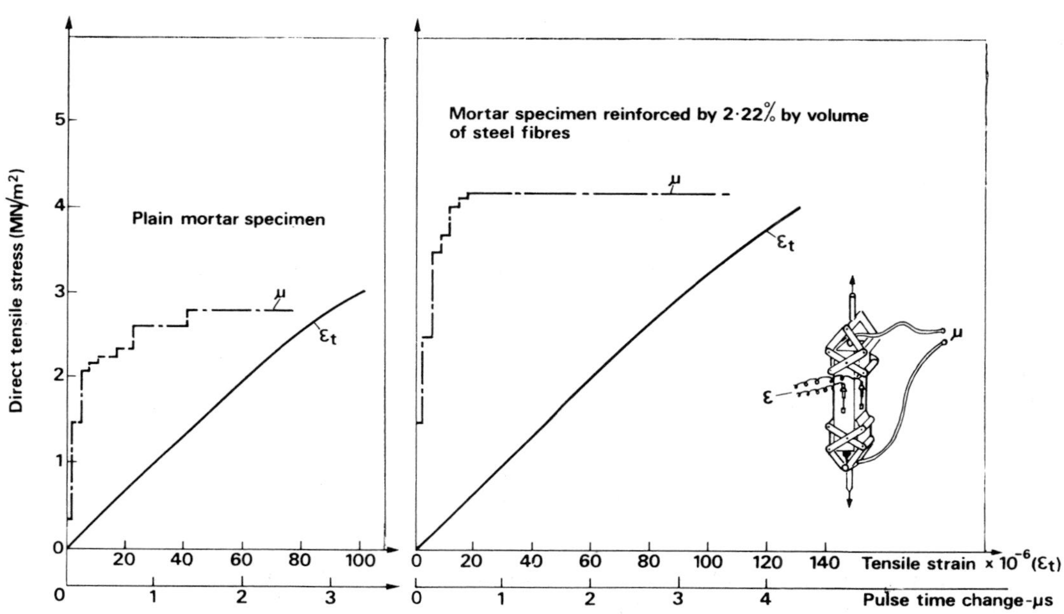

Figure 9 Cracking under direct tensile loading

(b) The plain specimen exhibits the characteristics expected from a semi-brittle material such as concrete in that the load-strain and load-deflection curves remained virtually linear up to failure, the neutral axis position was relatively stable with increasing load and only small changes in pulse time occurred before sudden failure. In the case of the fibre reinforced specimen, however, considerable quasi-plastic behaviour occurs once macro-cracking initiates, in this instance at a load of approximately 15 kN.

One of the most significant features of the fibre reinforced specimens is the progressive movement upwards of the neutral axis from the centre of the specimen as load is increased from 15 kN to failure at about 30 kN. Thus attempts to use the modulus of rupture[4] of fibre reinforced specimens as a measure of the maximum tensile stress at failure will result in a considerable overestimate. This is because the modulus of rupture, as defined in BS 1881[4], assumes that the neutral axis remains at half the depth of the specimen until failure. This is an important finding and one which must be carefully considered when judging the claims made regarding the tensile properties of steel fibre reinforced concrete.

A set of results obtained from the more limited programme of tests in uniaxial tension is shown on Figure 9.

It can be seen that pulse time changes initiated at a stress level less than 2 MN/m^2 for both specimens and in no case were time changes detected in other specimens at stresses greater than 2.6 MN/m^2. This evidence does not support the claims made by the patent[2] that the tensile cracking stress of concrete is elevated to beyond 6.9 MN/m^2 by the incorporation of closely spaced steel fibres.

However, an important point is to consider the techniques and terminology used to define the cracking stress. If only a visual assessment or a load-deflection curve is used then there is no doubt that, in testing small beams in flexure, the load at which cracks are judged to form can be increased by the addition of fibres.

7 STRENGTH

7.1 Influence of fibre orientation

It has been found that under table vibration there is a tendency for the steel fibres to align in planes at right angles to the direction of vibration or gravity. The results of this study have already been published[6] and the main conclusions drawn from the work were as follows:

(a) Steel fibre reinforced concrete or mortar products which are nominally randomly reinforced in three dimensions, can exhibit anisotropic behaviour due to fibre orientation during compaction;

(b) The direction of preferential fibre alignment should be made clear when the properties of steel fibre reinforced composites are quoted;

(c) From the practical point of view of manufacturing steel fibre reinforced cement products, this type of anisotropic behaviour could be put to good effect by arranging the compaction procedure so that the fibres are aligned in the most beneficial direction relative to the stress field. On the other hand, if the effects of vibration on fibre alignment are not fully appreciated, the strengths of steel fibre reinforced concrete products could be much lower than predictions based on laboratory tests using different compaction procedures.

7.2 Flexural[4], uniaxial tensile and torsional strengths

The strengths were determined from 100 mm x 100 mm x 500 mm prisms using the same materials detailed in Tables 1 and 2.

The strengths obtained[3,7] are shown in bands in Figure 10. From this figure it can be seen that the direct tensile and torsional strengths of concretes are only marginally increased. The flexural strengths, however, are considerably increased, the plain concrete strength being doubled by the addition of certain fibre types at 2 per cent by volume. Thus it may be concluded that the addition of steel fibres to concrete enables small flexural members to sustain considerably increased loading whilst, at the same time, the actual tensile strength of the material is only marginally increased.

It has been shown[3] that these marginal increases in tensile strength can be adequately predicted using the conventional theory of mixtures, ie reinforced concrete theory, when efficiency factors for the fibre reinforcement are introduced to allow for fibre orientation and to take account of bond strength.

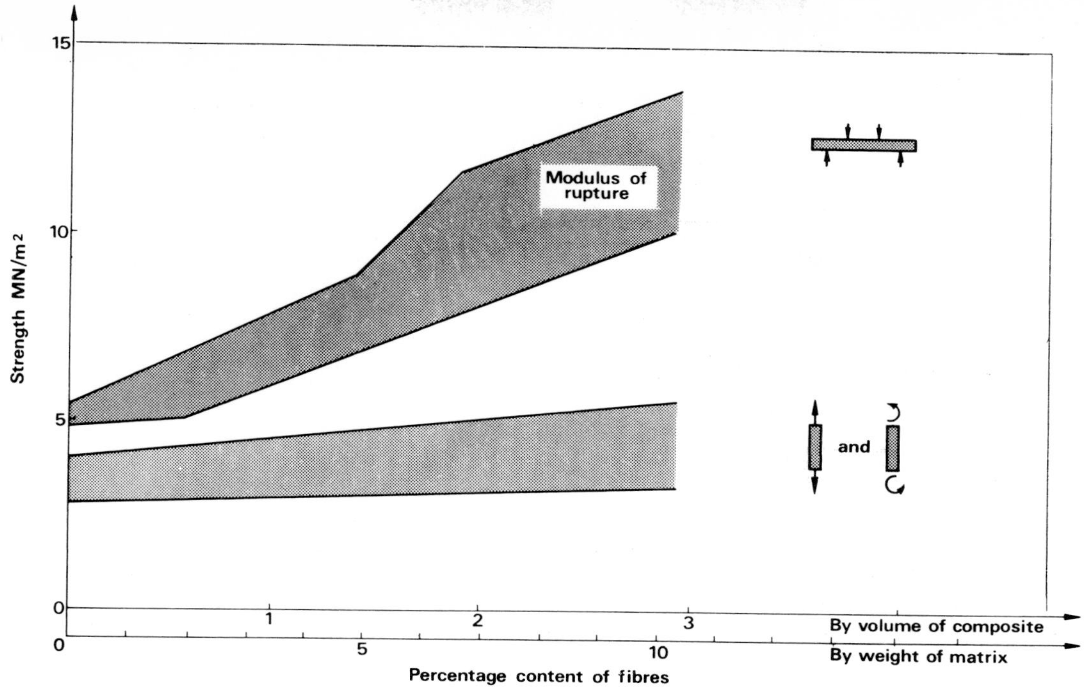

Figure 10 Flexural, direct tensile and torsional strengths of fibre reinforced mortar and concrete

The theoretical prediction of increased tensile strength with decreasing fibre spacing by Romualdi and Batson[1] is shown on Figure 11. The experimental results from this present investigation and those of Shah and Rangan[8] and of Johnston and Coleman[9] are also shown on Figure 11 from which it can be seen that for a constant volume of steel the measured tensile strengths only marginally increase with decreasing fibre spacing when compared with the large increases predicted theoretically[1]. Thus the theoretical fibre spacing concept does not in reality predict the cracking strength of fibre reinforced concretes and therefore some of the claims made in the patents for steel fibre reinforced concretes have not been substantiated by this investigation. The shortcomings of this aspect of the claims relate principally to a lack of understanding of the major part played by the method of test in determining the tensile strength of concrete.

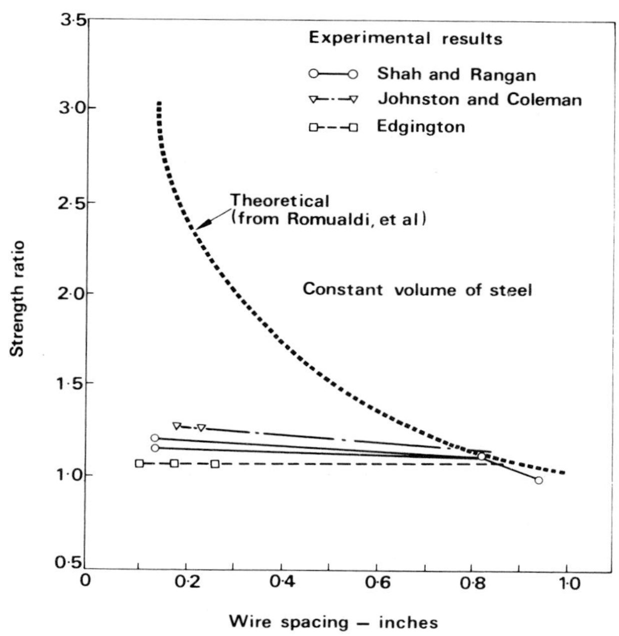

Figure 11 Effect of spacing of reinforcement on cracking strength of concrete

The inclusion of fibres having different aspect ratios revealed that for a constant volume of steel the loads carried by flexural specimens increased as the aspect ratio of the fibres increased and this is attributed to the increased effective bond length. It has been shown in Section 3, however, that for a constant volume of fibres, the workability of the composite decreases as the aspect ratio of the fibres increases. In design, therefore, it is necessary to optimise between considerations of strength and workability, increasing one at the cost of the other. Figure 12 illustrates, for a mortar matrix, the way in which aspect ratio and fibre content are interrelated to achieve the same load carried by a beam expressed as modulus of rupture.

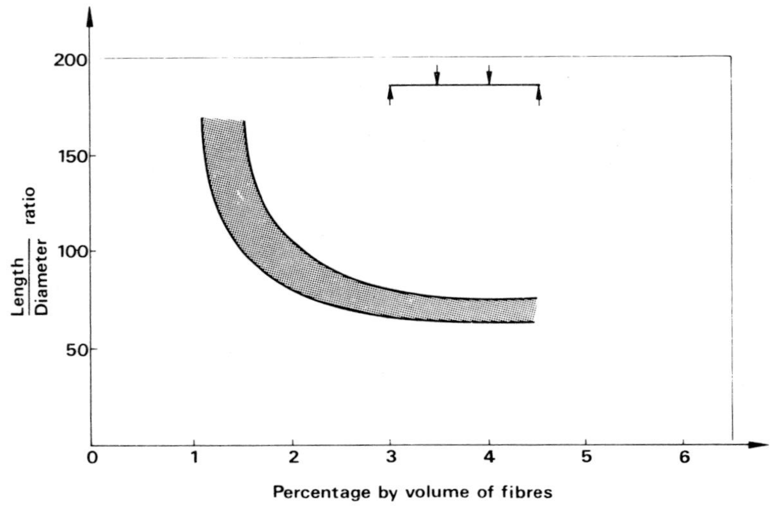

Figure 12 Fibre volume to achieve a modulus of rupture of 9 MN/m^2 in mortar

Although the detailed results show that the modulus of rupture of specimens containing crimped or indented fibres was slightly higher than those containing plain round wire at similar fibre concentrations and aspect ratios, no single fibre type proved to be significantly better than the others at all concentrations for improving the tensile or flexural strengths.

7.3 Compressive strength

Only small increases in the uniaxial compressive strength of mortar and concrete prisms and cylinders were achieved by the addition of steel fibres. The small increases in strength could have been more cheaply achieved by decreasing the water/cement ratio although the reinforced specimens have the merit of pronounced ductility.

7.4 Effect of age

It has been established from tests carried out at specimen ages ranging from 7 to 112 days that the strength of steel fibre reinforced concrete increases at a rate similar to that of unreinforced concrete.

8 MECHANISM OF FIBRE STRENGTHENING IN FLEXURE

It has been stated in Section 7 that the maximum flexural load sustained by a concrete or mortar beam may be considerably increased by the inclusion of steel fibres whilst there is only a marginal increase in the uniaxial tensile strength of the material. It is suggested that the increased load-carrying capacity is due to the partially plastic stress blocks developed within the tensile zone of fibre reinforced beams, as compared with the predominantly linear elastic stress blocks thought to exist within unreinforced beams.

8.1 Hypothesis

Consider a fibre reinforced concrete beam subjected to an increasing load ($P + \Delta P$) as shown in Figure 13a. As the tensile strain increases, cracks are formed but, unlike plain concrete, a proportion of the load is maintained across the crack by those fibres spanning the crack and hence equilibrium is maintained. Due to the formation of these cracks the measured tensile strains are increased and hence the neutral axis moves upwards to some new value. When further load is applied to the beam, the measured tensile strains increase at a greater rate than the compressive strains, see Figure 13b, and the values, d_n', increase until there is no simple relationship between the measured strain and the apparent stress sustained across the crack. The stress block in the tensile zone will then probably be represented by Figure 13c with a limit of 13d as the maximum tensile area which can be developed.

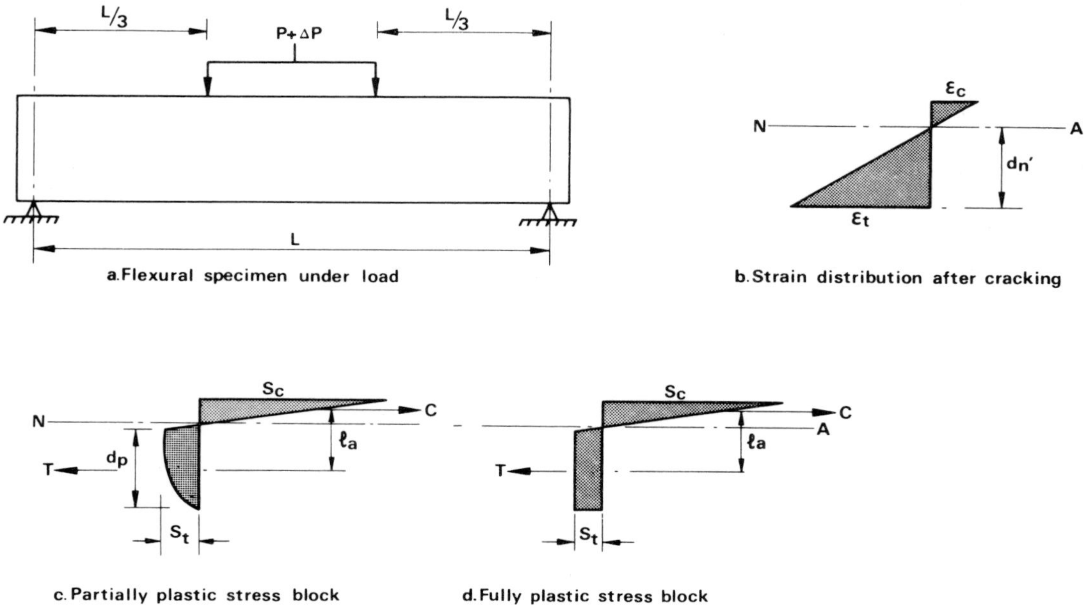

Figure 13 Suggested mechanism of fibre strengthening in flexure

8.2 Procedure adopted to check hypothesis

The compressive and tensile strains, and hence the neutral axis position, were measured up to failure for a wide range of plain and fibre reinforced specimens. The maximum tensile strength of each composite was known from the strength investigation.

A triangular stress distribution in the compressive zone of the beams was assumed because in general the maximum measured compressive strains were less than the strain at which the material becomes significantly inelastic.

Using values obtained from the tests the following were calculated at 95 per cent of the ultimate load.

(a) The compressive force C

Where $C = \dfrac{\epsilon_c \times E \times b \times (d - d_n)}{2}$

in which ϵ_c = compressive strain

E = modulus of elasticity

b = breadth of beam

d = depth of beam

d_n = height of neutral axis from tensile face.

(b) From various assumed shapes of the tensile stress block, ie between partially plastic (Figure 13c) and fully plastic (Figure 13d)

(i) the maximum tensile stress, (St), by equating T = C

(ii) the lever arm (la)

(iii) the moment of resistance = la x T

(iv) the theoretically sustained flexural load (P_{th}) by equating moment of resistance to bending moment, from which $P_{th} = \frac{6 \times la \cdot T}{L}$

The derived values of P_{th} and St were then compared with the measured values to identify a shape of stress block which would accurately predict the specimen failure load when the maximum tensile stress, St, was within the range determined from the direct tension tests. In addition to the assumptions made regarding the shape of the stress block, some error may have been introduced in the strain measurements due to the size of the lvdt supports but the results given in Table 4 suggest that reasonable agreement is obtained between predicted and observed behaviour. Of the 100 specimens tested, the failure load was predicted in each case to within an accuracy of ±20 per cent. It is therefore concluded that the improved load-carrying capacity of steel fibre reinforced beams is due to the formation of plastic or partially plastic tensile stress blocks and not to an increase in the tensile strength of the composite.

Table 4 Comparison of theoretical and experimental failure loads

Matrix type	Percentage of fibres by volume	Theoretical values		Experimental results		
		$0.95P_{th}$ (kN)	St (MN/m^2)	$0.95P_{actual}$ (kN)	Range of measured direct tensile strength (MN/m^2) minimum	maximum
Cement paste	0.82	17.0	4.96	18.1	4.77	5.15
	1.50	18.2	5.44	18.7	4.54	5.58
	2.70	19.9	5.33	19.4	5.03	5.51
	3.61	25.2	5.56	25.6	5.02	6.27
Mortar	0	11.7	3.57	12.1	3.13	4.00
	1.12	13.1	3.95	14.8	3.91	4.04
	1.81	25.9	4.84	25.5	4.50	4.92
	2.76	30.6	5.21	28.6	5.00	5.36
10 mm concrete	0	10.5	3.21	10.3	2.97	3.55
	1.18	17.6	4.05	17.7	3.87	4.14
	1.75	20.3	4.18	20.5	3.76	4.43
20 mm concrete	0.60	11.9	3.15	11.7	2.89	3.21
	1.19	15.6	3.16	15.9	2.61	3.21

It should be emphasised that although the areas and shapes of the calculated tensile stress blocks may be accurate for the purpose of calculating the moment of resistance of the beams, the tensile stresses are not real quantities. This is because the 'real' quantities are the forces in the individual fibres spanning cracks and these are effectively integrated, averaged and divided by the beam cross-sectional area to give a quantity which is convenient for engineering design, known as the average tensile stress in the composite. This is the same convenient quantity as is measured in a direct tensile test after matrix cracking and must not be confused with the modulus of rupture.

9 IMPACT

The impact toughness of various fibre reinforced composites was assessed using 100 mm x 100 mm x 500 mm specimens. A pendulum type impact machine was modified as shown in Figure 14 so that an estimate could be made of the energy required to displace and rotate the fractured specimen halves.

The results obtained for mortar and for 10 mm and 20 mm maximum sized concretes are shown on Figure 15. It is clear that the impact toughness of concrete is considerably increased by incorporating steel fibres, the improvements being especially favourable with the 0.50 mm dia x 50 mm high tensile crimped fibre, but further work is required to establish whether this is due to the shape or the strength of this type of fibre.

Figure 14 Impact machine

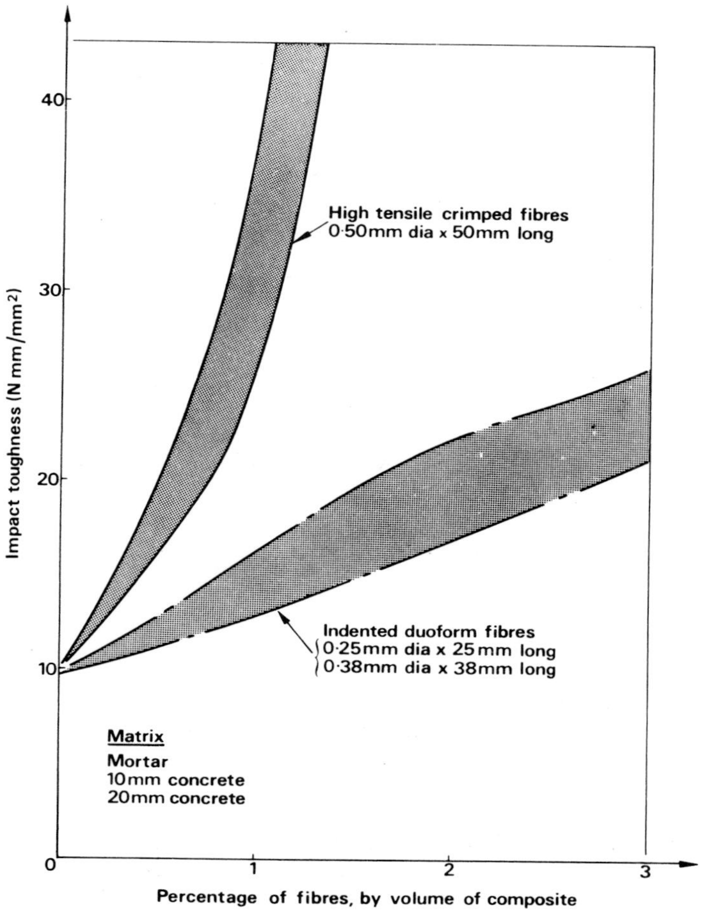

Figure 15 Impact toughness of fibre reinforced mortar and concrete

10 TIME-DEPENDENT PROPERTIES

10.1 Creep

From the results of compressive creep tests carried out over a loading period of 12 months it has been found that the addition of steel fibres in concrete does not reduce the creep strains of the composite. This behaviour is consistent with the low volume concentration of fibres when compared with an aggregate volume of approximately 70 per cent.

10.2 Shrinkage

The shrinkage of concrete over a period of 3 months on specimens subjected to various curing environments was unaffected by the presence of steel fibres. This again is as expected for the same reasons given for creep.

10.3 Durability

The weathering of plain and fibre reinforced normal and lightweight aggregate concretes was observed over a period of 3 years. For this study, cylindrical specimens were exposed at BRE, Garston, and also to aggressive environments on Department of the Environment exposure sites at Hurst Castle in the Solent, and at Beckton gas works, London. Apart from rust staining on the outer faces, the specimens exhibited no significant deterioration over the period of exposure, the rate of carbonation being the same for the plain and for the fibre reinforced specimens. The depth of penetration of carbonation was greater for the lightweight aggregate concretes.

The specimens used however were in an uncracked condition and further work is in progress to judge the long-term structural integrity of specimens containing cracks.

11 APPLICATIONS AND ECONOMICS

Having identified the stress systems for which concretes may be improved by the addition of steel fibres, it is now possible to consider suitable applications. The most beneficial characteristics of fibre reinforced systems are those of increased flexural capacity, toughness, post-failure ductility and crack control. The following are applications referred to in published papers:

> Airfield and highway pavements both for new construction and as overlays, bridge nosings and bridge decks, loading bays, heavy duty floors and chutes; tunnel linings, machine bases, security safes and strong rooms; thin precast components subjected to flexural or impact loading such as pile shells, paving slabs, fence posts, steps, manhole covers, duct covers and cast iron replacements; domes, shells, concrete boats and marine structures; regions of high shear stress such as prestressed, post-tensioned concrete end blocks; structures requiring resistance to thermal shock such as refractory linings, and fire-resistant coatings for beams and columns; more exotic applications such as explosives stores, pads for vertical take-off aircraft and tank turning pads.

The most advantageous outlets are still rather uncertain but the authors' experiences suggest that the main practical applications will be in paving, precast components and in those situations where toughness and post-failure ductility are important. The rate of progress will depend on economic considerations and may be influenced by future work in which the durability of cracked sections is examined.

At the current price of 25 to 30p/kg for steel fibres, the addition of 1 per cent by volume to the concrete more than trebles the material costs. At first sight this increase seems prohibitive but when fabrication and manufacturing costs are taken into account the materials costs represent only a fraction of the total cost. Furthermore, if conventional reinforcement can thereby be reduced or eliminated, the savings in cost of detailing, receiving, cutting, bending, and fixing conventional reinforcement may offset the increased materials cost.

12 CONCLUSIONS

Mixing and fibre distribution

1 A mixer with a power-driven pan and a power-driven paddle, which rotates in the same direction as the pan, provides a good mixing action for the production of steel fibre reinforced concrete.

2 The sequence in which the mix constituents are added to the mixer has no apparent influence on the resultant degree of fibre dispersion.

3 In order to ensure good fibre distribution, the fibres should reach the concrete individually and be immediately removed from the point of entry by the mixing action. This may be readily achieved by passing the fibres through a mechanically operated sieve into the type of mixer described in 1 above.

4 Fibre dispersion is more easily achieved when incorporating fibres of a low aspect ratio. As a result, higher volume fractions of wire can be used in a given matrix when the aspect ratio is low.

5 Uniform dispersion of a given fibre type within a concrete matrix becomes more difficult to achieve as the proportion of aggregate particles greater than 5 mm increases.

Workability and compaction

6 In general, the slump test provides very little indication of either the workability or the ease with which fibre reinforced concrete can be compacted.

7 The V-B test is the best of the 3 standard workability tests for assessing the behaviour of fresh fibre reinforced concrete subjected to compaction by vibration.

8 The workability of a composite is decreased as the fibre content is increased and there is a critical fibre volume above which the rate of decrease in workability is very rapid.

9 The most important single fibre characteristic which influences workability is the aspect ratio.

10 When the fibre concentration is held constant then the workability of a composite is decreased as the aspect ratio of the included fibres is increased.

11 When the fibre concentration and aspect ratio are held constant then the workability of the composite is decreased as the ratio:

$$\frac{\text{volume of cement paste + aggregate particles less than 5 mm in size}}{\text{total volume of the matrix}}$$

decreases.

Strength

12 The increase in the direct tensile, torsional and compressive strengths of fibre reinforced concretes, when compared with their unreinforced counterparts, are relatively small even at a fibre concentration of up to 5 per cent by volume. Hence there is likely to be little practical merit in including short random steel fibres in concrete to increase any of these strengths.

13 Significant increases in the modulus of rupture of fibre reinforced concretes may be achieved when the maximum size of the aggregate particles is not greater than 10 mm. These increases can exceed 100 per cent when 2 per cent by volume of certain fibre types are used.

Cracking

14 The cracking of all direct tensile fibre reinforced specimens initiated at a tensile stress less than 2.6 MN/m^2, this being approximately 1/3 of the value claimed in the Patent [2].

15 The onset of micro-cracking within cementitious matrices subjected to increasing flexural load is apparently unaffected by the presence of steel fibres.

16 The flexural load required for crack propagation is increased as the fibre content increases.

Mechanism of strengthening in direct tension

17 The theory of fibre strengthening, based on fibre spacing [1], grossly overestimates the true direct tensile strength of such composites.

18 A reinforcing theory, based upon the laws of mixtures, enables satisfactory prediction to be made of the direct tensile strength of fibre reinforced concretes. It may, therefore, be concluded that the direct tensile behaviour of fibre reinforced concrete characterises the performance of conventional reinforced concrete in which the reinforcing bars are inefficiently orientated, with respect to the direction of applied stress, and poorly bonded to the concrete matrix.

Mechanism of strengthening in flexure

19 The large increases in the load-carrying capacity of small fibre reinforced concrete beams is due to the formation of plastic or partially plastic stress blocks within the tensile zones of such beams, as a result of the forces maintained by the fibres after matrix cracking. Attempts to interpret the increases in modulus of rupture as being indicative of increased material strength will result in a considerable overestimate of the true tensile strength of the material.

Toughness under impact loading

20 Of the four fibre types investigated, the 0.50 mm diameter x 50 mm long high tensile crimped fibre proved to be the most beneficial at improving the impact toughness of the cementitious matrices. Increases in impact toughness of more than 400 per cent, at less than $1\frac{3}{4}$ per cent by volume, were measured with this fibre.

Dimensional changes

21 The addition of steel fibres to concrete at volume fractions of up to 3 per cent provides only marginal increases in the elastic moduli.

22 The shrinkage of mortar and the creep deformations of gravel aggregate concrete are not significantly reduced by the inclusion of 2 per cent by volume of steel fibres.

ACKNOWLEDGEMENTS

The authors wish to thank the Department of the Environment for financing the project, Mr F J Grimer of the Building Research Establishment for his useful guidance and the laboratory staff of the Construction Materials Research Group at the University of Surrey for their willing help at all times.

13 REFERENCES

1 **Romualdi, J P** and **Batson, G B.** Mechanics of crack arrest in concrete. Proceedings of the American Society of Civil Engineers, Vol 89, No EM3, June 1963, pp 147-168.

2 **Battelle Development Corporation.** Concrete and steel material. British Patent No 1068163, December 1963.

3 **Edgington, J.** Steel fibre reinforced concrete. Research report submitted to the Department of the Environment, January 1974. Also available as PhD thesis, University of Surrey, 1974.

4 **British Standards Institution.** BS 1881:1970: Methods of testing concrete.

5 **Grimer, F J** and **Ali, M A.** The strength of cements reinforced with glass fibres. Magazine of Concrete Research, Vol 21, No 66, March 1969, pp 23-30.

6 **Edgington, J** and **Hannant, D J.** Steel fibre reinforced concrete. The effect on fibre orientation of compaction by vibration. Matériaux et Construction, Vol 5, No 25, 1972, pp 41-44.

7 **Edgington, J.** Steel fibre reinforced concrete. Intermediate report submitted to the Department of the Environment for the period ending 1st March 1972.

8 **Shah, S P** and **Rangan, B V.** Effects of reinforcement on ductility of concrete. Proceedings of the American Society of Civil Engineers, Journal of the Structural Division, June 1970, pp 1167-1184.

9 **Johnston, C D** and **Coleman, R A.** Strength and deformation of steel fibre reinforced mortar in uniaxial tension. To be published in the Journal of the American Concrete Institute, 1974.

A stress-strain relationship for concrete at high temperatures (CP 40/74)

R. Baldwin and M.A. North

Notation

σ stress
σ_{max} maximum stress
ε strain
ε_{max} strain at maximum stress
E tensile modulus of elasticity
T temperature

Introduction

This paper reviews some of the data in the literature relating to the effect of temperature upon the elastic properties of concrete under compression at high temperatures. The only complete stress-strain curves appear to be those of Furamura[1], and these are the main concern of this paper. There are, however, other data: Malhotra[2] has measured the effect of temperature upon the compressive strength of concrete, whilst Saemann and Washa[3], Philleo[4] and Cruz[5] have investigated the effect of temperature upon the modulus of elasticity. These data will be used for the purposes of comparison with those of Furamura where possible.

Malhotra, Philleo and Cruz have all shown that the physical properties of concrete are strongly influenced by various factors such as the aggregate and age. In view of this, the numerical value of the elastic constants derived from Furamura's data apply only to the particular specimens tested, and hence the emphasis in this paper will be upon qualitative results, and in particular the way in which the stress, σ, and strain, ε, vary with temperature.

Furamura's data

Some typical stress-strain curves of Furamura are shown in Figure 1 for temperatures ranging from 30° to 600°C. These experiments were conducted with cylindrical specimens placed under compression in a furnace, the strains being derived from measurements of the change in length of the specimen. The loads were cycled during the course of each measurement and although, when stressed specimens were unloaded, the strain did not return to zero, when they were restressed, the original stress-strain curve was resumed from the point at which it had been left. This implies that the concrete acquires a permanent set during stressing and that the stress-strain curves in Figure 1 are not reversible; that is, they apply only in the direction of increasing strain.

Normalization of data

The stress-strain curves of Figure 1 exhibit a rather complicated relationship with temperature, non-linear over most of the range. However, some important conclusions may be drawn by imposing the following transformation on the data. For each temperature, let σ_{max} be the maximum stress (i.e. the peak of the stress-strain curve) and let the strain at this point be ε_{max}. We now plot σ/σ_{max} against $\varepsilon/\varepsilon_{max}$ for each temperature and obtain Figure 2. It is clear that the effect of this transformation is to bring all the curves for different

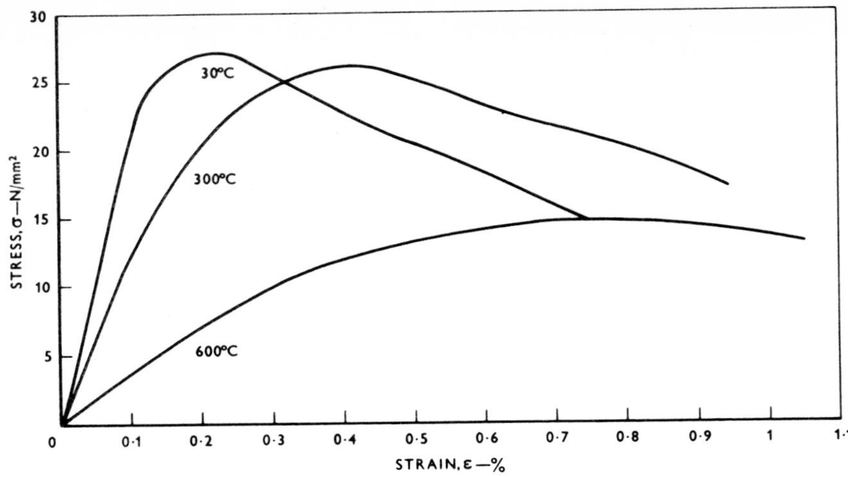

Figure 1: Typical stress-strain curves of Furamura at different temperatures.

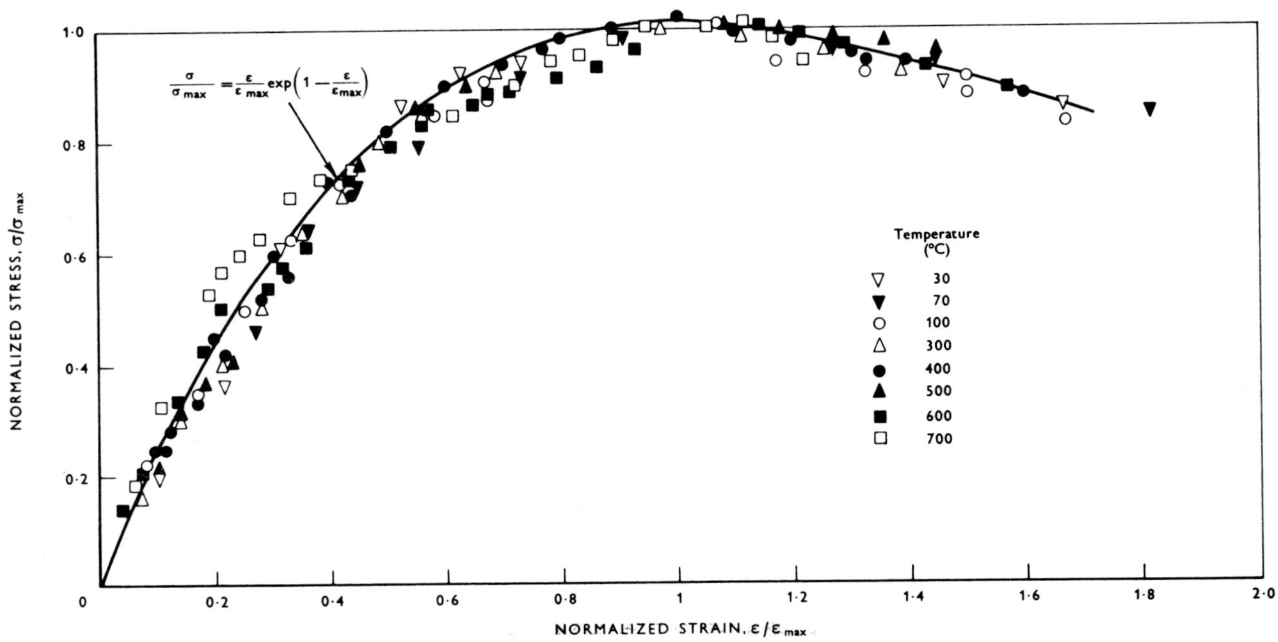

Figure 2: Normalized stress-strain curves (data from Furamura).

temperatures together so that they lie along the same course.

This result is significant because it implies that the stress-strain curves for high temperatures can be derived entirely from the stress-strain curves measured at room temperatures together with the variation of the compressive strength of the material with temperature, corresponding to the peak of the curve, and the strain at this point.

Compressive strength at different temperatures

The compressive strength of the specimens at different temperatures, derived by finding the maximum stress of each stress-strain curve, is plotted in Figure 3. This shows that the compressive strength remains nearly constant until around 300 to 400°C, when there is a dramatic decrease in strength. Malhotra[2] has investigated the compressive strength of concrete at high temperatures, and a typical curve from his paper is plotted in Figure 3 for the purpose of comparison. It should be noted that this curve may undergo considerable modification, both in location and slope, for example, for varying aggregates or ages of the material tested.

The strain, ε_{max}, at the peak of the stress-strain curves is plotted for different temperatures in Figure 4. These are the only data of their kind available in the literature, so no comparisons are possible.

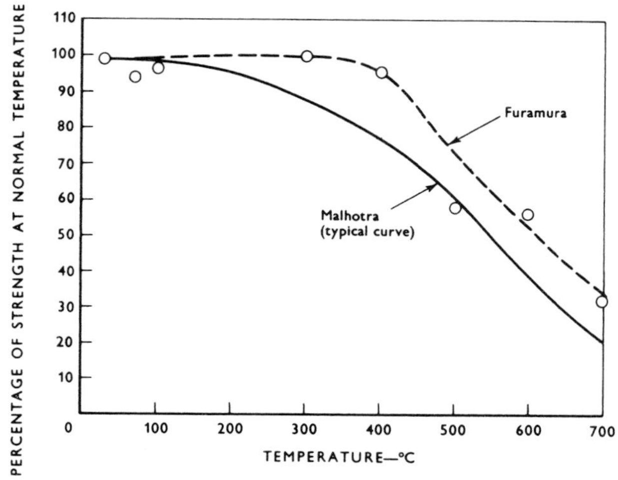

Figure 3: Compressive strength at higher temperatures as a percentage of strength at normal temperatures.

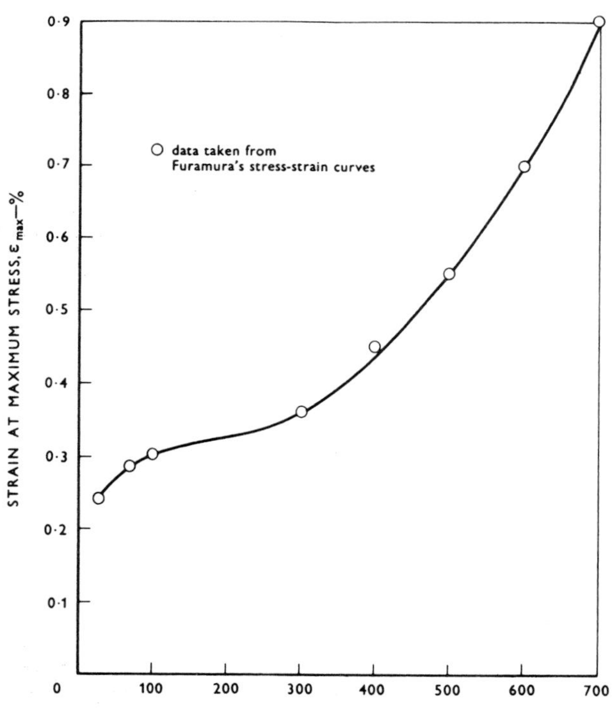

Figure 4: Variation of ε_{max} *with temperature.*

Tensile modulus of elasticity

It has been shown in Figure 2 that σ/σ_{max} is a function of $\varepsilon/\varepsilon_{max}$ only, to a very good approximation, and certainly within the experimental accuracy of the data. The stress-strain curve is distinctly non-linear and may be represented by the following equation (communicated personally by P. H. Thomas)

$$\frac{\sigma}{\sigma_{max}} = \frac{\varepsilon}{\varepsilon_{max}} \exp\left(1 - \frac{\varepsilon}{\varepsilon_{max}}\right)$$

over the whole range of $\varepsilon/\varepsilon_{max} > 0$.

Over the early part of the curve, for small values of $\varepsilon/\varepsilon_{max}$, the curve is approximately linear and then

$$\frac{\sigma}{\sigma_{max}} = K\frac{\varepsilon}{\varepsilon_{max}}$$

where K is constant, so that

$$\sigma = \left(K\frac{\sigma_{max}}{\varepsilon_{max}}\right)\varepsilon$$

Writing E for the tensile modulus of elasticity, where $\sigma = E\varepsilon$, we now have

$$E = K\frac{\sigma_{max}}{\varepsilon_{max}}$$

The constant K in this equation can be determined from the values of E and σ/σ_{max} at room temperature, with E derived from the slope of the stress-strain curve for small values of ε. This gives

$$K = 2{\cdot}18 \text{ and } E = 2{\cdot}18\frac{\sigma_{max}}{\varepsilon_{max}}$$

By using this equation and deriving σ_{max} and ε_{max} from the stress-strain curves, the variation of E with temperature can be determined. This is plotted in Figure 5. Also plotted on this curve are the corresponding values of E derived by measuring the initial slope of the stress-strain curves where, for the sake of uniformity, the part of the curve lying between zero and one-quarter compressive strength has been taken to be linear. This is a reasonable approximation for all temperatures. The agreement between the two sets of points is remarkable.

Measurements of E for concrete at different temperatures have been reported by Cruz[5] and the average of his curves is plotted in Figure 6, together with the data of Furamura. Cruz remarks that an exponential function will fit the plot of E against T. However, the evidence of this in Furamura's data is slight and a straight line is adequate over most of the range, particularly for $T > 100°C$, implying that E decreases linearly with temperature. The difference in slope in the region $T < 100°C$ is presumably due to the presence of moisture. The data of Furamura lie well within the limits derived by Cruz[5] for concrete of different materials.

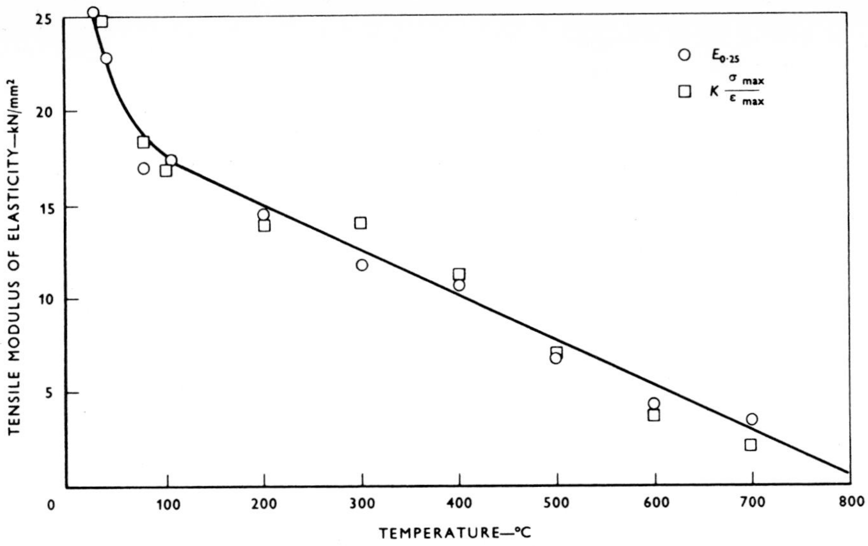

Figure 5: Comparison between two estimates of tensile modulus of elasticity (data from Furamura's stress-strain curves). $E_{0\cdot 25}$ = E derived from initial slope of stress-strain curve between 0 and 0·25 compressive strength.

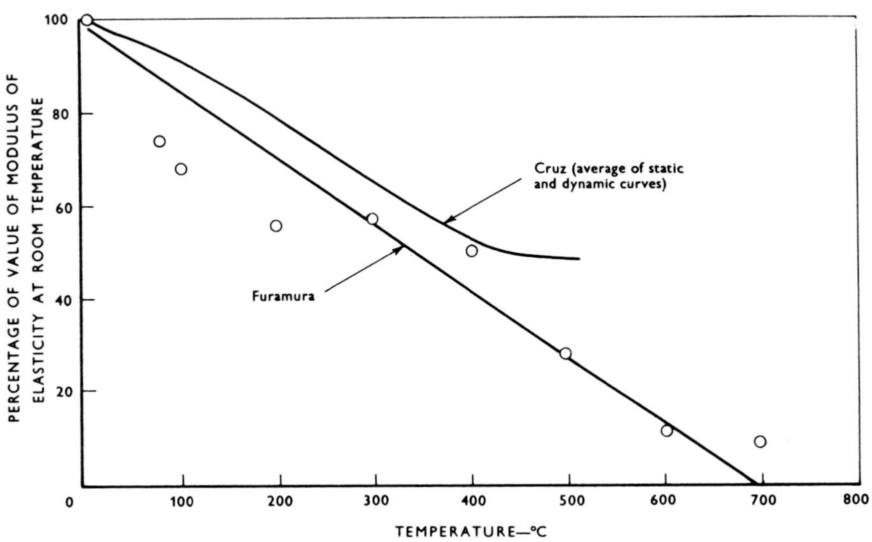

Figure 6: Variation of tensile modulus of elasticity with temperature.

Conclusions

(1) The stress-strain curves for concrete under compression at high temperatures as measured by Furamura show that the curves are of the form

$$\frac{\sigma}{\sigma_{max}} = f\left(\frac{\varepsilon}{\varepsilon_{max}}\right)$$

where f is a functional form independent of temperature, and σ_{max} and ε_{max} are the strain and stress at the peak of the stress-strain curve for a given temperature, and are themselves, therefore, functions of temperature. The data may be represented by the equation

$$\frac{\sigma}{\sigma_{max}} = \frac{\varepsilon}{\varepsilon_{max}} \exp\left(1 - \frac{\varepsilon}{\varepsilon_{max}}\right)$$

(2) This result implies that the stress-strain curve for concrete at high temperatures can be derived from the stress-strain curve at room temperature, together with the location of the maximum of the stress-strain curve at high temperatures—the compressive strength of the material.

(3) The tensile modulus of elasticity at temperature T

can be derived from the expression

$$E = K\frac{\sigma_{max}}{\varepsilon_{max}}$$

where K is a constant which can be derived from the stress-strain curve at room temperature. For Furamura's data, $K = 2{\cdot}18$. This method is in good agreement with values of E derived from measuring the slope of stress-strain curves at different temperatures.

(4) The modulus of elasticity appears to decrease approximately linearly with temperature, particularly when $T > 100°C$.

(5) The data of Furamura are in agreement with the measurements of Cruz[5].

REFERENCES

1. FURAMURA, F. The stress-strain curve of concrete at high temperatures—No. 1. Paper No. 7004 for Annual Meeting of the Architectural Institute of Japan, 1966. (In Japanese. Copy held by Fire Research Station Library, references D17.CTF67/23. 1966.) Summary (also in Japanese but with three Figures) printed in extra number of *Transactions of the Architectural Institute of Japan* (Summaries of technical papers of Annual Meeting of A.I.J., 1966). October 1966. p. 686.
2. MALHOTRA, H. L. The effect of temperature on the compressive strength of concrete. *Magazine of Concrete Research.* Vol. 8, No. 23. August 1956. pp. 85–94.
3. SAEMANN, J. C. and WASHA, G. W. Variation of mortar and concrete properties with temperature. *Journal of the American Concrete Institute. Proceedings* Vol. 54, No. 5. November 1957. pp. 385–395.
4. PHILLEO, R. Some physical properties of concrete at high temperatures. *Journal of the American Concrete Institute. Proceedings* Vol. 54, No. 10. April 1958. pp. 857–64.
5. CRUZ, C. R. Elastic properties of concrete at high temperatures. *Journal of the PCA Research and Development Laboratories.* Vol. 8, No. 1. January 1966. pp. 37–45. Reprinted as: Research Department Bulletin 191.

Deflections of reinforced concrete beams (CP 3/73)

R.F. Stevens

A series of sustained loading tests on reinforced concrete beams is described, in which the major factors affecting the development of deflexion and cracking were varied. The results of the tests are discussed and methods for estimating deflexion of reinforced concrete beams are proposed.

Introduction

There have been considerable increases in the strengths of reinforcing steels and structural concretes since the early days of reinforced concrete. For example, in the UK permissible stresses in reinforcement have risen from 110 MN/m^2 to 350 MN/m^2 and permissible stresses in concrete have risen from 4·3 $MN/^2$ to 10·5 MN/m^2 for nominal mixes. There has been no increase in elastic modulus of steel, and the increase in stress (hence the use of higher working stresses in steel and concrete), although correctly maintaining an adequate factor of safety against collapse, has given rise to local damage due to increased deflexion.

2. In discussion of the practical consequences of deflexion of reinforced concrete beams it has been suggested that long-term movements of a structure should be considered if the structure is to fulfil its purpose satisfactorily.[1] The increasing importance of deflexion has been recognized by the Draft Unified Code for Structural Concrete[2] in which the concept of providing an adequate factor of safety against collapse is extended to include adequate factors of safety against excessive deflexion.

3. This Paper describes a study of the behaviour of reinforced concrete beams at the Building Research Station with the aim of providing information for design for deflexion.

Current methods of estimating deflexions

Short-term deflexions

4. When subjected to load, a section of a beam undergoes compression on one side and tension on the other, which gives rise to a local curvature at that section. The local curvature ϕ may be expressed by

$$\phi = \frac{\epsilon_c + \epsilon_t}{d}$$

where ϵ_c is the maximum compressive strain at the section, ϵ_t is the maximum tensile strain at the section and d is the depth of the section. For a beam consisting of an elastic material, the local curvature ϕ is

$$\phi = M/EI$$

where M is the moment applied to the section, E is the elastic modulus and I is the second moment of area of the section.

5. The deflexion of the beam Δ may be found by solving the differential equation based on the local curvature

$$\Delta = k\phi l^2$$

where k is a constant, depending on the variation of curvature along the beam, and l is the length of the beam.

6. At low loads a typical load/deflexion relationship for a reinforced concrete beam coincides with that calculated by the elastic behaviour of the uncracked section. At the onset of cracking, there is a major change in the slope of the load/deflexion curve, due to a reduction in flexural rigidity at the cracked section. With larger load more cracks form, and the cracks propagate in height, causing further gradual reductions in the flexural rigidity of the beam. In the range of working loads observed deflexions are less than those calculated on the cracked section, due to the contribution of concrete in tension. The assessment of the initial deflexion of a reinforced concrete beam is influenced considerably by the estimation of when cracking occurs.

7. A major analysis of American tests on concrete beams has been made by Yu and Winter,[3] who found good agreement with test results and two simplified engineering methods of calculation. One method was an elastic calculation with a cracked section; the other was a refinement of the first method to allow for the influence of concrete in tension.

8. The European Concrete Committee (CEB) has given[4] a simplified method for determining deflexions under short-term loads, in which the value of the instantaneous deflexion was considered equal to the sum of the deflexion of the uncracked section under the moment at which cracking is produced, and the deflexion of the cracked section under a moment equal to the working moment less the moment at which cracking is produced. The moment at which cracking takes place is calculated from the tensile strength of concrete in bending. This is a sound and logical method, which can allow for the application of a load lower than the working load, but there is some uncertainty in assessing the tensile strength of concrete, as this can be reduced by differential shrinkage. In fact, the CEB suggested that 'where there is considerable shrinkage, this strength must be halved'.

Long-term deflexions

9. When subjected to sustained load, the compressive strains in a reinforced concrete beam increase due to creep and shrinkage of the concrete. The depth of the neutral axis increases with time, and there is a concurrent reduction in maximum compressive stress in the concrete. Concrete which is just below the neutral axis under short-term loading is subjected to tension; under the action of sustained loads this concrete will be finally subjected to compression. Little attention has been paid to the tensile force and stress distribution in the concrete under short-term loads. Even less is known about the changes with time under sustained loads, although this has been observed in a study of floor beams by Hollington.[5]

10. One of the earliest analyses of long-term deflexions, which was supported by tests carried out by Glanville and Thomas[6] was the use of an effective or reduced modulus of elasticity of concrete. This allowed for elastic and creep strains, and was used with the procedures of structural mechanics, i.e. that the section was assumed to be fully cracked to the neutral axis over the full length of the beam, and deflexions were calculated using the second moment of area of the cracked section and

$$E'_c = \frac{E_c}{1 + cE_c}$$

where E'_c is the effective concrete modulus, E_c is the initial concrete modulus and c is the specific creep. The effects of shrinkage are ignored in this method.

11. In their study of deflexions in America, Yu and Winter[3] also developed two simplified methods of calculation for long-term deflexions, one using an effective modulus of elasticity and the other considering the long-term deflexion as a simple multiple of the instantaneous deflexion.

12. The CEB[4] recommended that long-term deflexions should be calculated by trebling that part of the deflexion which corresponds to that part of the load which may be considered long-term. This implies only one particular value of the creep coefficient, a value of 2, and relates to an uncracked section. Shrinkage is ignored. This is a crude approach to the calculation of long-term deflexions, which is made even less reliable by the difficulty of assessing the initial cracking moment.

13. As a result of sustained loading tests on concrete beams, Hajnal-Koyni[7] found that the CEB estimate of long-term deflexion was a reasonable overall approximation, but better agreement between measured and calculated deflexions was obtained with an 'effective modulus of elasticity'.

14. The proposals for calculation of deflexions in the Draft Unified Code are based on the American approach of considering that long-term deflexion is directly related to initial deflexion.

Sustained loading tests

15. The major factors affecting the development of deflexions under sustained loading appeared to be strength of reinforcing steel, maximum compressive stress in concrete and atmospheric environment. A series of sustained loading tests was carried out at the Building Research Station to include these major factors.

Description of beams

16. The main series of beams comprised 24 pairs of beams, the variables being three strengths of reinforcing steel, two sizes of concrete cover, two proportions of reinforcement and two types of environment, one controlled and one natural. The beams were 5·2 m long, 200 mm wide and 385 mm deep. They were cast in gravel concrete designed to have a cube crushing strength of 34 MN/m^2 at 28 days.

17. In addition to the main series of beams, further pairs were cast to investigate additional features: type of concrete, compression reinforcement and size of specimen. The additional specimens consisted of two pairs of beams made with lightweight concrete, one pair of beams with no compression reinforcement, one pair of beams with compression reinforcement equal to tension reinforcement, and one pair of beams with dimensions twice those of other beams. Details of the combination of the main variables are given in Table 1, together with the system of lettering and numbering of the beams.

18. Beams undergoing long-term loading in the Structures Laboratory are shown in Fig. 1.

Results of tests

Creep measurements

19. Two creep specimens of one quarter of the beam section 100 mm × 190 mm × 710 mm long were spring loaded to 10 N/mm^2 and 5 N/mm^2 to observe the creep characteristics of the gravel concrete used. These specimens were provided with accompanying shrinkage prisms.

Table 1. Details of beams

Beam type	Type of steel	Amount of steel	Number and size of bars, *mm*	Concrete cover, *mm*	Exposure
A	Mild steel	Balanced	2 × 32	25	Laboratory
A	Mild steel	Balanced	2 × 32	25	Natural
B	Mild steel	Half balanced	2 × 22	25	Laboratory
B	Mild steel	Half balanced	2 × 22	25	Natural
C	Mild steel	Balanced	2 × 32	51	Laboratory
C	Mild steel	Balanced	2 × 32	51	Natural
D	Mild steel	Half balanced	2 × 22	51	Laboratory
D	Mild steel	Half balanced	2 × 22	51	Natural
E	High strength	Balanced	2 × 25	25	Laboratory
E	High strength	Balanced	2 × 25	25	Natural
F	High strength	Half balanced	2 × 19	25	Laboratory
F	High strength	Half balanced	2 × 19	25	Natural
G	High strength	Balanced	2 × 25	51	Laboratory
G	High strength	Balanced	2 × 25	51	Natural
H	High strength	Half balanced	2 × 19	51	Laboratory
H	High strength	Half balanced	2 × 19	51	Natural
J	Very high strength	Balanced	2 × 22	25	Laboratory
J	Very high strength	Balanced	2 × 22	25	Natural
K	Very high strength	Half balanced	2 × 16	25	Laboratory
K	Very high strength	Half balanced	2 × 16	25	Natural
L	Very high strength	Balanced	2 × 22	51	Laboratory
L	Very high strength	Balanced	2 × 22	51	Natural
M	Very high strength	Half balanced	2 × 16	51	Laboratory
M	Very high strength	Half balanced	2 × 16	51	Natural
Ra	High strength	Balanced	2 × 38	51	Laboratory
Rb	High strength	Balanced	6 × 22	51	Laboratory
X	High strength	Balanced	2 × 25	25	Laboratory
X	High strength	Balanced	2 × 25	25	Natural
Y	High strength	Balanced	2 × 25	25	Laboratory
Z	High strength	Balanced	2 × 25	25	Laboratory

Beams R were double size.
Beams X were made with lightweight aggregate (medium Lytag).
Beams Y had neither stirrups nor carrier bars in central zone.
Beams Z had compression steel similar to tensile.

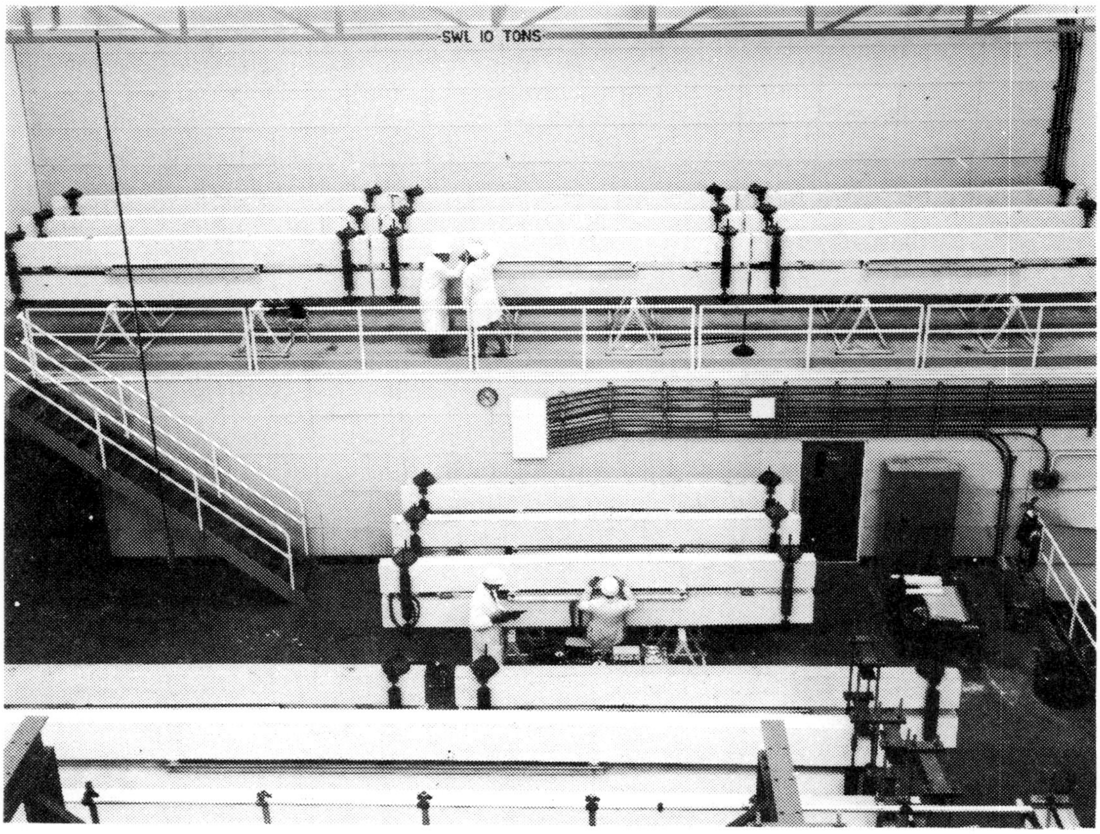

Fig. 1. Beams in Structures Laboratory

20. The creep coefficients (C is creep strain/elastic strain) were 2·4 for the specimen loaded to 10 N/mm^2 and 3·0 for that loaded to 5 N/mm^2. The mean value of 2·7 was adjusted according to the 'fictitious thickness' concept of the CEB resulting in a coefficient of 2·6 for the 205 mm × 385 mm beams and 2·2 for the 410 mm × 770 mm beams.

21. The results for the beams stored in the natural environment indicated a reduction in creep compared with that observed in the laboratory. Such a reduction might be expected to result from exposure at a mean relative humidity of 83% compared with that at a relative humidity of 65%.

Strains and deflexions

22. The changes in the distribution of strain with time for a typical beam are shown in Fig. 2. The distribution of a strain through the depth of the beam remained linear throughout sustained loading.

23. The depth of the neutral axis increased with time because of the growth of compressive strain in the concrete. The maximum observed compressive strain in concrete, the tensile strain in steel and the deflexion increased with time at a reducing rate, and observations for a typical beam are shown in Figs 3–5 for laboratory and natural exposure.

24. The average values of the maximum observed compressive strain in concrete, the tensile strain in the steel and the central deflexion for the constant moment region of the beams are given at initial loading, after one year and after two years for all beams in Tables 2 and 3. Table 2 gives the results for beams in the Structures Laboratory where the atmosphere was controlled at 19 C and 65% relative humidity. Table 3 gives the results for beams under natural conditions at the exposure site, although protected from direct rainfall.

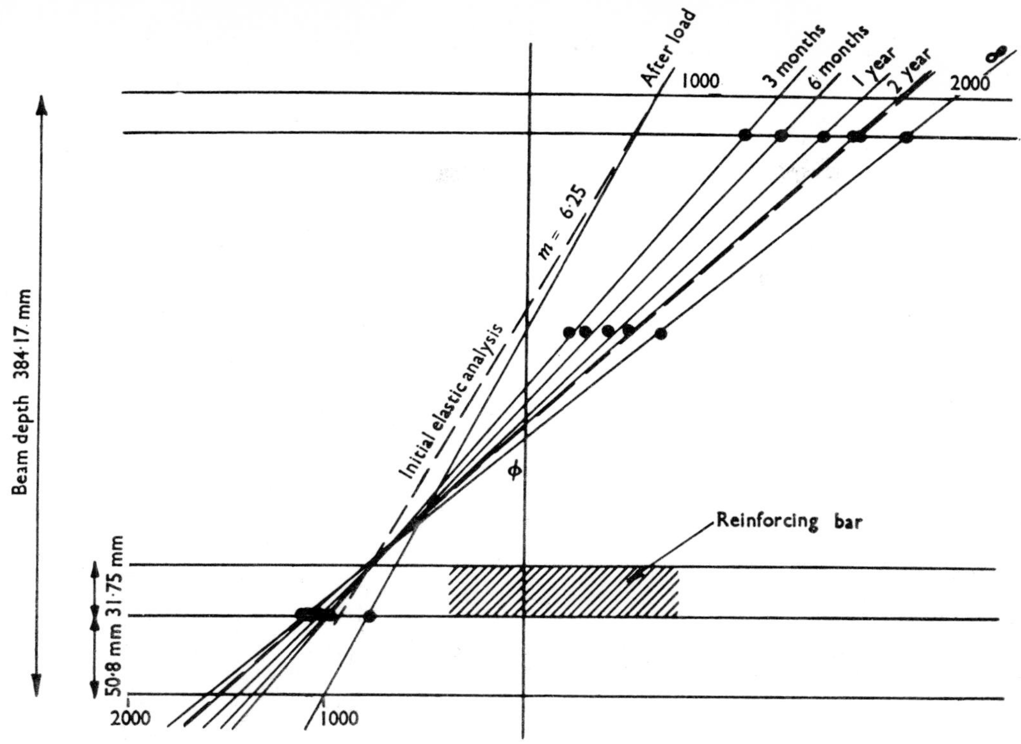

Fig. 2. Distribution of strain with time

25 The initial maximum compressive strain in the concrete was close to that estimated by elastic analysis assuming that concrete does not resist tension, using the values of modulus of elasticity for concrete given by tests on cylinders. In contrast the initial tensile strain in the reinforcing steel was appreciably less than was anticipated by this analysis (Fig. 2), which must indicate that tensile forces were developed in the concrete.

26 A simple approximation to the tensile force in the concrete may be obtained by considering that the initial compressive force in the concrete is equal to or slightly greater than that estimated by elastic analysis, so that the initial tensile force in the steel plus the initial tensile force in the concrete is equal to or slightly greater than the tensile force in the steel estimated by elastic analysis. Hence, if T is the tensile force in the steel estimated by elastic analysis, T' is the initial tensile force in the steel calculated from observed strains and T_c is the tensile force in the concrete, then $T_c \geqslant T - T'$.

Table 2. Results of tests—Structures Laboratory

Beam type	Age, *years*	Strain $\times 10^{-6}$		Deflexion, *mm*
		Concrete	Steel	
A	0	546	769	1·93
	1	1452	948	3·51
	2	1610	968	3·96
B	0	215	496	1 02
	1	687	582	1 80
	2	756	597	2 16
C	0	533	778	2·16
	1	1416	988	3 99
	2	1600	1050	4 32

Table 2—continued

Beam type	Age, *years*	Strain × 10^{-6}		Deflexion, *mm*
		Concrete	Steel	
D	0 1 2	178 697 777	315 423 508	0·79 1·80 2·16
E	0 1 2	515 1304 1480	1056 1319 1340	2·29 3·81 4·14
F	0 1 2	334 831 936	828 998 1069	1·42 2·46 2·72
G	0 1 2	483 1311 1450	976 1301 1310	2·26 4·17 4·39
H	0 1 2	287 811 900	730 927 990	1·65 2·85 3·12
J	0 1 2	534 1228 1341	1262 1495 1535	2·72 4·06 4·32
K	0 1 2	266 728 811	808 1083 1221	1·58 2·54 2·85
L	0 1 2	546 1493 1653	1261 1611 1650	2·87 4·93 5·21
M	0 1 2	276 818 927	873 1192 1257	1·85 3·35 3·61
Ra	0 1 2	270 655 722	708 832 874	3·10 4·57 4·95
Rb	0 1 2	365 778 850	831 933 952	3·18 4·78 5·33
X	0 1	783 1700	1122 1280	2·85 4·55
Y	0 1	633 1700	1150 1470	2·62 4·65
Z	0 1	453 920	1006 1200	2·21 3·12

Table 3. Results of tests—natural exposure

Beam type	Day and month of loading	Age, *years*	Strain × 10^{-6}		Deflexion, *mm*
			Concrete	Steel	
A	7 December	0 1 2	507 1236 1263	752 786 916	1·96 2·69 2·87
B	31 August	0 1 2	247 617 —	515 636 —	1·45 1·52 —
C	13 June	0 1 2	545 1164 —	772 889 —	2·11 3·05 —
D	19 October	0 1 2	184 565 —	356 414 —	0·81 1·49 —
E	7 July	0 1 2	519 1031 —	1061 1313 —	2·36 3·30 —
F	7 September	0 1 2	334 747 804	827 884 1017	1·55 2·44 2·69
G	23 June	0 1 2	470 — —	952 — —	2·18 — —
H	26 October	0 1 2	289 740 —	742 852 —	1·60 2·36 2·41
J	23 November	0 1 2	519 1090 1057	1223 1320 1433	2·69 3·33 3·43
K	14 September	0 1 2	268 588 627	819 1051 1088	1·70 2·46 2·54
L	23 December	0 1 2	575 1420 1370	1241 1318 1499	2·97 4·22 4·42
M	2 November	0 1 2	278 824 817	807 1061 1203	1·65 2·97 3·02
X	8 November	0 1	765 1258	1115 1258	2·82 3·79

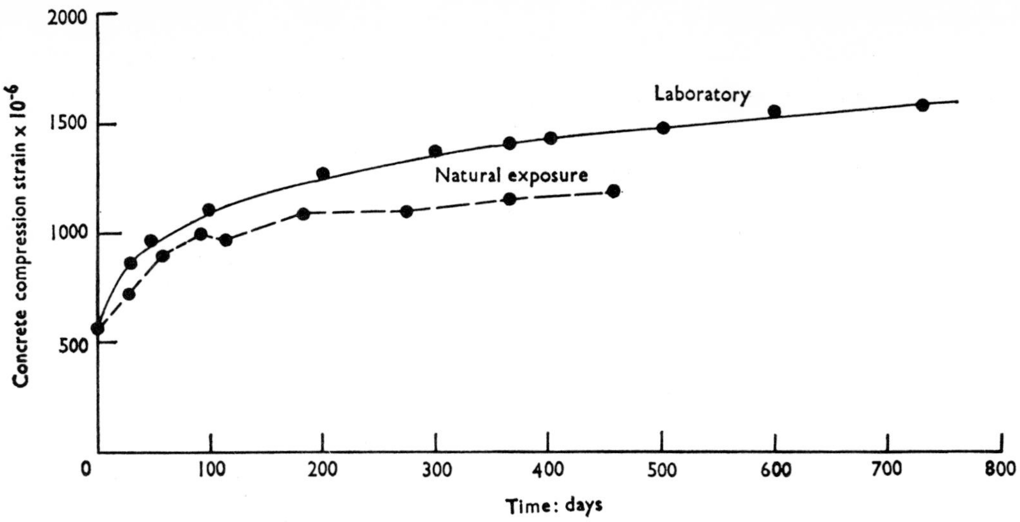

Fig. 3. Change in maximum concrete compressive strain

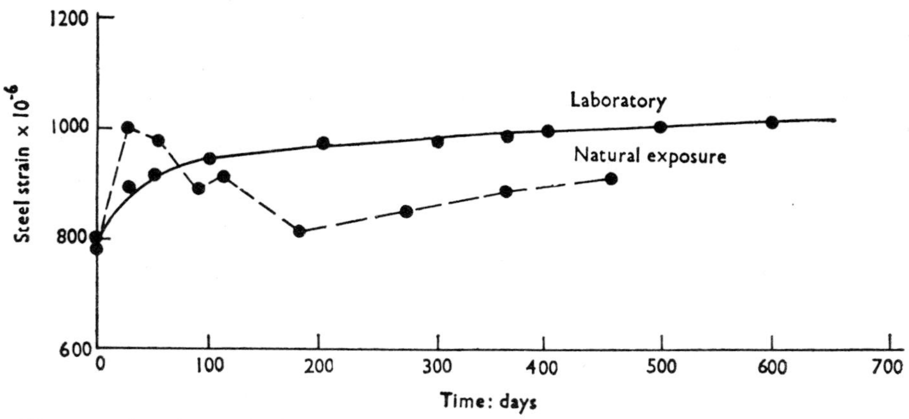

Fig. 4. Change in steel strain

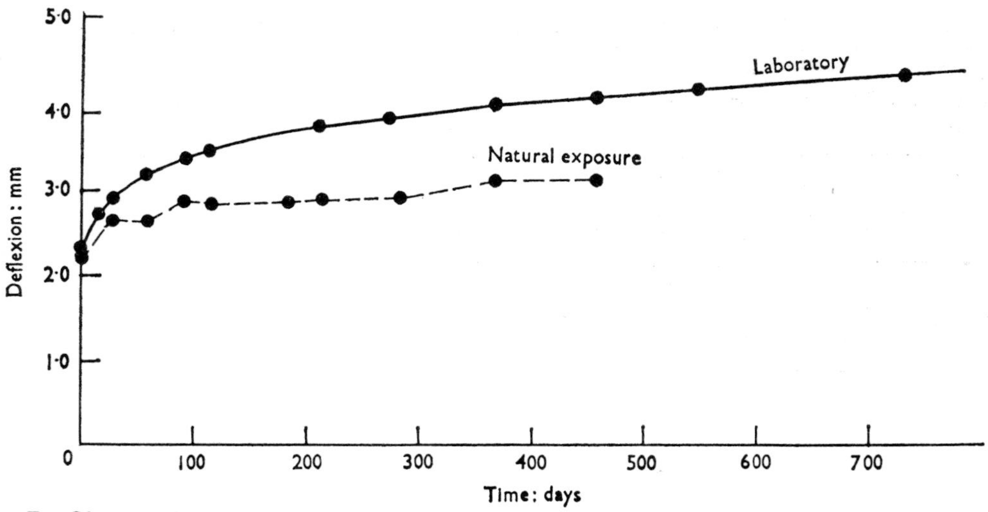

Fig. 5. Change in central deflexion

Table 4

Beam	Duration of loading, *years*	Calculated deflexion from observed strains, *mm*	Observed deflexion, *mm*
A	0 1	1·98 3·61	1·93 3·51
B	0 1	1·09 1·88	1·01 1·80
C	0 1	2·14 3·91	2·16 3·96
D	0 1	0·79 1·80	0·79 1·80
E	0 1	2·41 3·91	2·29 3·81
F	0 1	1·70 2·69	1·55 2·46
G	0 1	2·34 4·21	2·26 4·17
H	0 1	1·63 3·07	1·65 2·85
J	0 1	2·67 4·04	2·72 4·09
K	0 1	1·57 2·67	1·57 2·54
L	0 1	2·90 4·98	2·87 4·93
M	0 1	1·83 3·28	1·85 3·35

Discussion of results

Beams stored in laboratory

27. Observed deflexions and deflexions calculated from local curvatures determined from observed strains are compared in Table 4. The good agreement indicates the high reliability of the experimental results, and suggests that if the strain distribution were known deflexions could be accurately calculated.

28. The values of T_c for all beams under initial loading are shown in Table 5. These values are considerable, ranging from 12% to 54% of the calculated force in the reinforcement. There was no consistent difference between the tensile force developed in the concrete beams reinforced with round bars (beams A–D) and in the beams reinforced with deformed bars (D–M, R_a, R_b), there was no consistent difference in tensile force in the concrete between beams with 25 mm concrete cover (A, B, E, F, J, K) and those with 50 mm concrete cover (C, D, G, H, L, M) and there was no consistent difference in tensile force in the concrete between beams with differing steel perimeters. (The perimeters of steel for beams A, C, E, G, J, L were approximately 40% greater than those for beams B, D, F, H, K, M.) The most apparent relationship emerging from Table 5 is between the tensile force in the concrete and the size of the beam. The tensile force in the concrete before cracking would be estimated as

$$\frac{bd_0}{2}\frac{f'_c}{2}$$

where f'_c is the tensile strength of concrete in bending. This is about 61 kN for the beams A–M, and 240 kN for beams R_a and R_b.

29. The average tensile force in concrete after cracking T_c ranged between 43% and 86% of the tensile force just before cracking, with an average value of about 75%. The average tensile force in the concrete after cracking may be estimated as

$$T_c = \frac{3}{16} bd_o f'_c$$

The changes with time of the average tensile force in the concrete have been analysed, assuming that the distribution of compressive stress in the concrete

Table 5. Tensile force in concrete

Beam	T, kN	T', kN	T_c, kN	T_c/T, %
A	274	222	53	19
B	101	72	29	28
C	275	223	52	19
D	98	45	53	54
E	225	199	26	12
F	125	87	37	30
G	214	183	31	15
H	121	78	42	35
J	228	184	44	21
K	110	56	55	49
L	227	183	44	19
M	113	66	45	41
R_a	451	314	147	33
R_b	531	372	161	30

Table 6. Changes in tensile force

Beam	Tensile force, *kN*		
	Initial	At 1 year	At 2 years
A	53	36	34
B	29	27	24
C	52	23	18
D	53	48	37
E	26	3	0
F	37	26	10
G	31	0	0
H	42	32	25
J	44	24	22
K	55	37	26
L	44	13	10
M	45	30	26
R_a	147	130	110
R_b	161	140	134

remains linear. Although this assumption may not be strictly valid for those portions of concrete beams which are initially in tension and in time become subjected to compression, it is thought to give the broad picture of behaviour of the beam. The results of the analysis are shown in Table 6. In all cases the value of the tensile force reduced with time, and for two of the beams reduced to zero. After two years' sustained loading, the average value of tensile force in the concrete was about one third of that developed just before cracking.

30. The changes in the average tensile force with time are shown for the typical beam in Fig. 6, indicating that the tensile force in concrete of a cracked beam reduces with time at a reducing rate, a pattern akin to that of relaxation of stress of concrete under sustained load.

31. The rates of increase of maximum compressive strain in concrete, tensile strain in steel and central deflexion with time have been analysed following a method used by Ross[8] for hyperbolic functions, and the good agreement found with the rectangular hyperbolic form is shown in Fig. 7. Using this analysis, it has been possible to predict ultimate values of compressive strains in concrete, tensile strains in steel and central deflexions, which are given in Table 7.

32. Estimated ultimate values of compressive strain for the beams obtained by adding shrinkage strains and the product of the creep coefficient and the initial strain were greater than those predicted by hyperbolic analysis of observed strains. Comparison of creep strains in the concrete obtained by subtracting shrinkage strains from ultimate values of compressive strain showed that these strains were equal to the product of two thirds of the creep coefficient from control specimens and the initial concrete compressive strain. The reduction factor of two thirds is due to the reduction of maximum concrete compressive stress with time under sustained loading.

Calculation of deflexions

33. Initial deflexions of the beams calculated by two different methods are set out in Table 8.

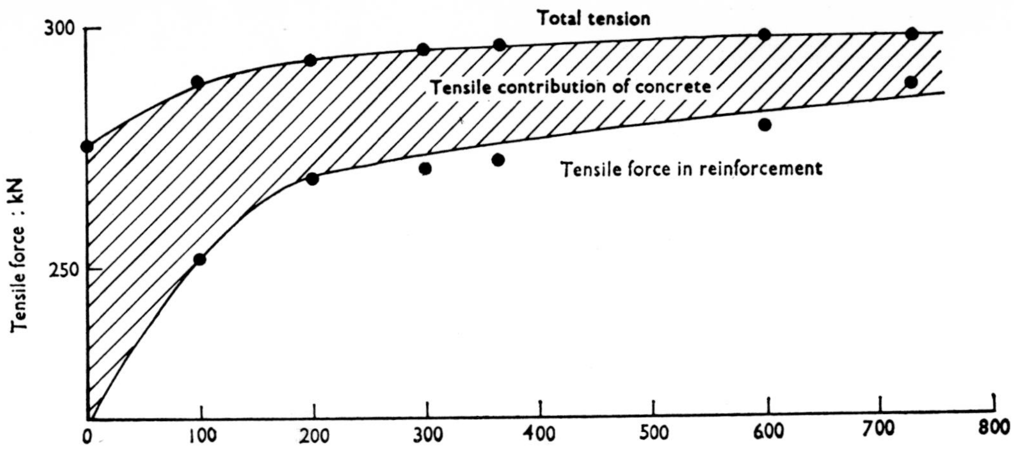

Fig. 6. Loss of tensile resistance of concrete

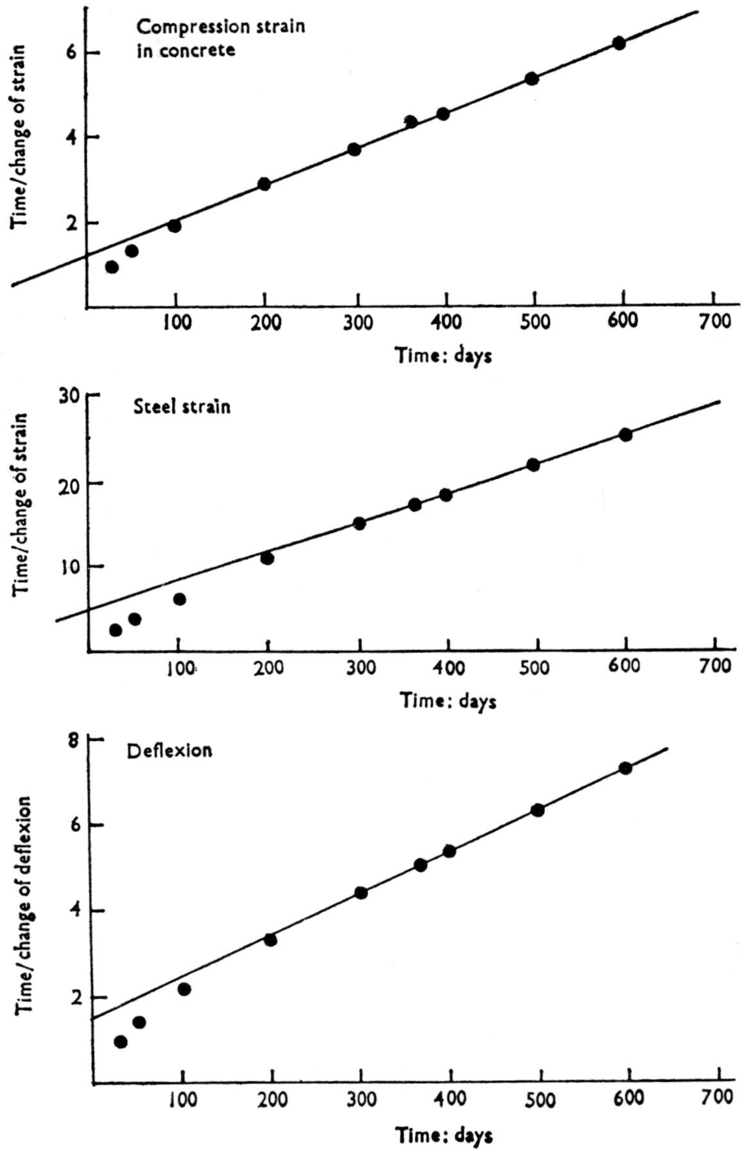

Fig. 7. Hyperbolic analysis

Table 7. Ultimate values

Beam type	Strains × 10^{-6}		Deflexion, *mm*
	Tensile	Compressive	
A	1000	1800	4·45
B	615	880	2·16
C	1080	1800	4·90
D	935	835	2·80
E	1360	1710	4·57
F	1200	1120	3·99
G	1300	1620	4·67
H	1010	1030	3·68
J	1610	1480	4·78
K	1460	900	3·51
L	1700	1900	5·64
M	1730	1130	4·01
X	1280	2180	5·26
Y	1460	2160	5·84
Z	1200	1040	3·30
R_a	950	860	5·64
R_b	990	980	7·42

34. In the first method an elastic analysis is used to determine the maximum compressive strain in the concrete and the tensile strain in the steel, assuming that concrete does not resist tension. In the second method a modified elastic analysis is used in which concrete is assumed to resist tension between cracks, so that the tensile strain in the steel is less than that given by conventional elastic analysis.

35. The immediate deflexions calculated by the first method overestimated the actual deflexions, the ratio of calculated deflexions to observed deflexions being 1·47, with a standard deviation of 0·14. The second method gave good agreement with actual deflexions (ratio of calculated deflexions to observed deflexions being 0·96, and the standard deviation being 0·08). Calculations of final deflexion of the beams are shown in Table 8, using two methods. Table 8 also shows observed deflexions at two years, and values of estimated final deflexions based on observations over two years.

36. In the third method an elastic analysis is used to determine the strains in concrete and steel, but an effective modulus of elasticity of concrete is chosen so that

$$E'_c = E_c \frac{1}{1 + C_t}$$

where E'_c is the effective modulus of concrete, E_c is the initial modulus of concrete and C_t is the creep coefficient.

37. In the fourth method the maximum compressive strain in the concrete is determined from the initial strain, by allowing for creep and shrinkage in the relationship

$$\epsilon_{ct} = \epsilon_{ci}\left[1 + \frac{2}{3} C_t\right] + \epsilon_s$$

where ϵ_{ct} is the maximum compressive strain in concrete, ϵ_{ci} is the maximum initial compressive strain in concrete and ϵ_s is the shrinkage strain.

Table 8. Calculation of deflexion

Beam	Observed on loading, *mm*	Calculated method 1, *mm*	Ratio Calculated/observed	Calculated method 2, *mm*	Ratio calculated/observed	Beam	Observed deflexion and 2 years, *mm*	Estimated final deflexion from observations, *mm*	Calculated method 3, *mm*	Ratio calculated/established	Calculated method 4, *mm*	Ratio calculated/established
A	1·93	2·08	108	1·85	96	A	3·96	4·45	3·51	79	4·01	90
B	1·01	1·32	130	0·86	85	B	1·96	2·16	1·98	92	2·49	115
C	2·16	2·26	105	2·05	95	C	4·32	4·90	3·99	81	4·47	91
D	0·79	1·35	172	0·86	109	D	2·16	2·79	2·06	74	2·59	93
E	2·29	2·39	105	2·03	89	E	4·14	4·57	3·73	82	4·27	94
F	1·55	2·08	134	1·45	94	F	2·82	3·99	2·92	73	3·51	88
G	2·26	2·49	110	2·10	93	G	4·39	4·67	3·94	84	4·50	96
H	1·65	2·21	133	1·52	92	H	3·12	3·68	3·15	85	3·76	102
J	2·72	2·92	107	2·59	95	J	4·32	4·78	4·45	93	4·80	100
K	1·57	2·57	163	1·65	105	K	2·85	3·51	3·43	98	4·04	115
L	2·87	3·30	115	2·79	98	L	5·21	5·64	4·95	88	5·59	99
M	1·85	2·82	153	1·85	100	M	3·61	4·01	3·84	95	4·47	112
Mean			128		96	Mean				85		100
Coefficient of variation			14		8	Coefficient of variation				9		8

38. The third and fourth methods gave reasonably good agreement with estimated final deflexions. The third method had an average ratio of calculated deflexion to estimated deflexion of 0·85, with a standard deviation of 0·09, thus tending to underestimate the final deflexion. The fourth method had an average ratio of calculated deflexion to estimated deflexion of 1·00, with a standard deviation of 0·08.

Beams on exposure site

39. Analysis of the behaviour of beams on the exposure site was complicated by variations in temperature and relative humidity. Changes of strain and deflexion were much affected by the time of year the beams were exposed. This is illustrated in Fig. 8, which shows the observed central deflexion for beam A. Beam A was placed out of doors in December, and there was little increase in deflexion until April. Deflexions continued to grow through the summer, followed by a stationary period from September to the following April. Between May and July the deflexion again increased, but the growth was about one third of that of the previous summer.

40. The beams on the exposure site showed less creep and shrinkage than companion beams in the controlled environment of the Structures Laboratory. The observed creep of concrete on the exposure site was about 80% of that observed in the laboratory, and the shrinkage was about 60% of that in the laboratory.

41. In the CEB recommendations[4] charts are given for estimating the effect of relative humidity on creep and shrinkage. Using these charts for concrete exposed to an average relative humidity of 82%, creep would be about 75% and the shrinkage 60% of concrete exposed at a relative humidity of 65%. These values are close to observed values, and would suggest that out of door deflexions may be calculated using the average relative humidity and the CEB charts.

42. In relation to the additional beams which were subjected to sustained loads for one year it was observed that

(*a*) creep in the beams without carrier bars, i.e. with no steel in the compression zone, was slightly greater than creep of beams with carrier bars

(*b*) the addition of compression steel equal to that provided in the tension zone approximately halved the amount of creep

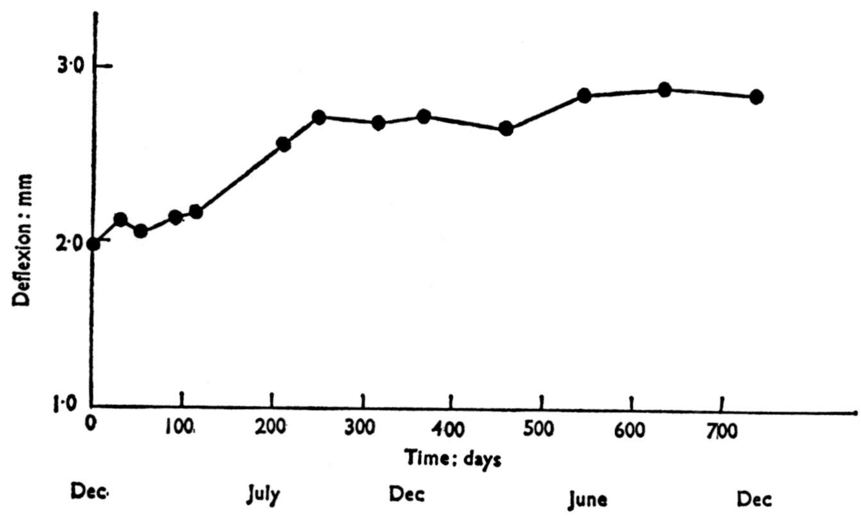

Fig. 8. Deflexion of beam A on exposure site showing seasonal variation

(c) the initial deflexion of the lightweight concrete beams was greater than that of corresponding gravel concrete beams, as would be anticipated from a lower elastic modulus for the concrete; subsequent changes in deflexion were similar to those of gravel concrete beams.

Conclusions

43. The following conclusions are drawn for reinforced concrete beams in bending under sustained working loads.

Strains and deflexions

44. The average tensile force in the concrete diminishes with time at a reducing rate, in a manner akin to relaxation of stress of concrete.

45. The maximum compressive strain in the concrete, the maximum tensile strain in the steel and the maximum central deflexion increase with time at a reducing rate. The rate of increase of these functions with time is of the form of a rectangular hyperbola, which permits ultimate strains and deflexions to be deduced.

46. Concrete beams exposed in a natural environment display about three quarters of the deflexions of companion beams in the controlled environment of the laboratory, which supports the concept of less creep at higher relative humidity.

47. The addition of compression steel equal in amount to that in the tension zone approximately halves the amount of creep.

Calculation of deflexion

48. The initial deflexion may be calculated with reasonable accuracy from local curvature by means of a modified elastic analysis in which concrete is assumed to resist tension between cracks, so that the average tensile strain in the steel is less than that given by conventional analysis.

49. The average tensile force in the concrete is given by

$$T_c = \frac{3}{16} b d_0 f'_c$$

where T_c is the average tensile force in concrete, b is the breadth of the beam, d_0 is the overall depth of the beam and f'_c is the tensile strength of concrete in bending.

50. The long-term deflexion may be calculated with reasonable accuracy by using an elastic analysis with an effective modulus of elasticity to allow for creep

$$E'_c = E_c \left[\frac{1}{1 + C_t}\right]$$

where E'_c is the effective modulus of concrete, E_c is the initial modulus of concrete and C_t is the creep coefficient, or from local curvature assuming the tensile strain in the steel is that appropriate to the working stress, and determining the maximum compressive strain in the concrete in terms of the initial maximum compressive strain, the creep coefficient of the concrete and the shrinkage coefficient of the concrete

$$\epsilon_{ct} = \epsilon_{ci}\left[1 + \frac{2}{3} C_t\right] + \epsilon_s$$

where ϵ_{ct} is the maximum compressive strain in concrete, ϵ_{ci} is the maximum initial compressive strain in concrete and ϵ_s is the shrinkage strain.

Acknowledgements

51. The work described forms part of the programme of research of the Building Research Station of the Department of the Environment, and is published by permission of the Director. The Author wishes to acknowledge the assistance given by his colleagues and in particular Mr D. W. Bryden-Smith with the experimental work.

References

1. *Design for movement in buildings.* Concrete Society, London, 1969.
2. British Standards Institution. *Draft British Standard Code of Practice for the structural use of concrete.* British Standards Institution, London, 1969.
3. Yu W. W. and Winter G. Instantaneous and long-time deflections of reinforced concrete beams under working loads. *Proc. Am. Concr. Inst.*, **57**, 1960, No. 1, July, 29–50.
4. European Concrete Committee. *Recommendations for an international code of practice for reinforced concrete.* Cement and Concrete Association, 1964.
5. Hollington M. R. *The long-term deformation of reinforced concrete members.* PhD thesis, Queen Mary College, London, 1969.
6. Glanville W. H. and Thomas F. G. *Further investigations on the creep or flow of concrete under load.* Building Research Technical Paper 21, HMSO, London, 1939.
7. Hajnal-Koyni K. Tests on beams with sustained loading. *Consult. Engr, Lond.*, 1963, Oct., 437–444.
8. Ross A. D. Concrete creep data. *Struct. Engr*, 1937, Aug., 314–326.

The deflexion of reinforced concrete beams under fluctuating load with a sustained component (CP 14/74)

P.R. Sparks and J.B. Menzies

Introduction

Today practising engineers are using reinforced concrete more efficiently than ever before and flexural members are becoming relatively more slender. Design stresses for the materials in structural reinforced concrete have increased substantially over the years. For example the DSIR Code of Practice 1934[1] specified a permissible stress for concrete in compression due to bending of 5·15 N/mm² (750 lbf/in²) for a 1 :2 :4 nominal mix and in the Code of Practice CP114 (1965)[2] this stress had been increased to 6·89 N/mm² (1000 lbf/in²) for the same mix. Similarly the permissible stress for mild steel reinforcement in tension was 124 N/mm² (18 000 lbf/in²) in 1934; in 1965 the permissible stress for this material was 138 N/mm² (20 000 lbf/in²) and stresses up to 228 N/mm² (33 000 lbf/in²) could be used for high-yield steel. These increased stresses have in effect, been adopted for limit-state design in the new Code of Practice CP110[3] although the greater flexibility of the limit-state method allows some variation in the design stresses in different situations. The increases in stress have been justified either by better quality control of existing materials or by the use of new materials. They have not been accompanied however by corresponding increases in material stiffnesses. Consequently the deflexions of reinforced concrete structures are becoming more important in design for serviceablity and a more detailed knowledge of the factors which influence deflexions is needed.

Fluctuations of imposed loads may influence deflexions. At present deflexions are estimated in design for the serviceability limit state[3] by summation of the immediate deflexions under load and those due to creep. The code states that the 'creep' deflexion depends on the dead-load and the imposed load of long duration and allows this deflexion to be calculated on the assumption that only dead-load and that part of the imposed load likely to be permanent, is effective. The èffect of the fluctuating part of the imposed load on 'creep' deflexion is assumed to be negligible except that attention is drawn to the situation where the imposed loads are predominantly cyclic in character. This guidance will influence designers to ignore the effects of the fluctuating part of the loading although most service loads do not remain constant and, in fact, many have cyclic components. Experience suggests that this simplification allows the design of serviceable structures although the effects of fluctuation of loading on deflexions may be considerable. Research in Belgium[4] has been described in which beams were subjected to fluctuating service loading and others to conditions which were likely to cause fatigue failure in the constituent materials. Increases of deflexion ranging from 18 per cent to more than 40 per cent were observed as the number of fluctuations of loading increased. Similarly, in step-loading tests at the Building Research Station, Snowdon[5] found that deflexions increased by 20 to 25 per cent in the early stages of fatigue loading. Research in America[6] has concentrated mainly on assessing the ability of reinforced concrete structures to resist a few near-ultimate overloads such as may occur in earthquakes and hurricanes. Under service loading, increases of deflexion ranging from 18 per cent to 48 per cent of the initial deflexion were found and up to ten times the initial deflexion at near ultimate loading.

There have been few comparisons made between the deflexions of reinforced concrete structures under loading regimes which have both a fluctuating component and a sustained component such as is likely to occur in service and those under sustained loading only. The design criteria in structural codes of practice have evolved principally from study of the behaviour of members loaded incrementally to failure and others subjected to sustained constant loads. This paper therefore describes an investigation undertaken specifically to determine the influence of fluctuation of loading on the deflexions of reinforced concrete beams and whether the fluctuation causes greater deflexions over the long-term than will occur in any case because of creep movement under a constant load. Twenty beams were made and subjected to a range of loading regimes usually with a fluctuating component.

Test beams and loading apparatus

The reinforced concrete beams were 3·12 m (10·23 ft) long and 200 mm (7·86 in) × 200 mm (7·86 in) square in cross-section. The concrete was made with crushed-limestone aggregate graded one third by weight 4·75 mm–9·5 mm (³⁄₁₆ in–⅜ in) and two-thirds 9·5 mm–19 mm (⅜ in–¾ in). The mix was made with ordinary Portland cement, Ham River sand and Cheddar limestone in the proportions 1 :2·4 :3·6 by weight. The water/cement ratio was 0·55 and the nominal cube strength 45 N/mm² (6500 lbf/in²).

The type of reinforcement was varied from beam to beam. Beams A to I and beam K contained only tensile reinforcement of three 22 mm (⅞ in) diameter high-tensile deformed bars. Beam L had only two of these bars, beam M had three similar bars of 19 mm (¾ in) diameter and beam N contained four 19 mm (¾ in) diameter plain mild-steel bars. Beam J was reinforced with three 22 mm (⅞ in) diameter high-tensile deformed bars which were greased over the middle half of their length to destroy the bond. This beam was included specifically to show, by removing the possibility of changes in bond, the influence of deformational changes in the concrete on the increase in deflexions under load. In addition to tensile reinforcement of three 19 mm (¾ in) diameter high-tensile deformed bars, beams O to T had shear stirrups of 6 mm (¼ in) diameter plain mild-steel bar at 125 mm (5 in) centres and nominal compressive reinforcement of two 10 mm (⅜ in) diameter plain mild-steel bars in the top of the beams.

The beams were cast in wooden moulds in two batches of five beams, one batch of four beams and one batch of six beams. Three control cubes were cast from each mix of concrete. The beams were cured under damp sacking for at least three days before being demoulded and moved to the test laboratory. The control cubes were demoulded after twenty-four hours and then stored under water at 18·3°C (65°F) until tested at an age of 28 days (Table 1). Tension tests were made on samples of each type of tensile reinforcement (Table 1). To reduce the effect of ageing of the concrete during testing each of the beams was at least two months old before it was tested.

The test load was applied to each beam by means of a servo-controlled hydraulic actuator. Each beam was simply-supported over a span of 3 m (9·84 ft) and the centrally applied load was transferred to it through a spreader girder to form two equal vertical loads applied 250 mm (9·84 in) on either side of midspan. Vertical deflexions were measured at midspan using a dial gauge reading to 0·01 mm (4×10^{-4} in) mounted beneath the beam (Fig 1).

Test procedure

The loads applied to each beam are shown diagrammatically in Fig 2. Three beams A, O and Q were subjected to an incrementally increasing load until failure occurred. Beams C, D, G, I–N and R–T were all subjected to fluctuating loads made up of a fluctuating component and a sustained component. For each beam the sustained component was between about 20 per cent and 40 per cent of the load required to produce bending failure of the beam and the fluctuating component increased the total load until the maximum value was between about 35 per cent and 90 per cent of the same failure load. The fluctuating component had a haversine (squared sine) wave form and was usually applied at a rate of 1 Hz. For some beams the first few cycles of load were applied more slowly to enable detailed readings of deformations to be obtained. The numbers of cycles of load applied to each beam is given in Fig 2. Sustained loads were applied to beams B, E, F, H and P; the length of time the load was held constant on each beam is also given in Fig 2. The deflexions of the beams at midspan were measured initially on first application of the load and subsequently at intervals throughout the periods the beams were under fluctuating and sustained load.

At the end of the periods of fluctuating or sustained loading the load was removed from each beam and the recovery of deflexion under zero load was observed for some days. Beams F, G, J–N, P, R and S were then subjected to a steadily increasing load until failure occurred.

Fig 1. Test arrangements

The initial deflexions of the beams

The midspan deflexions of the beams during the first application of load are given in Table 2 together with deflexions predicted by three methods of analysis sometimes used in design. The bi-linear method assumes that the stiffness of the beam has a constant value until the applied bending moment reaches the level which cracks the concrete. Where the bending moment exceeds this value a lower constant stiffness is employed. The other two methods use a constant stiffness throughout the length of the beam. In the transformed-section method this stiffness is assumed to be that of a fully cracked beam while the effective-stiffness method[7] reduces the stiffness in the fully cracked beam by the use of a factor which depends upon the concrete strength and maximum steel stress.

The accuracy of the predictions of the initial deflexions of the beams varied with the method of analysis employed. The bi-linear method was generally the most accurate, predicting in most cases to within 10 per cent of the measured deflexion. However, this method was cumbersome to use and little accuracy was lost by employing one of the other two methods. Of these the transformed-section method was simpler to use but the effective-stiffness method could be adapted to predict long-term deflexions. Two other methods are commonly used to determine beam stiffnesses; they are the concrete-section method and the gross-section method. In the former method the presence of the reinforcement is ignored and the section is assumed to consist entirely of concrete. In the latter method the steel area enhanced by the modular ratio is also included. Whilst these methods are suitable for determining the distribution of moments in redundant structures they gave deflexions of only about half of those measured and they are unsuitable therefore for predicting deflexions of beams.

As expected the area of tensile reinforcement in the sections had a considerable effect on the initial stiffness of the beams. The destruction of the bond in beam J reduced the initial stiffness by about 30 per cent and produced a corresponding decrease of about 30 per cent in ultimate strength compared with beam K which was similarly reinforced but had full bond. Beam N reinforced with mild-steel bars contained a greater cross-sectional area of steel for a given beam strength than one reinforced with high-tensile steel bars and was thus stiffer initially.

Increase in deflexions

The deflexions of all the beams subjected to fluctuating or sustained loading increased with the number of cycles of loading and with the length of time under load. In common with the results reported in the literature[7] the relationship between the maximum deflexion and the number of cycles of fluctuating load or the length of time under sustained load was found to be non-linear. However for beams subjected to fluctuating load a linear relationship was found between the logarithm of the increase in deflexion above the initial deflexion and the logarithm of the number of cycles of loading, i.e. a relationship of the form:

$\Delta = AN^b$ applies where
Δ is the increase in deflexion in mm,
N is the number of cycles of load, and
A and b are constants

The values of A and b obtained from the regression analyses of the logarithms of the measured increases of deflexion on the logarithms of the number of cycles are given in Table 3.

For the beams subjected to a sustained load a similar logarithmic relationship was found between the increase in deflexion and the time under load, i.e. a relationship of the form:

$\Delta = ET^f$ applies where
Δ is the increase in deflexion in mm,
T is the time under load in seconds, and
E and f are constants

This parabolic form of relationship corresponds to that which is commonly used to describe the primary stage of creep of many materials under constant sustained load[8]. Table 4 gives the values of E and f obtained from regression analyses of the logarithms of the measured increases of deflexion on the logarithms of time.

No measurements of deflexions were made on beams A–I between the initial deflexion measurements and those made at about 5000 cycles of loading or the equivalent length of time in the sustained-load tests. Consequently more confidence can be placed in the regression analyses on the observations from the later beams than those from the tests on beams A–I. For the later beams under fluctuating loads (J, K, L, M, N, R, S and T) where the measurements are available throughout the period of testing, the rates of increase in deflexion 'b' were remarkably constant (Table 3). This applies even for beam J where the bond between the reinforcement and concrete had been deliberately destroyed in the central portion

TABLE 1
Properties of beam materials

Beam identifier	Type of specimen	Material property	Measured value of property[1] N/mm²
A – E F – I J – N O – T	Concrete cube; 100 mm size, water-stored 28 days old	Cube strength	55·1 51·6 47·2 45·7
A – L	Steel reinforcing bar; GK 60 22 mm diameter	Yield stress Modulus of elasticity Ultimate stress	415 $2{\cdot}02 \times 10^5$ 691
M and O – T	Steel reinforcing bar; GK 60 19 mm diameter	Yield stress Modulus of elasticity Ultimate stress	401 $1{\cdot}93 \times 10^5$ 653
N	Steel reinforcing bar; mild steel 19 mm diameter	Yield stress Modulus of elasticity Ultimate stress	271 $2{\cdot}01 \times 10^5$ 437

1. Mean of measurement on three specimens

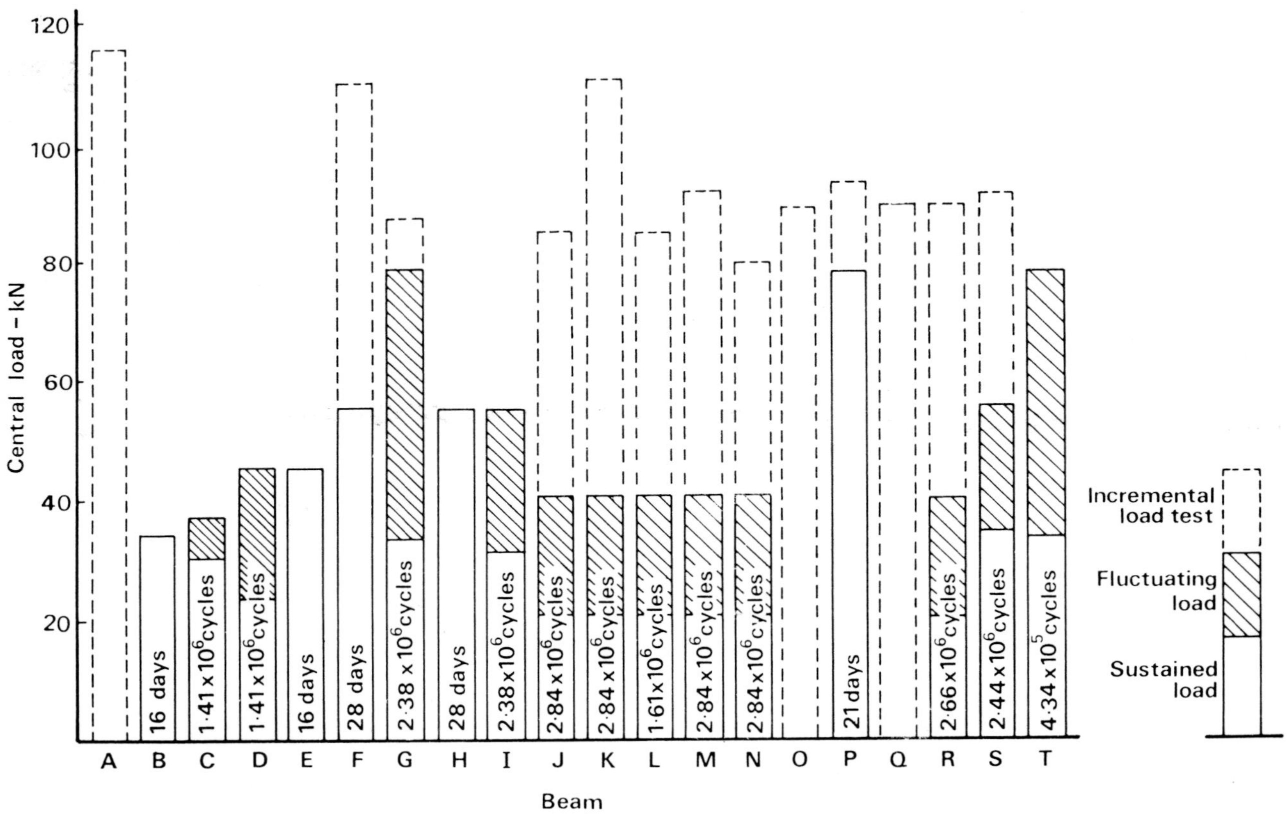

Fig 2. The nature of the loads on each beam

TABLE 2
Initial deflexions at midspan

Beam identifier	Load[1] kN	Measured initial deflexion (mm)	Predicted deflexion (mm) by		
			Bi-linear method	Transformed-section method	Effective-stiffness method
B	33·5	5·24	6·09	6·05	5·37
C	30·0 36·8	4·81 6·11	5·34 6·78	5·41 6·64	4·74 6·15
D	23·4 43·5	3·99 8·68	3·96 8·20	4·22 7·85	3·40 8·25
E	43·5	8·18	8·20	7·85	8·25
F	55·5	11·30	10·73	10·02	9·35
G	33·5 77·5	6·31 16·34	6·09 15·50	6·05 14·08	5·37 13·00
H	55·5	11·72	10·73	10·02	9·35
I	35·6 55·5	7·21 12·01	6·55 10·73	6·44 10·02	5·95 9·35
J	20·0 40·0	4·35 9·12	3·25 7·46	3·61 7·22	2·58 6·80
K	20·0 40·0	3·23 7·48	3·25 7·46	3·61 7·22	2·58 6·80
L	20·0 40·0	4·15 9·91	3·70 8·73	4·32 8·64	3·60 8·25
M	20·0 40·0	4·14 9·83	3·64 8·55	4·21 8·43	3·42 8·50
N	20·0 40·0	2·99 7·08	3·28 7·55	3·66 7·32	3·16 7·10
P	77·5	18·42	17·84	16·36	16·80
R	20·0 40·0	4·13 9·40	3·64 8·55	4·21 8·43	3·42 8·50
S	35·6 55·5	7·51 12·36	7·47 12·40	7·50 11·70	7·35 12·10
T	33·5 77·5	7·25 18·30	6·92 17·84	7·04 16·36	6·80 16·80

1. Minimum and maximum levels are given for beams under fluctuating load and maximum constant level for beams under sustained load.

TABLE 3
Regression constants for beams under fluctuating loads

Beam identifier	10 'A'	'b'
C*	0·39	0·28
D*	0·70	0·24
G*	0·86	0·30
I*	1·24	0·21
J	2·22	0·16
K	2·32	0·15
L	1·58	0·19
M	2·72	0·14
N	1·80	0·18
R	2·42	0·15
S	2·75	0·14
T	4·81	0·16

*Values based on measurements after 5000 cycles

TABLE 4
Regression constants for beams under sustained loads

Beam identifier	10 'E'	'f'
B*	0·44	0·26
E*	0·84	0·22
F*	1·62	0·21
H*	2·63	0·19
P	0·99	0·26

* Values based on measurements after 1·5 hours

of the beam. The differences in their increases in deflexion were due mainly to the amount of increase in deflexion in the first cycles of loading, i.e. the 'A' value in Table 3. At high load levels the value of 'A' depended upon the load level but at the lower levels which are directly comparable with working loads currently used in service in reinforced concrete structures the value of 'A' was insensitive to changes in load level and type of reinforcement. For design purposes a mean relationship would give acceptable accuracy (Fig 3). In the case of beams with sustained loads the rate of increase in deflexion 'f' also appears to be approximately constant but with the value of E depending upon the load level (Table 4). However, a full set of results was available only for beam P and thus this conclusion is much more tentative than that for beams under fluctuating loads.

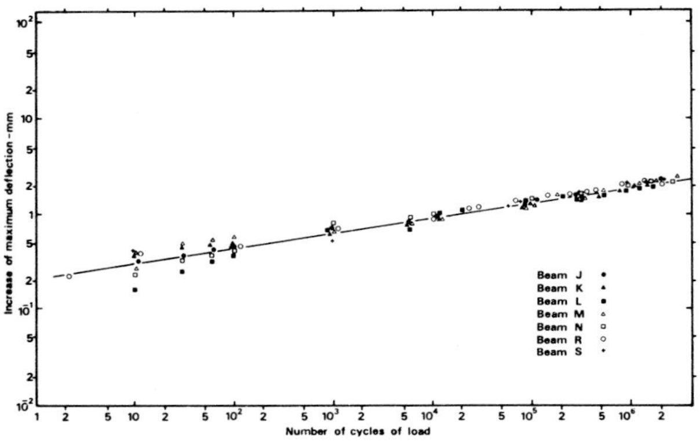

Fig 3. Increase of deflexions for beams under fluctuating service loads

A fluctuating load may produce both time- and cyclic-dependent creep whilst a sustained load normally only produces time-dependent creep. The relationship between these two is not known and thus it is difficult to compare the effects of the two types of loading. However, at the rates of loading used in these tests it appears that at high loads the increase in deflexion was greater in the early stages of loading for the beam subjected to fluctuating load than for the beam under sustained load only (Fig 4). The rate of increase of deflexion was however greater for the beam under sustained load at the maximum level of the fluctuating load so that after about one million cycles of load, or the equivalent length of time, the actual increases in deflexion were similar. At lower loads this effect was less pronounced and the increases in deflexion of beams under fluctuating load were similar to those of beams under sustained load equal to the maximum level of the fluctuating load.

During the testing of the beams under fluctuating load the minimum deflexions in the cycle increased considerably, in some cases the increase being more than 100 per cent in one million cycles of load (Tables 2 and 5). Most of the increase occurred during the first few cycles of load. These increases were far in excess of those exhibited by beams which had been subjected only to a sustained load equal to the minimum load applied in a fluctuating load test. Fig 5 shows the increase in minimum deflexions for beams with approximately the same sustained component of loading but different fluctuating components. Table 5 also gives the values of deflexions predicted by a modified effective-stiffness method[7] for 1×10^6 seconds of loading. This method adequately predicted the maximum deflexions of normal beams under most conditions provided it was assumed that the effect of the fluctuating component of load was the same as if a sustained load equal to the maximum load level was applied for the whole period. The minimum deflexions were determined assuming that the fluctuating component of load had no effect upon the long-term deflexions of the beams, an assumption which clearly leads to a gross underestimate of the actual deflexions.

Cracking

Visible tensile cracks in the concrete formed on the first application of fluctuating or sustained load to each beam. These cracks were usually perpendicular to the soffit of the beam and reached to the neutral axis near midspan (Fig 1). The lengths of the cracks diminished towards the ends of the beam. The pattern of cracks in beam J was influenced as expected by the destruction of the bond in the central portion of the beam; only two very large cracks appeared in this portion and a few smaller tensile cracks occurred towards each end of the beam where the bond was intact. With the exception of beam G and beam T there was almost no visible change in the crack pattern during the periods of loading. In the case of beam G a considerable amount of damage occurred during loading in the areas of high shear with the growth of diagonal tension cracks through the concrete. Beam T failed during the application of its fluctuating load due to a fatigue fracture in the tensile reinforcement. This failure brought about considerable concrete cracking in the region of the fatigue fracture.

Recovery of residual deflexions

The residual deflexions of each of the beams subjected to fluctuating or sustained load reduced with time after the removal of their respective loads. The values of deflexions at various times after the removal of these loads are given in Table 6. In general the residual deflexions of beams subjected to fluctuating loads were greater than those subjected to a sustained load equal to the maximum of the fluctuating load. The amount of recovery however was not significantly different for the two types of loading. For both types of loading most of the recovery had taken place within five to ten days when the deflexions had been reduced to between 80 per cent and 90 per cent of their value on removal of the load.

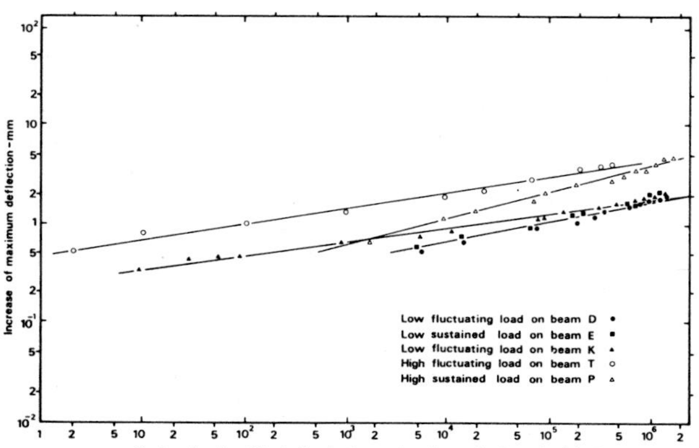

Fig 4. Increase of deflexions for beams under fluctuating load or sustained load

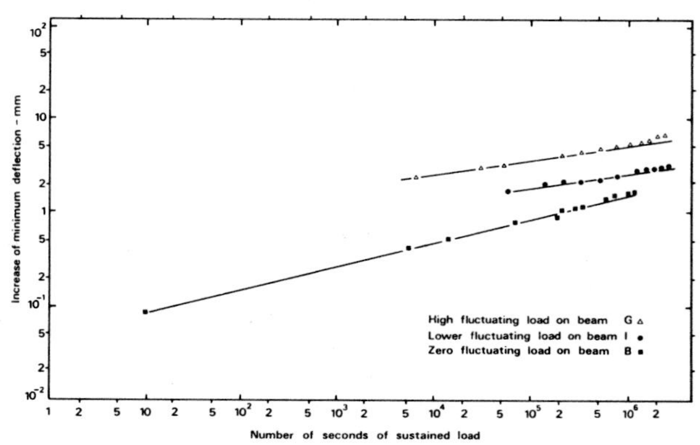

Fig 5. Increase of deflexion for beams under a sustained load with different superimposed fluctuating loads

TABLE 5

Long-term deflexions at midspan

Beam identifier	Load kN	Measured deflexion after 1×10^6 cycles or 1×10^6 seconds of loading (mm)	Deflexion predicted by the effective stiffness method (mm)
B	33·5	6·86	7·26
C	30·0 36·8	7·17 7·86	6·46 8·18
D	23·4 43·5	8·01 10·52	4·84 9·78
E	43·5	10·18	9·78
F	55·5	15·26	12·65
G	33·5 77·5	13·20 22·18	7·26 17·15
H	55·5	16·19	12·65
I	35·6 55·5	13·38 16·01	8·15 12·65
J	20·0 40·0	7·32 11·06	3·34 7·90
K	20·0 40·0	6·24 9·28	3·34 7·90
L	20·0 40·0	8·01 11·65	4·50 10·00
M	20·0 40·0	7·64 11·78	4·50 10·30
N	20·0 40·0	6·15 9·00	3·74 8·15
P	77·5	21·93	20·20
R	20·0 40·0	7·54 11·22	4·50 10·30
S	35·6 55·5	10·83 14·38	9·00 14·50
T	33·5 77·5	12·18[1] 20·14	8·55 20·20

1. Beam T failed in fatigue before 1×10^6 cycles; the deflexion at 1×10^5 cycles is quoted here. The predicted values are for 1×10^6 cycles.

Bending strength of the beams

The maximum load carried by those beams which were eventually subjected to a steadily increasing load to failure is shown in Fig 2. Only in the case of beam G and beam T did the earlier fluctuating load appear to affect the strength of the beam. The shear cracking which developed in beam G during the application of the fluctuating load was probably the main reason for its strength being about 20 per cent less than that of beam A. The fatigue fracture in the reinforcing steel of beam T occurred whilst the fluctuating load was being applied up to a maximum load of about 85 per cent of its assumed static strength, i.e. the strength which would be determined in an incremental-load test to failure.

Implications for design

Deflexions should not jeopardize serviceability by adversely affecting the appearance or efficiency of a structure. To achieve this a limit of deflexion for the serviceability limit state of span/250 is suggested in the current code for the structural use of concrete[3]. This is the total deflexion including the effects of temperature, creep and shrinkage and the limit is deemed to be satisfied if the span/effective-depth ratio is less than the appropriate value set out in the code. These values depend upon the support conditions of the beam, the span, the amount and type of reinforcement and the stress in the tension reinforcement. In this investigation only the more heavily loaded beams F–I, P, S and T failed to satisfy the span/effective-depth ratio limits. A span/250 limit represents a deflexion of 12 mm (½ in) in a span of 3 m (9·84 ft) and during these tests, of the beams which satisfied the code's span/effective-depth requirements, only beam M exceeded this deflexion (after $1{\cdot}8 \times 10^6$ cycles of load). However if the rate of creep is taken as that suggested in the CEB recommendations[9] the increase in deflexion observed in these tests represents only about one third of the final increase in deflexion. On this basis only beam B would still have satisfied the span/250 requirement when the terminal value was reached.

The procedure for estimation of total deflexion in design has already been mentioned; it is usually made by summation of the immediate deflexions due to dead-load and imposed load and the time-dependent deflexions due to creep, shrinkage and temperature. This procedure neglects the 'creep' effects produced by the transient components of load and the present tests throw some doubt on its validity. Prediction of immediate deflexions using the bi-linear, transformed-section or effective-stiffness method of analysis usually estimates within ±20 per cent; the concrete-section method or the gross-section method seriously underestimate actual de-

TABLE 6

Midspan deflexions (mm) after removal of load (days)

Time Lapse / Beam	0	2	5	19	22
B	1·55	1·24	1·03	0·94	1·01
C	2·05	1·74	1·58	1·41	1·46
D	3·40	3·23	3·12	3·07	3·01
E	2·96	2·62	2·39	2·28	2·38
Time Lapse / Beam	0		5	21	48
F	5·18		4·59	4·47	4·45
G	7·63		7·10	6·90	6·85
H	5·28		4·65	5·22	5·36
I	5·32		4·72	4·57	4·63
Time Lapse / Beam	0	2	4	9	57
J	3·93	3·63	3·61	3·55	3·38
K	3·22	2·93	2·88	2·83	2·72
L	3·13	3·05	3·04	3·04	3·00
M	4·01	3·67	3·60	3·54	3·36
N	3·05	2·75	2·71	2·66	2·54
Time Lapse / Beam	0	1	7	16	158
P	5·01	4·45	4·21	4·11	4·00
R	3·41	3·11	2·99	2·91	2·85
S	3·77	3·48	3·31	3·24	3·16

flexions. The increase of the immediate deflexions to allow for time-dependent effects should be made on the basis of the creep deflexion associated with a sustained load equal to the maximum load likely to be applied; the use of a method such as the modified effective-stiffness method[7] can predict total deflexions with acceptable accuracy. Deflexions predicted on the basis of 'creep' deflexion caused by permanent loads plus elastic deflexions due to transient loads will underestimate actual deflexions.

Conclusions

The following conclusions may be drawn from the investigation:

1. A load with both a fluctuating and sustained component or with a sustained component only causes the deflexion of a beam to increase with the continuing application of the load. A linear relationship exists between the logarithm of the increase in maximum deflexion and the logarithm of the number of cycles of load or the length of time under load.

2. At service-load levels and rates of loading used in these tests a beam subjected to a load with a fluctuating component increases its deflexion by about the same amount as a similar beam subjected to a sustained load at the level of the maximum of the fluctuation. At greater load levels the increase of deflexion is more rapid in the early stages for a beam under a load with a fluctuating component. After about one million cycles of fluctuation, however, a beam under sustained load at the maximum level of the fluctuation and for the same length of time has developed a similar increase in deflexion.

3. Within the limits of the present tests the rate or the amount of increase in deflexion under fluctuating load does not appear to depend upon the type, amount or bonding of the reinforcement but merely upon the duration and level of the loading.

4. Both initial and long-term deflexions of reinforced concrete beams can be predicted with acceptable accuracy using available methods of analysis provided it is assumed that any fluctuating component of loading acts as if it were a sustained load at the maximum level of the fluctuations. An assumption that only the sustained component of load contributes to the 'creep' of a beam leads to a considerable underestimate of the true deflexion.

Acknowledgements

The authors wish to thank their colleague Mr. D. Redfearn for assistance with the experimental work. The investigation was made as part of the research programme of the Building Research Establishment and this paper is published by permission of the Director.

References

1. Scott, W. L. and Glanville, W. H. *Explanatory handbook on the Code of Practice for Reinforced Concrete*, Concrete Publications, 1934.
2. Scott, W. L., Glanville, W. H. and Thomas, F. G. *Explanatory handbook on the BS Code of Practice for Reinforced Concrete CP114: 1957 with amendments to 1965*, Concrete Publications, 1965.
3. *British Standard Code of Practice CP110: The structural use of concrete:* London, British Standards Institution, 1972.
4. Lambotte, H., Claude, G. and Motten, H. *Essais de flexion sur poutres en béton armé;* Bruxelles, CSTC, 1969.
5. Snowdon, L. C. 'The static and fatigue performance of concrete beams with high strength deformed bars', Current Paper CP7/71: Building Research Station, Department of the Environment.
6. Ruiz, W. M. and Winter, G. 'Reinforced concrete beams under repeated loads': American Society of Civil Engineers, *Journal of the Structural Division*, June 1969, p. 1189.
7. Neville, A. M. *Creep of concrete: plain, reinforced and prestressed:* Amsterdam, North Holland Publishing Co. 1970.
8. Conway, J. B. *Numerical methods for creep and rupture analysis:* New York, Gordon & Breach, 1967.
9. Comité Européen du Béton. *Recommendations for an international code of practice for reinforced concrete:* Detroit, American Concrete Institute; London, Cement and Concrete Association, 1968.

Long-term cracking in reinforced concrete beams (CP 14/73)

J.M. Illston and R.F. Stevens

It is currently thought desirable that surface cracks in reinforced concrete should not be greater than a certain value. Formulae exist for estimating such widths under short-term loading but it is not always realized that considerable increases in width occur if the load is sustained. A study is presented of the progress of cracking in 60 beams kept under load for over two years. The influence of the major variables such as concrete cover, type of steel and conditions of storage was investigated and a method is suggested for estimating the surface widths of cracks in the long term. A further important feature of the study was the investigation, using a technique of resin injection, of the changes in internal crack widths and of the breakdown of adhesion bond.

Introduction

The strengths of reinforcing steels, and hence the working stresses and strains, have increased over the years and seem likely to continue to do so. Although the tensile strengths of concrete are increasing also there is no corresponding increase in tensile strain capacity to match that of the steel. It follows that the numbers and the widths of cracks in reinforced concrete members are also increasing and there is a risk of a critical deterioration in the resistance to corrosion and of composite action between steel and concrete.

2. This danger has been recognized by the adoption of limit state design in the draft unified code for structural concrete,[1] in which provision is made for adequate factors of safety against excessive cracking. The normal approach to this problem is that of the European Concrete Committee[2] which recommends that the widths of surface cracks should not exceed 0·1 mm in corrosive atmospheres, 0·2 mm in external exposures and 0·3 mm inside buildings.

3. At the level of the steel, the average width of crack at the surface of a beam is the average elongation of the steel between cracks minus the average elongation of the concrete. Thus, if s_m is the average spacing of cracks, ϵ_{ta} is the average strain at the level of the steel and ϵ_{tca} is the average concrete strain between cracks at the level of the steel, the average width of crack w_m is given by

$$w_m = s_m(\epsilon_{ta} - \epsilon_{tca}) \quad . \quad . \quad . \quad . \quad . \quad . \quad . \quad . \quad (1)$$

In the short term, the average concrete strain is often small and can be neglected so that

$$w_m = s_m \epsilon_{ta} \quad . \quad . \quad . \quad . \quad . \quad . \quad . \quad . \quad (2)$$

and the widths of cracks depend on the tensile strain in the steel and the crack spacing.

4. Broms[3,4] considered crack spacing in terms of effective cover, which he defined as the minimum distance from the concrete surface to the centre of the reinforcing bar. He found that crack spacing was a linear function of effective cover and that no other variable was significant. This unexpected conclusion was supported in a comprehensive series of tests carried out by Base *et al.*[5] in which a number of variables were investigated in a programme involving the testing of 120 beams. Little difference was found between the cracking in beams reinforced with deformed bars and round bars; neither the number of bars nor the proportion of steel was found to be a major influence; there was no evidence that the concrete strength or the condition of curing had a significant effect on cracking.

5. The final conclusion was that, in the short term, the maximum crack width could be expressed simply in the form

$$w_{max} = kc\epsilon_{ta} \quad \text{. . . . . (3)}$$

where c is the concrete cover to the nearest bar and k is a constant equal to 3·3 for deformed bars and 4·0 for plain round bars.

6. It is a matter of observation that the deflexions of beams and the widths of cracks increase under sustained load and the short-term measurements and formulae do not represent the worst state of cracking. To investigate the development of cracking with time a programme of tests was mounted at the Building Research Station in which 60 beams were kept under constant load for over two years. It was thought important to discover whether the variables found to be of little influence in the short term would intervene in the long term. In addition the conditions of storage are important and two groups of beams were tested: one in the internal atmosphere of the Structures Laboratory and the other outside, but under cover.

7. It is usually assumed that the likely corrosion is related to the surface crack width. It follows from the work of Base *et al.*[5] and Broms[3,4] that beams with greater cover will be the most subject to corrosion. This conclusion might be valid if the cracks were of uniform width between the steel and surface but this has not been established and Broms,[6] using a technique of resin injection, has shown that in the short term the width of flexural cracks near the steel is much smaller than at the surface.

8. It would thus appear that the width of surface cracks may not be of itself a sufficient measure of potential risk of corrosion. More information is needed on the form of internal cracks, particularly in the long term. Additional tests, using Broms's resin injection technique, were carried out at the Building Research Station to determine the shape of the internal cracks in a small number of the beams at the end of their period under sustained load.

Table 1

Type of reinforcement	Round mild steel, yield stress 248 MN/m^2 High tensile steel, yield stress 414 MN/m^2 Very high tensile steel, yield stress 552 MN/m^2
Proportion of reinforcement	Proportion given by balanced design using a load factor of 1·75 One half of that given by balanced design
Concrete cover	25 mm
Soffit cover/side cover	50 mm
Atmospheric environment	Controlled at 19°C and 65% relative humidity Natural, but shielded from direct rain (exposure site at Garston)—average annual conditions were temperature of 10°C and relative humidity of 83%

Surface cracking under sustained loading

Description of beams

9. The main series of beams comprised 24 pairs of beams, the variables being three strengths of reinforcing steel, two sizes of concrete cover, two proportions of reinforcement and two types of environment: one controlled and one natural. The beams were 5·2 m long, 200 mm in breadth and 385 mm in depth.

10. The variables are summarized in Table 1. The beams were cast in gravel concrete designed to have a cube crushing strength of 34 MN/m^2 at 28 days.

11. In addition to the main series of beams, further pairs were cast to investigate additional features: type of concrete, compression reinforcement and size of specimen. The additional specimens consisted of

(*a*) two pairs of beams made with lightweight concrete
(*b*) one pair of beams with no compression reinforcement
(*c*) one pair of beams with compression reinforcement equal to tension reinforcement
(*d*) one pair of beams with dimensions twice those of other beams.

The beams were simply supported in pairs, and were loaded at about the third points so that the central zone was subjected to uniform working movement.

12. The series of beams is the same as that used for a parallel investigation of deflexions. Further details of the beams, including details of manufacture and testing are given by Stevens.[7]

13. All visible cracks at the level of the tensile reinforcing steel on the side faces of the beams along the length of the central constant moment section were grouped into main structural cracks, and their widths were measured with microscopes magnifying 60 times. Where it appeared that the aggregate particles or other local effects had caused a structural crack to form more than one branch in its progression from its origin at the soffit of a beam, these were grouped as one crack.

Results of the tests

14. From the measurements the values of the quantities of average crack spacing, average crack width and maximum crack width were obtained. The maximum crack width was the greatest single crack width observed and therefore the most likely to allow corrosion.

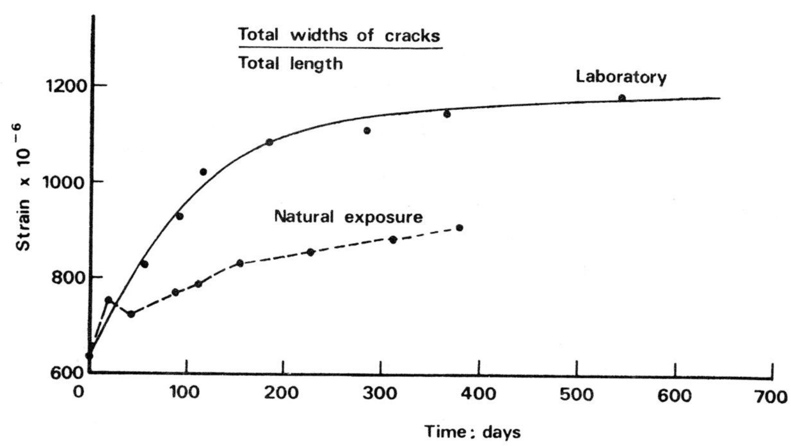

Fig. 1. Variation of cracking with time

Table 2. Results of tests—Structures Laboratory

Beam type	Type of steel	Concrete cover, *mm*	Age, *years*	Average spacing, *mm*	Average crack width, *mm*	Maximum crack width, *mm*
A	Mild	25	0 1 2	104 102 102	0·05 0·11 0·12	0·10 0·20 0·23
B	Mild	25	0 1 2	111 106 106	0·04 0·05 0·06	0·08 0·13 0·13
C	Mild	51	0 1 2	160 157 157	0·10 0·18 0·19	0·15 0·28 0·31
D	Mild	51	0 1	198 180	0·05 0·10	0·10 0·23
E	High tensile	25	0 1 2	111 104 104	0·09 0·13 0·13	0·18 0·25 0·25
F	High tensile	25	0 1 2	101 99 99	0·06 0·09 0·11	0·10 0·23 0·25
G	High tensile	51	0 1 2	147 139 139	0·13 0·18 0·21	0·23 0·38 0·38
H	High tensile	51	0 1 2	177 170 170	0·09 0·17 0·18	0·15 0·33 0·38
J	Very high strength	25	0 1 2	93 93 93	0·07 0·14 0·14	0·18 0·31 0·31
K	Very high strength	25	0 1 2	109 102 102	0·08 0·10 0·13	0·13 0·23 0·23
L	Very high strength	51	0 1 2	152 152 152	0·15 0·27 0·27	0·25 0·46 0·46
M	Very high strength	51	0 1 2	154 142 142	0·12 0·19 0·21	0·25 0·36 0·38
R_a	High tensile	51	0 1 2	203 170 170	0·13 0·13 0·15	0·25 0·28 0·31
R_b	High tensile	51	0 1 2	162 160 157	0·12 0·14 0·15	0·28 0·33 0·33

Table 2—*continued*

Beam type	Type of steel	Concrete cover, *mm*	Age, *years*	Average spacing, *mm*	Average crack width, *mm*	Maximum crack width, *mm*
X	High tensile	25	0 1	101 99	0·10 0·16	0·15 0·25
Y	High tensile	25	0 1 2	109 96	0·08 0·11	0·15 0·20
Z	High tensile	25	0 1 2	124 111	0·08 0·11	0·15 0·23

Table 3. Results of tests—natural exposure

Beam type	Age, *years*	Average spacing, *mm*	Average crack width, *mm*	Minimum crack width, *mm*
A	0 1	119 131	0·05 0·08	0·13 0·18
B	0	109	0·05	0·10
C	0 1	159 156	0·10 0·15	0·15 0·20
D	0 1	176 169	0·05 0·08	0·08 0·13
E	0 1	121 117	0·09 0·12	0·18 0·25
F	0 1 2	112 115 115	0·07 0·08 0·08	0·13 0·18
G	0	176	0·14	0·25
H	0 1	180 176	0·09 0·15	0·15 0·25
J	0 1	102 100	0·06 0·11	0·13
K	0 1 2	117 116 109	0·08 0·09 0·09	0·23 0·20
L	0 1 2	135 135 135	0·13 0·19 0·20	0·25 0·30
M	0 1	176 172	0·13 0·18	0·25 0·38
X	0	107	0·10	0·18

15. The mean spacing of the flexural cracks, computed average crack widths and maximum crack widths are given in Table 2 for the beams stored in the Structures Laboratory controlled at 19°C and 65% relative humidity and in Table 3 for beams stored in the natural environment in which the mean annual conditions were 10°C and 83% relative humidity. The growth in cracking, expressed in terms of the sum of all crack widths over the constant moment regions of beams, is shown in Fig. 1 for a typical beam in the laboratory and natural exposure.

Relations for short-term cracking

16. The emphasis of the tests reported was on the long-term effects of sustained load on cracking. To this end two covers only were used and the tests cannot be regarded as a basic investigation into cover. However, the results provide further confirmation of the dominance of cover among the variables that affect the spacing of cracks.

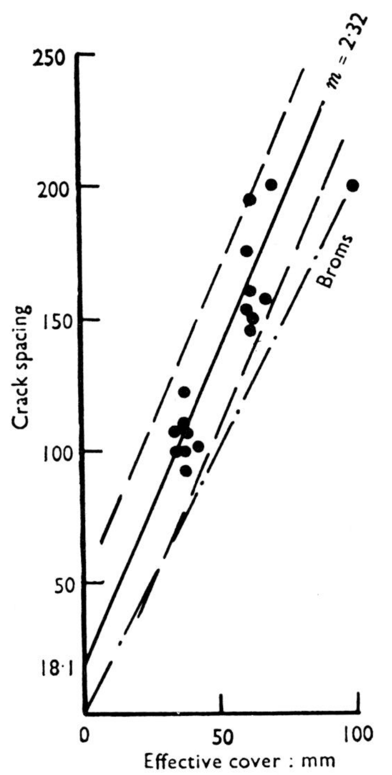

Fig. 2. Relationship between crack spacing and effective cover at initial load (m slope of regression line)

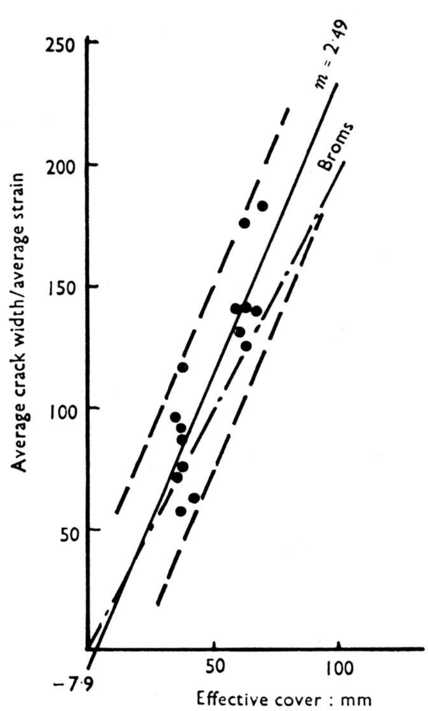

Fig. 3. Relationship between ratio of mean crack width to average strain and cover at initial loading

17. Two versions of cover have been suggested: actual cover as used by Base *et al.*[5] and effective cover as used by Broms.[3,4] On the assumption that the relationship between crack spacing and cover is linear, better agreement was found if effective cover was adopted. Crack spacing is plotted against effective cover in Fig. 2. Crack slope, defined as

$$\frac{\text{average crack width at steel level}}{\text{average strain at steel level}}$$

is plotted against effective cover in Fig. 3. In both Figs 2 and 3 Broms's expressions are given for comparison.

18. The relationships by linear regression analysis are

$$s_m = 2{\cdot}3c_e + 18 \qquad (4)$$

$$\frac{w_m}{\epsilon_{ta}} = 2{\cdot}5c_e - 8 \qquad (5)$$

where c_e is effective cover, equal to actual cover plus bar radius. The crack spacings and crack widths are distinctly greater than those reported by Base *et al.*[5] This is attributable to differing test conditions and systems of measuring cracks, rather than to intrinsic differences in the phenomena under investigation. The same comment can be made with reference to Broms's[4] results, but they are in reasonable agreement as can be seen from Figs 2 and 3.

19. If c_e is eliminated between equations (4) and (5)

$$\epsilon_{ta}s_m = w_m + (25\,\epsilon_{ta} - 0{\cdot}1w_m) \qquad (6)$$

The term in brackets represents the extension of the concrete between the cracks. The simplified expression $s_m = w_m/\epsilon_{ta}$, neglecting the effect of the concrete extension, applies when $s_m = 250$ mm.

20. In these tests s_m is always less than 250 mm and there is a significant extension of the surface concrete between the cracks. This can be seen in the typical curve of Fig. 4 at zero on the time scale.

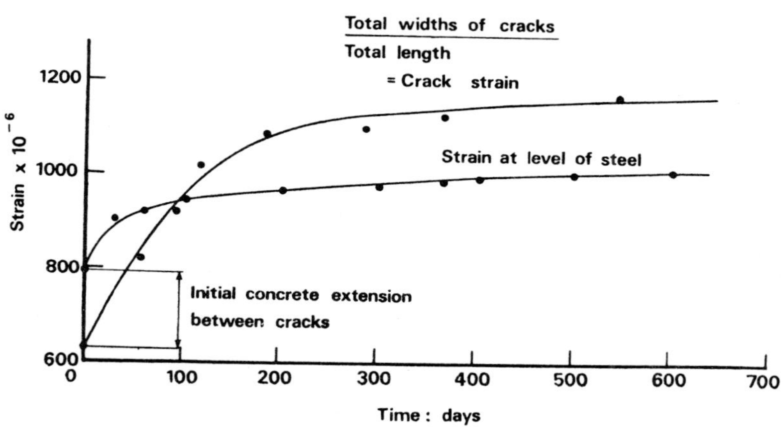

Fig. 4. Growth of cracking

Long-term cracking

Beams in Structures Laboratory (controlled environment)

21. The average spacing of cracks in the beams as given in Table 2 changed little over the two year period of sustained loading, so that the number of cracks was virtually unchanged with time. The development of widths of cracks with time is shown for the beams in Fig. 4, which shows the ratio of total widths of cracks at the level of the steel over constant moment region to total length of constant moment region and the observed surface strain at the level of the steel. The ratio of total width of cracks to total length is identical to the ratio of average width of crack to average crack spacing (r) and, as the value average crack spacing does not change with time, the ratio represents the growth in average crack width.

22. The average crack grows in width with time at a reducing rate and, significantly, at a greater rate than the average observed surface strain at the level of the steel. Initially, the ratio r is less than the observed surface strain, showing that concrete is subjected to an extension between cracks, as already deduced.

23. There are two causes of the increase with time of the crack widths at the steel level. First, as discussed by Stevens,[7] the creep of the concrete in the compression zone of the beam results in a redistribution of stress, a drop in the neutral axis and an increase in the tensile stress in the steel. This effect is shown by the curve of strain at the level of the steel in Fig. 4.

24. Second, the concrete contracts between the cracks because of the combined effect of shrinkage and the loss of tensile stress which is also shown by Stevens.[7] In Fig. 4 the ratio average width of crack to average crack spacing increases more rapidly than the strain at the level of the steel, which indicates a loss of concrete tension between cracks. After two years the ratio r is greater than the observed average surface strain, indicating that concrete has contracted between cracks. The amount of the contraction at two years—200×10^{-6}—is close to the observed free shrinkage strain for the concrete on the upper face of the beams, which suggests for long-term cracking

$$\frac{w_m}{s_m} = (\epsilon_{ta} + \epsilon_s) \quad . \; . \; . \; . \; . \; . \; . \; . \quad (7)$$

where ϵ_{ta} is the average surface strain at the level of the steel and ϵ_{cs} is the free shrinkage strain of concrete.

25. The relationships between the ratio of average width of crack to average surface strain at steel level plus observed shrinkage strain and effective cover are computed at initial loading, after one year of sustained loading and after two years' sustained loading.

$$\frac{w_m}{\epsilon_{ta} + \epsilon_{cs}} = 2{\cdot}5c_e - 15, \text{ at initial loading in mm.} \quad . \; . \quad (8)$$

$$\frac{w_m}{\epsilon_{ta} + \epsilon_{cs}} = 2{\cdot}3c_e - 3 \text{ after 1 year in mm.} \quad . \; . \; . \; . \quad (9)$$

$$\frac{w_m}{\epsilon_{ta} + \epsilon_{cs}} = 2{\cdot}0c_e + 13 \text{ after 2 years in mm.} \quad . \; . \; . \; . \quad (10)$$

These equations indicate the slight change in average crack spacing with time.

26. The observed maximum width of cracks was almost exactly twice the observed average width of crack. This also did not change during the period of sustained loading.

Estimation of cracking

27. A simplified estimate of cracking in the constant moment region of reinforced concrete beams is given by

average crack spacing	$2{\cdot}3c_e$
average crack width	$2{\cdot}3c_e(\epsilon_{ta} + \epsilon_{cs})$
maximum crack width	$4{\cdot}6c_e(\epsilon_{ta} + \epsilon_{cs})$

The strain at the level of the steel ϵ_{ta} may be taken as f_s/E_s where f_s is the stress in the steel and E_s is the modulus of elasticity of the steel. Methods of estimating f_s have been given by Stevens[6] for initial working load and working load sustained for some years.

28. The time-dependent increase in curvature means that the crack widths at the soffits of the beams increase proportionately more than those at the level of the steel. It is a reasonable approximation to consider that the crack widths increase linearly with the depths from the neutral axis. This is shown in Fig. 5, in which the sums of all the crack widths in the constant moment region are plotted against the position of measurement on the surface of the section, for two similar beams, one under initial load and the other subjected to sustained working load for two years.

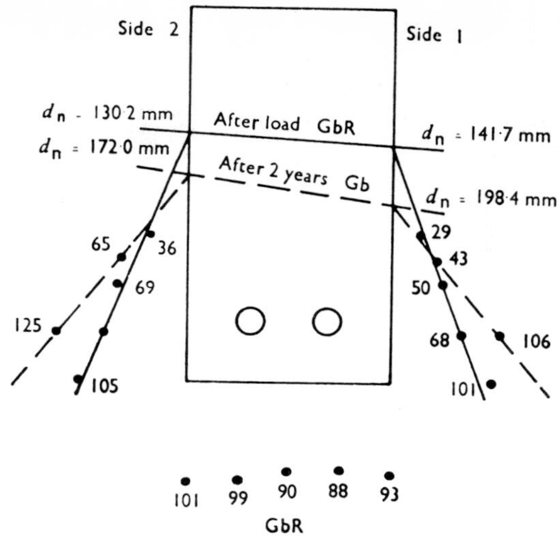

Fig. 5 . Total crack width (d_n depth to neutral axis)

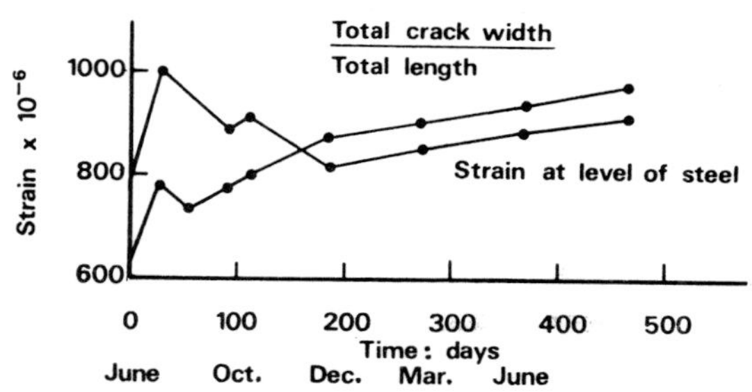

Fig. 6. Cracking with time for typical exposure site beam

Beams on the exposure site

29. The observations of cracking in beams on the exposure site proved difficult after about one year's exposure, and it was not always possible to obtain reliable readings.

30. The initial crack spacing again remained virtually constant. Changes with time in the rates of total widths of cracks/observation span and average strain at steel level are plotted in Fig. 6, which follows a similar overall pattern to that for beams in the Structures Laboratory in that initially the average strain due to cracking was less than the average surface strain, showing that concrete between cracks was carrying tension. At a later time the average strain due to cracking was greater than the average surface strain, showing that the shrinkage of the concrete between cracks had occurred.

31. Cracking developed at a slower rate for beams on the exposure site compared with beams in the laboratory; after one year the increase in crack width was about 40%, compared with a value of about 60% for beams in the laboratory.

32. The development of cracking for beams on the exposure site followed the same pattern of behaviour as those in the laboratory, although less shrinkage of concrete took place, and the formulae developed for cracking of laboratory-stored beams are equally applicable to the beams stored on the exposure site.

Fig. 7. Specimen showing almost complete loss of bond along the bottom of the reinforcement (set 1B central section, vacuum filled)

Internal cracking

33. Previous work by the Authors[8] was concerned with resin injection of axially reinforced cylinders. These tests confirmed the findings of Broms,[6] and Broms and Lutz[9] that in the short term the crack widths at the steel/concrete interface, although usually measurable, are much less than the widths of the same cracks on the surface. Sectioning and separation on a longitudinal plane through the steel revealed a resin stain on the steel/concrete interface at each crack. This is a clear indication of the breakdown of adhesion bond and it is interesting to observe whether the length of stained interface increases with time under load. This is an additional objective to the main one of measuring changes in the widths of internal cracks.

34. Preliminary tests were mounted on four pairs of small beams.[10] One of each pair was injected with resin immediately after loading and the second after five months under sustained load. The results confirmed the findings of the prism tests that in the short term cracks narrowed towards the steel, and that loss of adhesion bond occurred and was shown by resin staining. In the long term the trend was towards an extension of the length of bond breakdown; a particularly dramatic example, in which the staining covers nearly the whole length of the bar, is shown in Fig. 7. The internal widths of the cracks increased with time just as they have been observed to do on the surface.

35. The beams used in the preliminary tests were small and the period of

Table 4. Details of beams

Beam type	Type of steel	Size of bars, *mm*	Concrete cover, *mm*	Type of aggregate	Period of sustained loading	Loading
Cb	Round M/S	32	51	River gravel	2 years	Overloaded $1{\cdot}7\times$ working load: steel yielded
Gb	Square twist HT	25	51	River gravel	2 years	Working load
GbR	Square twist HT	25	51	River gravel	7 hours	Working load
Xb	Square twist HT	25	25	Medium lytag	$1\frac{1}{4}$ years	Working load

sustained loading was short and it was decided to extend the resin injection programme to a limited number of full-size beams in the main series.

36. Four of the beams in the main series, which had been stored in the laboratory, were resin impregnated over a portion of the constant moment zone. These were the upper beams Cb, Gb and Xb. Beam type Gb was then repeated, loaded at an age of 28 days after casting and filled as quickly as possible (reference mark GbR). Table 4 gives the relevant details of these beams, which all contained two tensile reinforcing bars.

37. The method adopted was one of vacuum filling. As the beams were not provided with a duct, a method was adopted of drilling 13 mm dia. holes from each crack on the side faces near the level of the reinforcement on both sides of the beam. Plates with an attached tube were then fixed to the surfaces of the beams over each hole. The cracks and connexion plates were then covered with a surface coating of resin, to which strips were clamped to form a reservoir to contain the resin on the upper soffit surface as shown in Fig. 8. The tensile area of the beam was then coated and effectively sealed around and beyond the section about to be filled. When the coating had hardened, a vacuum pump was applied to the linked connectors and resin was drawn through the cracks from the reservoir.

Fig. 8. Vacuum filling a beam, showing resin reservoir, pump and pressure gauge

38. Two days after resin impregnation each beam was unloaded and the filled section removed. Four cuts were made with a track saw using diamond or carborundum blades from the outer surfaces towards the bars as indicated in Fig. 9, to allow the removal of a concrete prism including the surface which had been in contact with the bar. The crack widths at the positions marked A, B and C on each side outwards from the bars, and at the positions marked 1, 2 and 3 downwards from the bar to the extreme tension face were then measured, together with the length of resin stain at each crack.

39. Table 5 gives the average values of the internal cracks, the average length of resin stain and the corresponding percentage of bond breakdown.

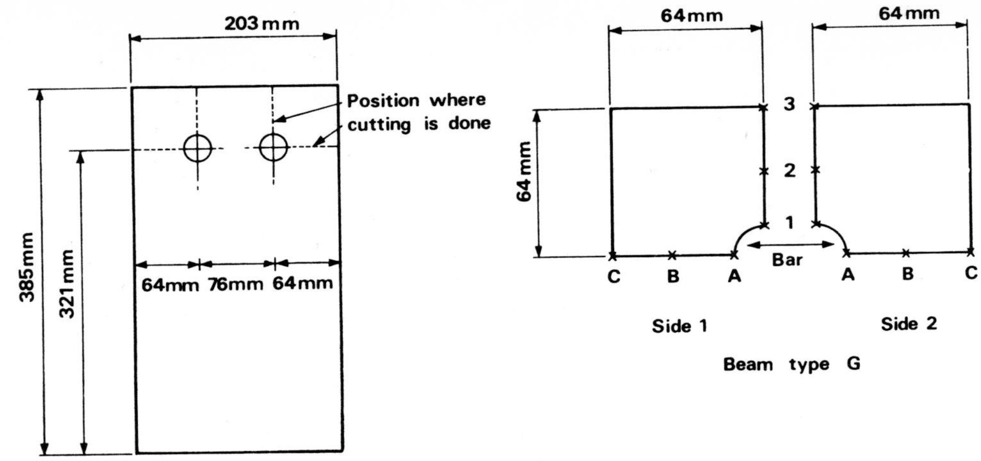

Fig. 9. Position of cutting and internal crack width measurement

Time-dependent changes in crack widths and bond breakdown

40. The observations made with the full-size beams revealed the same major features which had been observed previously,[7] namely that cracks were continuous from the soffit to near the neutral axis and that cracks had finite width at the bar. Each crack was accompanied by local breakdown of adhesion bond at the bar. These beams permitted a further investigation of the variation of crack width from the surface of the beam to the bar, and the interesting comparison of cracking under initial loading and after two years' sustained loading. The increase in crack width from reinforcing bar to soffit was regular for all the beams. The ratio of crack width at the soffit to crack width at the bar ranged from about 1·2 to 1·4, which was approximately in direct agreement with the distribution of strain from soffit to bar.

41. Crack widths from the bar to the side face of the beam at the level of the bar usually showed an increase in width from the bar to the surface, this increase being most marked in the case of the beam subject to initial loading, where the width of crack at the face of the beam at the level of the steel was about 2·5 times the crack width at the steel. However, for the companion beam after two years' sustained loading the crack width at the surface was about only 20% greater than the crack width at the steel. This is further confirmation that, under initial load, steel bars cause cracks to narrow towards the bars. This effect tends to disappear with time.

42. The length of breakdown in adhesion bond was observed to increase from about 20% of the total length of bar for the beam under initial loading to about 50% of the total length for the companion beam after two years' loading. The beam made with lightweight aggregate concrete showed a breakdown of adhesion bond over about 30% of the total length after 1¼ years' sustained loading. These results point to an appreciable increase in breakdown of adhesion bond with time. The beam subjected to overloading also showed an increase in breakdown of bond, suggesting that the breakdown also increases with stress in the bar.

Mechanism of cracking in beams

43. It is thought that cracking in reinforced concrete beams subjected to bending originates at the soffit. The initial crack is generated when the local tensile strength of the concrete is less than the maximum tensile stress produced by bending. The initial crack is likely to run continuously through the beam

Table 5. Average crack widths

Beam detail		Cb	Gb	Gbr	Xb
		Sustained load for 2 years, then overloaded × 1·7	After 2 years	After loading	After $1\frac{1}{4}$ years
		Internal, *mm*	Internal, *mm*	Internal, *mm*	Internal, *mm*
Variation from reinforcement to soffit	1	0·32	0·19	0·08	0·16
	2	0·36	0·21	0·1	0·18
	3	0·39	0·27	0·12	0·20
Variation from reinforcement to side face	A	0·37	0·17	0·05	0·13
	B	0·31	0·19	0·1	0·16
	C	0·31	0·20	0·14	0·16
Average stain, *mm*		91	71	36	31
Percentage breakdown		57	47	21	29

from the soffit to close to the neutral axis. Where the path of the crack crosses a steel reinforcing bar, a large increase of stress in the bar occurs, causing a slight reduction of diameter of the bar and helping to generate local breakdown of adhesion bond from the crack. The breakdown of adhesion bond occurs to a similar extent both for ribbed bars and smooth bars. At the surface of the crack the stress in the concrete is zero, but tension is developed in the concrete from the interface between the perimeter of the steel bars and the concrete. The distance along the soffit from the crack required for the concrete to redevelop the tensile stress at which the initial crack formed is linearly related to the distance from the centre of the steel/concrete interface (which is the centre of the bar) to the soffit. Further cracking may then occur in random fashion, save that there is an inhibited length from the initial crack required to redevelop the previous tensile stress at the soffit, and further inhibited lengths at subsequent cracks so that the average crack spacing will be in proportion to the effective cover.

44. The width of cracks is determined by the crack spacing and the difference between the strain in the steel and the tensile strain in the concrete. No difference in cracking between beams reinforced with ribbed or plain bars occurs under initial loading, as the value of adhesion bond is the same for both surfaces and little relative movement between the steel and concrete occurs where adhesion bond breaks down.

45. The effect of sustained load on concrete beams produces changes in cracking. There is no appreciable change in the number, and hence the spacing of cracks, but the widths of cracks show appreciable increases, because of time-dependent change of curvature and contraction of the concrete. The contraction of the concrete is caused by shrinkage and by loss of tension due to the breakdown of adhesion bond with time.

46. The pattern of widths of cracks on initial loading is likely to show a narrowing in widths of cracks from the face of the concrete to the steel bars, but under the action of sustained loading local changes in widths are smoothed out and cracks are produced of uniform width across the beam with a constant increase in width from about the neutral axis to the soffit.

47. It may be possible to misjudge the effects of corrosive atmospheres on cracked reinforced concrete when merely the surface widths of cracks are

considered. Where cracked reinforced concrete is observed after some years' exposure there could be good correlation between surface widths of cracks and the widths of cracks through the beam, but where short-term accelerated corrosion tests are carried out in laboratories, the narrowing of cracks from the surface of the beam to the bar and the limited breakdown of bond could give deceptively good resistance to corrosion. It would be even more misleading to consider the effect of corrosive environments on initial widths of cracks, rather than the larger widths which exist in the longer term.

Conclusions

Surface cracking

48. The initial spacing of cracks is dependent mainly on the effective cover to the steel reinforcement, defined as the distance from the centre of a reinforcing bar to the face of the beam. The initial spacing of cracks is not affected by different surfaces to the reinforcing steel, such as round, square twisted or ribbed. The average initial width of crack is dependent on the effective cover to the steel reinforcement and the difference between average strain in the steel and average strain in the concrete. The spacing of cracks does not change with time under sustained loading. The average crack grows in width with time under sustained loading at a reducing rate. In the tests crack widths doubled in two years. The increase in widths is caused by reduction in average tensile strain in concrete between cracks due to shrinkage and by time-dependent change of curvature.

Estimation of cracking

49. The average crack spacing in the constant moment region of a reinforced concrete beam is given by $s_m = 2{\cdot}3c_e$, where s_m is the average spacing of cracks and c_e is effective cover to reinforcing steel. The average width of crack in the constant moment region of a reinforced concrete beam subjected to sustained loading for a long time is given by $w_m = 2{\cdot}3c_e(\epsilon_{ta} + \epsilon_{cs})$, where w_m is the average width of crack, ϵ_{ta} is steel strain associated with working stress and ϵ_{cs} is the shrinkage strain of the concrete. The maximum observed width of crack was approximately double the average width of crack, both under initial loading and under sustained loading.

Internal cracking

50. Cracking in reinforced concrete beams subjected to bending is continuous from the soffit through the beam around any reinforcement to near the neutral axis. Where the path of a crack crosses a reinforcing bar, a breakdown of adhesion bond is produced along the perimeter of every reinforcing bar at each side of the crack. Similar lengths of breakdown of bond occur for both ribbed and plain bars. Under initial loading steel bars in a concrete beam produce a restraint on the widths of cracks, in particular the widths of cracks at the level of the steel. The length of breakdown of adhesion bond increases under sustained loading. Under sustained loading the restraint afforded by steel bars to the widths of cracks is reduced, and the cracks tend to become of triangular form, increasing regularly in width from neutral axis to soffit, and being of uniform width across the beam.

Acknowledgements

51. The research work represents a collaboration between the Department of Civil Engineering, King's College, University of London and the Building Research Station, Department of the Environment. The Authors wish to thank Mr D. W. Bryden-Smith for his assistance with the work.

References

1. BRITISH STANDARDS INSTITUTION. *Draft British Standard Code of Practice for the structural use of concrete.* British Standards Institution, London, 1969.
2. EUROPEAN CONCRETE COMMITTEE (CEB). *Recommendations for an international Code of Practice for reinforced concrete.* Cement and Concrete Association, 1964.
3. BROMS B. B. Stress distributions in reinforced concrete members with tension cracks. *J. Am. Concr. Inst.*, 1965, **62**, No. 9, Sept., 1095–1108.
4. BROMS B. B. Crack width and crack spacing in reinforced concrete members. *J. Am. Concr. Inst.*, 1965, **62**, No. 10, Oct., 1237–1255.
5. BASE G. D. *et al. Crack control in beams.* CERA Research Report 6, 1966.
6. BROMS B. B. Technique for investigation of internal cracks in reinforced concrete members. *J. Am. Concr. Inst.*, 1965, **62**, No. 1, Jan., 35–43.
7. STEVENS R. F. Deflexions of reinforced concrete beams. *Proc. Instn Civ. Engrs*, Part 2, 1972, **53**, Sept., 207–224.
8. ILLSTON J. M. and STEVENS R. F. Internal cracking in reinforced concrete. *Concrete*, 1972, **6**, No. 7, July, 28–31.
9. BROMS B. B. and LUTZ L. A. Effect of arrangement of reinforcement on crack width and spacing in reinforced concrete members. *J. Am. Concr. Inst.*, 1965, **62**, No. 11, Nov., 1395–1409.
10. STEVENS R. F. The time-dependent deflections of reinforced concrete beams under sustained loading and the development of cracking. PhD thesis, submitted to the University of London, 1970.

Effects of various factors on the extensibility of concrete (CP 15/76)

C.R. Lee and W. Lamb

INTRODUCTION

In 1970, at the Montreal meeting of the ICOLD Committee on Concrete, the British representative was asked to prepare a report on 'the effect of various factors on the extensibility of concrete for large dams'. The British Sub-Committee considered that this task would be much facilitated and a better report would result if each country would draw attention to its own publications and sources of information. A circular letter dated 19 April 1971 was accordingly addressed to the ICOLD representatives in the fifteen countries thought most likely to be interested, inviting each of them to provide a bibliography and brief review of publications dealing with the subject, primarily in their own country but also including information from the countries not included in the circulation of the letter. In the letter, it was pointed out that while the meaning of extensibility was usually well understood, an alternative definition might be tensile strain capacity up to failure taking account of creep and in relation to stress, shrinkage, etc.

At the Dubrovnik meeting in 1971 of the ICOLD Committee on Concrete, the British representative tabled a short report on the progress of the enquiry. The definition of extensibility as 'strain to failure' was confirmed.

At the Canberra meeting in April 1972 of the Sub-Committee on Concrete of the new ICOLD Committee on Materials, 'Dr Rosanov requested that the speed of applying load should be taken into account when compiling the report and that only tensile strain should be considered'.

At the Madrid meeting in June 1973 of the Sub-Committee on Concrete of the ICOLD Committee on Materials, a first draft report was presented consisting of a commentary and seven appendices, containing the replies to the circular from Great Britain, Australia, France, Japan, Portugal, USA and USSR.

Further useful comments were received from the USA representative both in Madrid and in later correspondence. The draft report was circulated for comment to the six countries other than Great Britain who had contributed to the appendices but no comments have been received.

Following approval by the Sub-Committee on Concrete a revised draft was presented to the Committee on Materials in Athens in 1974, who recommended publication. The draft was circulated to all National Committees under cover of Circular Letter No 711 and publication was approved by the Executive in Tehran in 1975. A late comment by Dr Rosanov (USSR) has been added as Appendix 1.

The present report is a revision of the first draft with the original 'British' bibliography expanded to include papers published in other countries, including some references given in the appendices to the first draft. This brief history may help to explain an apparent numerical bias in the bibliography in favour of British papers. It is hoped that the bibliography is now fully representative of the international literature.

In preparing the bibliography an attempt has been made to limit it strictly to the subject of extensibility defined as 'tensile strain capacity to failure'. Priority has been given to papers giving quantitative information. Some papers without such information but which contribute to an understanding of extensibility have, however, been included. Papers dealing with such subjects as 'creep', 'cracking', etc, but without information on extensibility, as strictly defined, have generally been excluded.

The library search has been concentrated on papers published since 1950 although it has been recognised that such is the general paucity of quantitative information on extensibility, reference might usefully be made to papers published before that date. For obvious reasons, the search has not been confined only to the field of dam construction. Also for obvious reasons, while papers dealing exclusively with effects in simple compression have generally been excluded, some papers dealing with complex states of stress have been held to qualify.

COMMENTARY

The extensibility of concrete is defined as 'tensile strain capacity to failure'. The property has a particular application for concrete dams in that there has been considerable interest in changes in strain, such as those due to temperature, creep or shrinkage, and measurements of strains are far easier than measurements of stresses. There are thus considerable attractions in studying the possibility of cracking in terms of extensibility rather than in terms of tensile strength.

A study of the literature shows a considerable difference in quantitative values according to circumstances and it may be as important to consider the factors governing failure as it is to consider earlier creep. The literature illustrates the very considerable effect of control of the rate of increase in strain and the substantial difference between extensibility determined as the tensile strain reached before failure in a **direct** tensile test at progressively increasing stress (eg of the order of 100×10^{-6} strain) and the appreciably greater values of extensibility (eg generally of the order of 200×10^{-6} strain but sometimes much higher) which can be reached when the rate of increase in strain is controlled. Experimentally, control can be achieved in especially stiffened testing machines or by the presence of reinforcement so that as the strain increases part of the maximum load carried by the concrete can be shed. This leads to the determination of the so-called 'complete' stress/strain curve incorporating a 'falling' component in which the load decreases as the strain increases towards eventual failure. For similar reasons greater strains can be achieved in flexural tests than in direct tensile tests.

Determinations with controlled rates of straining are very relevant to the conditions in dams where the effective rates of strain increase may be determined, for example, by the slow rates of temperature change and, in the case of surface cracking, by bond to large masses of unaffected interior concrete. In this concept it is, of course, not necessary that the concrete in question should actually change in dimensions. It can remain at constant or nearly constant size and an effective strain be imposed by temperature change or drying. Simulation of this condition forms the basis for the so-called 'restrained' tests referred to in the bibliography [9, 22, 39, etc].

With controlled rates of strain the precise stage at which failure occurs becomes less obvious and the study of extensibility merges into a study of the failure mechanism of concrete, eg of the development of micro-cracking within the concrete. Such cracking may exist well before any approach to failure, perhaps first between the paste and larger aggregate particles and then progressively extending across the paste to link up as failure is approached. Some of the more recent papers consider the application of the energy release concepts of fracture mechanics, a subject originally developed to explain, by consideration of crack propagation, the unexpectedly low tensile strengths of brittle elastic materials. The extension to concrete has not been without difficulty but the general interest in fracture and crack-arrest mechanisms has been helpful, eg in the development of fibrous concrete.

This view of the conditions under which cracking occurs in a dam and its important effect on the way in which the property of extensibility should be assessed, has not been accepted by dam engineers everywhere even in recent years. There still remains a tendency to equate high creep at lower loads to high eventual extensibility disregarding the changing internal conditions within the concrete which determine eventual failure.

The US Corps of Engineers appeared to have adopted a useful compromise when they compared the potential crack resistances of alternative concretes proposed for recent dams on the basis of very slow **flexural** tests on massive concrete beams under progressively increasing loads. Comparisons were based on 'strain capacity', ie the surface strains extrapolated from internal strain gauge readings corresponding to 90 per cent of the eventual failing load which occurred typically at 30 to 60 days. Although this approach to the problem appears to have been largely creep-orientated the slow flexural nature of the tests does ensure that the rate of straining is controlled, particularly in the surface layers where cracks may be initiated. It would be interesting to know what happens beyond the arbitrary 90 per cent of the failing load.

Another practical approach has been employed in the USSR where concretes have been compared by the ratios of tensile strength (by splitting test) to the dynamic elastic modulus. Although this has proved a convenient way of performing a large number of tests, it is some way from being a simulation of practical conditions. The emphasis given to tensile strength rather than to creep, except even more indirectly, is very interesting.

Looking at the published literature, it becomes evident, as already noted, that by far the most important factor affecting the measurement of extensibility is the limitation of the rate of straining so as to employ more or all of the 'complete' stress/strain curve. Differences in this respect make comparisons difficult between information from different sources and the applicability of some of the results to practical conditions questionable. An allied factor is the difference between flexural and direct tensile tests.

Another factor which might be expected to be critical but upon which evidence is scanty, is the moisture conditions of the concrete surface and interior. These might considerably affect micro-cracking through their effect on shrinkage and also affect other properties of the concrete. Ohno and Shibata's [20] results for immature concrete suggest a greater extensibility when wet than dry.

The rate of straining might also be expected to be important in assessing numerical values, whatever view is taken of extensibility. Ohno and Shibata [20] showed for immature concrete higher extensibility for lower rates of loading. Hansen[47] showed similar results in a study of the time of form removal. More direct confirmation for maturer concrete is given by the results of the US Corps of Engineers tests already referred to[115, 116]. It was found that the improvement in 'strain capacity' at slower rates varied from about 8 per cent to nearly a 100 per cent.

Another factor to be taken into account when assessing the experimental results is that of 'averaging'. In direct 'uncontrolled-strain' tests the specimen might fail as the result of a single flaw or continuous crack. This, of course, accounts for the variability of tensile **strength** tests in general. In 'controlled-strain' tests there is the possibility of evening out the effects of individual micro-cracks. Even here, care must be taken that the measurement of strain is not made on an unduly short gauge length spanning a crack which is itself insufficient to cause failure[38, 98]. Some of the freakishly high extensibilities reported, eg of the order 2000 x 10^{-6} strain or more, were probably associated with an exceptional juxtaposition of gauge length and crack. Flexural tests might be an improvement on direct tests in this respect but the effect needs watching.

The effects of other factors, such as the materials used and the mix proportions, tend to be rather smaller than the effects already noted and they are difficult to summarise in isolation from the conditions of test. However, the general picture emerges of a higher extensibility being associated with a lower water/cement ratio[94, 99, 115], and with aggregate of a smaller maximum size[94, 115], the aggregate being crushed[67, 99, 115] and of a rough texture[59, 115]. Less attractive to dam constructors are less definite suggestions that extensibility may be increased by a smaller proportion of coarse to fine aggregate[29] and a higher cement content[36, 57, 94, 99, 115]. There is, however, some conflicting evidence on the latter point[77].

Very tentatively, it could be suggested that the effect of these factors is compatible with a view of extensibility determined at least partly by the bond between paste and the larger aggregate particles. This places the emphasis rather differently than in a view expressed by the US Corps of Engineers[115] that a higher 'strain capacity' goes with an aggregate of a lower E value.

Other support for the 'bond' or 'micro-cracking' view is the mention of the possible importance of bond by Stolnikov[99] in explaining the improvement gained from the use of limestone sand. He considered that the type and grain size of the sand has a great influence on his measure of extensibility. The results by the US Corps of Engineers[115] showed great improvements in 'strain capacity' when all the aggregate, including sand, was manufactured (crushed or milled).

Johnston[94] found differences between various aggregates in tests at uncontrolled strain suggesting an improvement in extensibility in the ascending order of merit basalt, limestone, gravel, granite, sandstone. The last two were consistently first and second in merit even when the water/cement ratio and aggregate size were varied. Kaplan[29] also found little difference between limestone and gravel but found an improved extensibility as the amount of coarse aggregate was reduced.

Houk, Paxton and Houghton[100] reported that 'strain capacity' increased with age, a result consistent with a view of extensibility as related to bond strength providing no drying occurs. Hansen[47], on the other hand, reported that in tests related to the time of form removal, extensibility decreased with age.

Hughes and Ash[77, 95] found that the effect of anisotropy due to the direction of casting a specimen was significant.

An interesting result was reported by McDonald, Bombich and Sullivan[116] in describing tests for Trumbull Dam for US Corps of Engineers. Here, when comparing two possible concretes, a greater increase in 'strain capacity' with decrease in speed of loading was shown by the concrete with the lower creep. The authors considered the result anomalous but it could be that the effect of differences in creep are less important than they thought. Their results also indicated that the substitution of fly ash for part of the cement decreased 'strain capacity' while it increased creep.

In this study an attempt has been made to concentrate on the subject of extensibility and not to widen the study to the subject of cracking in dams. This is because there are, in most cases, other factors affecting the strains which lead to cracking, eg the differences in heat evolution due to different types of cement, differences in temperature change or drying, etc. It is worth noting, however, since strains in the different dams may not be seriously involved, that the US Corps of Engineers reported[115] a lower cracking incidence in dams in their North Pacific Division made with totally 'manufactured' aggregate, a result which correlates with their 'strain capacity' comparisons referred to above.

RECOMMENDATION

It is recommended that comparative tests should continue on the factors determing the extensibility of concrete defined as 'tensile strain capacity to failure' and that special attention should be paid in the tests to a condition of controlled strain simulating that in a dam and to the conditions within the concrete determining the stage at which visible cracking starts. Slow rates of loading should be employed and regard should be paid to the moisture condition of the concrete.

BIBLIOGRAPHY ON EXTENSIBILITY OF CONCRETE FOR LARGE DAMS

e indicates paper containing quantitative information on extensibility defined as 'strain capacity to failure'

c indicates paper from US Corps of Engineers containing information on strain capacity up to 90 per cent of failing load in slow flexural tests

a indicates paper from USSR with indirect methods of determination.

Other papers contain discussions relevant to extensibility.

REFERENCES

1 **Thomas, F G.** Cracking in reinforced concrete. Structural Engineer, Vol 14, 1936, pp 298-320.

2 **Rao, K L.** Structural Engineer, Vol 20, 1942, pp 44 and 68 (Discussion of paper by Marshall, W T and Tembe, N R. Experiments on plain and reinforced concrete in torsion. Structural Engineer, Vol 19, 1941, pp 177-191; Discussion, Vol 20, pp 38-44, 68-69.)

3 **Squire, R H.** Some important aspects of the elastic modulus of concrete. Structural Engineer, Vol 21, 1943, pp 211-239.

4 **Rao, K L.** Prestressed beams under direct sustained loading. Structural Engineer, Vol 22, 1944, pp 425-454.

5e **Evans, R H.** Extensibility and modulus of rupture of concrete. Structural Engineer, Vol 24, 1946, pp 636-659.

6 **Ramaley, D** and **McHenry, D.** Stress-strain curves for concrete strained beyond the ultimate load. US Department of the Interior, Bureau of Reclamation, Engineering and Geological Control and Research Division, Laboratory Report No Sp-12, 1947.

7e **Grassam, N S J** and **Fisher, D.** Tests on concrete with electrical-resistance strain gauges. Engineering, Vol 172, 1951, pp 356-358.

8e **Rao, K L.** Extensibility and cracking in concrete. International Commission on Large Dams, 4th Congress, New Delhi, 1951, Proceedings Vol 3, pp 249-269.

9 **Lee, C R.** Creep and shrinkage in restrained concrete. International Commission on Large Dams, 4th Congress, New Delhi, 1951, Report R46.

10 **Roberts, C M.** Cement variations as affecting cracking in large dams. International Commission on Large Dams, 4th Congress, New Delhi, 1951, Proceedings Vol 3, pp 129-153.

11e **Blakey, F A.** Mechanism of fracture of concrete. Nature, Lond, Vol 170, 1952, p 1120.

12 **Blakey, F A** and **Beresford, F D.** Tensile strains in concrete. CSIRO, Melbourne, Reports C2, 2-1, 1953 and C2, 2-2, 1955.

13e **Blakey, F A** and **Beresford, F D.** A note on strain distribution in concrete beams. Civil Engineering and Public Works Review, Vol 50, No 586, 1955, pp 415-416.

14e **Todd, J D.** The determination of tensile stress-strain curves for concrete. Proceedings of the Institute of Civil Engineers, Part 1, Vol 4, 1955, pp 201-211.

15 **Saemann, J C, Warren, C** and **Washa, G W.** Effect of curing on the properties affecting shrinking and cracking of concrete block. American Concrete Institute Journal, Proceedings Vol 51, No 9, 1955, pp 833-852.

16e **Blakey, F A.** Some considerations of the cracking or fracture of concrete. Civil Engineering and Public Works Review, Vol 52, No 615, 1957, pp 1000-1003.

17 **Jones, R** and **Kaplan, M F.** The effect of coarse aggregate on the mode of failure of concrete in compression and flexure. Magazine of Concrete Research, Vol 9, No 26, 1957, pp 89-94.

18e **Blakey, F A.** Influence of water-cement ratio on mortar in which shrinkage is restrained. American Concrete Institute Journal, Proceedings Vol 55, No 5, 1958, pp 591-604.

19e **Blackman, J S, Smith, G M** and **Young, L E.** Stress distribution affects ultimate tensile strength. American Concrete Institute Journal, Proceedings Vol 55, No 6, 1958, pp 679-684. Discussion by Bredsdorff, P K and Kierkegaard-Hansen, P. American Concrete Institute Journal, Proceedings Vol 55, No 6, pp 1421-1426.

20e **Ohno, K** and **Shibata, T.** On extensibility of fresh concrete under slowly increasing tensile load. RILEM Bulletin, Vol 4, 1959, pp 24-31.

21 **Neville, A M.** Some aspects of the strength of concrete. Civil Engineering and Public Works Review, Vol 54, 1959, pp 1153-1156, 1308-1310, 1435-1439.

22 **Blakey, F A** and **Lewis, R K.** The deformation and cracking of hardened cement paste when shrinkage is restrained. Civil Engineering and Public Works Review, Vol 54, 1959, pp 577-579, 759-762.

23 **Blakey, F A** and **Beresford, F D.** Cracking of concrete. Constructional Review, Vol 32, No 2, 1959, pp 24-28.

24 **Waugh, W R** and **Rhodes, J A.** Control of cracking in concrete gravity dams. American Society of Civil Engineers Proceedings, Journal of Power Division, Vol 85, No PO5, 1959, pp 1-19.

25e **Hatano, T.** Dynamical behaviour of concrete under impulsive tensile load. Technical Laboratory, Central Research Institute of Electric Power Industry, Tokyo. Technical Report: C-6002, Nov 5, 1960.

26 **Jones, R.** The development of microcracks in concrete. RILEM Bulletin, No 9, 1960, pp 110-114.

27 **Blakey, F A** and **Lewis, R K.** Effect of sand characteristics on the cracking of small bars of mortar in which shrinkage is restrained. Civil Engineering and Public Works Review, Vol 55, 1960, pp 389-393.

28 **Blakey, F A** and **Lewis, R K.** The effect of curing on the cracking of small bars of mortar in which shrinkage is restrained. Civil Engineering and Public Works Review, Vol 56, No 657, 1961, pp 473-477; No 658, pp 639-643.

29e **Kaplan, M F.** Strains and stresses of concrete at initiation of cracking and near failure. American Concrete Institute Journal, Proceedings Vol 60, No 7, 1963, pp 853-880. Discussion by Abeles, P W. American Concrete Institute Journal, Proceedings Vol 61, 1963, pp 1937-1943.

30e **Rusch** and **Hilsdorf, H.** Verformungseigenschaften von beton unter zentrischen zugspannungen (Deformation properties of concrete under concentric tensile stresses). Teil 1, Bericht Nr 44, des Material-prufungsamtes fur das Bauwesen der Technischen Hochschule Munchen, May 1963.

31e **Namiki, M, Kiuchi, A** and **Oshio, A.** On the elongation of concrete in the local rupture region. Review of the 17th General Meeting of the Japan Cement Engineering Association, Tokyo, May 1963, pp 195-198.

32 **Hsu, T T C, Slate, F O, Sturman, G M** and **Winter, G.** Microcracking of plain concrete and the shape of the stress-strain curve. American Concrete Institute Journal, Proceedings Vol 60, No 2, 1963, pp 209-224.
Discussion by 13 contributors and closure by authors, American Concrete Institute Journal, Proceedings Vol 60, No 2, 1963, pp 1787-1824.

33 **Romualdi, J P** and **Batson, G B.** Mechanics of crack arrest in concrete. American Society of Civil Engineers Proceedings, Journal of Engineering Mechanics Division, Vol 89, No EM3, 1963, pp 147-168.
Discussion by Broms, B B, Shah, S P and Abeles, P W. American Society of Civil Engineers Proceedings, Journal of Engineering Mechanics Division, Vol 90, No EM1, 1964, pp 167-173.

34 **Glucklich, J.** Fracture of plain concrete. American Society of Civil Engineers Proceedings, Journal of Engineering Mechanics Division, Vol 89, No EM6, 1963, pp 127-138.

35e **Evans, R H** and **Kong, F K.** The extensibility and microcracking of the in-situ concrete in composite prestressed beams. Structural Engineer, Vol 42, No 6, 1964, pp 181-189.

36e **Oladapo, I O.** Cracking and failure in plain concrete beams. Magazine of Concrete Research, Vol 16, No 47, 1964, pp 103-113.

37e **Oladapo, I O.** Extensibility and modulus of rupture of concrete. Technical University of Denmark, Structural Research Laboratory Bulletin, No 18, 1964.

38e **Mamiki, M, Kiuchi, A** and **Oshio, A.** Tensile deformation of concrete. The Journal of Research of the Onoda Cement Co (in Japanese with summary in English):
Vol 15, No 55, 1963, (On the tensile performance of green concrete).
Vol 15, No 57, 1963, (On the elongation of concrete in the local region for rupture).
Vol 15, No 58, 1963, (On the relation between strength and localised plastic strain of concrete).
Vol 16, No 59, 1964, (On the stability of cracks in concrete).
Vol 16, No 60, 1964, (On the microscopic theory of compressive failure of concrete).
Vol 16, No 61, 1964, (Theory of creep deformation).
Vol 16, No 62, 1964, (Crack formation due to restraint of drying shrinkage).

39e **Mamiki, M** and **Oshio, A.** On the meanings and utility of restrained drying shrinkage test of concrete. Review of the 18th General Meeting of the Japan Cement Association, Tokyo, May 1964, pp 156-163.

40 **Lee, C R.** Temperature and other factors influencing the cracking of concrete in a dam. International Commission on Large Dams, 8th Congress, Edinburgh 1964, Proceedings Vol 3, pp 179-191.

41 **Newman, K.** The structure and engineering properties of concrete. International Symposium on Theory of Arch Dams, Southampton University, April 1964. Proceedings pp 683-712. Pergamon Press, 1965.

42 **Robinson, G S.** The influence of microcracking and state of stress on the elastic behaviour and discontinuity of concrete. International Symposium on Theory of Arch Dams, Southampton University, April 1964. Proceedings pp 713-721. Pergamon Press, 1965.

43 **Coutinho, A de S.** The influence of the type of cement on its cracking tendency. Ministério das Obras Públicas, Laboratório Nacional de Engenharia Civil, Technical Paper No 216, Lisbon, 1964.

44 **Coutinho, A de S.** A fissurabilidade dos cimentos, argamassas e betoes por efeito da sua contraccao (Cracking tendency in cements, mortars and concretes due to shrinkage). Ministério das Obras Públicas, Laboratório Nacional de Engenharia Civil, Publicacao No 57, Lisbon, 1964 (Summary in English pp 119-122).

45e **Illston, J M.** The creep of concrete under uniaxial tension. Magazine of Concrete Research, Vol 17, No 51, 1965, pp 77-84.

46e **Jones, R.** Cracking and failure of concrete test specimens under uniaxial quasi-static loading. International Conference on Structure of Concrete, London 1965, Paper B5.

Proceedings, pp 125-130. Cement and Concrete Association, 1968.

47e **Hansen, T C.** Surface cracking of mass concrete structures at early form removal. RILEM Bulletin, No 28, 1965, pp 145-153.

48e **Sturman, G M, Shah, S P** and **Winter, G.** Effects of flexural strain gradients on micro-cracking and stress - strain behaviour of concrete. American Concrete Institute Journal, Proceedings Vol 62, 1965, pp 805-822.
Discussion by Swamy, N. American Concrete Institute Journal, Proceedings Vol 63, 1966, pp 1717-1719.

49 **Shah, S P** and **Slate, F O.** Internal microcracking, mortar-aggregate bond and the stress-strain curve of concrete. International Conference on the Structure of Concrete, London, 1965. Proceedings pp 82-92. Cement and Concrete Association, 1968.

50 **Kaplan, M F.** The application of fracture mechanics to concrete. International Conference on the Structure of Concrete, London 1965. Proceedings pp 169-175. Cement and Concrete Association, 1968.

51 **Romualdi, J P.** The static cracking strength and fatigue strength of concrete reinforced with short pieces of thin steel wire. International Conference on the Structure of Concrete, London, 1965. Proceedings pp 190-201. Cement and Concrete Association, 1968.

52 **Newman, K.** Criteria for the behaviour of plain concrete under complex states of stress. International Conference on the Structure of Concrete, London 1965. Proceedings pp 255-274. Cement and Concrete Association, 1968.

53 **Vile, G W D.** The strength of concrete under short-term static biaxial stress. International Conference on the Structure of Concrete, London, 1965. Proceedings pp 275-288. Cement and Concrete Association, 1968.

54 **Marshall, A L.** Review of some of the problems involved in the early-age cracking of concrete. Civil Engineering and Public Works Review, Vol 60, No 709, 1965, pp 1169-1175.

55 **Alexander, S.** A single equation for the stress-strain curve of concrete. Indian Concrete Journal, Vol 39, No 7, 1965, pp 274-277.

56 **Yerlici, V A.** Behaviour of plain concrete under axial tension. American Concrete Institute Journal, Proceedings Vol 62, No 8, 1965, pp 987-992.

57e **Welch, G B.** Tensile strains in unreinforced concrete beams. Magazine of Concrete Research, Vol 18, No 54, 1966, pp 9-18.

58e **Hughes, B P** and **Chapman, G P.** The deformation of concrete and micro-concrete in compression and tension with particular reference to aggregate size. Magazine of Concrete Research, Vol 18, No 54, 1966, pp 19-24.

59e **Hughes, B P** and **Chapman, G P.** The complete stress-strain curve for concrete in direct tension. RILEM Bulletin, No 30, 1966, pp 95-97.

60 **Becker, G** and **Weigler, H.** Investigations on the failure and deformation of concrete under combined biaxial stresses. Translation by Commonwealth Scientific and Industrial Research Organisation, Melbourne, Translation No 7282, 1966.

61e **Davies, J D** and **Nath, P.** Complete load-deformation curves for plain concrete beams. Building Science, Vol 2, No 3, 1967, pp 215-221.

62e **Heilman, H G, Hilsdorf, H** and **Finsterwalder, K.** Festigkeit und Verformung von Beton unter Zugspannungen (Strength and deformation of concrete under tensile stress). Materialprufungsamt fur das Bauwesen der Technischen Hochschule Munchen 1967, Bericht Nr 64.
(Same title - Deutscher Ausschuss fur Stahlbeton 1969, Heft 203).

63e **Komlos, K.** The determination of the tensile strength of concrete. Indian Concrete Journal, Vol 41, No 11, 1967, pp 429-436; Vol 42, No 2, 1968, pp 68-76; Vol 42, No 11, 1968, pp 473-478, 482; Vol 43, No 2, 1969, pp 42-49, 54.

64 **Clark, L E, Gestle, K H** and **Tulin, L G.** Effect of strain gradient on the stress-strain

curve of mortar and concrete. American Concrete Institute Journal, Proceedings Vol 64, No 9, 1967, pp 580-586.

65 **Meyers, B L.** Time dependent strains and microcracking of plain concrete. PhD Thesis, Cornell University, 1967.

66 **Barashikov, A Y A.** Creep of concrete under cyclic deformations. Beton I Zhelezobeton, Vol 13, No 12, 1967, pp 28-30. (In Russian.)

67e **Haroun, W A.** Uniaxial tensile creep and failure of concrete. PhD Thesis, University College London, 1968.

68e **Evans, R H** and **Marathe, M S.** Microcracking and stress-strain curves for concrete in tension. Matériaux et Constructions (RILEM), Vol 1, No 1, 1968, pp 61-64.

69e **Hughes, B P** and **Ash, J E.** Short-term loading and deformation of concrete in uniaxial tension and pure torsion. Magazine of Concrete Research, Vol 20, No 64, 1968, pp 145-154.

70e **Isenberg, J.** Properties of concrete change when microcracking occurs. American Concrete Institute Special Publication SP 20, 1968, pp 28-41.

71 **Illston, J M.** Components of creep in mature concrete. American Concrete Institute Journal, Proceedings Vol 65, No 3, 1968, pp 219-227.

72 **Mitzel, A.** The hypothesis of equal creep deformation in compression and in tension, and the principle of superposition of creep and shrinkage Annales Institut du Bâtiment, Vol 21, No 241, 1968, pp 96-100. (In French.)

73 **Shah, S P** and **Chandra, S.** Critical stress, volume change and microcracking of concrete. American Concrete Institute Journal, Proceedings Vol 65, No 9, 1968, pp 770-781. Discussion by Sidenbladh, T. American Concrete Institute Journal, Proceedings Vol 66, 1969, pp 227-229.

74 **Shah, S P** and **Winter, G.** Inelastic behaviour and fracture of concrete. American Concrete Institute Journal Special Publication SP 20, 1968, pp 5-28.

75 **Orr, D M F.** A note on the ductility of concrete in tension. Constructional Review, Vol 41, No 6, 1968, pp 22-24.

76 **Welch, G B** and **Haisman, B.** The application of fracture mechanics to concrete. University of New South Wales, UNICIV Report 1.12, 1968.
Also similar title in Matériaux et Constructions (RILEM), Vol 2, No 9, 1969, pp 171-177.

77e **Hughes, B P** and **Ash, J E.** The effect of water gain on the behaviour of concrete in tension, torsion and compression. The Concrete Society, Technical Paper PCS 54, November 1969.

78e **Facaoaru, I, Tannenbaum, M** and **Stanculesou, G.** Studies on the deformation and cracking of concrete under uniaxial stress fields. International Conference on Structure, Solid Mechanics and Engineering Design, Southampton University 1969. Proceedings, Part I, pp 653-666. John Wiley, 1971.

79e **Johnson, R P** and **Lowe, P G.** Behaviour of concrete under biaxial and triaxial stress. International Conference on Structure, Solid Mechanics and Engineering Design, Southampton University 1969. Proceedings, Part II, pp 1039-1051. John Wiley, 1971.

80e **Ward, M A** and **Cook, D J.** The development of a uniaxial test for concrete and similar brittle materials. Materials Research and Standards, Vol 9, No 5, 1969, pp 16-20.

81e **Carvalho Franca, G de** and **Pincus, G.** The distribution of concrete strains in the split cylinder test. Journal of Materials, Vol 4, No 2, 1969, pp 393-407.

82e **Kupfer, H** and **Hilsdorf, H K.** Behaviour of concrete under biaxial stresses. American Concrete Institute Journal, Proceedings Vol 66, 1969, pp 656-666.
Discussion by Pandit, G S and Zimmerman, R M. American Concrete Institute Journal, Proceedings Vol 67, 1970, pp 194-197.

83c **Houghton, D L.** Concrete volume change for Dworshak Dam. American Society of Civil

Engineers Proceedings, Journal of Power Division, Vol 95, No PO2, 1969, pp 153-166.

84 **Ward, M A** and **Cook, D J**. The mechanism of tensile creep in concrete. Magazine of Concrete Research, Vol 21, No 68, 1969, pp 151-158.

85 **Franklin, R E**. The effect of weather on early strains in concrete slabs. RRL Report No LR 266, 1969.

86 **McCreath, D R, Newman, J B** and **Newman, K**. The influence of aggregate particles on the local strain distribution and fracture mechanism of cement paste during drying shrinkage and loading to failure. Matériaux et Constructions (RILEM), Vol 2, 1969, pp 73-84.

87 **Grimer, F J** and **Hewitt, R E**. The form of the stress-strain curve of concrete interpreted with a diphase concept of material behaviour. International Conference on Structure, Solid Mechanics and Engineering Design, Southampton University, 1969, Proceedings.

88 **Hatano, T**. Failure of concrete and similar brittle solids on the basis of strain. International Journal of Fracture Mechanics, Vol 5, 1969, pp 73-79.

89 **Moavenzadeh, F** and **Kuguel, R**. Fracture of concrete. Journal of Materials, Vol 4, 1969, pp 497-519.

90 **Popovics, S**. Fracture mechanics in concrete. How much do we know? American Society of Civil Engineers Proceedings, Journal of Engineering Mechanics Division, Vol 95, No EM3, 1969, pp 531-544.
Discussion by Marathe, M S, Chandra, S, Cowan, H J and Cook, D J. American Society of Civil Engineers Proceedings, Journal of Engineering Mechanics Division, Vol 96, No EM1, 1970, pp 95-102.

91 **Béres, L**. Relaxation of stresses in concrete. Bauplanung Bautechnik, Vol 23, No 6, 1969, pp 286-288. (In German.)

92 **Welch, G B** and **Haisman, B**. Fracture toughness measurements of concrete. University of New South Wales, UNICIV Report R42, 1969.

93 **Naus, D J** and **Lott, J L**. Fracture toughness of Portland cement concretes. American Concrete Institute Journal, Proceedings Vol 66, No 6, 1969, pp 481-489.

94e **Johnston, C D**. Strength and deformation of concrete in uniaxial tension and compression. Magazine of Concrete Research, Vol 22, No 70, 1970, pp 5-16.

95e **Hughes, B P** and **Ash, J E**. Anisotropy and failure criteria for concrete. Matériaux et Constructions (RILEM), Vol 3, No 18, 1970, pp 371-374.

96e **Imbert, I D C**. The effect of holes on tensile deformations in plain concrete. Highway Research Record, No 324, 1970, pp 54-65.

97e **Chen, W F**. Extensibility of concrete and theorems of limit analysis. American Society of Civil Engineers Proceedings, Journal of Engineering Mechanics Division, Vol 96, No EM3, 1970, pp 341-352.

98e **Raju, N K**. Strain distribution and microcracking in concrete prisms with a circular hole under uniaxial compression. Journal of Materials, Vol 5, No 4, 1970, pp 901-908.

99a **Stolnikov, V V** and **Litvinova, R E**. Factors influencing crack resistance of concrete for large dams. Communication by Soviet National Committee to 10th Congress on Large Dams, Montreal, 1970, pp 45-76.

(See also contribution without quantitive data on extensibility: Stolnikov, V V. Discussion on question 38, 10th Congress on Large Dams, Montreal 1970, Proceedings Vol 6, pp 719-721.)

100c **Houk, I E, Paxton, J A** and **Houghton, D L**. Prediction of thermal stresses and strain capacity of concrete by tests on small beams. American Concrete Institute Journal, Proceedings Vol 67, No 3, 1970, pp 253-261.

101e **Johnston, C D**. Deformation of concrete and its constituent materials in uniaxial tension. Highway Research Record, No 324, 1970, pp 66-76.

102 Strain capacity of mass concrete. Department of the Army, Office of Chief of Engineers, Washington, Engineer Technical Letter No 1110-2-89, April 1970.

103 **Venuat, M.** Deformation of hardened concrete under load. Revue des Matériaux de Construction, 1970, No 655, pp 87-97; No 656, pp 121-134; No 657-8, pp 167-196. (In French.)

104 **Campbell-Allen, D** and **Holford, J G.** Stress and cracking in concrete due to shrinkage. Institution of Engineers Australia, Civil Engineering Transactions, Vol CE12, No 1, 1970, pp 33-39.

105 **Popovics, S.** A review of stress-strain relationships for concrete. American Concrete Institute Journal, Proceedings Vol 67, No 3, 1970, pp 243-248.

106e **Hunt, J G.** A laboratory study of early-age thermal cracking of concrete. Cement and Concrete Association, Technical Report 42.457, London, 1971.

107e **Birkimer, D L** and **Lindemann, R.** Dynamic tensile strength of concrete materials. American Concrete Institute Journal, Proceedings Vol 68, No 1, 1971, pp 47-49 and supplement No 68-8.

108e **Chen, W F** and **Carson, J L.** Stress-strain properties of random wire reinforced concrete. American Concrete Institute Journal, Proceedings Vol 68, No 12, 1971, pp 933-936.

109 **Marshall, A L.** The nature of early-age cracking in concrete. Sunderland Polytechnic, May 1971, Report No CR/1.

110 **Buyukozturk, O, Nilson, A H** and **Slate, F O.** Stress-strain response and fracture of a concrete model in biaxial loading. American Concrete Institute Journal, Proceedings Vol 68, No 8, 1971, pp 590-599.

111 **Cook, D J.** Factors affecting the tensile creep of concrete. University of New South Wales, School of Engineering, UNICIV Report R-64, April 1971.

112 **Yerlici, V A.** Stresses and cracking in reinforced concrete members under axial tension. Matériaux et Constructions (RILEM), No 23, 1971, pp 313-322.

113 **Shah, S P** and **McGarry, F J.** Griffith fracture criterion and concrete. American Society of Civil Engineers Proceedings, Journal of Engineering Mechanics Division, Vol 97, No EM6, 1971, pp 1663-1676.
Discussion by Brown, E T, Hudson, J A and Marathe, M S. American Society of Civil Engineers Proceedings, Journal of Engineering Mechancis Division, Vol 98, No EM5, 1972, pp 1310-1313.

114 **Magnas, J P** and **Audibert, A.** Criteria for the strength of concrete subjected to multiaxial strains. Annales de l'Institut Technique du Bâtiment et des Travaux Publics, No 287, 1971, pp 21-43.

115c **Houghton, D L.** Concrete strain capacity tests - their economic implications. Conference on Economical Construction of Concrete Dams, Proceedings of the Engineering Foundation Conference, Asilomar Conference Grounds, Pacific Grove, California, May 1972. Proceedings published by American Society of Civil Engineers.

116c **McDonald, J E, Bombich, A A** and **Sullivan, B R.** Ultimate strain capacity and temperature rise studies, Trumbull Pond Dam. US Army Engineer Waterways Experiment Station, Vicksburg, Miss, Miscellaneous Paper C-72-20, August 1972.

117 **Brown, J H.** Measuring the fracture toughness of concrete paste and mortar. Magazine of Concrete Research, Vol 24, No 81, 1972, pp 185-196.

118 **ACI Committee 224.** Control of cracking in concrete structures. American Concrete Institute Journal, Proceedings Vol 69, No 12, 1972, pp 717-753.

119 **Liu, T C Y, Nilson, P H** and **Slate, F O.** Stress-strain response and fracture of concrete in uniaxial and biaxial compression. American Concrete Institute Journal, Proceedings Vol 69, No 5, 1972, pp 291-295.
Discussion by Selvam, V K M, Aldridge, W W and Hsu, C T. American Concrete Institute Journal, Proceedings Vol 69, No 11, 1972, pp 710-712.

120 **Cook, D J.** Some aspects of the mechanism of tensile creep in concrete. University of New South Wales, UNICIV Report No R97, November 1972.

121e **Swamy, R N.** Fibre reinforced concrete is here to stay. Civil Engineering and Public Works Review, Vol 68, No 809, 1973, pp 1075-1081. This paper is a review of the First International Symposium on Fibre Reinforced Concrete, Ottawa, October 1973, sponsored by the American Concrete Institute.

122c **McDonald, J E.** Ultimate strain capacity tests, Clarence Cannon Dam, St Louis District. US Army Engineer Waterways Experiment Station, Vicksburg, Miss, Miscellaneous Paper C-73-5, March 1973.

123 **ACI Committee 207.** Effect of restraint, volume change and reinforcement on cracking of massive concrete. American Concrete Institute Journal, Proceedings Vol 70, No 7, 1973, pp 445-470.

124 **Rostam, S** and **Byskov, E.** Cracks in concrete structures: a fracture mechanics approach. Danmarks Tekniske Høјskole, Afdelingen for Baerende Konstructioner, Rapport Nr R34, 1973.

125 **Naus, D J.** Fracture mechanics applicability to Portland cement concrete. Army Construction Engineering Research Laboratory, Champaign, Illinois, Technical manuscript M42, March 1973.

126 Third International Congress on Fracture, Munich, April 1973. Thirteen papers concerning fracture of concrete given at the Congress form a special issue of Cement and Concrete Research, Vol 3, No 4, 1973, pp 343-494.

127 **Radjy, F** and **Hansen, T C.** Fracture of hardened cement paste and concrete. Cement and Concrete Research, Vol 3, No 4, 1973, pp 343-361.

128 **Zaitsev, J W** and **Wittmann, F H.** Fracture of porous viscoelastic materials under multiaxial state of stress. Cement and Concrete Research, Vol 3, No 4, 1973, pp 389-395.

129 **Swamy, R N** and **Kameswara Rao, C V S.** Fracture mechanism in concrete systems under uniaxial loading. Cement and Concrete Research, Vol 3, No 4, 1973, pp 413-427.

130 **Brown, J H** and **Pomeroy, C D.** Fracture toughness of cement paste and mortars. Cement and Concrete Research, Vol 3, No 4, 1973, pp 475-480.

APPENDIX 1

Dr Rosanov (USSR) has commented:

'The procedure for the comparative evaluation of the extensibility of concrete using the dynamic elasticity modulus and the tensile strength at splitting is but one of several approaches used in the USSR for this purpose. Moreover in a number of studies under way in the USSR on the extensibility of concrete in particular to obtain complete stress-strain diagrams the moving boundary method or similar approaches are used. Information on those methods and on data received when applying them is available in the communication by A A Borovoi, K A Maltsov and N S Rosanov, which was distributed at the 11th ICOLD Congress[131].

It is an accepted practice in the USSR to subdivide the complete stress-strain diagram into a number of sections with different physical characteristics of the materials when tested at a prescribed deformation speed. In investigating the effect of different factors on the extensibility of concrete, not only the variation in the general shape of the diagram is studied, but also that of the individual sections and the distribution of the points representative for the transition of the material from one state into another. Such an approach seems to us to be the most appropriate and promising one.

In enumerating the characteristics and parameters of the concrete which affect its extensibility it should be pointed out that according to the experimental research conducted in the USSR one of the major factors is the type of rock used for the coarse aggregate. Namely, concretes with carbonate aggregate possess, as a rule, greater extensibility than those with igneous aggregate.

We consider it expedient to emphasise in the recommendations adduced in the paper not only the necessity of developing procedures for the estimation of the extensibility of concrete for large dams fitting the actual service conditions of the structures, but also the need to define clearly the state of the material corresponding to different sections of the complete stress-strain diagram. The latter is required for the substantiation of employing the experimental findings in question in engineering design and in solving problems encountered in engineering practice while constructing large dams.'

Additional reference

131a **Borovoi, A A, Maltsov, K A** and **Rosanov, S N.** Critères de sécurité des barrages en béton. Communication by Soviet National Committee to 11th Congress on Large Dams, Madrid, 1973.

High alumina cement concrete in buildings (CP 34/75)

Building Research Establishment

1 INTRODUCTION

In February 1974 two roof beams made with high alumina cement concrete collapsed into a school swimming pool at Stepney. An investigation was immediately started by the Building Research Establishment at the request of DOE/DES who issued a joint Circular[1] in February suggesting that local authorities identify buildings with similar roof construction, in particular those likely to have high temperatures at roof level, eg swimming baths, assembly halls, laundries, etc and arrange for an appraisal of the structural design.

During the investigation, it was found that similar roof beams in the nearby gymnasium of the same school were also highly converted and had lost strength, indicating that the deterioration of the concrete was not primarily due to the high temperatures and humidity in the swimming pool roof. A further DOE/DES Circular was issued in May[2] which extended the initial warning to all buildings with roofs of a type similar to the Stepney School.

The report of the BRE investigation[3], published in June 1974, attributed failure to a loss of strength of the high alumina cement concrete due to conversion followed by chemical attack.

During this period, the Department had also received a number of reports of structural inspections and appraisals which indicated loss of strength of high alumina cement concrete in varying degrees and in many forms of construction over the country. This led to the issue of a further joint DOE/DES Circular in July 1974[4] in which it was concluded that:

1 all existing buildings incorporating high alumina cement concrete should be regarded as suspect, at least in the longer term

2 a further programme of action would be required starting immediately with those buildings with the greatest risk

3 high alumina cement concrete should not be used for structural work in buildings until further notice.

It recommended that local authorities should extend appraisal to all precast prestressed non-composite roof or floor members and also to columns of high alumina cement concrete, dealing in the first place with those buildings with roof or floor members exceeding 5 metres. This Circular also included advice, prepared by the Building Research Establishment, on testing which described procedures for visual examination, chemical analysis, core testing, ultrasonic pulse velocity measurement and performance examination.

By then the Building Research Establishment had started field studies and laboratory tests to examine the degree of risk likely in buildings with precast prestressed non-composite members, particularly those with spans shorter than about 5 metres. Investigations of other situations in which high alumina cement concrete had been used were also to be made.

2 DEVELOPMENT OF THE STRUCTURAL USE OF HIGH ALUMINA CEMENT

The manufacture of high alumina cement commenced in the United Kingdom in 1925 following its introduction from France where it had been developed to provide concrete, which would resist chemical attack, particularly for marine use. From the structural engineering viewpoint, the capability of the cement to develop a high early strength in concrete offered advantages, since strengths reached one day after casting were very appreciably greater than could be obtained with normal Portland cements. Its high cost however prevented extensive use.

In 1933, Davey [5] published the results of a study of the influence of temperature upon the strength development of concretes including concrete made with high alumina cement. Most of his specimens had a total water/cement ratio of 0.6 with mix proportions of 1:6 by weight. The results showed that the concrete suffered some reduction in strength after curing at 25°C for one month and a much more serious reduction in strength after curing for seven days at 35°C. The majority of his tests were completed in 90 days but he also stated that there was a slight regression in strengths at one year although no details were given.

Few results of long-term tests are available from this period but two series of cylindrical specimens were prepared at BRS in 1925 with mix proportions of 1:6 by weight. The total water/cement ratio was 0.6; the storage conditions were not at first controlled but later were maintained at 18°C and 65% relative humidity. Results over a period of 20 years are given in Figure 1, and show a residual strength of about 70% of the strength at one year with little difference between air and water storage.

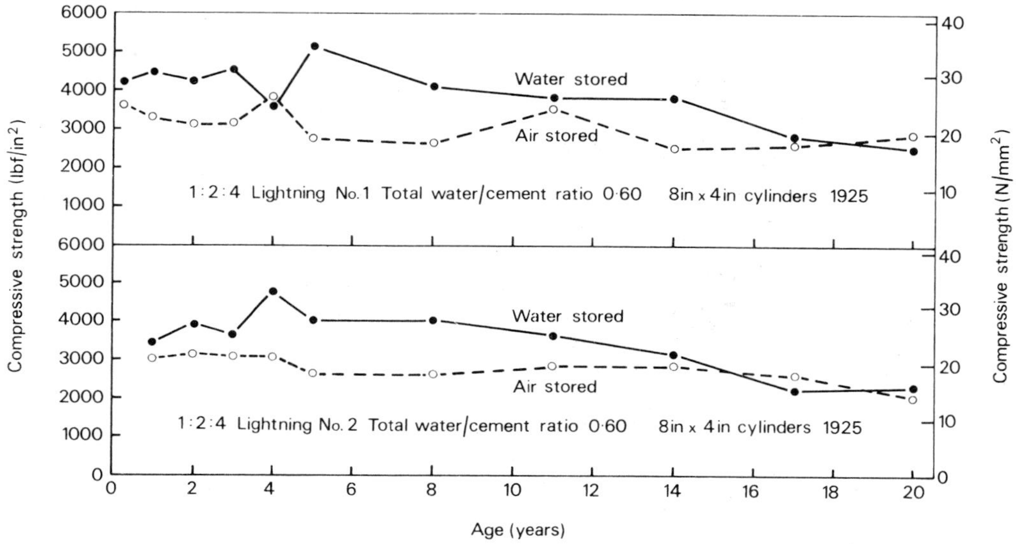

Figure 1 Long age tests on early HAC 'Lightning' concrete. (After 1932 the specimens were stored in a laboratory with air controlled at 18°C and 65% RH. Before this date conditions were variable, but temperature was usually 12-14°C).

Research, commenced in 1929 and reported by Lea and Watkins [6] in 1960 on the durability of reinforced concrete in sea water, also examined the performance of high alumina cement concrete. The results showed high residual strengths of cylinders after 10 years immersion in water or concentrated sea water; the strengths ranged upwards from 36 N/mm^2 for concrete with a total water/cement ratio of 0.53 and mix proportions by weight of 1:5.

By the time therefore that the BS Code of Practice, CP 114, 'The Structural Use of Reinforced Concrete in Buildings' [7] was published in 1948, there was some experimental evidence on the use of high alumina cement and by then, its quality was controlled by a BS Specification [8] which had been issued in 1940. This particular code permitted its use for reinforced concrete and recommended a 1:2:4 mix with a total water/cement ratio of 0.58, this figure being chosen mainly to ensure a not too rapid rise in temperature during curing. In the explanatory handbook [9] to the Code, the manufacturers of the cement were said to recommend that the formwork should be kept wet and struck at eight hours, the concrete being kept wet for at least 24 hours in order to avoid the strength from being impaired and objectionable surface dusting. A warning was also given against placing high alumina cement concrete at high ambient temperatures or in large masses which would result in high temperatures being developed due to the evolution of heat in the cement. The sub-Code of CP 114, CP 114.100 - 114.105 [10] issued in 1950 added a warning against the use of high alumina cement with aggregates which might liberate soluble alkalis or lime.

In 1951, BRS Digest No 27 [11] was published giving the general characteristics of concrete made with high alumina cement. It recommended the use of mixes from 1:5 to 1:7 with just enough water to enable the concrete to flow easily.

A total water/cement ratio of 0.5 was regarded as suitable for a 1:2:4 mix and it was stated that the total water/cement ratio should not be less than 0.4. Reference was made to loss of strength of concrete where exposed to moist and hot conditions at any time in its life but it was stated that the residual strength was adequate for structural purposes. Specific recommendations for use were:

(a) Concrete should not be mixed and placed when the atmospheric temperature was likely to remain above 30°C for any length of time.

(b) The maximum thickness of concrete to be placed in one operation should not exceed 450 mm.

(c) Formwork should be stripped as soon as possible and the concrete sprayed with water.

It was at about this time that high alumina cement was being introduced into the manufacture of prestressed concrete members on pretensioning beds. The main reason for this development was that, by using high alumina cement, the precast concrete manufacturer was able to achieve a daily production cycle, since he was able to stress the concrete at the age of about 18 hours. If however he used Portland cement, the plant was only able to operate on a four or five day cycle, and on this basis it was found more economic to use the much more expensive high alumina cement.

During the next few years, the precast concrete industry carried out experiments with the use of high alumina cement concrete and concluded that, provided the water/cement ratio was kept low and the concrete was maintained, during curing and in service, at temperatures below about 30°C, the concrete would not lose strength. If however conversion did occur, concretes with low water/cement ratios would still retain substantial strength. By this time, however, high alumina cement concrete with water/cement ratios somewhat higher than was being recommended had already been incorporated in some construction.

The BS Code of Practice, CP 115, 'The Structural Use of Prestressed Concrete in Buildings'[12] was published in 1959. It restricted its scope to the use of concrete made with Portland cement complying with BS 12 and Portland blastfurnace cement complying with BS 146. High alumina cement could only be used with the Engineer's approval.

In 1959, K Newman[13] published the results of his research on the design of mixes with high alumina cement in which he examined the properties of concretes made with low water/cement ratios and the effect of conversion at 38°C on their strength. This work established the importance of water/cement ratio in relation to the residual strength at a high level of conversion.

In 1963, Neville[14] produced a paper on the deterioration of high alumina cement concrete, which presented the results of a review of numerous laboratory investigations. These showed an overall pattern of gradual loss of strength of high alumina cement concrete. The paper also contained detailed evidence of practical difficulties and failures that had occurred in structures. The lengthy discussion on the paper included much evidence from the cement manufacturers and manufacturers of prestressed concrete beams on the adequacy of the product.

Following the publication of this paper, the Institution of Structural Engineers set up a committee to report on the use of the cement in structural engineering. The Committee studied all published data and other material made available to it regarding reported cases in which high alumina cement had been used for structural and civil engineering purposes. They reported in 1964[15] that there was ample evidence to show that this cement could be used satisfactorily to produce sound and durable concrete when proper precautions were taken during mixing, placing and curing, and due regard was paid to the temperature and humidity conditions to which it might subsequently be subjected. The report gave information on strengths obtained with the cement and on the effect of conversion, and made specific recommendations for limiting water/cement ratio, controlling the temperature of the concrete during its early life and adopting a reduced strength where the rate of conversion might be high.

Publication of the BS Code of Practice, CP 116, 'The Structural Use of Precast Concrete'[16] followed in 1964 and dealt in some detail with use of high alumina cement using the information in the Report of the Institution of Structural Engineers on relationships between strengths and water/cement ratio for strengths at one day when normally cured and for concrete fully-converted at 38°C. It included the following recommendations for reducing the harmful effects of conversion by:

1 Using a total water/cement ratio which is never greater than 0.5 for reinforced concrete and never greater than 0.4 for prestressed concrete.

2 Keeping the concrete cool and moist during the first 24 hours to ensure proper curing and to assist the dissipation of heat due to hydration which is so detrimental at early ages.

3 Keeping the concrete reasonably dry and cool after its initial curing.

It advised that the specified cube strength at one day should not be less than 52 N/mm^2 (7500 lb/in^2) for prestressing and gave details for the control of concrete strength. Limits were set on the thickness of concrete cast and cooling of the concrete with water sprays was required for at least 24 hours. Particular care was required to keep materials and concrete cool in hot weather. It was stated that if temperatures much above 25-30°C were allowed to persist in the concrete during the first two or three days of hardening, strength and durability might be greatly reduced. It was also said that if concrete, even after proper curing, is exposed to hot wet conditions above 35°C conversion could also occur with a considerable degree of strength reduction. This, to some extent, contradicted the introductory warning that in general high alumina cement concrete was not recommended for structural units which would be maintained in wet or humid conditions at temperatures above about 27°C and should not be used in such situations without prior consultation with the cement manufacturer. Permissible stresses were related to two sets of conditions, Class 1 where the precautions to minimise conversion were taken as required by the Code and permissible stresses were then based on the strength for normal curing at one day, and Class II where these precautions were not taken or where the service conditions were adverse and permissible stresses were then based on the fully converted strength.

This Code is still current but likely to be superseded by the BS Code of Practice, CP 110, 'The Structural Use of Concrete' [17] issued in 1972, which also contained recommendations on the use of high alumina cement. In these, it did not differ substantially from CP 116. Clauses were more specific in recommending that high alumina cement concrete should not be used in wet or humid conditions at temperatures above 27°C unless designed for full conversion, but less so in respect to water/cement ratio which was limited to a value of free water/cement ratio of 0.4 for all structural concrete, whereas previously the requirement had been for a total water/cement ratio of 0.4 for prestressed with 0.5 for reinforced concrete.

Since the development of precast prestressed concrete beams for buildings had taken place in the 1950's and CP 116 had been prepared in the mid 1960's in the light of the experience gained, it is against the recommendations of CP 116 that the present state of high alumina cement concrete needs to be viewed. In this respect CP 110 is of secondary importance although its provisions for design provide a more appropriate basis for the appraisal of construction as discussed later in the report. It should be noted that all BS Structural Codes which contained recommendations for the use of high alumina cement have now had these clauses removed and a note added to the effect that, although these Codes do not give guidance, they do not prohibit its use. It should however be observed that, at the time of the Stepney collapse, more than three quarters of the production of this cement was used for refractory concrete, flue linings and non-structural purposes. Of the remainder about 10 000 tons were used annually in the production of prestressed concrete floor and roof beams. Apart from use in lintels, other uses were few.

A substantial number of failures have been experienced in the development of the use of high alumina cement for structural purposes in France, Germany and other countries. High alumina cement was used in a minor fashion for structures and maritime works in France in the inter-war period. Following a number of failures, the structural use of this material was banned for public works in 1943. A committee was set up in 1963 to consider its reintroduction for government building, and this led to a Ministerial Circular in 1970 which permitted its use with severe restrictions on mix design and control. A major restriction was the requirement for a high minimum cement content of 400 kg/m^3 of concrete which was greater than that used by British manufacturers. Since the issue of the Circular no applications have been made to use this material in buildings.

The manufacture of high alumina cement was introduced into Hungary about 1930, and it was subsequently used in construction of reinforced concrete. Its composition was similar to, but not the same as the British cement. Its manufacture was stopped in the mid 1950's when it became known that the strength of the concrete decreased with time. In 1967 a survey was made by the Ministry of Buildings of all buildings containing the material and subsequently a number of buildings were found to require strengthening. It is understood that building designs have been recalculated on the basis of a fully converted strength.

High alumina cement was used in Germany in the manufacture of prestressed concrete units, and their experience is therefore of particular interest. A series of structural collapses occurred in agricultural buildings in 1961, due to corrosion of the prestressing steel. The high alumina cement concrete was found to be highly-converted, porous and permeable. There were special features about these failures which would suggest that such failures would not occur in this country. Firstly the German steels were formed by hotrolling and quenching as opposed to British practice of cold-drawing and were more susceptible to stress corrosion cracking; secondly the German cement was not of the same composition as the British cement, but contained sufficient sulphide to cause embrittlement of the steel; and thirdly the cover provided to the steel was 10 mm (0.4 in) - whereas in this country the minimum cover for internal construction is 15 mm (0.6 in). One important feature of the German failures was the discovery that the highly-converted concrete had lost considerable strength. In the investigation that followed at the Technical University of Munich, the strength of concrete at about 15 years of age was found to range from 17 N/mm^2 to 50 N/mm^2, and structural calculations for certain structures in a dry environment were subsequently re-checked on the basis of an assumed fully converted strength of 21 N/mm^2 (compared with the one day strength of about 45 N/mm^2) with a slightly lower factor of safety.

3 DESIGN CONSIDERATIONS

Until 1959 when the BS Code of Practice, CP 115, for prestressed concrete was issued, the design of prestressed concrete tended to conform with the guidance given in the First Report on Prestressed Concrete[18] published by the Institution of Structural Engineers in 1951. It laid down the principles for design which are still followed and in some ways was the forerunner of the limit state design approach adopted in BS Code of Practice, CP 110:1972, for the Structural Use of Concrete. It required consideration of behaviour at four stages, the transfer of prestress to the concrete, handling, stresses under service loading and conditions at failure. During these first three stages, it was assumed that the materials behaved elastically with appropriate allowance for losses of prestress due to permanent deformation of the materials, whilst in the fourth stage the mode of failure of the materials was considered. Thus two separate sets of conditions were dealt with, for the first the permissible stresses were defined, and for the second, the levels of strength to be used and the mechanism of failure to be adopted were given. The latter were not however specifically described until CP 115 appeared eight years later.

Nevertheless, the permissible stress given in the First Report and in CP 115 were almost identical so that designs for service loading, which tended to control, are little different for either document. The Code, CP 115, gave a method for calculating the ultimate strength of beams and recommended values for load factors, and these clauses were also adopted without change in, CP 116, the Code for precast concrete, which appeared in 1965. Compressive stresses under service loading were however increased in CP 116 by 10%. The BS Code, CP 110, for structural concrete which was issued in 1972 has as yet had little impact on the design of roof and floor units of prestressed concrete. Thus there have been no major changes in the levels of stress or safety adopted in design throughout the whole period that high alumina cement has been used in prestressed units. The changes that have occurred have been mainly in the way in which the cement has been used and possibly in the character of the cement itself.

Whilst design was generally in accordance with CP 116, the newer approach to considerations of serviceability and safety given in CP 110 are now more appropriate to the appraisal of existing construction. When compared with the earlier Codes, CP 115 and CP 116, which are still current, the new Code permits slightly higher tensile stresses at the serviceability limit state. Thus units designed under the older codes could now be designed to carry slightly higher loads when satisfying serviceability requirements.

Comparison of requirements for ultimate strength cannot be made readily. The Codes, CP 115 and CP 116, require the load which must be considered in ultimate strength calculations to be:

$$1.5\ G + 2.5\ Q$$

or $$2.0\ (G + Q) \text{ whichever is less}$$

where G is the dead load and Q is the imposed load as given in CP 3, Chapter V, Part I[19]. CP 110 requires this load for ultimate conditions to be:

$$1.4\ G + 1.6\ Q$$

(The values 1.4 and 1.6 are termed partial safety factors for loads and load effects in CP 110).

In resisting these loads, CP 115 and CP 116 give a method of calculation which assumes that the steel may be stressed to its tensile strength for beams which would fail by tension failure of the tendons, ie where the proportion of steel in the section is small, or that the concrete is stressed to 4/5 of its strength in beams which would fail as a result of crushing of the concrete, ie where the proportion of steel is large. The comparable figures for CP 110 are that, at the limits, the steel may be stressed to its tensile strength divided by 1.15 and the concrete may be stressed to its potential in a beam divided by 1.5. (The figures 1.15 and 1.5 are termed partial safety factors for materials in CP 110). Thus again the newer code is slightly more conservative in design than the older codes, particularly where the ratio of dead load to imposed load is less than about 1.5.

Review of the different types of prestressed concrete beam used in building construction shows that the majority have conformed to certain standard types adopted by the manufacturers. There has been a tendency for each manufacturer to concentrate on a limited number of sizes. To achieve economy in the range of sizes adopted, each size has been manufactured with different numbers of tendons to meet the needs of standard floors of different spans. Examples of some of the sections and tendon arrangements are shown in Figures 2 and 3 for X5$\frac{1}{4}$ and X7 joints made by Pierhead Limited; other manufacturers adopted a similar approach. The design service loads and maximum stress under these loads are tabulated for these joists in Table 1. The particular significance of this practice in terms of appraisal is that the sections with few tendons suffer substantially smaller losses of strength through deterioration of the concrete than those with larger numbers of tendons. Figures 4 and 5 give the relationships, obtained using the method of calculation and stress-strain relationships for the materials advocated in CP 110 (taking partial safety factors for the materials as unity), between cube strength of the concrete and the ultimate strength of the joists, while Figures 6 and 7 show the safety factors for floor beams with different numbers of tendons for different cube strengths. At a cube strength of 20 N/mm^2, the beam with six wires in Figure 6 has a safety factor of about one whilst that with three wires has a safety factor in excess of 1.5. Thus for the latter beams, the effect of a loss of strength in the concrete due to conversion is less important. The Code, CP 116, permits higher tensile stresses in roof than in floor units, since the imposed load on the former seldom occurs. Hence the safety factors for roofs are lower by about 7 - 15%, the smaller reduction in value relating to beams with the larger number of tendons as shown by Table 1. This feature is important in relation to isolated beams, usually in roof construction, and particularly so for long span beams where failure would have very serious consequences.

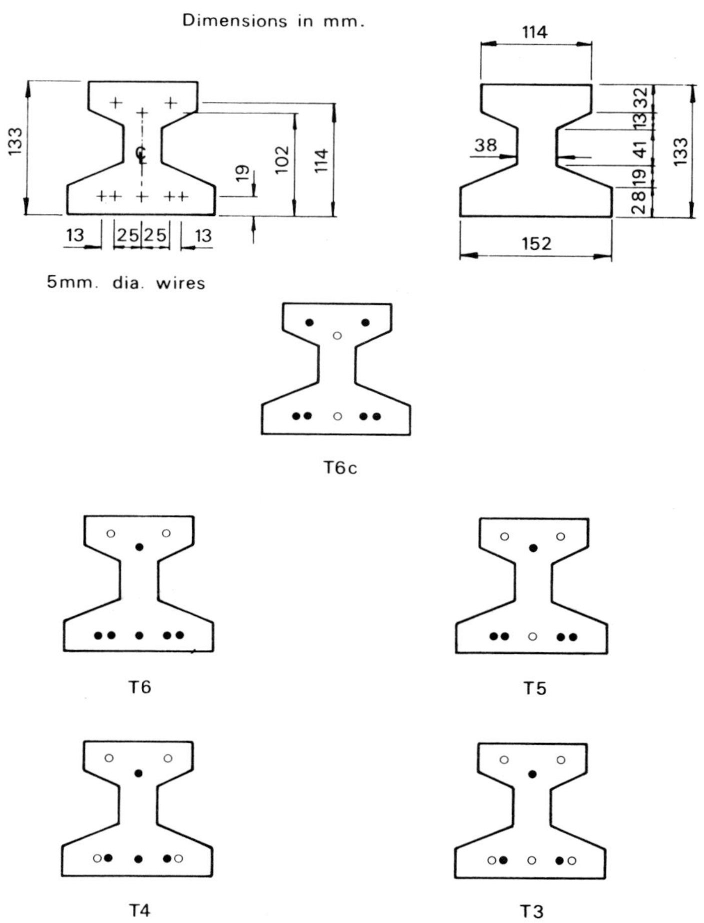

Figure 2 X5$\frac{1}{4}$ Pierhead joists

In floor construction, joists were usually used with hollow blocks and until 1965 when Building Regulations required improved levels of resistance to sound transmission, it was common practice to cast the finishing screed in direct contact with the beams and hollow blocks forming the floor. From 1965 on however, the screed was separated from the beams and hollow blocks by a quilt. In neither case was any allowance made for any structural contribution from the screed. Other aspects of construction, which are not taken into account in design, may also increase the strength and stiffness of roofs and floors; these include end restraint, two-way slab action, membrane action and non-loadbearing partitions.

Where composite action between the beams and a structural topping was deliberately provided in construction, the problems of conversion of high alumina cement concrete are much less likely to provide difficulty. Evidence from the examination of the roof collapse at Camden [20] suggested that there was no major loss of strength in the composite roof and no signs of undue deflection. Shear might however need special consideration for beams of I or X section in composite construction.

In assessing the safety of existing buildings, it is not only necessary to consider the direct effects of the loss of strength in the concrete on the strength of components and to take account of the way in which the component is built into the structure, but it is also necessary to consider the real loads imposed on the structure since these may be different from those assumed in design. If required, the dead load can be determined with considerable accuracy. For special occupancies, such as industrial premises and hospitals, the imposed loading may be found from the way in which the building is used. The results of loading surveys in offices [21] and retail premises have been published from which a realistic assessment of load may be made. This may lead to the assessment of a load in appraisal lower than that assumed in design.

Data are now becoming available from a similar survey in housing. For bays spanning one-way over 3 to 5 m span with an aspect ratio of span to width less than 1.2 and an assumed 18 changes of occupier in the remaining life of the building, the intensity of loading including occasional loading from groups of people is unlikely to exceed $1.2 kN/m^2$ on more than one occasion in 20. This figure is based on a survey of 567 bays with an area between 20 and 40 m^2. It allows for non-uniform distribution of loading and assumes that mid-span bending moment is critical and that slab action develops in the floor system.

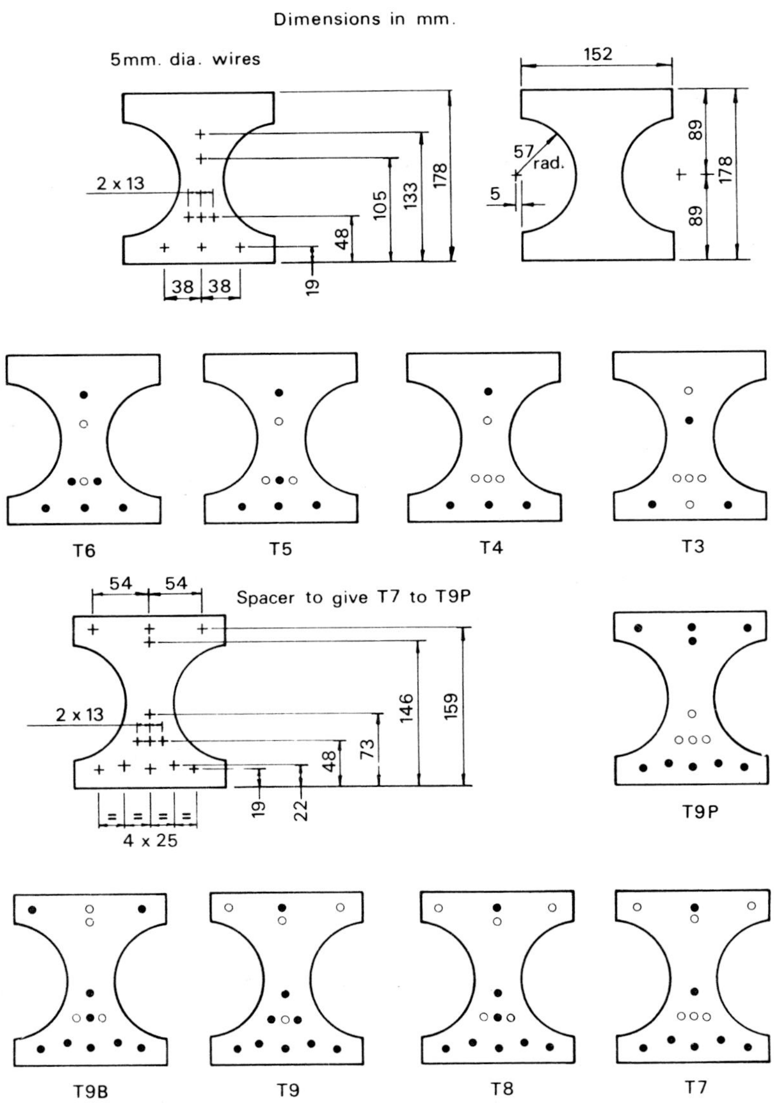

Figure 3 X7 Pierhead joists

Table 1 Design service moments for some pierhead X joists and corresponding stresses in the concrete

Beam ϕ	Design service moment: kN.m		Maximum prestress in the concrete	Maximum compressive stress in the concrete under service loads : N/mm²	
	Floors+	Roofs+	N/mm²	Floors	Roofs
X5¼ T3	3.57	4.06	6.54	11.12	12.55
T4	4.81	5.32	9.50	13.97	15.42
T5	5.96	6.46	12.14	16.59	18.05
T6	7.00	7.03*	14.50	18.88	18.95
X7 T3	5.96	6.85	5.87	7.43	8.65
T4	7.62	8.51	8.02	9.29	10.52
T5	8.95	9.85	9.77	11.12	12.35
T6	10.21	11.11	11.41	12.85	14.08
T7	11.67	12.68	13.17	14.41	15.66
T8	12.80	13.72	14.60	15.97	17.22
T9	13.87	14.79	15.94	17.44	18.69
X9 T14	28.33	30.24*	15.23	17.70	18.96

ϕ Designates size of joist eg X5¼ T3 is 5¼ in deep with 3 No 5 mm wires

\+ Permissible tensile stress under service loads - floors 2.2 N/mm²
- roofs 3.4 N/mm²

* Design service moment limited by permissible compressive stress, elsewhere permissible tensile stress controls.

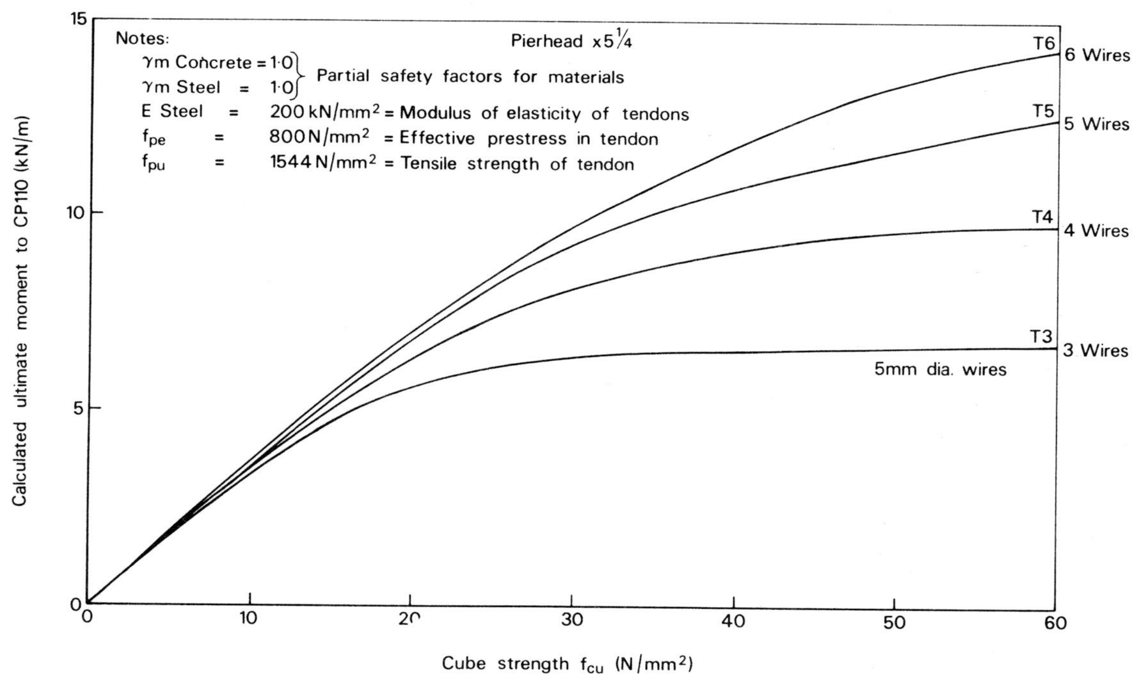

Figure 4 Relationship between calculated ultimate moment and cube strength - Pierhead X5¼

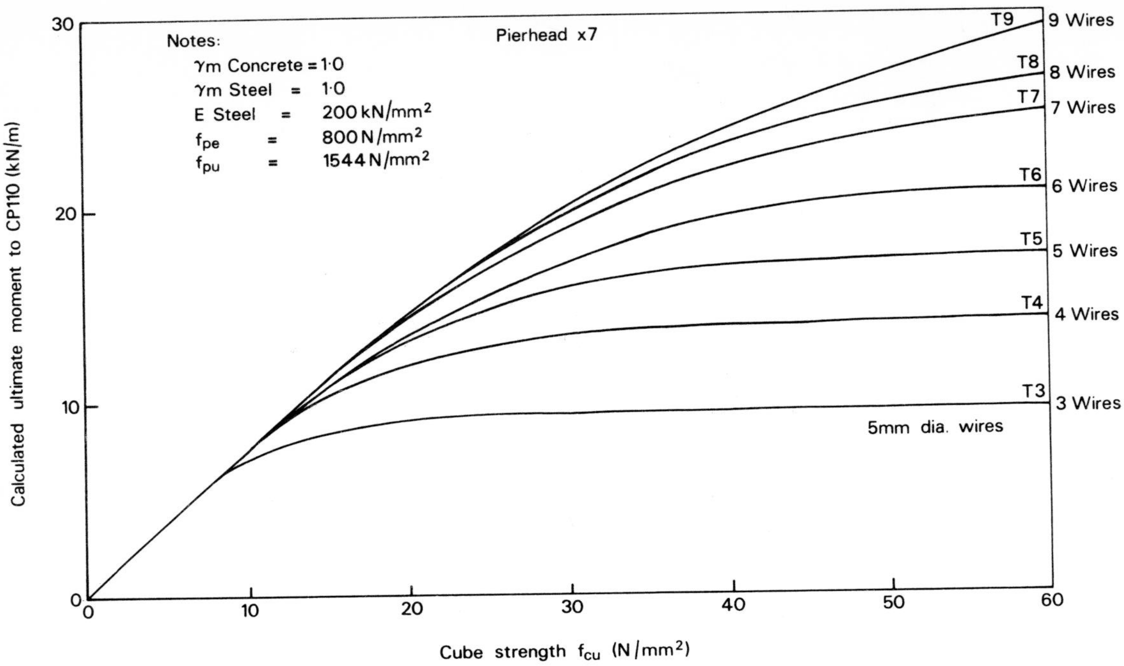

Figure 5 Relationship between calculated ultimate moment and cube strength - Pierhead X7

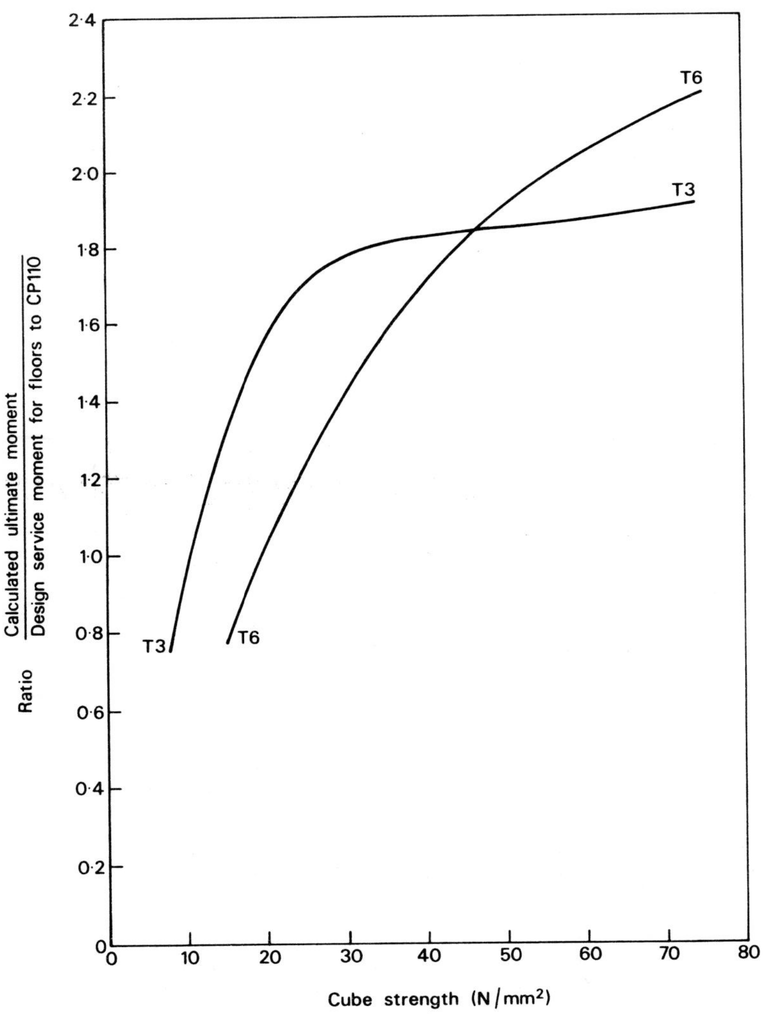

Figure 6 Relationship between calculated overall factor of safety and cube strength from figure 4 - Pierhead X5¼

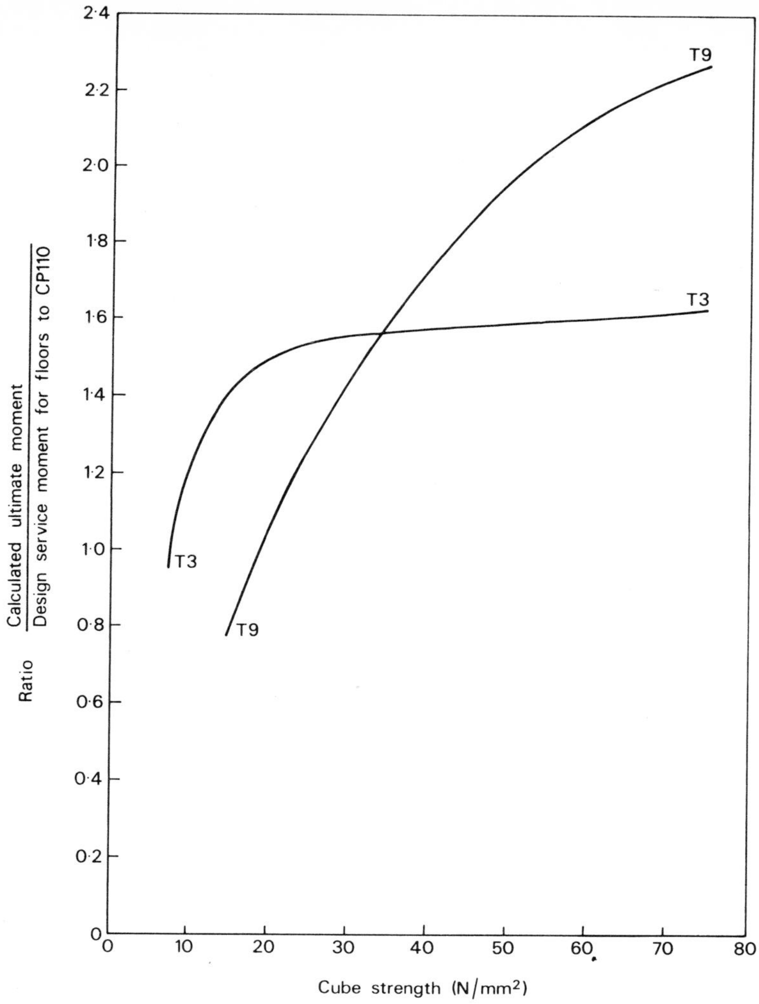

Figure 7 Relationship between calculated overall factor of safety and cube strength from figure 5 - Pierhead X7

4 OUTLINE OF THE LABORATORY AND FIELD INVESTIGATIONS

Fundamental studies of the hydration and conversion processes in high alumina cement and of the consequences of conversion on the engineering properties of high alumina cement concrete had been intermittently active at the Building Research Establishment following the early work of Davey, and Lea and Watkins, to which reference has already been made. These studies in themselves had not indicated that a serious situation was likely to develop in normal buildings where quality of concrete was properly controlled, but provided a basis for developing an understanding of the failures which have recently been investigated. Some details of this work are given in this report to provide the background for the current appraisal of high alumina cement concrete in existing buildings.

After completion of the investigation of the collapse at Stepney, the current investigation of high alumina cement concrete in existing buildings commenced. At that stage reports had been received of a number of buildings, assessed as being distressed by their owners, mostly local authorities. In order to gauge the size of the problem, a survey was conducted of some 38 buildings from which it was possible to deduce some broad descriptions of what could readily be recognised as distressed. Most were school buildings with isolated roof beams.

The Department of the Environment in association with the Department of Education and Science asked local authorities to make available to the Building Research Establishment the results of their structural appraisals; the Institution of Structural Engineers also invited their members to do the same. As a result information has been received on the condition of high alumina cement concrete in over 400 buildings. This feedback has been collated and analysed and has provided a valuable fund of data to supplement the laboratory investigation. A summary of the types of structure included in this review is given in Table 2.

Table 2 Structure of sample provided by information received from outside sources

Group		No of buildings or part of buildings in group
Age (in whole years)	Not reported	4
	0 to 4 + years	47
	5 + to 9 +	49
	10 + to 14 +	45
	15 + to 19 +	16
	20 + and over	5
	Total	166
Use of building	School, college, etc	100
	Hospital	4
	Industrial	5
	Office	24
	Residential	18
	Others	15
	Total	166
Floor or roof	Not known	61
	Floor	60
	Roof	54
Type of beam (pre-tensioned unless stated)		
Shape	I or X	59
	Rectangular	2
	Trough or tee	0
	Lattice	15
	Special	6
Depth	133 mm ($5\frac{1}{4}$ in)	5
	179 mm (7 in)	40
	229 mm (9 in)	27
	254 mm (10 in)	4
	305 mm (12 in)	2
	381 mm (15 in)	2
	457 mm (18 in)	5
	152 mm (6 in)	1
Includes 4 buildings with some post-tensioning		

During this investigation more than two hundred prestressed concrete beams of high alumina cement concrete from existing buildings were subjected to a variety of tests by BRE including:

1 Measurements of conversion in concrete members;

Differential thermal analyses were made on samples taken from nearly every beam received, and a calibration service was provided to ensure uniformity of results between test consultants from whom results were received.

2 Strength tests on concrete, including concrete subjected to accelerated conversion;

Cores or beam sections were cut from a large number of beams to determine compression strength in relation to beam strength and degree of conversion. Methods of test of cores were examined in relation to the standard method and a considerable number were also subjected to tensile tests. Other tests were made to determine carbonation and porosity.

3 Bending tests on concrete beams, including beams subjected to accelerated conversion*;

*Accelerated conversion was induced by heating beams and cores in the kilns of the Princes Risborough Laboratory for one month at 50°C and 75% relative humidity.

Between 40 and 50 beams of X type as well as a number of non-standardised types were subjected to flexural tests. Some were found during the tests to have been damaged in such a way that it interfered with the results and these results were rejected. Other tests were made on beams to estimate shear strength, the prestress remaining in the tendons and the transmission length.

4 Loading tests on floors made of beams and hollow blocks using beams subjected to accelerated conversion;

In all more than 60 beams were used in these tests which were made to determine the degree of composite action and distribution of concentrated load where standardised forms of beams were used in floor construction.

In order to complete the investigation within the shortest possible time it was necessary to make use of considerable research resources - tests have been made at the Building Research Station, the Fire Research Station, the Princes Risborough Laboratory, the Transport and Road Research Laboratory, and the Polytechnic of Central London.

Although all this information has been obtained from only a small part of the total population of buildings in which the cement was used, it is to some extent biased, since it was derived from structures judged to be at risk.

5 CONVERSION AND CHEMICAL ASPECTS

High alumina cement, which derives its name from the high proportion of alumina in its composition, is made in the UK by fusing together bauxite ore and limestone in a process originally developed in France at the beginning of the century. Manufacture commenced here in Northern Ireland in 1925 and followed at Grays in Essex where it is still made by the sole present manufacturer, the Lafarge Aluminous Cement Company.

The chemical and mineralogical changes that take place during conversion have been the subject of research at the Building Research Establishment for many years. The results of this work have contributed to the general understanding of the complex processes involved. These results have been drawn together and reviewed in this report, and are described in greater detail in a paper now awaiting publication.

The important reaction during the initial setting of the cement is the formation on hydration of monocalcium aluminate decahydrate (CAH_{10}), dicalcium aluminate octahydrate (C_2AH_8) and alumina gel (AH_n). These aluminates give the high strength to high alumina cement concrete but they are metastable and at normal temperature convert gradually to tricalcium aluminate hexahydrate (C_3AH_6) and gibbsite (AH_3) which are more stable. The change in composition is accompanied by a loss of strength and by a change in crystal form from hexagonal to cubic and rhobhedral with the release of water which results in increased porosity of concrete. The precise way in which these changes take place depends on the temperature, water/cement ratio and chemical environment.

Experimental evidence suggests that the important reaction in conversion is the change from monocalcium aluminate decahydrate to tricalcium aluminate hexahydrate and alumina hydrate. Temperature affects the rate of decomposition, the higher the temperature the faster the rate of conversion. Since the products of the conversion reaction have a higher density, stress should increase the rate but there is no clear evidence that the stresses normally induced in concrete by prestressing or under load are of sufficient magnitude to produce a significant effect. External chemical agents such as alkali soluble sodium and potassium which are freeable from certain aggregates, as for example those used at Stepney, are known to increase the rate. Experimental study has also shown that the higher the water/cement ratio the greater is the rate of conversion. Examination of the crystal size of C_3AH_6 shows that it increases with water/cement ratio and with the rate of conversion. Crystal size, and hence the number of contacts between grains has a considerable effect on strength, the smaller crystals lead to greater strength.

At temperatures below 55^oC, the relationship between mineralogy and strength is that initially the formation of monocalcium aluminate decahydrate leads to a rise in strength to a maximum. The fall to a minimum is then due to conversion and the small subsequent rise is the result of the gradual formation of Stratlings compound (C_2ASH_8). The minimum strength is reached quicker at higher temperatures and the amount of conversion at minimum strength is also affected by temperature, it can be as little as 45% for the lower water/cement ratios and lower temperatures.

It has been stated in the past that moist as well as warm conditions were necessary to accelerate conversion, but since the conversion reaction releases the water needed, it is only in exceptionally dry conditions that conversion is likely to be delayed.

For laboratory work, the degree of conversion is expressed as:

$$\frac{\text{weight of } C_3AH_6 \times 100}{\text{weight of } C_3AH_6 + \text{weight of } CAH_{10}}$$

Since, in the concrete in structures, the C_3AH_6 may become decomposed by carbonation and, fortuitously, the quantities of C_3AH_6 and AH_3 are not very different, it may also be defined with a difference of not more than 4% at 50% conversion as:

$$\frac{\text{weight of } AH_3 \times 100}{\text{weight of } AH_3 + \text{weight of } CAH_{10}}$$

A number of different methods may be used in determining the quantities of different minerals present but differential thermal analysis offers a rapid method readily calibrated for different laboratories.

For the appraisal of high alumina cement in existing structures it was recommended in the DOE/DES Circular of 20 July 1974, that differential thermal analyses should be made of the concrete. The results were then to be compared with the curves given in Figure 8. Curve A was derived from the long-term laboratory study, referred to later, and represents the relationship obtained between conversion and age for a high alumina cement concrete with a free water/cement ratio of 0.4 stored in water at 18°C. This particular concrete and storage condition was chosen since the minimum strength as a result of conversion was unlikely to reduce much below the strength at one day. Concretes in buildings with a degree of conversion less than that given by curve A were therefore defined as 'good' and were unlikely to have suffered a serious loss in strength. Appraisal of buildings with concrete in this category was not therefore of immediate concern.

Curve B was drawn parallel to Curve A in such a way that all concrete from buildings where failures had occurred together with laboratory concretes which had strength less than ½ their one-day strength were in the region designated as 'highly converted'. It is now considered more appropriate to use the terms 'low-level', 'medium-level' and 'high-level' to describe the conversion of concrete in relation to age as shown in Figure 8.

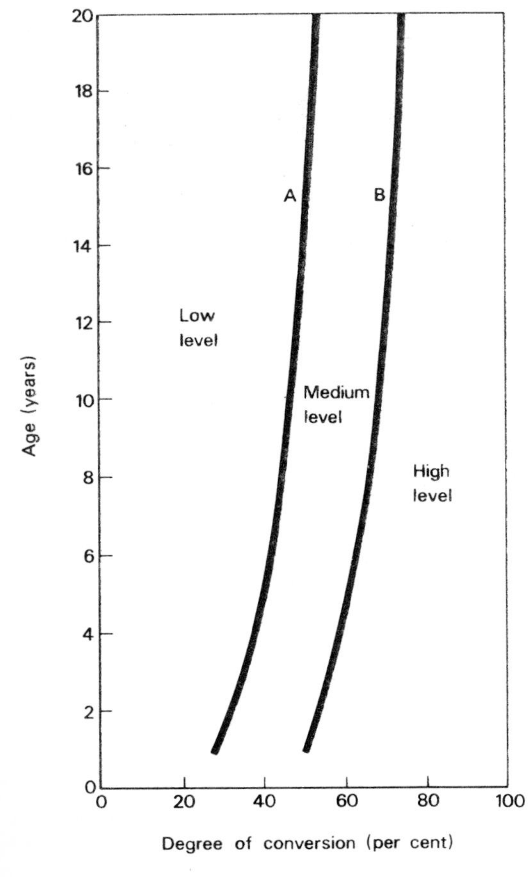

Figure 8 Relationships between age and degree of conversion

Failure of high alumina cement concrete may result from attack by alkalis external to the concrete, which is often described as alkaline hydrolysis and is readily detectable by the growth of white crystalline deposits on the surface with erosion of the concrete. It is essentially attack by carbon dioxide from the air in which the alkali acts as a carrier. If the source of alkali is internal due to the nature of the aggregate, then the deterioration may not be apparent from a simple external examination.

Susceptibility of concrete to chemical attack is not regarded as entirely a function of the minerals present but of the porosity which is dependent on the rate of conversion and the water/cement ratio. Thus, if conversion has taken place rapidly and the water/cement ratio is moderate or high, the cement will be more liable to chemical attack. This view was substantiated by tests on neat cement pastes with high and low water/cement ratios. Specimens were converted at different rates and immersed in sulphate solutions for periods up to one year. The rapidly converted series only showed chemical attack with the formation of ettringite, which was one of the primary causes of the failure of the beams in the swimming pool roof at Stepney.

The environment to the steel tendons or reinforcement provided by high alumina cement concrete is alkaline although less so than for Portland cement concrete. Atmospheric carbon dioxide attacks and decomposes the predominant calcium aluminate products and results ultimately in the formation of calcium carbonate and hydrated alumina. With normal dense concretes stored out of doors such carbonation only occurs in the surface layer and the main body of the concrete remains unaffected. If concrete adjacent to the steel becomes carbonated, the alkalinity of the concrete is reduced and the normal protection afforded to the steel is impaired. These comments apply to high alumina cement manufactured in France and in the UK. The cement made in Germany contains a high proportion of sulphide, not present in the French and British cement, which caused stress-corrosion fractures of the prestressing tendons in humid conditions.

6 LONG-TERM STUDY OF ENGINEERING PROPERTIES

Samples of high alumina cement have been received at the Building Research Establishment for research purposes since at least 1925 and routine tests have been made to determine their strength in concrete since 1931. These tests have consisted of compression tests on concrete with proportions by weight of 1:2:4 with a total water/cement ratio of 0.6, corresponding to a free water/cement ratio of about 0.5. The specimens were cubes and they were tested at the ages of 1, 3, 7 and 28 days and at three months. In certain cases tests were also made at greater ages.

The results obtained in these routine tests are shown in Figure 9, and suggest that there has been little change in quality as expressed by the tests at these ages over the past 40 years.

During the discussions in the course of the preparation of the Institution of Structural Engineers' Report[15] and of the Code of Practice, CP 116:1965,[16] for precast concrete, it was appreciated that further information on the long-term strength and durability of high alumina cement concretes with low water/cement ratios was desirable. A research project was therefore commenced at

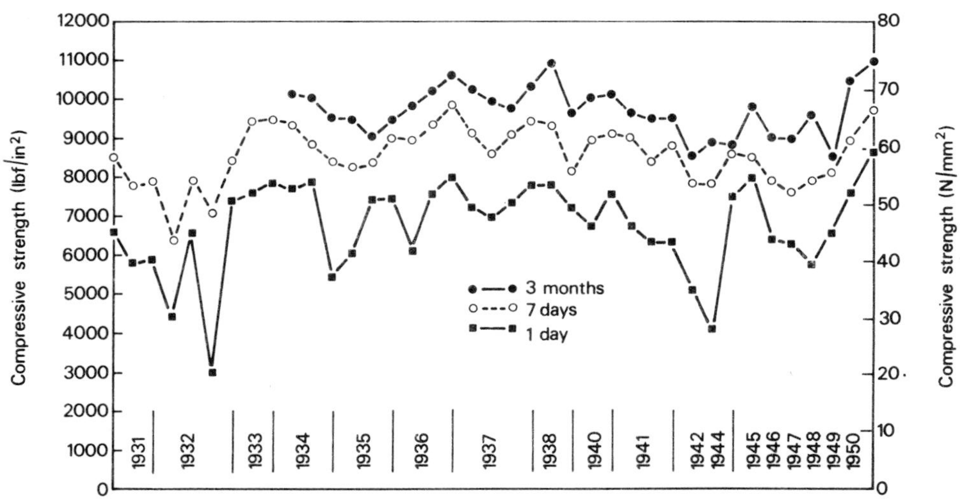

Compressive strength 1:2:4 concrete, total w/c 0·60, water-stored 4 in cubes

Figure 9 (part one) Check tests on high alumina cement from 1931 to 1950

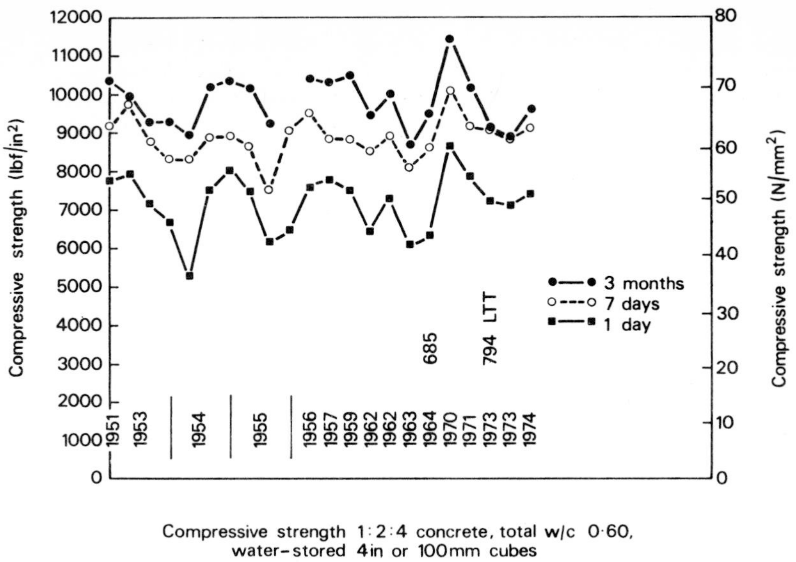

Figure 9 (part two) Check tests on high alumina cement from 1951 to 1974

the Building Research Station with advice and financial support from the Lafarge Aluminous Cement Company to obtain comprehensive data on the effect of time and temperature on high alumina cement concrete with a range of water/cement ratios covering the extreme limits of what could be used in practice.

At the time of the Stepney collapse, the five-year tests had been completed and some of the specimens due for test at 10 years were already being tested at $8\frac{1}{2}$ years as a consequence of the Camden collapse. This study provided valuable background information to the Stepney investigation and has particular relevance to the problems of appraisal of existing buildings in high alumina cement concrete. The main results of this project are presented here, and details of the materials and test methods will be published elsewhere.

There were two main series of tests; the first (A1), dealt with the effect on strength of storage of high alumina concrete, after one day in air at 18°C, in water at 18°C and at 38°C for periods of up to 20 years; the second series (A2) examined the influence of various temperature conditions during the first day on the subsequent strength of concrete stored in water at 18°C. Two smaller test series were later carried out, one (B) to determine the minimum converted strength at 38°C on immediate immersion in water as compared with that obtained at 38°C following one day in air at 18°C, and the other (C) to provide information on the modulus of elasticity of the concrete.

The strength development at the two levels of temperature over the period of $8\frac{1}{2}$ years is shown for concretes of different water/cement ratios in Table 3, where the data are given as strengths of cubes and as percentages of the one day strengths of cubes stored at 18°C in air. Some of the results are also plotted in Figure 10.

Table 3 shows that at 18°C the concrete gains in strength until a maximum is reached at between three months and one year for the concretes with the lower water/cement ratios and at between 1 month and 3 months at the higher water/cement ratios. The maximum values are between 50% and 33% greater than the 1-day strengths, the greatest increase being at the lowest water/cement ratio. After passing the maximum, the strength reduces with time and reaches a minimum for the concretes with the lower water/cement ratios in about 5 years; for the concretes with the higher water/cement ratios, this minimum may not have been reached at $8\frac{1}{2}$ years. The drop in strength is small with low water/cement ratios and very substantial with the high water/cement ratios. So long as the free water/cement ratio is less than 0.4 the minimum strength is not less than the 1-day strength.

When the concrete is immersed in water 38°C at the age of 1 day, the strength increases, and at 7 days is greater with the lower than with the higher water/cement ratios, thereafter there is a reduction in strength and a minimum value is reached after about 3 months. Subsequently a slight increase occurs with time. For the values of water/cement ratio examined, the residual strength ranges from 68% of the 1-day strength for the lowest water/cement ratio to 26% for the highest. With a free water/cement ratio of 0.4, this minimum is 46% of the 1-day strength.

There is therefore a very substantial reduction in strength when the level of temperature is raised from 18°C to 38°C. The results of the tests on concretes stored at 18°C for $8\frac{1}{2}$ years and then stored for 3 months at 38°C reduced to the same minima experienced after 3 months when the concretes were stored continuously at 38°C, as shown in Figure 10. Thus the stage at which the temperature is increased has little effect on the minimum strength reached. The main results of these long-term tests are summarised in Figure 11 which gives the relationships between compressive strength and water/cement ratio for the different storage conditions set against background curves which are typical of relationships for Portland cement concrete.

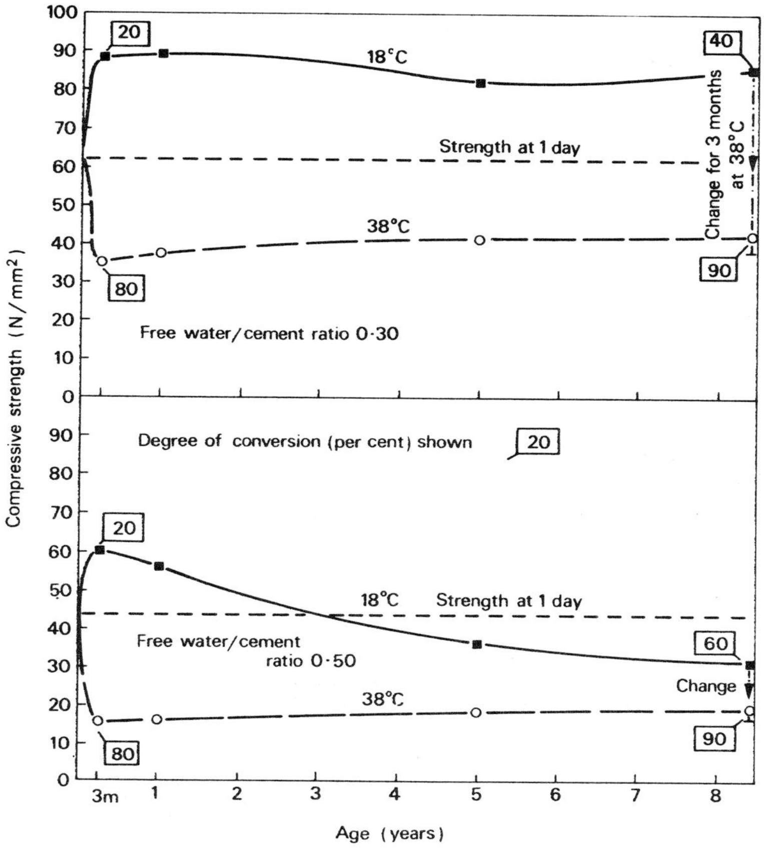

Figure 10 Typical strength development curves

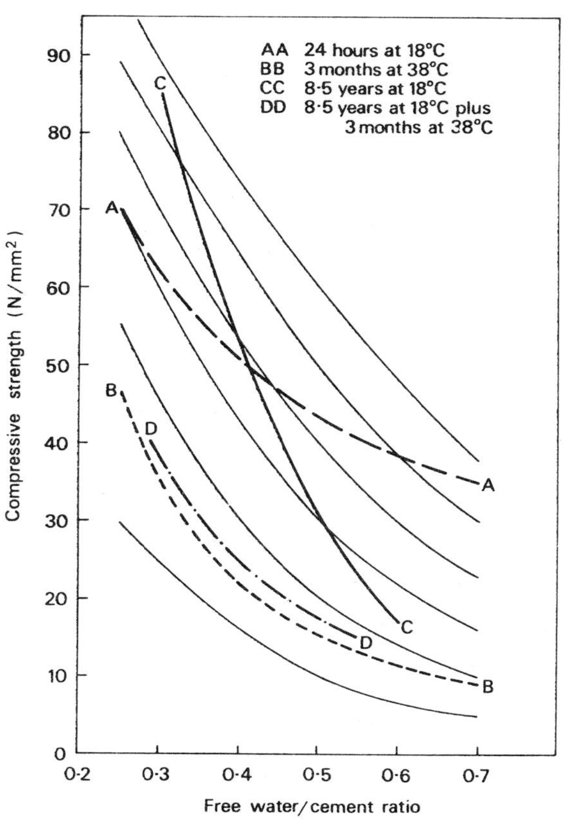

Figure 11 Relationship between strength and water/cement ratio from test series A1

Table 3 Strength development of concretes of different free water/cement ratios (calculated from data for test series A1)

Free water/cement ratio by weight	Compressive strength of 101.6 mm cubes N/mm^2												
	Stored in water at 18°C							Stored in water at 38°C					
	24 hours	7 days	28 days	3 months	1 year	5 years	8.5 years	7 days	28 days	3 months	1 year	5 years	8.5 years
0.25	70.0	86.0	94.5	102.0	105.0	-	-	91.5	76.0	47.0	51.0	55.0	51.0
0.30	62.0	76.5	83.0	88.5	89.0	82.0	83.0	78.0	55.5	35.0	37.5	41.0	42.0
0.35	55.5	69.0	75.0	78.5	77.5	64.5	67.0	68.5	42.5	27.5	29.0	32.0	34.5
0.40	51.0	63.0	68.0	71.0	68.5	52.0	53.0	61.0	34.0	22.0	23.0	25.5	28.0
0.45	47.0	58.0	62.5	65.0	61.5	43.0	40.5	55.5	27.5	18.0	19.0	21.0	23.0
0.50	43.5	54.5	58.0	60.0	56.0	36.5	31.0	50.5	23.0	15.5	16.0	18.0	19.0
0.55	41.0	51.0	54.5	55.5	51.5	31.0	23.5	46.5	19.5	13.0	13.5	15.5	16.0
0.60	38.5	48.0	51.5	52.0	47.5	27.0	17.0	43.0	17.0	11.5	11.5	13.5	13.5
0.65	36.5	45.5	48.5	49.0	44.5	24.0	-	40.5	15.0	10.0	10.0	11.5	-
0.70	35.0	43.5	46.0	46.5	41.5	21.0	-	38.0	13.0	9.0	9.0	10.5	-

Table 3 **Strength development of concretes of different free water/cement ratios (calculated from data for test series A1)**

Free water/cement ratio by weight	Compressive strength expressed as a percentage of the 24 hour strength												
	Stored in water at 18°C							Stored in water at 38°C					
	24 hours	7 days	28 days	3 months	1 year	5 years	8.5 years	7 days	28 days	3 months	1 year	5 years	8.5 years
0.25	100	123	136	146	150	-	-	131	109	68	73	79	73
0.30	100	123	135	144	144	133	137	127	90	57	61	66	68
0.35	100	124	135	142	139	116	121	124	77	49	52	57	62
0.40	100	124	134	140	135	102	104	121	67	43	45	51	55
0.45	100	124	133	138	131	92	85	118	59	38	40	45	49
0.50	100	124	133	137	128	83	71	116	53	35	36	41	39
0.55	100	124	133	136	126	76	57	114	48	32	33	38	39
0.60	100	124	133	135	123	70	44	112	44	29	30	34	35
0.65	100	125	133	134	121	66	-	111	41	27	28	32	-
0.70	100	125	132	133	119	61	-	109	37	26	26	30	-

The investigation did not examine the effect of intermediate temperatures over the long-term. Results of tests reported by French, Montgomery and Robson[23] for concrete with a free water/cement ratio of about 0.32 indicated that a major reduction in strength occurred when the storage temperature was between 25°C and 30°C and that, at higher temperatures, the reduction only increased slightly. In terms of the 1-day strength at 20°C the minima reached in their study were approximately as given in Table 4 in comparison with results derived from this investigation for the same free water/cement ratio.

Table 4 Effect of temperature on minimum strength

Temperature: °C	Minimum strength as a percentage of the 1-day strength	
	French, Mongomery and Robson*	BRE +
18	-	125
25	131	-
30	75	-
38	70	54
50	61	-
60	60	-

* 1-day strength at 20°C
\+ 1-day strength at 18°C

The extent of conversion during these long-term tests at the Building Research Establishment is indicated in Figure 10, which shows that after 3 months at 18°C the concretes have converted by 20% but after $8\frac{1}{2}$ years the level of conversion has increased by a smaller amount for the concrete with the lower water/cement ratio. At 38°C, conversion is very rapid and is 80% at the minimum strength. Some slight increase in strength takes place as conversion increases from 80% to 90%. Measurements of conversion obtained throughout the long-term investigation are plotted against strength of the concrete in Figure 12 which shows the wide scatter obtained. If however results for conversion are plotted against age of concrete for the different temperatures of storage and grouped by values for water/cement ratio as in Figure 13 which also shows strengths similarly grouped, it may be seen that increases in levels of conversion tend to be accompanied by reductions in strength.

The second series of tests (A2) dealt, as already mentioned, with the effect of exposure to various temperatures during the first day after casting on the subsequent strength when stored at 18°C. The results are not relevant to the appraisal of existing construction except in so far as they provide one explanation for rapid conversion and serious loss of strength. For example, exposure of concrete in water to a temperature of 35°C for 3 hours, after 3 hours at 18°C, will reduce the strength of concrete after 1 year of further storage in water at 18°C by an average of 25% where the free water/cement ratio is in the range of 0.27 to 0.4 and by an average of nearly 50% when the free water/cement ratio was in the range of 0.45 to 0.62. The extreme sensitivity of the concrete to loss of strength in this way is illustrated by the additional fact that if the concrete had been heated to 35°C for the first three hours instead of the second three hours after casting; the strength at 1 year would instead have been increased slightly for all water/cement ratios. For heating for three hours at 25°C for either regime, the effect on strength at 1 year would have been a slight increase in each case.

A comparison was made, as part of the third test series (B), between the strengths of different batches of cement in concrete with different water/cement ratios of the strengths at 1 day for storage at 18°C and the strengths obtained after 3 months at 38°C. The results, which are presented in Figure 14, show the variation in strength between different batches and the variation in the reduction in strength for the different batches. Within the range for free water/cement ratio of 0.35 to 0.45, which may have been used for prestressed concrete, these results provide confirmation of the limits of characteristic strength given in the Handbook for the Unified Code[24] for the use of structural concrete. Over this range, the reduction in characteristic strength is nearly constant at about 30 N/mm^2. The significance of this feature is discussed later.

Values for the modulus of elasticity for both slowly and rapidly converted concretes, which were obtained from the fourth series of tests (C) are shown in Figure 15 and have direct relevance to appraisal. Comparisons with results obtained at one day are given in Table 5.

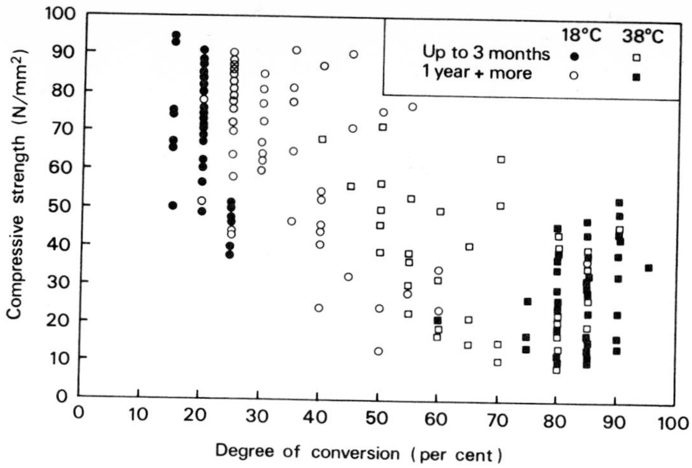

Figure 12 Compressive strength and degree of conversion from test series A1

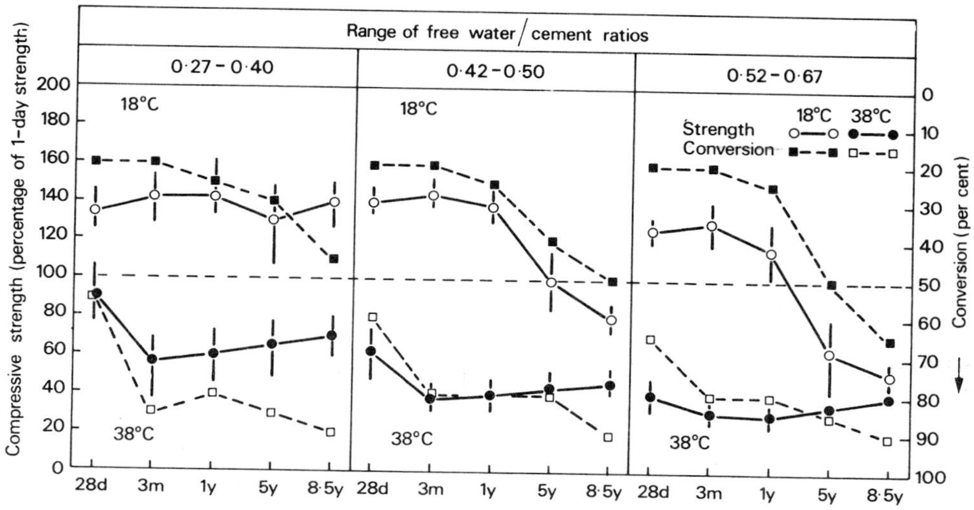

Figure 13 Development of conversion and strength expressed as a percentage of the 1-day strength from test series A1

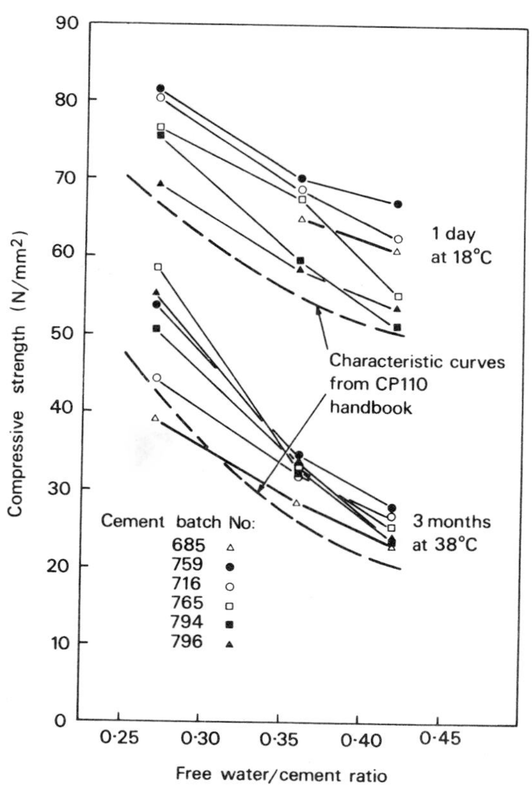

Figure 14 Relationship between 1-day strengths, minimum converted strengths and water/cement ratio for various batches of cement

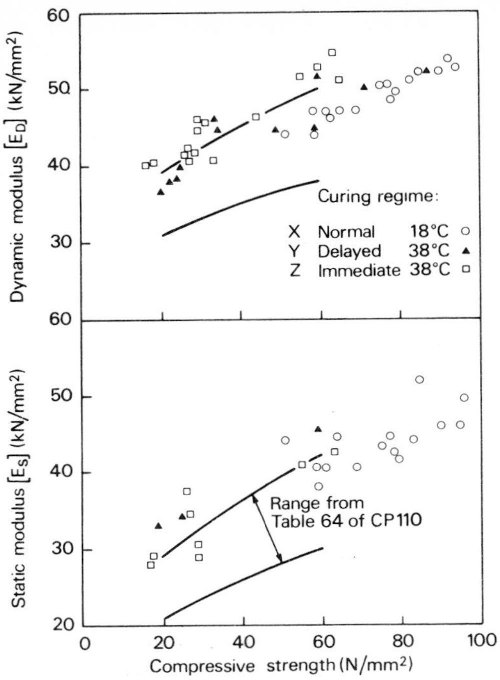

Figure 15 Relationship between modulus of elasticity and compressive strength

Table 5 Engineering properties of high alumina cement concrete after storage at 38°C as percentages of their values after storage for one day at 18°C

Free water/ cement ratio	Compressive strength %	Static modulus of elasticity %	Dynamic modulus of elasticity %
0.27	76	99	102
0.41	47	92	94
0.50	35	70	86

Appraisal of prestressed concrete beams may require estimates of losses of prestress, and therefore of the likely creep of the high alumina cement concrete. Few long-term data exist that are directly relevant to concretes with low water/cement ratios. Measurements made by Glanville and Thomas[25] over a variety of concretes showed that for mix proportions by weight of 1:2:4 with a free water/cement ratio of about 0.60, the creep with high alumina cement concrete over a period of 5 years was approximately 50% greater than with rapid-hardening Portland cement concrete for identical levels of stress applied when the concrete was 28 days old. Data obtained on unloaded prestressed beams with post-tensioned tendons, made in 1958, showed that, over a period of nearly 3 years of storage in air indoors, the creep of high alumina cement concrete was about 2½ times that of concrete made with rapid-hardening Portland cement.

The mix proportions by weight for the high alumina cement concrete were 1:2.4:3.3 with a free water/cement ratio of about 0.5 and stressing of the concrete to 14 N/mm^2 was effected at eight days. The strength of the concrete at the time of stressing was 70 N/mm^2, and the shrinkage and creep strains after three years were respectively 40 x 10^{-6} and 940 x 10^{-6}. Creep was then still continuing but at a reducing rate. These data are indicative only since the water/cement ratio was greater than was later advocated.

Figure 1 gives the results of tests over 20 years on concrete cylinders which were commenced in 1925 which have already been referred to briefly. The form of specimen accounts in part for the low strength but of particular interest is the relatively slow loss of strength for concrete with what would now be judged as a high water/cement ratio. Figure 16 shows the results of more recent tests commenced in 1945, 1964 and 1973. When compared with the results in Figure 1, it may be observed that the reduction in strength from one year takes place at a progressively greater rate as the manufacture of the cement becomes more recent. The recent increase in standard temperature for storage in the laboratory from 18°C to 20°C would not be expected to have had more than a small effect.

The evidence is sparse but the implication that the rate of loss of strength is increasing for newer cement may help to explain, for example, why the concrete at the Camden School for

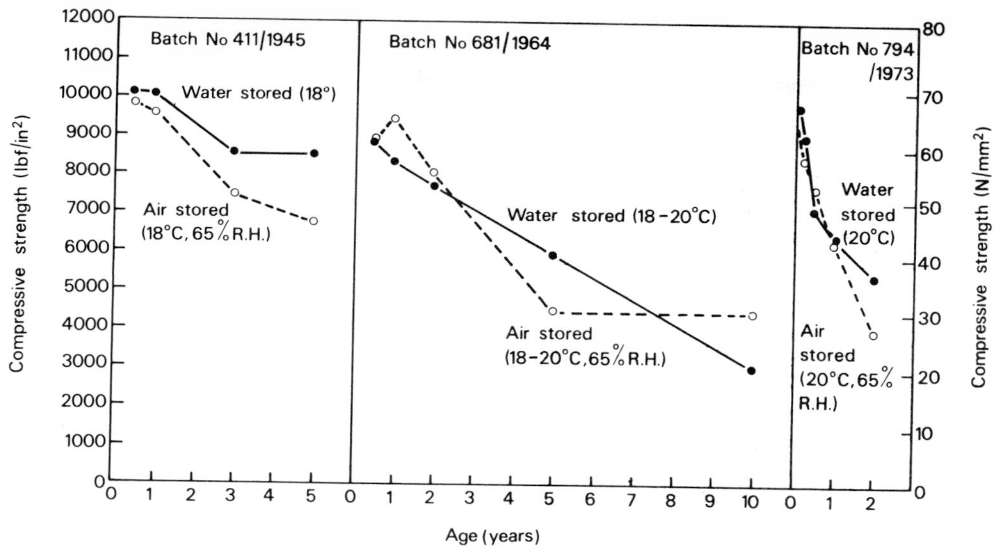

Figure 16 Long age tests on post-war HAC concrete. 1:2:4 Total w/c 0.60

Girls, which was about 20 years old and made before the need for a low water/cement ratio was appreciated, gave strengths from cores equivalent to cube strengths of between 24 and 42 N/mm^2 for degrees of conversion between 77 and 94% where the total water/cement ratio was probably about 0.6.

It should be noted that since 1973 (Batch No 794 in Figure 9) the cements received have shown an increase in strength up to 28 days but unlike previous cements there has been a decrease in strength between 28 days and three months. There is no reason to suppose however that the minimum strength attained at a high level of conversion would be unusual since the results of tests in water at 38°C are consistent with those obtained with other cements included in the main study already described. Nevertheless, this rapid reduction in strength at normal temperature might account for some of the unexpectedly high reductions found in some relatively new buildings.

Table 6 Analysis of specimens received by three testing laboratories for dta tests

Age	Low level no of examples	%	Medium level no of examples	%	High level no of examples	%
24	0	0	1	50	1	50
23	0	0	4	17	20	83
22	1	1	6	6	95	93
21	0	0	165	74	57	26
20	0	0	35	18	157	82
19	0	0	4	29	10	71
18	0	0	18	23	59	77
17	0	0	11	32	23	68
16	0	0	5	4	114	96
15	12	4	49	16	244	80
14	5	1	50	15	287	84
13	21	7	105	37	160	56
12	4	2	62	24	194	74
11	12	6	49	23	147	71
10	71	14	114	22	325	64
9	6	7	27	30	56	63
8	47	9	74	15	380	76
7	13	16	15	19	51	65
6	51	13	69	18	270	69
5	16	14	24	21	72	65
4	60	28	42	20	112	52
Average	319	8%	929	23%	2834	69%
3	64	50	39	30	24	20
2	149	41	112	31	104	28
1	372	63	185	31	38	6

7 CONDITION OF CONCRETE IN EXISTING BUILDINGS

The Department of the Environment's Circular of 20 July 1974 recommended that the degree of conversion should be determined for the concrete in existing structures as a guide to defining priorities in the appraisal of buildings.

Information from differential thermal analyses undertaken by three consulting firms is tabulated in Table 6 against the age of the concrete. These show that for concretes more than three years old, 69% are at a high level of conversion and 8% only are at a low level of conversion. A similar trend is clear from the results submitted by local authorities and structural engineers for appraised buildings, shown in Figure 17 which may well cover some of the same data. Some 2 800 individual readings are plotted from 148 buildings, of these the averages for 59% of buildings are in the high level range of conversion and 10% are in the low level range. The scatter of data obtained from one building is given in Figure 18. This range of values, ie from 30% to 85% at the age of seven years was not uncommon for buildings with concrete at a medium level of conversion. Examination of the results for conversion suggests that the higher levels obtained in a building were associated with higher ambient temperatures, indicating that conditions in service have influenced the rate of conversion.

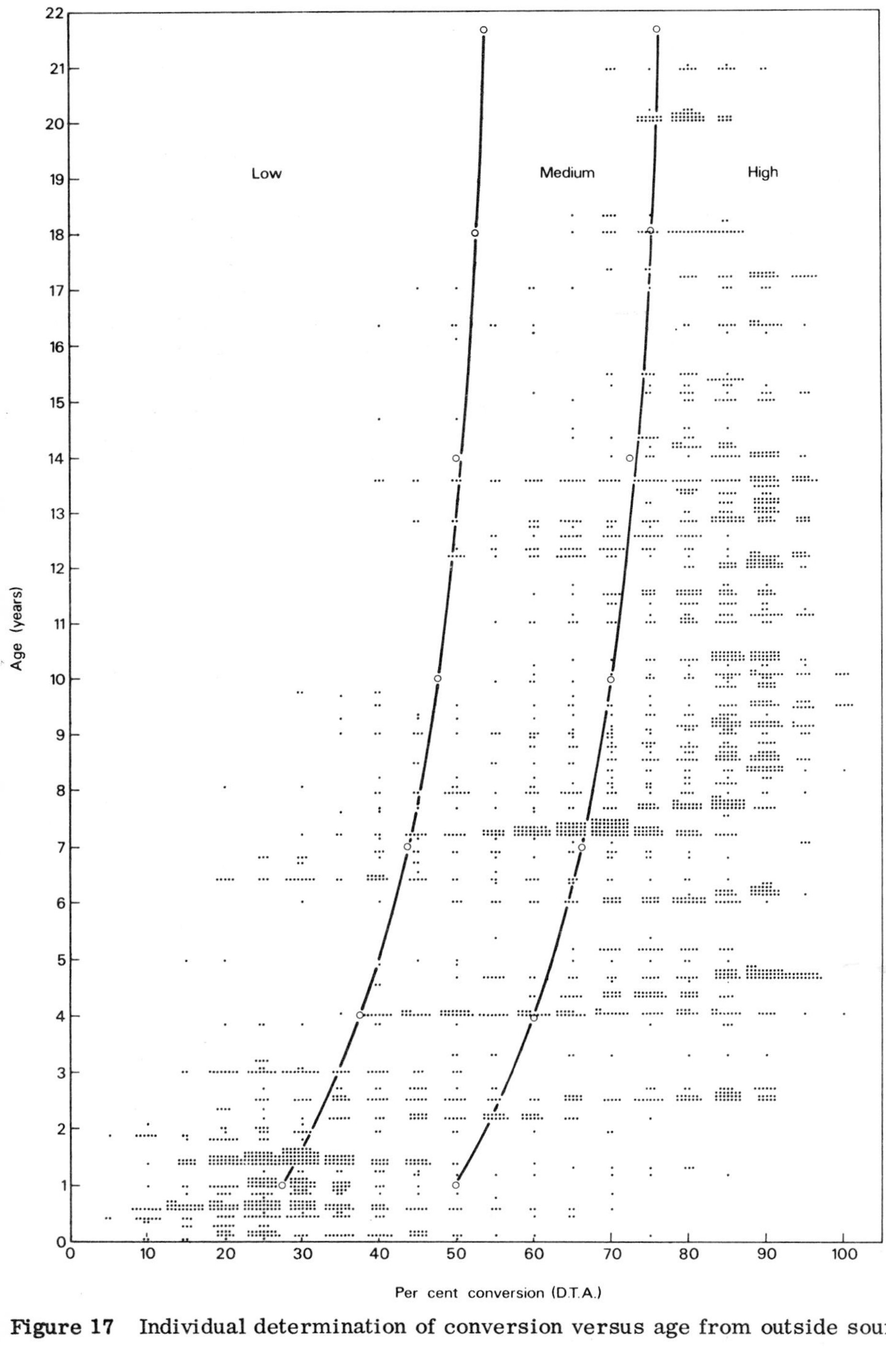

Figure 17 Individual determination of conversion versus age from outside sources

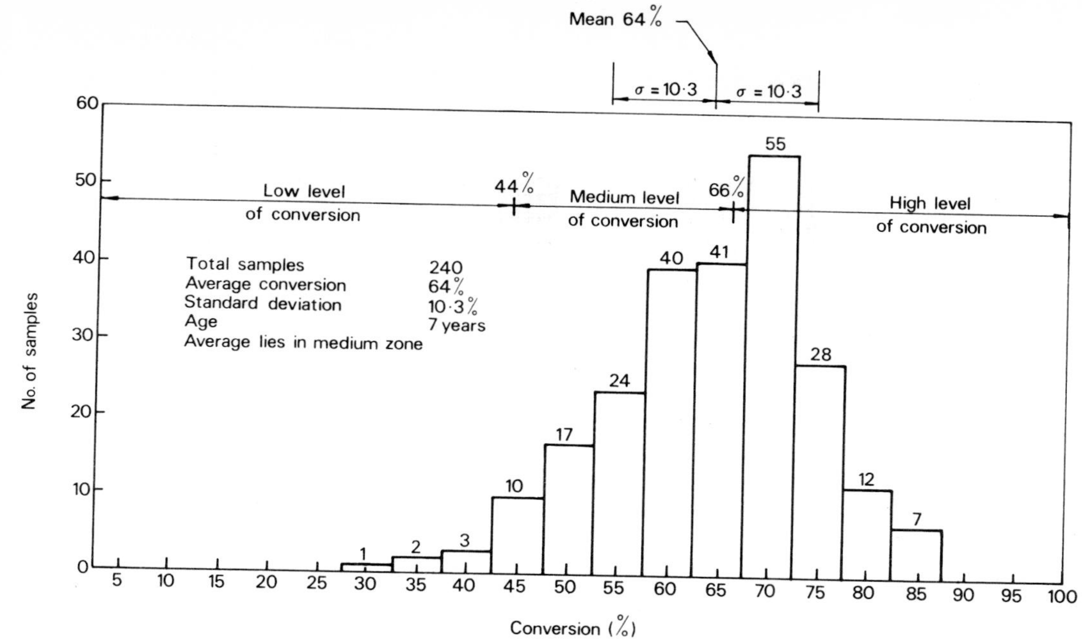

Figure 18 Histogram of percentage conversions from individual determinations for building No 132

Comparison between estimated cube strength and degree of conversion for data received from local authorities and structural engineers is illustrated in Figure 19. It will be seen that there is a general reduction in strength as the percentage of conversion increases and also for a given degree of conversion there is a considerable variability in strength. The pattern is broadly similar to that obtained in the laboratory tests shown in Figure 12 but of course there is increased representation in the highly converted region where most tests were made. These

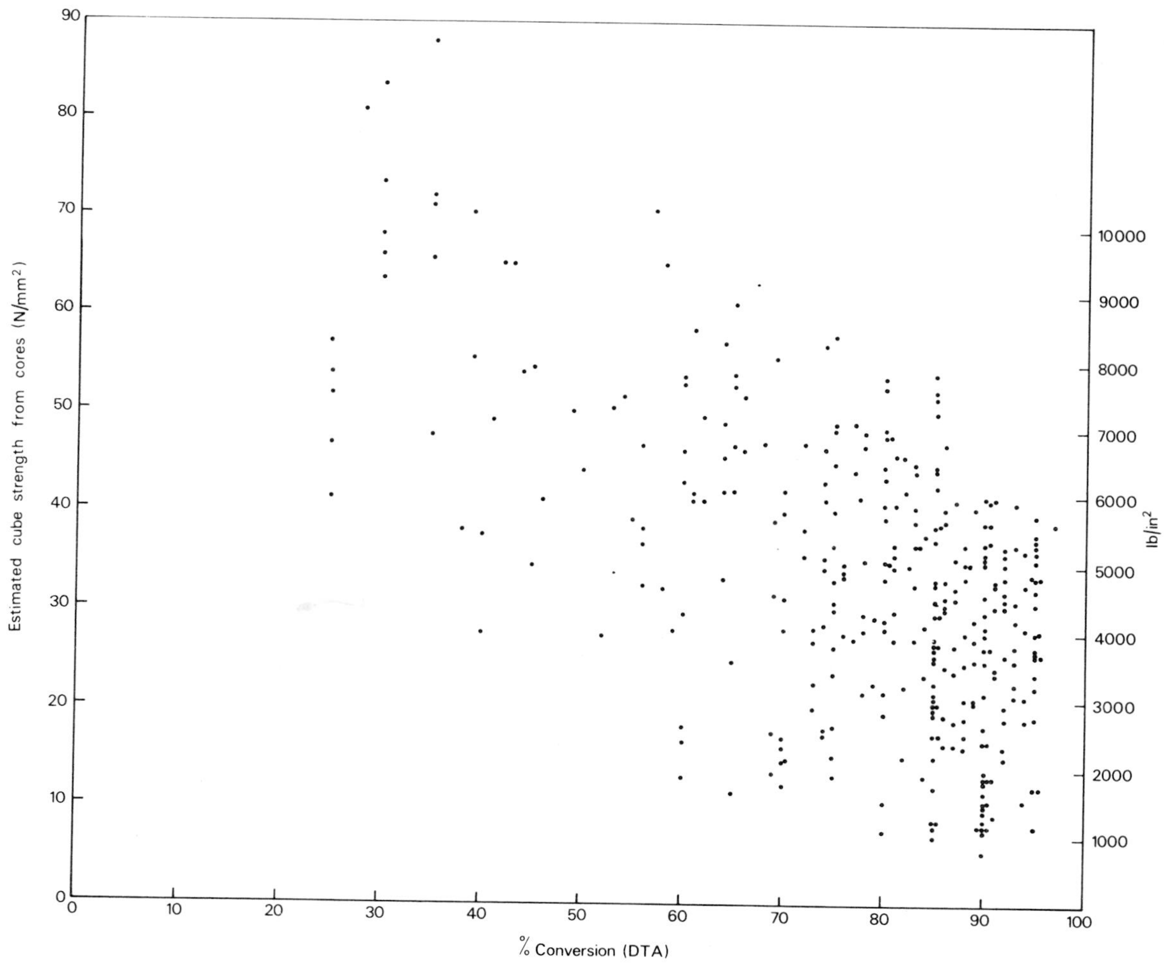

Figure 19 Cube strength versus conversion for individual cores

results confirm that the greatest loss of strength is likely to occur when the concrete is highly converted.

Before considering the significance of the estimated cube strengths obtained from core tests and their application in appraisal, the data on ultra-sonic pulse velocity measurements made at the Building Research Establishment and as received from outside sources are reviewed. Measurements of ultra-sonic pulse velocity were made across the flange, usually at ten positions, on a large number of the beams delivered to the BRE from sites including all those, from which cores were cut or sections prepared for crushing tests, as well as for most of the beams tested in flexure. The relationship between ultra-sonic pulse velocity for beams and the estimated cube strengths for cores or beam sections for individual beams from the sites dealt with in the BRE tests are shown in Figure 20(a), and the average values for each site in Figure 20(b). These results show that the limits of the relationship between cube strength and ultra-sonic pulse velocity given in the Institution of Structural Engineers' Guidelines[27] tend to underestimate strength where the ultra-sonic measurements are made in the laboratory. Closer agreement than this could not be expected however since it is known that the mix proportions and type of aggregate have an appreciable influence on pulse velocity.

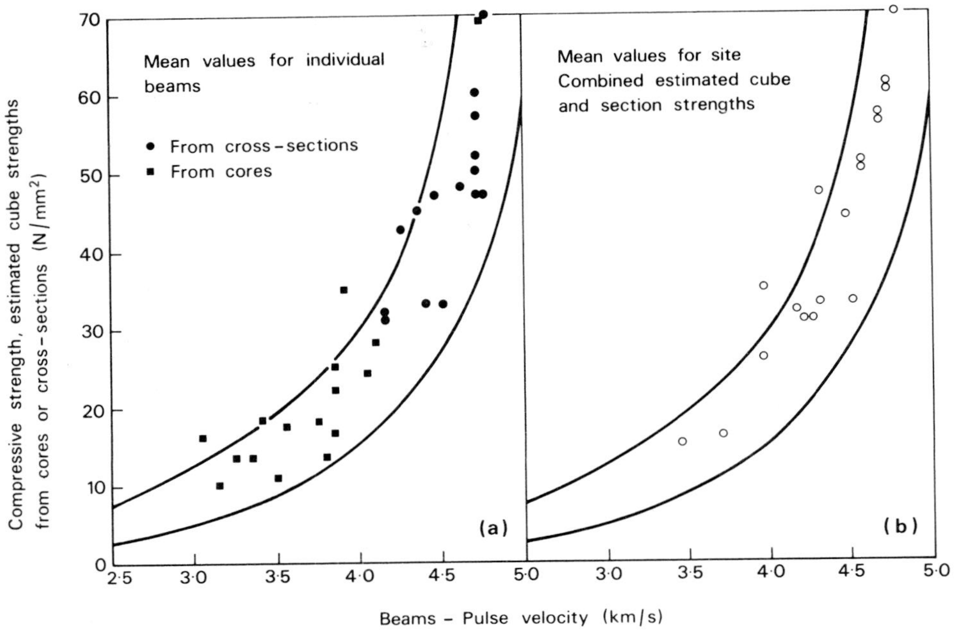

Figure 20 Relationships between cube strength and ultra-sonic pulse velocities - BRE tests

The relationship between ultrasonic pulse velocity and cube strength estimated from core tests for the data received from local authorities and structural engineers are plotted for individual strength tests in Figure 21 and averages for each of the 51 buildings are shown in Figure 22. These results show much more scatter than those obtained in the BRE tests, which may be due to some of the pulse velocity measurements being made on beams and others on cores and to the greater difficulty of coring and taking the pulse velocity measurements in the field. Both Figures 20(b) and 22 indicate that the curves in the Institution of Structural Engineers' Guidelines embrace the majority of the data averaged for each building but confirms that ultrasonic pulse velocities provide only a very coarse measure of concrete strength.

The overall data for 55 buildings received from local authorities and structural engineers are given in Table 7, which includes a brief description of their type and condition, estimated cube strengths from core tests, ultra-sonic pulse velocities and degree of conversion. The mean value for tests on all cores gave an estimated cube strength of 33 N/mm^2 with a standard deviation of 15 N/mm^2; the lowest strength was about 7 N/mm^2. The data have a considerable bias since cores were only cut in field investigations where a structure had shown some distress or had some concrete at a high level of conversion. The largest numbers of cores tended to be taken from construction with the weakest concrete; thus much of the concrete examined was not entirely representative of the 55 buildings which in themselves were not typical of all buildings with high alumina cement concrete. To make some allowance for the bias, the following analysis gives equal weight to each building.

A summary of the data in Table 7 is presented in Table 8 which shows the range of averages of strength for the buildings, the standard deviation of the average strength between buildings and

Table 7 Core strength data for individual buildings (outside sources)
Expressed as 'estimated cube strengths' in accordance with BS 1881, Part 4, 1970

Building or part building	Age (yrs)	Type of beam	Floor(F) or Roof (R)	Corresponding average dta %	Corresponding average dta Class	Whether visibly damaged	CORES: Diam of cores mm	CORES: No of cores	CORES: Average strength N/mm²	CORES: Standard deviation N/mm²	CORES: Coeff of variation %	CORES: Apparent characteristic strength N/mm²	Corresponding average UPV km/s
1	15		F	85	H	no		10	24.5	3.6	14.4	18.5	3.95
2	8	X12	R	82	H	no		10	41.5	9.0	21.6	26.5	4.25
3A	12	I10	R	75	H	yes		2	48.5	5.7	11.7	39.0	-
3B	12	I10	R	59	M	no		4	51.0	7.7	15.1	38.5	-
4	15	various - some post-tensioned	R and F	-	H	no	70	37	18.0	6.4	35.5	7.5	3.84
6A	14	non-proprietry	R	75	H	no	75	6	38.5	6.9	17.8	27.0	4.70
6B	14	non-proprietry	R	75	H	no	75	5	36.5	8.1	22.0	23.0	4.52
8	1	X15, X18	R and F	33	M	no	50	8	72.0	8.9	12.3	57.5	-
9	13		-	-	-	no		7	38.4	8.3	21.5	25.0	4.44
10	12	X7	R	-	-	no		4	41.8	7.5	17.9	29.5	4.37
11A	5	X15	R	72	H	yes		3	19.0	8.5	45.1	5.0	4.07
11B	5	X21	F	77	H	no		10	25.0	8.8	34.5	10.5	4.33
13	12		R	-	H	yes		5	8.5	1.2	13.8	6.5	3.86
14	9	X9	F	77	H	no		17	41.0	13.3	32.6	19.0	-
15	3	X9	F	25	L	no		4	52.5	4.4	8.4	45.5	4.66
16	4	X9	F	40	M	no		2	32.5	7.1	21.8	21.0	3.77
18	18	X7	R	90	H	no		2	30.0	16.3	54.7	3.5	4.01
19	13	X7	R	90	H	no		7	36.0	4.9	13.6	28.0	3.68
20A	11		R	79	H	no		5	38.5	11.7	30.3	19.5	
20B	11		F	89	H	no		3	28.5	7.0	24.5	17.0	
20C	11		?	81	H	no		12	32.0	7.6	23.6	19.5	4.61*
21	13		R	66	M	yes		24	16.5	4.8	29.6	8.5	-
22A	8		R	-	H	no		6	65.0	13.0	20.0	43.5	-
22B	8		F	-	H	no		14	39.5	26.0	65.9	(- 3.0)	-
22C	8		-	-	H	no		2	52.0	4.6	8.9	44.5	-

*The average UPV for building 20 corresponds to 6 cores with average strength of 38.1 N/mm²

Table 7 Core strength data for individual buildings (outside sources) (continued)

23	1	X5	F	35	M	no		2	38.0	4.8	12.7	30.0	-
24	16	X7	R	78	H	no		3	46.5	3.5	7.6	40.5	3.75
27	10	non-standard post-tensioned	R	83	H	yes	74 - 80	35	28.0	8.4	29.9	14.0	3.87
31	4	H9 and I10	F	89	H	-	-	30	22.5	4.2	18.8	15.5	4.21
33	8	X7	R	78	H	no		8	28.5	6.0	21.1	18.5	4.42
34	4	X9	F	75	H	no		2	40.5	1.8	4.5	37.5	-
37	9	T4	R	95	H	no		4	36.0	3.1	8.6	31.0	-
38A	7	T4	R	50	M	no		4	66.5	2.8	4.1	62.0	4.55
38B	7	T4	R	84	H	no		5	41.0	14.4	35.1	17.5	3.97
40A	9	T4	R	65	M	no		4	35.5	4.1	11.4	29.0	4.05
40B	9	T4	R	48	M	no		4	49.5	3.3	6.7	44.0	4.37
42	9	T4	R	95	H	no		6	32.0	6.0	18.8	22.0	3.83
43A	6	T4	R	50	M	no		4	66.6	11.4	17.1	48.0	4.39
43B	6	T4	R	34	L	no		4	70.0	10.6	15.2	52.5	4.37
45	9		R	47	L	no		4	42.5	4.7	11.1	35.0	4.27
46A	6		-	86	H	no	75	28	22.5	7.5	33.5	10.0	4.11
46B	6		F	77	H	no	75	12	25.0	8.3	32.8	11.5	4.24
46C	6		R	63	H	no	75	4	34.5	16.5	47.8	7.5	4.44
47A	-		R		H	no	48.5	3	32.5	8.0	24.9	19.5	4.52
47B	-		F		H	no	48.5	23	40.0	11.7	29.2	21.0	4.21
48	-		-	82	H	-		8	23.0	7.4	32.0	11.0	4.26
56A	14 approx		R	76	H	no	68	3	32.5	5.0	15.4	24.5	4.45
56B	"		F	92	H	no	68	7	26.0	6.5	25.3	15.5	4.31
57A	14 approx		R	79	H	no	68	11	40.0	7.3	18.1	28.0	4.60
57B	"		F	89	H	no	68	12	28.5	6.4	22.4	18.0	4.28
58A	8 approx		R	74	H	no	68	9	28.0	11.7	44.4	9.0	4.47
58B	"		F	83	H	no	68	13	31.0	12.1	39.1	11.0	4.50
59	-		-	85	H	no	68	25	33.0	8.7	26.4	18.5	4.35
60A	19 approx		R	73	M	no	68	17	43.5	8.3	19.1	30.0	4.54
60B	"		F	86	H	no	68	8	28.0	10.3	36.9	11.0	4.31

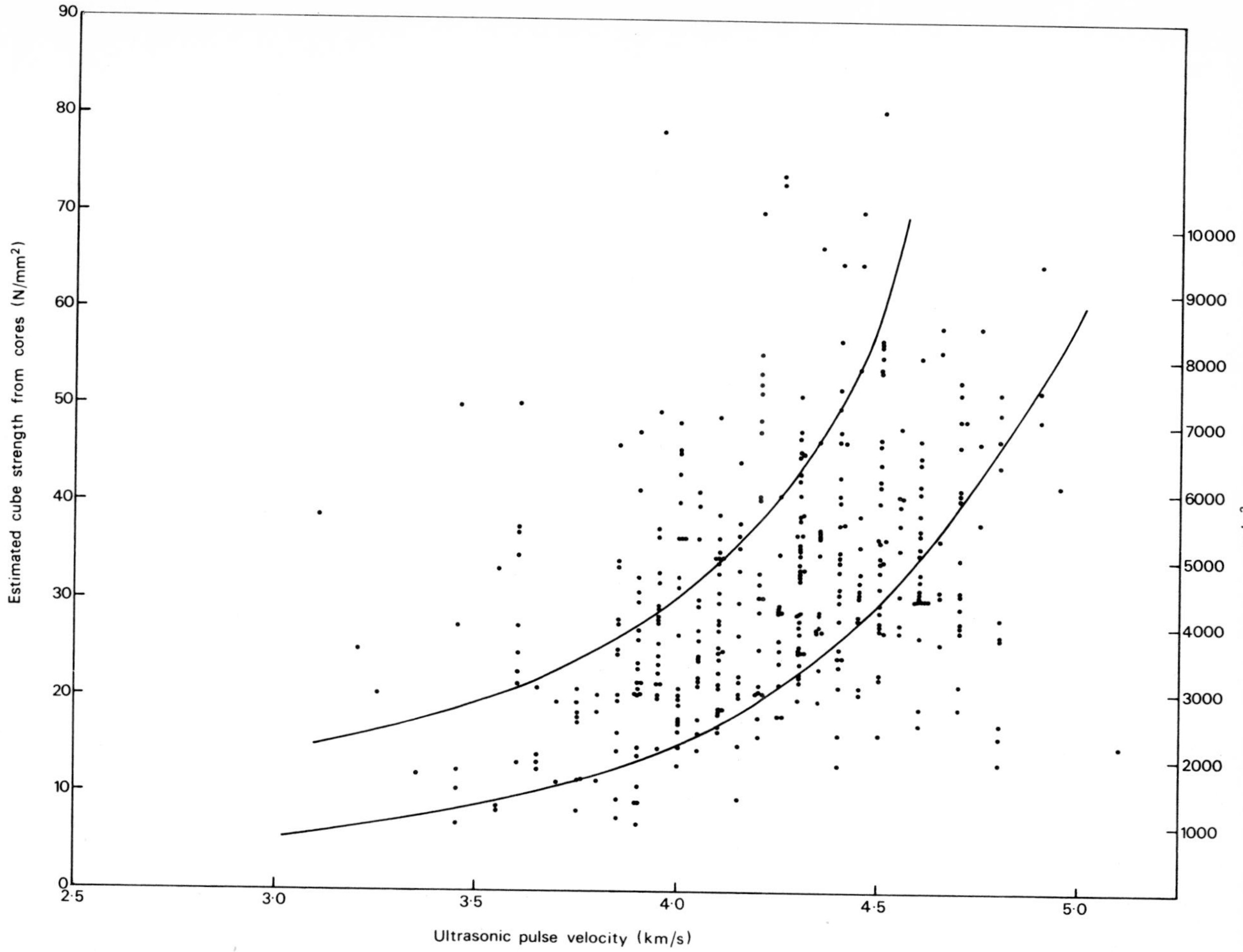

Figure 21 Ultrasonic pulse velocity versus estimated strength for individual cores

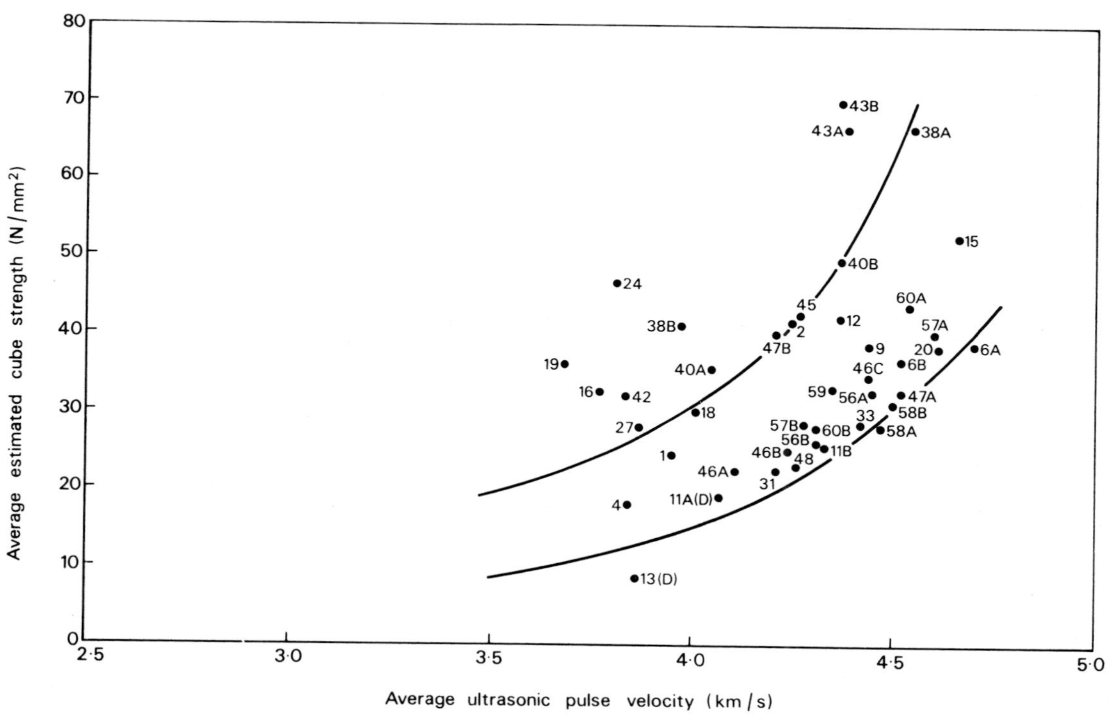

Figure 22 Average ultrasonic pulse velocity (km/s)

the mean standard deviation in buildings for a variety of groupings of the buildings for both the data from outside sources and from the BRE tests. If all the results in Table 8 are considered irrespective of the condition of the buildings and degree of conversion, it may be seen that the average cube strength for the buildings was 37 N/mm^2 with a range between $8\frac{1}{2}$ N/mm^2 and 72 N/mm^2, distributed as shown in Figure 23. This illustrates the tendency for buildings with

Table 8 Overall averages of core strength data (core strengths expressed as 'estimated cube strengths' in N/mm^2)

Category	Outside sources only					Combined sources (outside plus BRE tests)			
	Distribution of average strengths per building or part building				Mean standard deviation within within buildings N/mm^2 +	Distribution of average strengths per building or part building			
	No of buildings or part buildings	Overall mean strength N/mm^2 *	Range of averages N/mm^2	Standard deviation of distribution N/mm^2 o		No of buildings	Overall mean strength N/mm^2 *	Range of averages N/mm^2	Standard deviation of distribution N/mm^2 o
All results	55	37.0	$8\frac{1}{2}$ to 72	13.4	7.9	74	38.0	$8\frac{1}{2}$ to 72	14.1
All results less 'visibly damaged' buildings	52	38.5	18 to 72	12.5	8.1	64	40.0	18 to 72	13.2
All results 'high conversion' only	40	33.0	$8\frac{1}{2}$ to 65	10.0	8.4	50	33.0	$8\frac{1}{2}$ to 65	10.5
All results for 'high conversion' less 'visibly damaged' buildings	38	34.0	18 to 65	9.2	8.6	43	35.0	18 to 65	9.7
All results, roof beams only	31	39.5	$8\frac{1}{2}$ to 70	14.0	7.8	-	-	-	-
All results, floor beams only	16	32.5	$22\frac{1}{2}$ to $52\frac{1}{2}$	8.3	8.5	-	-	-	-
All results X5 and X7 beams only	6	36.5	$28\frac{1}{2}$ to $46\frac{1}{2}$	6.9	7.2	13	41.0	$28\frac{1}{2}$ to 70	11.7
All results less X5 and X7 beams	44	36.5	$8\frac{1}{2}$ to 72	13.7	7.7	56	37.0	$8\frac{1}{2}$ to 72	14.3
All 'visibly damaged' buildings □	3	14.7	$8\frac{1}{2}$ to $19\frac{1}{2}$	5.5	4.8	10	22.0	$8\frac{1}{2}$ to 35	8.5

* Mean value of average strength per building

\+ Mean value of standard deviation (σ) of individual core strengths per building

o Standard deviation of distribution of average strengths - not to be confused with the standard deviation (σ) of core strengths 'within' buildings

□ For this table, buildings 3A and 27 have been classed as <u>not</u> 'visibly damaged'

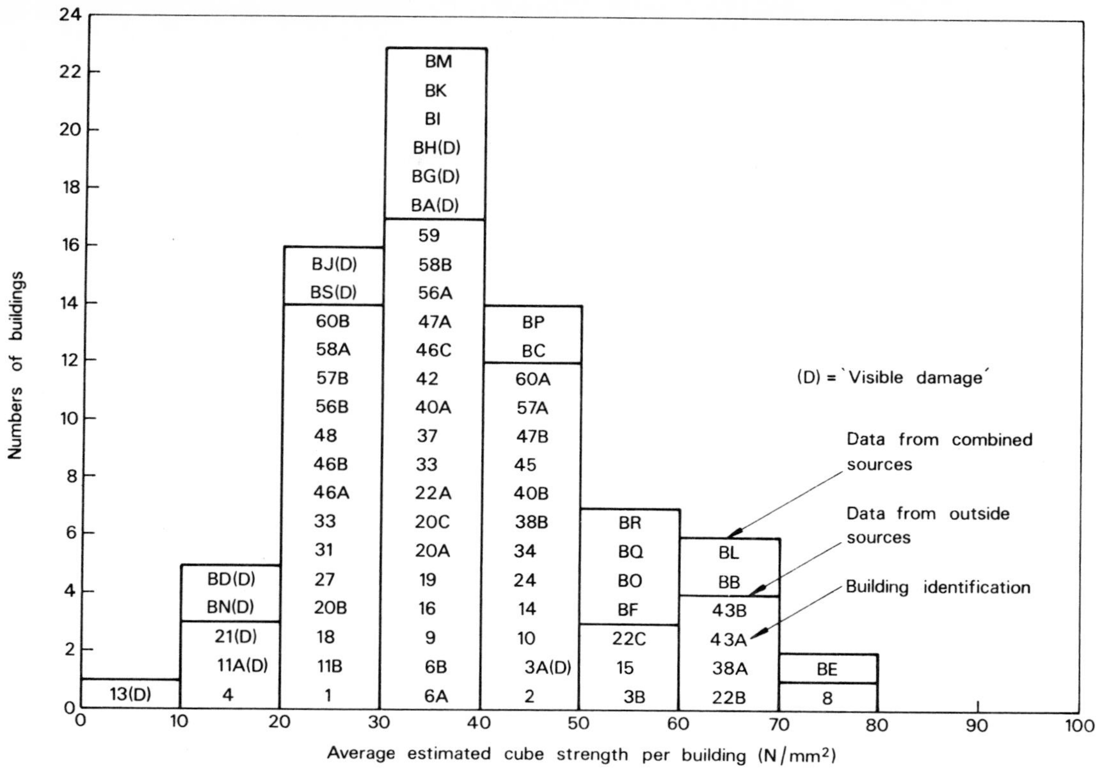

Figure 23 Histogram of average estimated cube strengths per building (expressed as 'estimated cube strengths')

visible damage to have a low strength for the concrete. Thus inspection can assist in identifying some of the buildings which require remedial measures.

Since the survey shows that most concrete tends to reach a high level of conversion, it is necessary to consider buildings in this condition when examining the significance of the strength tests on cores. The results for 50 buildings which were highly converted, given in Table 8, show that the overall mean strength for buildings was 33 N/mm^2. A slightly higher figure for the mean strength of 35 N/mm^2 was obtained when visibly damaged buildings were excluded. Under the present provisions of CP 110 and on the assumption that the recommendations for strength at one day given in CP 116:1965 were followed of requiring originally a one day strength of 52 N/mm^2, the strength which would now be assumed as a minimum for ultimate strength calculations would be 52 N/mm^2 divided by 1.5, ie 35 N/mm^2. Although this would be regarded as a lower acceptable limit for design to CP 110, it may nevertheless be concluded that much of the construction reviewed would conform with current requirements for safety in design. If consideration is also given to the feature considered later in analysing the tests on beams, that strength derived from cores tends to underestimate the strength of the concrete in the beam, more of the construction reviewed would meet present design requirements. The adverse bias in the sample of buildings reviewed would suggest that the proportion of construction as a whole which would conform with existing design requirements would be appreciably higher.

The results of the laboratory study showed that within the range of practical concrete mixes, concretes with a certain characteristic strength at one-day would have a characteristic strength after conversion at 38°C which was 30 N/mm^2 lower. Thus for concrete designed to meet the requirement of CP 116:1965 for a strength of 52 N/mm^2 at one day, the highly converted characteristic strength might be expected to be 22 N/mm^2. This concept would be particularly convenient to designers in assessing existing construction. The data were therefore examined to determine whether this approach was valid.

The concrete as originally specified under good control conditions might have given a standard deviation of 6 N/mm^2 and on this basis a comparison can be made between what might have been specified and what was actually obtained for the buildings covered in Table 8 from outside sources:

Original specification:	Mean	62 N/mm^2:	Characteristic	52 N/mm^2*	
All results:	Mean	37 N/mm^2:	Characteristic	11 N/mm^2*	(5%)
Highly converted results:	Mean	33 N/mm^2:	Characteristic	12 N/mm^2*	(5%)

Clearly these figures are likely to be of little assistance to the structural engineer in his appraisals, since the characteristic values also need to be reduced by dividing by a partial safety factor, which is considered further in Section 9.

If more specific applications are considered, such as the use of standardised beams, then the data from outside sources in Table 8 gives:

X5 and X7 beams: Mean 36.5 N/mm^2 : Characteristic 20 N/mm^2* (5%)

These data are derived from tests relating to six buildings for which there were in all 26 cores, 20 of them coming from highly converted concrete. If all the data obtained from outside sources and from the BRE tests is considered for standardised beams, which were either received as highly converted or were subjected to heat treatment for one month at 50°C on arrival, then, for the 32 beams concerned, the following are found:

X5, X7, and T4 beams: Mean 33.7 N/mm^2: Characteristic 22 N/mm^2*

Again, the characteristic strength given would need to be divided by a partial safety factor.

In reviewing these figures it has to be borne in mind that the core tests have tended to give low estimates of strength of the concrete in the structures for the following reasons:

(a) the samples were biased, as already discussed;

(b) difficulties in taking cores, particularly in the field lead to damage to the concrete;

(c) the need to take small diameter cores from some of the common types of beam tends to exaggerate the effect of aggregate size and minor defects in the concrete;

(d) the core may not represent the strength of the whole region in which the ultimate resistance of the beam is likely to develop (this difficulty is partly overcome by taking the mean strength from the cores in estimating beam strength).

The result for characteristic strength obtained for highly converted standardised beams of under 230 mm in depth encourage the further examination of these and other data to determine the applicability of an assumed value for characteristic strength of say 20 N/mm^2.

An extensive number of tests were made in the BRE investigation on material from prestressed concrete beams from 21 buildings some of which had already been drawn on in reviewing the strength of concrete in standardised beams. Where possible, the compressive strength was measured by drilling cores 75 mm (3 in.) in diameter. These were tested in accordance with BS 1881 Part 4, 1970[26] and correction factors applied to obtain the 'estimated cube strength'. Where this was not possible - that is for the X5 and X7 beams and some of the other types of beam - the compressive strength was measured on section sawn from the beam, crushed with the longitudinal axis in a vertical position as illustrated in Figure 24. No correction factors for shape were used, but a comparison was made between estimated cube strengths determined from cores and the section strength for beams where both cores and ends could be sampled. The relationship obtained is shown in Figure 25. Because there were few results, a simple relationship of equality between section strength and estimated cube strength has been assumed. The results of these tests from 21 sites are given in Table 9 and have been plotted against degree of conversion in Figure 26 with the results for Stepney and Camden. They show, as with the results from outside sources, that a number of individual equivalent cube strengths for cores were below the limit of 20 N/mm^2 being considered, but in fact none of this concrete came from standardised beams in the range of size up to X9.

The beams received from various sites for test at the BRE were at various levels of conversion and information was required on how much additional conversion might be expected to occur before it was complete and hence what further loss of strength might occur. Beams, cores and beam sections were therefore subjected to one month's exposure in the kilns of the Princes Risborough Laboratory at 50°C and 75% relative humidity. The test results for concrete from these beams are given in Table 10 which may be compared with those in Table 7. For the concrete at a low level of conversion, the average strength before treatment was 64 N/mm^2 and the average after treatment was 35 N/mm^2 equal to an average reduction of 29 N/mm^2, which is similar to the drop in characteristic strength from the one-day strength to that when

* All these results apply to individual estimated cube strengths from cores or from beam sections (referred to later) in the case of the BRE tests. Generally it is more convenient to deal with beam strength in terms of the mean estimated cube strength for the beam. The distribution of strength within the beam has a standard deviation of up to 4 N/mm^2 which has little influence on the calculations and so no correction has been made.

Figure 24 Mode of failure of beam section after compression test

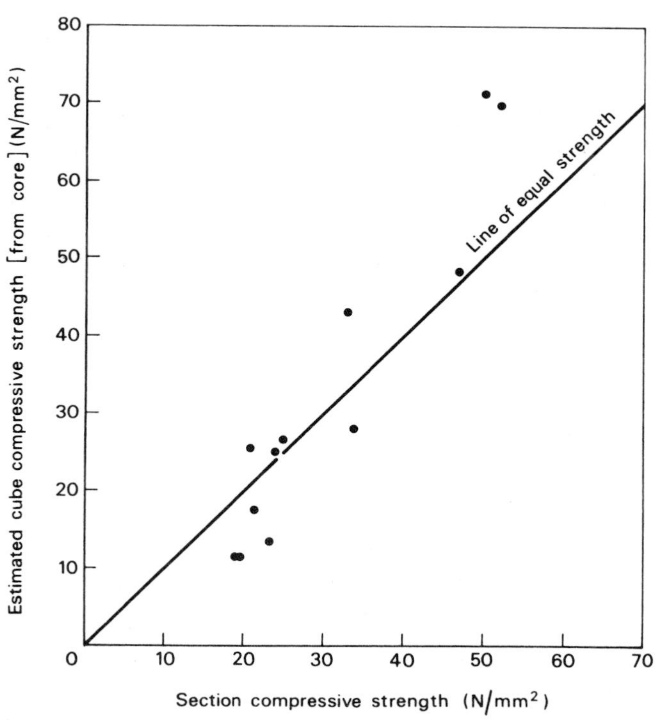

Figure 25 Relationship between strengths of sawn sections and drilled cores

Table 9 Mean values for compressive strength and pulse velocity on beams as received from sites

Source	Age of concrete (years)	Degree of conversion (high/low)	Beams		No of beams from which cores/sections were cut	Cores mean pulse velocity km/s	Mean compressive strength (cores/sect) N/mm^2
			No tested for pulse velocity	Mean pulse velocity km/s			
QM	7	high	6	4.65	2	4.35	49
HO	1	low	52	4.55	3	-	51
BF	8	high	2	4.50	1	-	33
HG	4	high	2	4.30	1	-	33
RE	1	low	5	4.70	1	-	71
LC	10 - 15	high	2	4.30	5	3.95	47
BA	10	high	1	4.15	1	-	32
FP	3	low	10	4.65	1	-	56
MF	1	low	2	4.75	1	-	70
RG	17	high	3	4.20	1	-	31
WC	1	high	5	4.55	3	4.35	50
BO	5	low	2	4.70	1	-	60
PS	8	high	3	3.95	2	4.35	35
MS	1	high	6	4.45	2	-	44
BF	8	high	4	3.95	3	3.95	26
P	20	high	14	3.45	8	3.55	15 +
BW	10	high	2	4.25	1	4.00	31
GW	1	high	2	3.70	2	3.70	16*
BU	5	high	1	4.20	None	-	-
WSW	6	high	3	4.70	None	-	-
PH	21	low	1	4.65	None	-	-
Totals for 21 sites	Max Mean Min	high	56	4.70 4.24 3.45	32	4.35 4.05 3.55	49 35 15
	Max Mean Min	low	72	4.75 4.67 4.55	7		71 62 51
	Max Mean Min	combined	128	4.75 4.35 3.45	39	4.35 4.05 3.55	71 42 15

+ Distressed building - excessive deflection. * Distressed building - cracking and collapse during erection

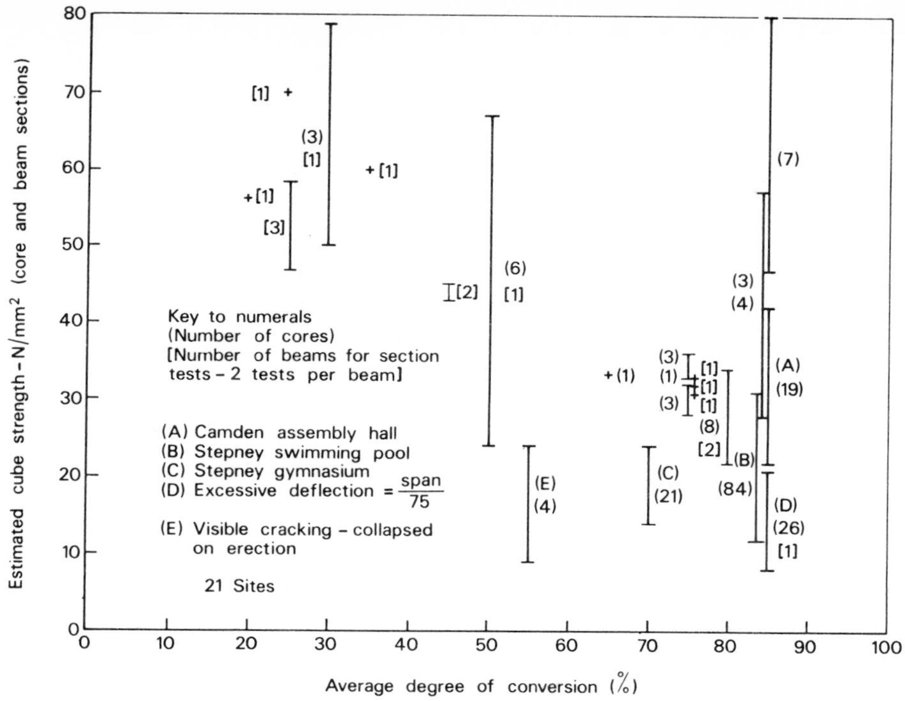

Figure 26 Ranges of individual strength test results for beams tested at BRE grouped by sites of origin

highly converted at 38°C. It should be noted that a number of the beams which were received in a highly converted condition lost strength when subjected to the heating regime; the average strength before heating was 35 N/mm^2 and after heating was 30 N/mm^2, the change however varied between a loss of strength of 13 N/mm^2 and a gain of 6 N/mm^2. The results were very variable, nevertheless the reduction in strength on heating of concrete at a high-level of conversion was, on average, about the same in proportion as that shown in Table 3 for a much stronger concrete when heated at 50°C instead of 38°C. This suggests that the lower temperature may provide a more realistic value for minimum strength.

The estimated individual cube strengths from tests on cores which had been heated in the kiln at 50°C ranged from 10 N/mm^2 to 59 N/mm^2. The majority of the low results came from one school roof with non-standard beams which had been demolished as a result of excessive deflection. The remainder of the treated cores came from 12 beams, three cores being cut from each beam. Of these, eight cores from three beams gave lower estimated cube strengths than 20 N/mm^2, and of these six cores were in the range of 10 to 15 N/mm^2, but these six cores were of 50 mm diameter and were obtained from beams which had a lower specified strength in manufacture than was usual for the X type beams and were of different section. The equivalent cube strengths from tests on heat-treated beam sections ranged from 18 to 61 N/mm^2 and referred to 21 beams for most of which there were two tests. Of the 33 beams from which these samples were cut, the individual estimated cube strengths were less than 20 N/mm^2 in four beams, the lowest two of which were for beams of unusual form. Thus these results are not inconsistent with a characteristic strength for standardised beams such as X beams of 20 N/mm^2.

Further information on the strength of concrete in existing buildings can be obtained from the considerable number of beams tested at the Building Research Establishment. Most were of the widely used X5, X7 and X9 types and analysis of their results also provides information on the relationship between beam strength and concrete strength. The beams were loaded by increments to failure with the load being applied at two equally spaced points in the span. Deflections were recorded at mid-span and strains in the concrete were measured on the upper and lower flanges with a demountable strain gauge of 200 mm gauge length. The main results are summarised in Table 11. The majority of the beams had the maximum number of wires in the section adopted by the manufacturer so that the effect of conversion on the strength of the beams was a maximum in most cases.

For the groups of beams, which were at a low-level of conversion when received, the effect of accelerated conversion obtained by heating at 50°C, gave a ratio for ultimate strength after heating to that before heating, which ranged from 0.59 to 0.94 with a mean of 0.76.

Table 10 Mean value of compressive strength and pulse velocity on beams after accelerated conversion treatment

Source	Age of concrete (years)	Degree of conversion		Beams		No of beams from which cores/sections were cut	Cores mean pulse velocity km/s	Mean estimated cube strength after conversion (cores/sections) N/mm^2	Loss in strength on completion of conversion N/mm^2
		Before	After	No tested for pulse velocity	Mean pulse velocity km/s				
QM	7	87	87	3	4.55	3	-	37	12
HO	<1	38	80	38	4.25	3	-	28	23 *
RE	<1	24	83	1	4.20	1	-	61	10 *
LC	10 - 15	76	75	1	4.45	3	4.00	38	9
BA	10	80	75	1	4.25	1	-	38	-6
FP	3	27	80	12	4.05	1	-	23	33 *
MF	<1	26	74	1	4.55	1	-	30	40 *
WC	1	33	83	4	4.25	3	3.90	30	20
WSW	6	55	85	1	4.30	1	-	36	-
BO	5	12	77	1	4.05	1	-	26	34 *
P	20	85	80	3	3.45	3	3.55	18 +	-3
MS	<1	38	65	20	4.10	1	-	36	8
BF	8	80	80	1	3.75	1	-	20	13
PH	<1	79	80	16	4.20	1	-	31	-
Totals for 14 sites		Max			4.55		-	61	
		Mean		103	4.15	24	3.80	32	
		Min			3.45		-	18	

+ Building showing excessive deflection of span/75

* Originally converted to a low level when received

Table 11 Summary of test data on flexural tests on X joists

Ref: Type	Wires		Age in years	Conversion group	Cube strength from beam sections N/mm^2	Test span Initial camber	Position of load points	Test span -	Bending moment kN/m		Type of failure
	No	Dia mm							At first crack	At failure	
21HO X5$\frac{1}{4}$	6	5	1	LA	47	280	L/4	25.4	10.2	15.5	F.C
32HC X5$\frac{1}{4}$	6	5	1	LA	-	400	L/4	24.0	9.0	16.2	F.C
61HO X5$\frac{1}{4}$	6	5	1	HC	-	200	L/4	25.9	8.1	10.4	F.C
133MS X5$\frac{1}{4}$	6	5	1	HC	36	180	L/4	29.2	7.3	12.5	F.C
145MS X5$\frac{1}{4}$	6	5	1	LA	43	220	L/4	35.8	11.5	14.8	F.C
146MS X5$\frac{1}{4}$	6	5	1	LA	45	400	L/4	35.8	9.2	12.1	F.C
104BA X5$\frac{1}{4}$	6	5	10	HA	32	500	L/4	26.3	9.8	10.5	F.C
105BA X5$\frac{1}{4}$	6	5	10	HC	38	420	L/2.5	22.4	8.8	10.1	F.C
X7 X7	7	5	21	HA	-	330	L/3	25.7	14.8	25.0	F.C
HG1 X7	7	5	4	HA	33	600	L/4	27.1	15.6	21.0	F.C
76HO X7	9	5	1	HC	-	330	L/2.8	17.7	21.1	30.8	F.C
79HO X7	9	5	1	HC	-	450	L/2.8	19.4	23.7	34.0	F.C
93HO X7	9	5	1	LA	-	670	L/2.6	18.7	20.1	33.0	S
99HO X7	9	5	1	LA	-	570	L/2.8	18.6	25.4	36.2	F.C
106MF X7	8	5	1	HC	30	190	L/4	28.3	15.3	23.0	F.C
107MF X7	8	5	1	LA	70	270	L/4	33.7	16.9	33.1	F.C
108MF X7	8	5	1	LA	-	270	L/4	33.7	14.1	32.7	F.C
182PH X7	8	5	1	HC	30	190	L/4	31.0	15.8	25.5	F.C
36WC X7	8	5	1	LA	-	4000	L/4	33.6	16.8	33.6	F.C
18BC X7	8	5	8	HA	-	1200	L/4	33.7	13.8	27.2	F.C
WSW1 X9	14	5	6	HA	-	1630	L/4	42.7	42.8	67.6	F.C
WSW2 X9	14	5	6	HA	-	990	L/4	25.9	41.2	66.4	F.C
WSW3 X9	14	5	6	HC	33	1000	L/4	26.2	33.6	46.5	F.C
QM1 X9	14	5	7	HC	35	600	L/4	26.2	39.5	61.7	F.C
QM3 X9	14	5	7	HA	-	1190	L/4	25.9	31.4	65.2	F.C
QM4 X9	14	5	7	HA	-	450	L/4	25.8	37.2	65.6	F.C + S
QM5 X9	14	5	7	HC	34	750	L/4	26.2	36.5	57.5	F.C
QM6 X9	14	5	7	HA	-	590	L/4	35.9	36.7	50.5	F.C
QM7 X9	14	5	7	HC	42	600	L/4	26.2	36.6	62.6	F.C
8RE X9	14	5	1	LA	-	6700	L/3	26.2	31.7	50.6	F.C +S
10RE X9	14	5	1	LA	-	680	L/2.4	25.3	25.8	44.3	F.C + S
11RE X9	14	5	1	LA	-	400	L/4	26.2	29.8	60.4	F.C + S
199BO X9	7	7	5	HC	25	840	L/4	25.6	32.5	40.5	F.C
200BO X9	7	7	5	LA	-	-	L/4	24.0	39.0	68.3	F.C

LA Low level of conversion as received.
HA High level of conversion as received.
HC Converted at 50°C to a high level of conversion

Final column F.C = Flexural compression
S = Shear

This drop of 24% in mean strength corresponds approximately to a drop in the strength of cores or sections for concrete from the same source from an average of 60 N/mm^2 at a low level of conversion to 27 N/mm^2 when treated to obtain complete conversion, ie a drop of 55%. This result is indicative of the trend for the reduction in beam strength to be proportionately less than the reduction in strength of the concrete.

Further results of the beam tests are given in Table 12, these include the load factors against cracking and failure based on the design service moment for floors permitted by CP 116, they are therefore about 7% greater than they would be if the X7 and X9 beams were used in roofs. These results for the overall safety factors may be further analysed, as in Table 13, to show the effect of different states of conversion. From these results it is clear that the likelihood of the overall safety factor reaching 1.0 is remote, ie about five in 1000 floor beams for accelerated conversion; a slightly higher proportion, about eight in 1000 roof beams might drop to an overall safety factor of 1.0. Alternatively, these data may be used with the calculated relationships between cube strength and beam strength, based on CP 110, to derive cube strengths corresponding to the level below which 5% of beam strengths may be expected to lie. For the beams highly converted in the kiln at 50°C, this value of overall safety factor is 1.33. The corresponding cubes strengths for X5, X7 and X9 beams, using the CP110 method of calculation, are in the range of 27 to 28 N/mm^2. Whilst this result indicates a relatively high level for characteristic strength derived from tests on beams, an even higher value of about 33 N/mm^2 was obtained from tests on beams which were highly converted when received. It must be appreciated the sample was

Table 12 Analysis of test data

Ref	Initial elastic modulus of concrete from deflection kN/mm^2	Penultimate load stage		Design service moment kN/m	Overall safety factors	
		M_{Pen}/M_{ult}	Max comp strain		Against cracking	Against flexural failure
21HO	38	0.99	.0025	7.0	1.46	2.21
32HO	51	0.99	.0021	7.0	1.37	2.31
61HO	33	0.93	.0016	7.0	1.16	1.49□
133MS	27	0.97	.0023	7.0	1.04	1.79□
145MS	33	0.93	.0015	7.0	1.64	2.11
146MS	39	0.95	.0018	7.0	1.31	1.73
104BA	35	0.98	.0011	7.0	1.40	1.50+
105BA	39	0.99	.0016	7.0	1.26	1.44□
X7	39	0.95	.0018	11.7	1.26	2.14+
HG1	39	0.91	.0011	11.8	1.32	1.78+
76HO	41	0.97	.0028	13.9	1.52	2.22□
79HO	47	0.96	.0021	13.9	1.71	2.45□
93HO	39	0.95	.0016	13.9	1.45	2.37
99HO	45	0.98	.0024	13.9	1.81	2.58
106MF	43	0.93	.0012	12.8	1.20	1.80□
107MF	38	0.96	.0016	12.8	1.32	2.59
108MF	34	0.93	.0020	12.8	1.10	2.55
182PH	32	0.94	.0018	12.8	1.23	1.99□
36WC	46	0.88	.0019	13.4	1.25	2.51
18BC	43	0.94	.0016	13.4	1.03	2.03+
WSW1	52	0.83	.0010	28.3	1.51	2.39+
WSW2	50	0.91	.0015	28.3	1.46	2.35+
WSW3	40	0.98	.0014	28.3	1.19	1.64□
QM1	45	0.93	.0017	28.3	1.00	2.18□
QM3	43	0.75	.0009	28.3	1.11	2.30+
QM4	46	0.87	.0014	28.3	1.31	2.32+
QM5	45	0.89	.0014	28.3	1.29	2.03□
QM6	46	0.92	.0008	28.3	1.30	1.78+
QM7	46	0.96	.0018	28.3	1.29	2.21□
8RE	41	0.94	.0018	28.3	1.12	1.79
10RE	36	0.94	.0011	28.3	0.91	1.57
11RE	48	0.94	.0017	28.3	1.05	2.13
199BO	35	0.94	.0012	27.3	1.19	1.48□
200BO	50	0.96	.0019	27.3	1.43	2.50

+ Highly converted concrete as received
□ Highly converted concrete after treatment

Table 13 Effect of the state of conversion on the overall safety factor against failure and cracking

	State of conversion		
	Low level - as received	High level - as received	High level - accelerated at 50°C
No of beams	13	9	12
Failure			
Mean	2.23	2.07	1.89
Maximum	2.59	2.39	2.45
Minimum	1.57	1.50	1.44
Standard deviation	0.34	0.32	0.34
$\frac{\text{Mean} - 1.00}{\text{Standard deviation}}$	3.6	3.4	2.6
Cracking			
Mean	1.32	1.30	1.26
Maximum	1.81	1.51	1.71
Minimum	0.91	1.03	1.00
Standard deviation	0.25	0.15	0.19
$\frac{\text{Mean} - 1.00}{\text{Standard deviation}}$	1.3	2.0	1.3

very restricted and no beams were tested with known estimated cube strengths from tests on beam sections of less than 25 N/mm^2. Nevertheless the results give some further support for the adoption of a specific characteristic strength for highly converted concrete in X5, X7 and X9 beams.

The results for cracking are more erratic and suggest that variations in prestress rather than in the strength of the concrete may have been the cause, eg 1 in 10 beams at a low level of conversion may be expected to crack under the design service load if these results are representative.

The measurements of strain in the concrete reported in Table 12 at the penultimate increment of loading, demonstrate that the ultimate compressive strain of 0.0035% assumed in CP 110 is not reached irrespective of the state of conversion. It is nevertheless convenient to continue using the method of calculation recommended in CP 110, since it is important to be consistent and the precise form of the stress-strain relationship for the concrete is not critical in calculating the strength of beams (it would be important if calculations of deformation were being made).

For some of these beams, there were results of ultra-sonic pulse velocity tests and results from crushing beam sections. A comparison was therefore made between the experimental beam strengths and those obtained in calculations by the method recommended in CP 110, in which the concrete cube strength was derived both from the pulse velocity tests, using the mean of the relationships with equivalent cube strength in the Institution of Structural Engineers' Guidelines, and from those obtained by crushing tests on beam sections. The results are tabulated in Table 14, and tend to confirm that the crushed beam sections give a strength almost the same as that for the equivalent cube derived from core tests. They show that on average the strengths derived in this way underestimate the actual strength of the beam by 15 - 16%. If however the pulse velocity tests were used for estimating strength then 5% of beams would be weaker than 0.86 times the calculated strength to CP 110. For the crushed sections, the corresponding figure would be 0.90 times the calculated strength. Further calculations were made in which the experimental beam strengths were used to derive the estimated cube strength of the concrete. This was found to be on average 1.30 times the average strength obtained from crushed beam sections with a standard deviation of 0.29. For this factor, the scatter of results was such that 5% of beam strengths would have been over-estimated if the value of the factor had been reduced from 1.30 to 0.82. In an earlier analysis of the results for the tests on two beams in the gymnasium at Stepney, where the results of a large number of cores were available, this factor had a value of about 1.8. The value for this factor may well depend on how representative the cores are of the concrete in the beam. The beams at Stepney had a very hard outer skin of concrete which would have contributed to the strength of the beam but would not have influenced the strength of the cores.

Table 14 **Ratio of experimental to calculated beam strengths for estimates based on ultra-sonic pulse velocities and crushing tests**

	Ratio of experimental beam strength to calculated beam strengths for strengths of cubes estimated from:	
	Results of ultransonic pulse velocity measurement	Results of tests on beam sections
Number of beams	30	16
Mean	1.16	1.15
Maximum	1.52	1.43
Minimum	0.79	0.89
Standard deviation	0.18	0.15

The evidence from these tests on X joists suggests that the variation in the strengths of the different sizes of beam for different states of conversion are less than the variations that might be expected from those occurring in the assessment of the strength of the concrete. It also supports the view that the method used for determing the strength of the concrete and its incorporation in the calculations tends to underestimate the strength of the beam; the extent to which it is underestimated is however confused by the variation in results obtained. Nevertheless the combination of these two factors may permit the adoption in certain construction of a characteristic converted strength of concrete of 20 N/mm^2 for beams manufactured under conditions where it was ensured that the characteristic strength at one day was 50 N/mm^2. For X5, X7 and X9 sections consideration could be given to enhancing this strength by a factor of 1.3.

While the main attention in the laboratory work on the properties of the concrete received was directed to the determination of degrees of conversion, ultra-sonic pulse velocities, crushing strengths and beam strengths, a number of subsidiary tests were made to provide information needed for appraisals. These included determinations of tensile strength, prestress, shear strength, and carbonation and porosity. Results of measurements of modulus of elasticity were presented in the description of the long-term laboratory investigation in Table 4 and in Table 12 which gave some of the data on beam tests.

Tensile strengths of concrete were determined on cores, which were subjected to biaxial pressure on their curved surface. This caused a single cleavage plane to develop at right angles to the axis of the core. Results are given in Figure 27, which shows that there is little difference between high alumina cement concrete and Portland cement concrete over the range of equivalent cube strengths considered.

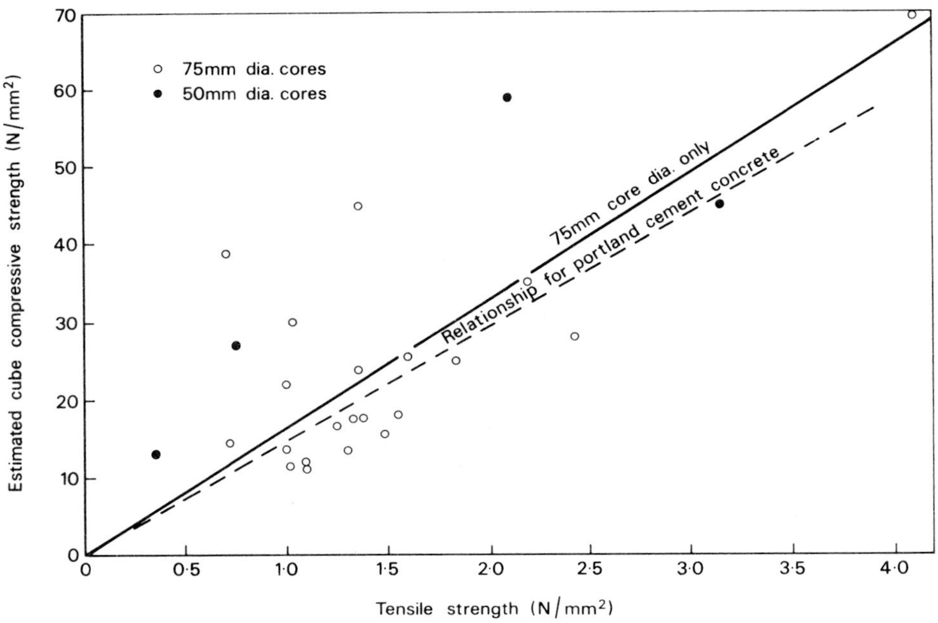

Figure 27 Relationship between estimated cube strength and tensile strength

Measurements of the residual prestress in the tendons of a number of prestressed concrete members stressed with 5 or 7 mm diameter wires were made by determining the length of tendons before and after removal from a cut length of beam. Allowance was made for the reduction of stress in the tendon at each end over the transmission length. The method only gave results to an estimated accuracy of ± 100 N/mm^2; the prestress in the tendons for beams with a low level of conversion was between 750 and 850 N/mm^2 and for those with a high level of conversion between 400 and 650 N/mm^2. These differences apparently due to the level of conversion are greater than was indicated in the determinations of cracking loads on beams where the differences due to the state of conversion were relatively small as shown in Table 13. The capability of the concrete to develop the stress in the steel through bond was determined by cutting sections from beams and measuring the change in the length of the concrete near the ends before and after cutting. For 5 mm and 7 mm wires the transmission length was between 30 and 50 diameters and was little affected by the degree of conversion. These values were about half those recommended for design in CP 115, it is not therefore anticipated that anchorage failures would be likely with 5 mm or 7 mm diameter wire.

In all eight beams were tested on short spans with loads centrally applied to determine their strength in shear. Bearings were between 25 mm and 75 mm with spans of 0.6 and 1.2 m for joists varying in size from X5 to X9. The shear resistance did not appear to be influenced by the effects of conversion and the failing loads were equivalent to shears of at least twice the design service shear adopted by the manufacturers. With these types of joist therefore, without composite action, shear does not present a serious problem.

The extent of carbonation due to exposure to atmospheric carbon dioxide was assessed for 25 of the beams received at the Building Research Establishment by breaking off pieces of concrete and spraying the freshly broken surfaces with phenolphtalein indicator solution. For samples at ages up to one year, the depth of carbonation was generally found to be less than 5 mm but occasionally it was as much as 10 mm. Depths of carbonation up to 10 mm were also observed with most of the older concretes aged from 5 to 17 years, although several samples were in the range of 15 to 30 mm. In one instance with concrete aged 20 years, extensive carbonation had occurred to a depth exceeding 60 mm. There was no apparent correlation between the degree of conversion and depth of carbonation. Since these concretes have been mainly exposed under relatively dry conditions indoors, the values for depth of carbonation obtained do not appear unduly high except in the case of the 20 year-old concrete.

The specimens tested for carbonation were also subjected to the capillary-rise test included in the Institution of Structural Engineers' Guidelines[17]. The results obtained were all appreciably higher than the values set as maxima for good quality well-compacted concrete. Furthermore the values obtained showed some tendency to increase with increasing degree of conversion of the cement.

The vacuum saturation method was also used to determine the porosity of the various concrete samples. The results of these tests are shown in Figure 28, which indicates some increase of porosity with degree of conversion. The highest values were experienced with 28 year-old beams.

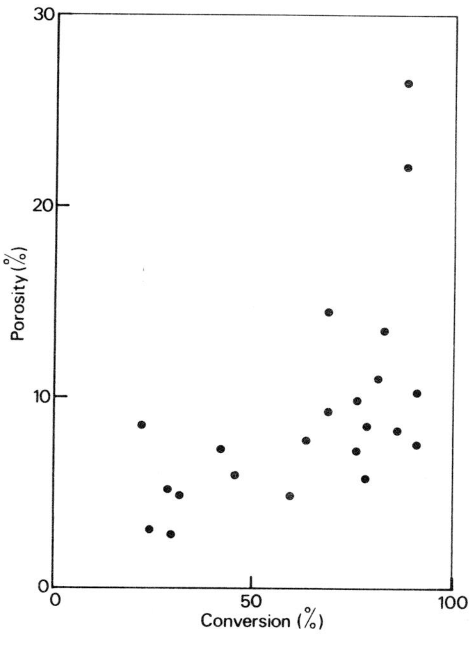

Figure 28 Relationship between degree of conversion and porosity

A visual examination was made of tendons removed from the beams. There was no significant corrosion of any of the embedded steel examined, only slight discolouration or superficial attack of the steel being observed, and some of this corrosion may well have been present initially on the steel surface. Furthermore, since no broken wires were ever found, there was no indication of the occurrence of any stress-corrosion cracking. The worst incidence of corrosion, with very slight localised pitting, was observed on the steel embedded in the 20 year-old concrete beams, where extensive carbonation had evidently reduced the alkalinity of the concrete adjacent to the steel so that the normal protective action of the concrete had become impaired. In this instance, some corrosion can take place, but even so the presence of both moisture and oxygen is necessary for progressive corrosion to occur and such carbonation is thus not necessarily accompanied by serious corrosion of the steel reinforcement. Hence, for concrete kept indoors in dry conditions the corrosion of the reinforcing steel may still not be significant, but if appreciable carbonation occurs the corrosion risk will be considerably increased for concrete exposed to the weather and to wet conditions.

8 STRENGTH OF ROOFS AND FLOORS

The first indications of the potential weakness of high alumina cement concrete in existing buildings came from the investigation of the collapse of the assembly hall at the Camden School for Girls in 1973. The concrete there had a minimum strength of 24 N/mm^2, but this was no more than contributory to the failure, which was due essentially to the inadequacy of the joint between the roof beams and the edge beams. Some other buildings of similar construction, which were inspected immediately afterwards, showed serious cracking at these joints which might well have led to failure. A number of these were built in Portland cement concrete.

At the University of Leicester, the collapse of prestressed beams of high alumina cement concrete in the roof the Bennett Building was due again to inadequacy of the connection between the roof beams and the edge beams, aggravated by loss of strength in the high alumina cement concrete in the precast edge-beam units. The roof beams were found by test to have adequate strength and have been retained in the reconstruction.

The failure at the Sir John Cass School, Stepney, of two beams in the roof of the swimming pool in 1974 was the first structural collapse in this country of prestressed concrete beams made with high alumina cement where the deterioration of the concrete was the primary cause of failure; this was not however due to conversion alone but was aided by sulphate attack, which led to final disruption of the concrete. In the adjacent gymnasium building of similar construction, a substantial loss of strength due to conversion had occurred but the residual strength was sufficient to give the beams an overall safety factor against failure of 1.4 as shown by loading tests.

Examination of the isolated roof beams in a 20-year old classroom block, in one of the many inspections following the Camden collapse, showed tensile cracking and deflections of up to span/75. The roof was immediately replaced and some of the beams were tested. They were of rectangular section being designed especially for that particular contract. On test, the overall safety factor was found to have approached close to unity. The likelihood of failure had increased as a result of ponding on the roof and the remedial measures to prevent roof leakage. Nevertheless adequate warning of failure was given and, once the staff of the local authority had been alerted, the appropriate remedial measures were taken. The beam in this case had been made before the need for controlling the water/cement ratio to the lowest possible level had been appreciated.

The warning of distress is unlikely to be as clear when isolated beams of I or X section are used. Once the flexural compressive strength of the concrete in the upper flange has deteriorated to such an extent that it cannot provide the compressive resistance necessary to support the load on the beam, there can be little or no redistribution of stress to lower levels in the section as is possible with rectangular sections, since the thin webs of I or X sections offer little flexural resistance. Thus little warning was given in the Stepney swimming pool before failure of the isolated X type beams. In the gymnasium roof at Stepney, the sag of the beams was as much as span/300, which increased for the two beams tested to span/145 and span/190 respectively just before failure. Deflections of this magnitude are not immediately obvious but are readily determined by measurement.

One other collapse of a roof beam of prestressed high alumina cement has come to the attention of the Building Research Establishment, this occurred during construction some months after the failure at Stepney. The concrete, although only a few months old, was converted to a high level suggesting that overheating might have occurred during the first few days. This was confirmed by the vertical cracking present in the upper part of the beam indicating weakness of the concrete at the time of transfer of the prestress.

Of the five failures or near failures associated with the use of prestressed concrete beams in roofs which have been reported to the Building Research Establishment, two did not directly involve the quality of the concrete in the roof, one was aggravated by chemical attack and two were due to seriously weak concrete. One of these was probably made with a high water/cement ratio and the other might well have been rejected on inspection. Thus of the many thousands of structures built during the past 20 or so years, the failures that have occurred, while representing an unacceptable level, have in some cases had contributory factors not due solely to the use of high alumina cement, but in part due to lack of supervision and control in manufacture, which might lead to similar failures if it were lacking in other forms of construction.

It is also significant that all the failures that have occurred were in roof construction. None were in floor construction where the manufacturers report that more than 17 x 10^6 m^2 have been built. This difference in experience between the behaviour of floors and roofs is probably due to several causes; higher levels of stress are adopted in the design of roofs for service loading; the proportion of service load permanently in place is higher for roofs; and the degree of interaction with other components is less than in floor construction.

Part of the Building Research Establishment's investigation was concerned with a survey of existing buildings regarded by their owners as being distressed. The majority were schools or educational buildings examined after the early DOE/DES Circulars had been issued. Of the 38 buildings, it had been decided to replace units in 15 and to strengthen in 17 cases. The reasons for strengthening or replacement were excessive deflection in nine instances, poor bearings at the ends of units in three buildings, a case of chemical attack and one where the design provided an inadequate margin of safety. Of the deflections reported, there were seven in excess of span/200 of which four were in excess of span/100. Whilst the study showed that decisions to strengthen or replace were being taken in many instances in adverse circumstances, there were indications in a few cases that the cost of immediate replacement was regarded as less of a liability than the cost of initial testing and of possible further inspections.

Chemical attack has figured in a number of reports received but its frequency is relatively insignificant. In all, fewer than 10 cases are known to have occurred and in some of these the attack was superficial with little effect on strength.

Results of between three and four hundred tests on floors and roofs have been received from local authorities and structural engineers. These were mostly carried out by applying test loads of 1.25 times the imposed load plus 0.25 times the dead load as in the recommendations of the DOE/DES Circular of 20 July 1974[4]. Results for 363 of these tests are summarised in Table 15 where the criterion passing the test is that given in CP 116 requiring 85% recovery after 24 hours although the loading was more severe than in CP 116. One of the roofs failed structurally and twelve roofs did not conform with the recovery requirement. Of these latter roofs, six would not be judged to fail under the newer provisions of CP 110, which removes the requirement for recovery when deflection is small. Of the remaining six roofs showing insufficient recovery, one comprised beams with hollow blocks and screed and subsequent tests showed that the beams alone gave an overall factor of safety of 2.21. The other five roofs were effectively of isolated beam construction. Beams, removed from one of the roofs which failed to show 85% recovery but were exempted from the requirement by the provisions of CP 110, gave an overall safety factor against failure of 1.44 The roof which failed structurally had a span of 6.8 m and consisted of X7 beams at 0.8 m centres supporting woodwool slabs and a screed. Failure occurred when the overall safety factor was 1.13.

One floor with X9 beams spanning 8 m with a monolithic cement-sand screed of about 50 mm in thickness was tested to failure. The floor was separated from adjacent construction and tested as a panel with a width of 2.3 m. The overall safety factor obtained was greater than 2.3 and failure took place by crushing of the screed showing that its bond with the beams was effective in enhancing the strength of the floor.

Testing of floor assemblies was included in the Building Research Establishment's investigation, four series each of three floors were tested using the resources of the Transport and Road Research Laboratory, the Princes Risborough Laboratory and Laboratory of the Polythechnic of Central London. Each floor consisted of five beams, which had previously been subjected to heating to 50°C for one month at 75% relative humidity to ensure their conversion to a high-level of between 70% and 80%. Most of the units used were Pierhead X5¼ and X7 beams but three floors were made with Francis Parker F6/4¾/7 beams. The different types of floor are listed in Table 15 and cross-sections of those with X5¼ and F6/4¾/7 beams are shown in Figure 29. For the Pierhead floors, the prewetted surface of the hollow blocks was brushed with 1:6 cement to sand slurry. Where floating screeds were provided, the sand-cement screed was cast on top of the glass fibre insulating quilt covered by polythene sheeting and wire mesh. The bonded screeds were cast directly on the slurry-grouted surface. The Francis Parker floors did not have a

Table 15 Results of site tests on floors and roofs supplied by consultants and local authorities

Type of construction load tested	Number of load tests carried out	Number of tests failing to meet recovery requirements (CP 116)
FLOORS:		
1 Beam and hollow blocks + bonded screeds	89	0
2 Beam and hollow blocks + floating screeds	2	0
3 Isolated floor beams	6	0
4 Construction not fully defined	25	0
	122	0
ROOFS:		
1 Beam and hollow blocks + bonded screeds	47	2
2 Beam + woodwool or similar + roof screed	75	7
3 Isolated roof beams	81	4
4 Roof construction not fully defined	38	0
	241	13

Total number of load tests - 363

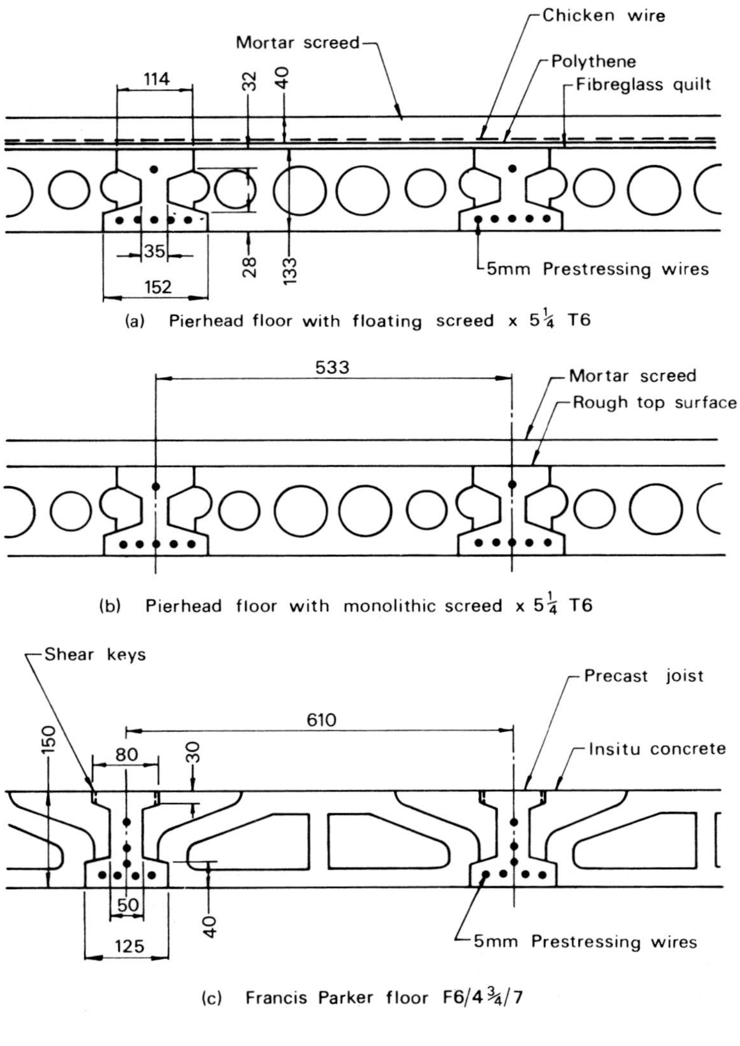

Figure 29 Cross-sectional details of test floors. Note dimensions given are typical values measured on the joists tested

screed, concrete was cast in the space between the joist and the hollow blocks, both of which had previously been wetted.

The floors are listed in Table 16. Those numbered 1 to 9 were tested with loads applied through transverse beams at four points in the span. The other three floors were subjected to uniform loading using bricks. The results of these tests are also shown in Table 16 from which it may be observed that the contribution from a bonded screed has ranged in four tests from an enhancement factor* of 1.26 to 2.02 whereas for floating screeds it ranged, again in four tests, from 1.04 to 1.10. The latter result, which was confirmed by the deflection measurements showed that there was little or no composite action between the floating screed and the pre-stressed concrete beams.

The floating screed was shown to play a much more effective role in distributing concentrated loading between adjacent members. In these tests which were preliminary to the main tests, a number of floors were loaded by two concentrated loads applied to the centre of the five beams in the floor at positions approximately a fifth of the span apart and placed centrally on the span. The deflections at mid-span in the loaded and adjacent beams are shown in Figure 30. Whilst it has been shown by Mayfield, Cairns and Miller [28] that measured distributions of deflections under concentrated loads do not bear a close relationship to the beam reactions, the fact that the adjacent beams have deflected to a high proportion of the deflection of the loaded beam indicates a substantial amount of load redistribution in each of the floors tested. It may be seen from Figure 30 (a) and (b) that the deflections of the outer beams in relation to those of the central

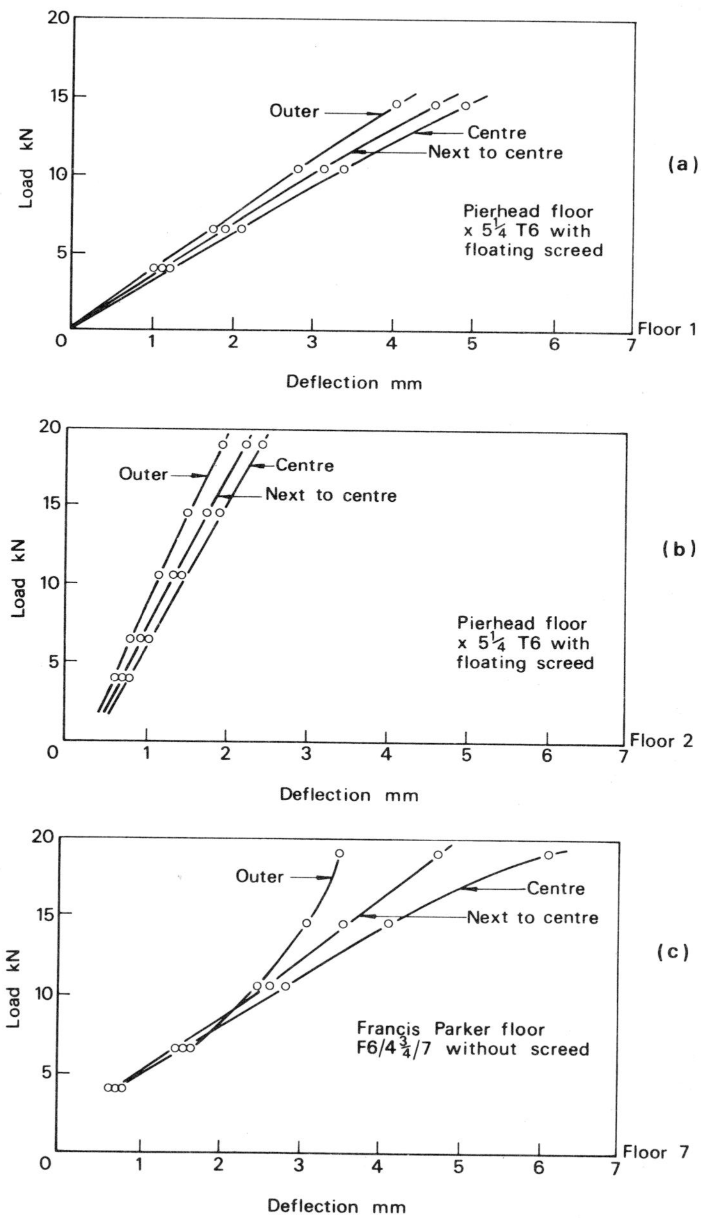

Figure 30 Central deflections of beams in floors with load applied to the central beam

*The enhancement factor is the ratio of the floor strength to that of the individual beams.

Table 16 Results of floor tests

Test no	Floor no (beams)	Screed thickness at mid-span mm	Screed floating or bonded	Span m	Spacing of beams mm	Strength of concrete N/mm^2	Strength of screed N/mm^2	Mean design moment single beams kN m	Mean ultimate moment single beams kN m	Ultimate moment floor kN m	Overall safety factor beams	Overall safety factor floor	Enhancement factor	Type of failure
1	3 X7 T9) 2 X7 T8)	50	B	5.72	600	31	18 - 20	13.5	26.5	166	1.96	2.46	1.26	Sudden delamination 2 beams failed in compression
2	5 X7 T8)	60	F	5.72	600	31	18 - 20	12.8	26.5	146	2.07	2.28	1.10	2 beams failed in compression
3	4 X7 T8) 1 X7 T9)	57	F	5.72	600	31	18 - 20	13.0	26.5	144	2.04	2.22	1.09	Compression failure
4	5 X5$\frac{1}{4}$ T6) Fig 29(a)	40	F	4.77	533	36	29	7.0	12.8	66.8	1.83	1.91	1.04	Compression failure
5	5 X5$\frac{1}{4}$ T6) Fig 29(b)	40	B	4.77	533	36	17	7.0	12.8	123	1.83	3.51	1.92	Delamination compression (3 beams) shear (bond slip 2 bms)
6	5 X5$\frac{1}{4}$ T6) Fig 29(b)	40	B	4.77	533	36	18	7.0	12.8	130	1.83	3.71	2.02	Shear (1 beam) compression (1 beam) bond slip (2 beams)
7	5 F6/48/7) Fig 29(c)	-	-	4.77	610	23	15$^+$			99		2.4	-	Compression spalling in compression zone, no damage to in situ concrete
8	5 F6/48/7) Fig 29(c)	-	-	4.77	610	23	18$^+$			61		1.5	-	Crushing and spalling of compression zone, longitudinal separation (dry mix, low bond of infill concrete)
9	5 F6/4$\frac{3}{4}$/7) Fig 29(c)	-	-	4.77	610	23	21$^+$			106		2.5	-	Some longitudinal separation. Crushing and spalling in compression zone
10	5 X5$\frac{1}{4}$ T6)	-	-	3.58	533	27	18 - 20	7.0	10.2	51.4	1.46	1.47	-	Compression (4 beams) Shear (1 beam)
11	5 X5$\frac{1}{4}$ T6) Fig 29(a)	60	F	3.58	533	27	18 - 20	7.0	10.2	54.1	1.46	1.55	1.06	Compression (5 beams) Shear (1 beam)
12	5 X5$\frac{1}{4}$ T6) Fig 29(b)	38	B	3.58	533	27	18 - 20	7.0	10.2	65.6	1.46	1.87	1.28	Compression (5 beams) Shear (1 beam)

$^+$ Strength of in-situ concrete

beams are in much the same proportion for the floors with floating and bonded screeds although the total deflection is less for the latter as would be expected. The importance of the inherent capacity in these floors for transverse transmission of load between beams is that this characteristic would provide for individual weak beams in contruction to receive support from adjacent stronger beams.

9 CONCLUDING COMMENTARY

In the earlier sections of this report and in this concluding commentary, account has been taken of the need to provide information on the Building Research Establishment's investigations to the Sub-Committee of the Building Regulations Advisory Committee which was appointed under the chairmanship of Mr C B Stone to make further recommendations on the structural appraisal of existing buildings with high alumina cement concrete. With this in mind, the report presents the results obtained and discusses their significance without presenting specific design recommendations.

The results of the surveys and the field and laboratory tests have been considered in previous sections without examining the levels of safety which might be regarded as satisfactory. Reference has already been made to the approach to safety given in CP 110 and it has been suggested that this might usefully be adopted for appraisal of existing buildings in which high alumina cement concrete has been used. The values for the partial safety factors for loads, for conditons at failure, which are used to enhance the values for the loads expected in normal service, and the corresponding partial safety factors for materials, by which their expected strengths are reduced, have already been mentioned. The values for these factors were originally chosen to give a level of safety in design which was consistent with that already established in the earlier codes of practice, and in some respects may be greater than is necessary for both adequacy and economy.

At the stage when a building is being designed there are a number of unknowns which are allowed for in design in the choice of the partial safety factors. Once the building has been constructed, these unknowns may be fewer and may justify the use of lower values for these factors in appraisal without reducing the level of safety accepted as adequate.

For dead loading, the partial factor of safety for failure conditions is taken in design as 1.4 to take account of the inaccuracies of construction and lack of knowledge about finishes. In the completed building these are no longer unknown and a value of possibly 1.2 or less might be more appropriate. When the factor for imposed loading is considered, it might be argued that, for normal design, the value in CP 110 of 1.6 is excessive. When more is known of the use of the building, the figure might be reduced substantially although consideration may need to be given to possible future changes in use. Appraisal of the actual imposed loads from surveys or published data may allow some reduction in loading from that assumed in design. In particular in housing, for example, an imposed floor load of 1.2 kN/m^2 for spans between 3 and 5 metres might be assumed instead of 1.5 kN/m^2 provided that the form of construction induced slab action. Loading due to partitions might also be regarded more realistically, since some contribute to the strength of the construction and the loading effect of others may be overestimated. Reductions in the factors for both dead and imposed loading may be further justified when it is known that the assumptions in design are conservative, eg floor slabs are nearly always designed as spanning in one direction only, when all test evidence shows that there is some transverse distribution of loading, membrane action and end fixity. It might be possible to adopt lower values for the partial safety factors without increasing the risk of failure to an unacceptable level for buildings where warning of the possibility of failure would be clear.

The choice of the partial safety factors for materials at failure must again be based on what more is known at appraisal than would be known at the design stage. Since the problem with high alumina cement concrete is that of dealing with a substantial loss of strength in the concrete, the strength of the steel is not critical and therefore a factor of unity might be appropriate. For concrete, it may be argued that if an inspection of the building for signs of distress has taken place that a lower factor than 1.5 as used in design might be acceptable. If inspection has included intensive sampling of the concrete for strength testing an even lower factor might be appropriate. It must however be appreciated that, when choosing a value for the partial safety factor for concrete, as the strength of the concrete is reduced by conversion there appears to be no reduction in the standard deviation of the variation in concrete strength. Thus when the concrete has a characteristic strength of 52 N/mm^2, the margin between the mean strength which might be 62 N/mm^2 and the strength used for ultimate design of 35 N/mm^2 corresponds to about 4σ (where σ is the standard deviation). This is approximately equivalent to accepting a level for failure of 1 in about 20 000. If however the strength had fallen to an average of 32 N/mm^2 with the same standard deviation the characteristic strength would be

22 N/mm^2; using this with a partial safety factor of 1.5 as before would give a design strength for ultimate strength calculations of 15 N/mm^2, which corresponds to less than 3σ or a failure level of about 2 in 1000. Thus the safety margin is appreciably less even though the value of the partial safety factor remains unchanged. It is however relevant to take account of engineering experience and judgement in making the choice of partial safety factor.

The main results of the BRE investigations are summarised in the following paragraphs in the light of the foregoing comments:

(i) Measurements of the degrees of conversion of the concrete in existing buildings indicate that most concrete has reached a high level of conversion in a few years. The results of crushing tests on highly converted concrete specimens cut from beams also indicate that strength is very variable and that some concrete suffers substantial losses of strength with respect to the strength at one day on which the design was usually based. In the appraisal of building therefore it is advisable to assume that all concrete has reached or will reach, during its expected life, a high level of conversion. Nevertheless, in the majority of construction the loss of strength is not sufficient to endanger the structure.

(ii) The long-term laboratory studies have shown that:

(a) if concrete with a free water/cement ratio of less than 0.4 is stored in water* at 18°C throughout its initial curing period and its subsequent life, a minimum strength will be reached after about 5 years and this minimum will not be appreciably less than the strength at one day; tests on a recent batch of cement shows, however, that the loss of strength at 18°C may take place more rapidly than this suggests;

(b) if concrete is stored in water at 38°C, after one day at 18°C, converts rapidly to a high level and reaches a minimum strength in about three months which is very substantially less than the strength at one day;

(c) if concrete is stored in water at 18°C for a long period (up to 8$\frac{1}{2}$ years) and is then immersed in water at 38°C it will rapidly convert and lose strength to the minimum level reached for continuous storage at 38°C;

(d) if concrete is subjected to temperature in excess of 25°C at any time during curing or subsequently, which in its early life need only be for a few hours, conversion becomes more rapid and a serious loss of strength occurs which may take some months to develop. Storage at 30°C appears to cause a loss of strength which is only slightly less than that at 38°C.

Since the temperature of 38°C represents the upper limit of what is likely to be reached during curing of these sections or in normally heated buildings, and the precise level is not critical, it is recommended that design appraisals should be based on the minimum strength at this temperature.

(iii) Appraisal of the results of crushing tests on specimens from beams of standardised sections which were likely to have been subjected to good control in manufacture and to have been required to conform with the one-day strength requirements of CP 116 has suggested that a characteristic strength for the concrete of 20 N/mm^2 might be adopted in conjunction with an appropriate partial safety-factor as discussed earlier. Tests on X5, X7 and X9 beams suggest that this strength might be increased by a factor of 1.3, since there was evidence that the method of calculation given in CP 110 tended to underestimate beam strength. The suggestion that the specified one-day strength of the concrete might drop from 52 N/mm^2 to about 20 N/mm^2 when a high level of conversion was reached is consistent with the findings of the long-term laboratory study. Here the characteristic strengths of concrete at one day for practical water/cement ratios might be expected to drop in value by about 30 N/mm^2.

Results obtained by Cusens and Jackson[29] in tests on highly converted prestressed concrete beams suggest that for more massive beams the equivalent cube strengths derived from cores do not underestimate the strength of the concrete, using the method of calculation in CP 110.

When sampling of concrete for strength tests is necessary, and the concrete has been shown to be highly converted, ie 80% converted, the specimens should be tested as sampled. Otherwise they should be tested after heating in water at 50°C for one month to obtain the fully converted strength or alternatively, if there is time, at 38°C

---

* The experimental evidence does not suggest that storage in air at 65% relative humidity would lead to appreciably different results.

in water for three months since the latter is a more realistic and apparently a slightly less onerous condition.

(iv) The results of tests to failure on floors indicate that where they consist of beams with hollow blocks there is a slight enhancement in strength over that of the beams alone and that the blocks are capable of distributing loads to adjacent beams; where the floors have beams and blocks with a floating screed separated by a quilt, the effect is greater. The greatest enhancement of strength and capability for distributing loads to adjacent beams is obtained where the screed is bonded to the beams and blocks, and composite action develops. The results of these tests taken together with those for proof loading tests made in existing buildings indicate that the risk of collapse is very small and may be no greater than is experienced with other forms of construction. This conclusion finds some confirmation from the fact that of about 17 x 10^6 m^2 of flooring, that is reported to have been built with prestressed concrete units of high alumina cement, no collapses have resulted from loss of strength in the concrete due to conversion

(v) The results of tests on roofs particularly those with isolated beams show that the position is less satisfactory and that the risk of collapse although small may not be acceptable. Remedial measures have been necessary for a small number of roof structures following the development of excessive deflection. The few collapses of roofs that have occurred have not all however been due solely to loss of strength due to conversion, but have, in several cases, been the result of a combination of conversion with defects such as chemical attack or inadequate detailing of reinforcement at joints or poor control of quality.

(vi) No specific experimental studies have been made of the performance of high alumina cement concrete in:

(a) precast prestressed composite members;

(b) precast concrete structural members;

(c) in-situ construction.

The data obtained in the investigation however suggest that once a basic strength has been established for appraisal, whether it is the result of accepting the validity of the control procedure and allowing for the effects of a high level of conversion, or whether it is as a result of tests on samples of concrete, this strength should be used in association with corresponding information for design in the Code of Practice, CP 110. Evidence on values for the modulus of elasticity, bond and tensile strength suggest that the properties of the concrete after becoming highly converted are not inconsistent with those of Portland cements of the same strength.

(vii) Highly converted high alumina cement concrete is vulnerable to chemical attack in the presence of long-term wetness and a chemically aggressive agent, which may be a more serious risk for concretes with greater water/cement ratios.

Long-term studies are currently in hand to determine the performance of high alumina cement in sulphate bearing soils in relation to the construction of foundations.

(viii) No adverse effects of corrosion have been detected on steel tendons removed from prestressed concrete beams in normal buildings. Neither the cement nor the prestressing steel used in this country have the characteristics which led to the failure of prestressed concrete beams with high alumina cement in Germany in 1961.

ACKNOWLEDGEMENTS

The investigations described in this report formed part of the research programme of the Building Research Establishment and the report is published by permission of the Director.

The initiation, direction and organisation of these extensive investigations was undertaken by Dr R F Stevens, Head of the Structural Properties Division, assisted by Mr A J Lockwood. Other members of staff who made major contributions to the work included Dr A J Chabowski, Mr F J Grimer, Mr J A Fielding, Mr K E Haine, Mr C R Lee, Dr H G Midgley, Mr D Redfearn, Dr M H Roberts, Mr E R Skoyles, Mr D C Teychenné and Mr G A Weeks. Mr J F Bowden was seconded from the Property Services Agency and prepared the design study.

The following organisations assisted in the laboratory research:

Princes Risborough Laboratory of BRE, Transport and Road Research Laboratory and the Polytechnic of Central London.

REFERENCES

1 Collapse of roof beams:
The Sir John Cass Foundation and Red Coat Church of England Secondary School - Stepney:
Department of the Environment, 28 February 1974.

2 Collapse of roof beams:
The Sir John Cass Foundation and Red Coat Church of England Secondary School - Stepney:
Department of the Environment, 30 May 1974.

3 Report on the failure of roof beams at Sir John Cass Foundation and Red Coat Church of England Secondary School, Stepney:
S C C Bate:
Building Research Establishment Current Paper, CP 58/74, June 1974.

4 High alumina cement concrete in buildings:
Department of the Environment, 20 July 1974.

5 Influence of temperature upon the strength development of concretes:
N Davey:
Building Research Technical Paper No 14, HMSO, 1933.

6 The durability of reinforced concrete in sea water:
F M Lea and C M Watkins:
Building Research Station National Building Studies, Research Paper No 30, HMSO, 1960.

7 BS Code of Practice, CP 114:1948, The structural use of reinforced concrete in buildings:
British Standards Institution, London, 1948.

8 BS 915:1940, High alumina cement:
British Standards Institution, London, 1940.

9 Explanatory Handbook on the BS Code of Practice for Reinforced Concrete:
W L Scott, W H Glanville and F G Thomas:
Concrete Publications Limited, London, 1950.

10 Code of Practice, CP 114.100 - CP 114.105:1950, Suspended concrete floors and roofs:
British Standards Institution, London, 1950.

11 High alumina cement:
Building Research Station, Digest 27, HMSO, 1951.

12 BS Code of Practice, CP 115:1959. The structural use of prestressed concrete in buildings:
British Standards Institution, London, 1959.

13 The design of concrete mixes with high alumina cement:
K Newman:
Reinforced Concrete Review, 1960, V (5) pp 269-94.

14 A study of deterioration of structural concrete made with high alumina concretes:
A M Neville:
Proc ICE, 1963, 25 (July), pp 287-324 and 1964, 28 (May), pp 78-84.

15 Report on the use of high alumina cement in structural engineering:
Institution of Structural Engineers, London, 1964.

16 BS Code of Practice, CP 116:1965, The structural use of precast concrete:
British Standards Institution, London, 1965.

17 BS Code of Practice, CP 110:1972, The structural use of concretes:
British Standards Institution, London, 1972.

18 First report on prestressed concrete:
Institution of Structural Engineers, London, 1951.

19 British Standard Code of Practice, CP 3, Chapter V, Part I: 1967:
British Standards Institution, London, 1967.

20 Report on the collapse of the roof of the assembly hall of the Camden School for Girls:
HMSO, London, 1973.

21 Floor loadings in office buildings:
G R Mitchell and R W Woodgate:
Building Research Station Current Paper, CP 3/71, January 1971.

22 Floor loading in retail premises:
G R Mitchell and R W Woodgate:
Building Research Station Current Paper, CP 25/71, September 1971.

23 High concrete strength within the hour:
P J French, R G J Montgomery and T D Robson:
Concrete, 1971, 5 (8), pp 253-8.

24 Handbook on the Unified Code for structural concrete (CP 110:1972):
Cement and Concrete Association, 1972.

25 Further investigations on the creep or flow of concrete under load:
W H Glanville and F G Thomas:
Building Research Technical Paper No 21, London, HMSO, 1939.

26 BS 1881: 1970, Methods of testing concrete:
British Standards Institution, London, 1970.

27 Guidelines for the appraisal of structural components in high alumina cement concrete:
Institution of Structural Engineers, London, October 1974.

28 The transfer of point-load effects on a typical pretressed, precast flooring panel:
B Mayfield, J E Cairns and R C Miller:
Civil Engineering and P W Review, 1969, 64 (754), pp 445-50.

29 Flexural behaviour of highly converted aluminous prestressed concrete beams:
A R Cusens and N Jackson:
Concrete, 1975, 9 (5), pp 30-2.

Long-term research into the characteristics of high alumina cement concrete (CP 71/75)

D.C. Teychenné

SYNOPSIS

The paper describes a research programme started in 1964 to study the characteristics of high alumina cement (HAC) concrete up to an age of 20 years. Experimental details are given of the main and two subsidiary programmes together with test results up to 8·5 years. Four test series were undertaken to establish: (1) the strength development of a wide range of concrete mixes stored continuously in water at 18°C and at 38°C for long periods; (2) the effect of changes in curing regime during the first 24 h upon the subsequent strengths up to 1 year; (3) the effectiveness of an immediate conversion curing regime to give minimum strengths at 5 days instead of 3 months; (4) the modulus of elasticity of HAC concrete, both unconverted and converted. This research includes the quantitative measurement of the degree of conversion, factors influencing it and its relation to the reduction in compressive strength. The results show that the strength after long periods of storage in water at 18°C is extremely sensitive to the water/cement ratio, and that the temperature conditions during the first 24 h are highly critical for strength at later ages. The modulus of elasticity of high alumina cement, converted or unconverted, is related to the compressive strength of the concrete.

Introduction

High alumina cement has been manufactured in the United Kingdom by the Lafarge Aluminous Cement Co. Ltd for about 50 years. The cement was developed in France from the earlier work of Bied[1] seeking a cement more resistant to sulphate attack than ordinary Portland cement. A great deal of research has been carried out into the characteristics of this type of cement and this is described in books by Robson[2] and Lea[3]. The first edition of the British Standard Specification, BS 915, *High alumina cement,* was published in 1940, and cement complying with this Standard was allowed for structural concrete in the first edition of CP 114 published in 1948[4].

The fact that high alumina cement could undergo chemical changes, now referred to as 'conversion', sometimes resulting in a decrease in compressive strength, had been known for many years, but in the early 1960s a number of papers was published resulting in a revival of interest in this phenomenon. K. Newman[5] published a paper dealing with the design of concrete mixes with high alumina cement along the same lines as the method described in Road Note No. 4[6]. This work involved using mixes with a range of cement: aggregate ratios from 1:3 to 1:7·5 and a range of water/cement ratios from 0·30 to 0·70, but unfortunately the specimens stored at 18°C were tested only at ages up to 7 days. Other specimens were stored in water at 38°C and tested at ages up to 2 months to obtain the minimum strength on conversion. In the same year, Lea and Watkins[7] published the results of long-term tests (23 years) on the effect of sea water upon experimental piles. The programme included tests on specimens made with high alumina cement and generally this type of cement gave the best over-all performance with the least cracking of the concrete and corrosion of the steel. In 1961 Masterman[8] published a paper stressing the importance of using low water/cement ratios and of restricting the temperature rise of high alumina cement (HAC) concrete during the first 24 h. A world-wide review of research on the strength reduction and the deterioration of structural concrete made with high alumina cement was published by Neville[9] in 1963.

During this period, 1960–65, a panel of the Institution of Structural Engineers was drafting the materials clauses for a new Code of Practice for precast concrete, which was published as CP 116[10] in 1965. To guide the panel, the Institution appointed a committee to examine the use of high alumina cement in structural concrete and its findings[11] were incorporated in the requirements of CP 116. From the papers published at this time, it was apparent that more information was required on the long-term behaviour of HAC concrete kept at normal temperatures of 18 to 20°C, and on the effect of temperature during the early life of the concrete. In 1963 a programme of research was proposed by the Building Research Station and accepted by Lafarge Aluminous Cement Co. Ltd, who gave advice and some financial support to the work. This programme started in 1964 and had three main objectives: (1) to extend work on the properties of HAC concrete by using different sand gradings and low water/cement ratios; (2) to study the behaviour of concrete stored for up to 20 years in water at 18°C and at 38°C; and (3) to examine the effect of different initial storage temperatures upon the strength of HAC concrete. This paper describes this work and the results up to an age of about 8·5 years together with the results obtained from two smaller, subsidiary series of tests on HAC concrete.

The conversion of high alumina cement

The mineralogy of high alumina cement is an extremely complex subject and beyond the scope of this paper. It is discussed in the books of Lea[3] and Robson[2] and more detailed research is described by Midgley[12], who contributes a new paper[13] to this present issue of *MCR*. In previous publications, high alumina cement has been described as being 'converted', 'fully converted' or 'partially converted', but the amount of conversion, or 'degree of conversion' has not been expressed quantitatively. In this paper, the degree of conversion is quantified and expressed as a percentage of the fully converted condition. It is convenient to use the cement chemists' shorthand notation as follows: C = CaO, A = Al_2O_3, S = SiO_2, F = Fe_2O_3 and H = H_2O. Thus C_3AH_6 is equivalent to $3CaO.Al_2O_3.6H_2O$.

The reaction between water and the anhydrous constituents of high alumina cement results in the formation of various aluminate hydrates depending upon the precise chemical composition of the cement and the hydration conditions. At ordinary temperatures (below 25°C), the most important reaction is the formation of the monocalcium aluminate hydrate CAH_{10} resulting from the reaction:

$$CA + 10H \longrightarrow CAH_{10} \quad \ldots\ldots\ldots\ldots (1)$$

Alumina gel, A_xH, is also formed, together with a small quantity of C_2AH_8. The hydrates CAH_{10} and C_2AH_8 are known as the hexagonal hydrates which are metastable at normal temperatures and convert to the stable cubic aluminate C_3AH_6 and either alumina gel or gibbsite, AH_3. The primary conversion reaction is indicated by the equation:

$$3CAH_{10} \longrightarrow C_3AH_6 + 2AH_3 + 18H \quad \ldots\ldots (2)$$

It should be noted that this reaction liberates all the water needed for the conversion process to continue. The specific gravities of these various compounds are CAH_{10} 1·72, C_3AH_6 2·53 and AH_3 2·44, and the conversion reaction must therefore result in a reduction in the volume of the solids and an increase in the porosity, since the over-all dimensions of specimens of cement paste or concrete remain sensibly constant.

Midgley[12] has shown how the presence of these various hydrates can be established by means of differential thermal analysis (DTA), since they produce peaks on the thermograms at different temperatures. The degree of conversion expressed as a percentage of full conversion is calculated from the expression:

$$\frac{\text{weight of } C_3AH_6 \times 100}{\text{weight of } C_3AH_6 + \text{weight of } CAH_{10}} \quad \ldots\ldots (3)$$

the relative weights of C_3AH_6 and CAH_{10} being derived from the measurement of the endothermic peaks obtained on the DTA thermogram. As, in this work, the specimens were stored in a CO_2-free system, measurement of the C_3AH_6 peak is satisfactory. Where carbonation can take place, the C_3AH_6 may become decomposed and the AH_3 peak is measured instead of C_3AH_6.

Description of the test programmes

The original test programme agreed in 1963 consisted of two series of tests. Series A1 dealt with the effect of mix parameters, particularly the water/cement ratio, upon the strength of concrete stored in water at 18°C and at 38°C, after initial curing for 24 h in moist air at 18°C, at ages up to 20 years. Series A2 dealt with the effect of storage at high temperatures during the first 24 h upon the strength of concrete stored subsequently in water at 18°C.

Later, two subsidiary test series were carried out. Series B compared the minimum strengths obtained from delayed immersion in water at 38°C with those obtained at 5 days from an immediate immersion regime used by Lafarge. Series C, using the same curing regimes, examined the static and dynamic modulus of elasticity of HAC concrete in the unconverted and converted condition.

Table 1 gives a summary of the materials used, the mix proportions, the curing regimes and the age at test in the four test series.

Experimental details and procedures

MATERIALS USED

Cements

For the main test series, A1 and A2, a single batch of cement was supplied in bags by Lafarge Aluminous Cement Co. Ltd. A number of different batches of

TABLE 1: **Summary of the experimental details of the four test series.**

Test series	Cement batch No.	Fine aggregate grading	Cement: aggregate ratio	Workability	Total water/ cement ratio	Curing regimes*	Testing ages
A1: Long-term	685	40% Zone 2 or 27% Zone 4 sand	1:3·0 1:4·5 1:6·0 1:7·5	Low, Medium, High Low, Medium, High Low, Medium, High Low, Medium, High		X, Y	1, 7 and 28 days 3 months 1, 5, 10 and 20 years
A2: Effect of first 24 h	685	40% Zone 2 sand	1:3·0 1:4·5 1:6·0 1:7·5	Low, Medium, High Medium Low, Medium, High Medium		A to K	1, 7 and 28 days 3 months 1 year
B: Accelerated conversion	685, 716, 759, 765, 794, 796	40% Zone 2 sand	1:4·5		0·35 0·44 0·50	X, Y, Z	1, 4, 5, 7 and 28 days 3 months
C: Modulus of elasticity	759	40% Zone 2 sand	1·45		0·35 0·50	X, Y, Z	1, 4, 5, 7 and 28 days 3 months
			1·60		0·50 0·60		

* X: 24 h in air at 18°C, then in water at 18°C
Y: 24 h in air at 18°C, then in water at 38°C
Z: immediately into water at 38°C for 24 h
A to K: various periods in water at 18°C and at 25, 35 or 50°C during first 24 h, then in water at 18°C (details in Table 8)

cement was purchased in bags for series B; one of these was used in series C. In each case, the separate delivery of cement was blended in a mechanical blender and then stored in airtight steel drums until required for use. The cements were tested in accordance with the requirements of BS 915[14]. With some batches, tests were made on a standard concrete (1:2:4 by weight, total water/cement ratio of 0·60). The results are given in Table 2. This shows that all the cements comply with the requirements of BS 915. (Although the fineness of batches 685 and 796 does not comply in terms of the residue on the 90 μm test sieve, both batches meet the alternative specific surface requirement. The strength of converted concrete made with these cements is given in Table 10.)

Aggregates

The aggregates used were Thames Valley sand and gravel of irregular shape, the maximum size being 19 mm. To maintain close control of the over-all grading, the aggregates were dried and seived into the size fractions shown in Table 3. The gravel fractions were combined 2 parts by weight of fraction G1 to 1 part of fraction G2 for all test series. The sand fractions were recombined to give combined gradings within the limits of grading Zones 2 and 4 of BS 882[15].

In series A1, two combined gradings were used, a continuous grading with the Zone 2 sand and a somewhat gapped grading with the Zone 4 sand. The proportions of sand and gravel were selected so that the surface modulus of the two combined gradings was the same. Details of the separate fractions and of the combined gradings used in series A1 are given in Table 3, which also includes the water absorption values used to estimate the free water/cement ratios.

As shown in Table 1, only the combined grading with the Zone 2 sand was used in test series A2, B and C. Similar combinations of the four aggregate fractions were used in series B and C but, since these tests were made at a much later date with different aggregates, the combined Zone 2 grading used in series B and C may differ slightly from that used in series A1 and A2.

DETAILS OF MIXES

In series A1 and A2, four cement:aggregate proportions were used giving a range of cement contents from about 550 to about 250 kg/m³. In the long-term strength series (A1), 24 different mixes were used, but only eight different mixes were used in series A2. In order to compare the results obtained from rapid conversion in series B with data obtained by Lafarge, one

TABLE 2: **Details of test results on the cements used.**

Test	Cement Batch No.					
	685	716	759	765	794	796
A: PHYSICAL TESTS IN ACCORDANCE WITH BS 915						
1: Fineness						
Percentage by weight retained on 90 μm sieve	9·5	—	2·7	6·3	—	9·8
Specific surface (m^2/kg)	250	—	336	—	280	319
2: Setting times						
Initial setting time (h-min)	4–25	—	3–45	2–20	2–15	2–40
Final setting time (h–min)	5–05	—	5–00	3–10	3–00	4–25
3: Soundness						
Expansion by Le Chatelier test (mm)	nil	—	0·5	1·0	nil	nil
4: Compressive strength (N/mm^2) measured on vibrated mortar cubes						
Average at 24 h	70·6	74·8	66·6	58·6	53·0	58·8
Average at 3 days	75·2	79·3	72·0	64·1	55·3	64·4
Average at 7 days	77·6	84·0	77·8	70·3	—	69·4
Average at 28 days	89·2	89·8	84·3	82·7	—	79·3
B: COMPRESSIVE STRENGTH (N/mm^2) OF 1:2:4 CONCRETE OF W/C 0·60						
Stored in water at 18°C						
Average at 24 h	43·5	—	—	53·7	49·2	49·0
Average at 3 days	52·3	—	—	57·9	54·2	54·0
Average at 7 days	60·0	—	—	63·3	62·5	60·4
Average at 28 days	65·6	—	—	69·5	67·1	67·7
Average at 3 months	65·2	—	—	70·0	62·5	60·9
Stored in air at 18°C and 65% R.H.						
Average at 3 days	51·1	—	—	61·6	56·0	58·1
Average at 7 days	59·4	—	—	65·2	62·0	61·8
Average at 28 days	64·5	—	—	67·9	65·0	63·1
Average at 3 months	69·7	—	—	65·9	58·0	60·3
C: CHEMICAL ANALYSIS (% BY WEIGHT)						
SiO_2	4·30	—	3·89	4·20	4·00	4·11
Fe_2O_3	10·98	—	12·56	13·07	11·99	17·17
FeO	5·28	—	3·26	2·77	3·75	—
Al_2O_3	38·28	—	38·76	37·83	39·47	39·01
TiO_2	1·90	—	2·02	2·55	1·50	1·12
P_2O_5	0·11	—	0·12	0·11	0·12	0·15
CaO	37·84	—	37·95	37·85	38·31	37·97
MgO	0·41	—	0·54	0·50	0·45	0·31
MN_2O_3	0·09	—	0·04	0·03	0·04	0·04
NA_2O	0·09	—	0·07	0·06	0·07	0·06
K_2O	0·16	—	0·12	0·14	0·08	0·07
SO_3	—	—	0·14	0·03	0·01	0·01
	99·62	—	99·47	99·17	99·79	100·02

mix with three different total water/cement ratios was used. In series C, four mixes were used, two cement: aggregate ratios at each of two total water/cement ratios, one of which was common to both mixes.

CURING REGIMES AND TESTING AGES

Table 1 shows that a number of different curing regimes and testing ages were used in the various test series. In series A1, two curing regimes were used in which the cubes were stored in the moulds for 24 h at 18°C covered by damp hessian and polythene. After demoulding, some cubes were tested at 24 h; the rest were stored in water, half at 18°C (regime X, normal curing) and half at 38°C (regime Y). These two regimes were used in series B and C and a third (regime Z) in which the test specimens were placed immediately in water at 38°C in their moulds, the top surface being covered by a steel plate.

TABLE 3: **Details of aggregates used in test series A1 and A2.**

	Individual fractions used				Calculated gravel and sand grades			Combined aggregate gradings	
	Gravel fractions		Sand fractions		Gravel	Zone 2 sand	Zone 4 sand	Aggregate with Zone 2 sand	Aggregate with Zone 4 sand
	G1	G2	S1	S2					
	19·00 to 9·50 mm	9·50 to 4·75 mm	4·75 to 0·60 mm	0·60 mm down	67% G1 33% G2	50% S1 50% S2	10% S1 90% S2	60% gravel 40% sand	73% gravel 27% sand
B.S. sieve size	Percentage (by weight) passing B.S. sieve size								
19·00 mm	100·0	—	—	—	100·0	—	—	100·0	100·0
9·50 mm	0	100·0	—	—	33·3	—	—	60·0	51·5
4·75 mm		0	100·0	—	0	100·0	100·0	40·0	27·0
2·36 mm			78·0	—		89·0	98·0	35·5	26·5
1·18 mm			43·0	—		71·5	94·5	28·5	25·5
0·60 mm			8·0	100·0		54·0	91·0	21·5	24·5
0·30 mm			1·0	44·0		22·5	39·5	9·0	10·5
0·15 mm			0	4·0		2·0	3·5	1·0	1·0
Water absorption (%)					1·90	1·35	0·75	1·70	1·60

In series A2, ten different curing regimes were used in order to simulate practical conditions of reaching various maximum temperatures at different times during the first 24 h after casting. The cubes were stored in water for 24 h immediately after casting, the top surface of each cube being covered by a steel plate.

Series A1 involved tests up to an age of 20 years, whereas in series A2 test ages did not go beyond 1 year. In series A1, there is a gap between tests at 5 years and at 10 years. However, following the collapse of the roof of the assembly hall at the Camden School for Girls[16], it was agreed to test some of the specimens due for test at 10 years at an earlier age. Since the cubes were cast over a period of about 1 year, there were differences in their age at test, the average being about 8·5 years. A few more cubes were transferred from water at 18°C to water at 38°C for 3 months to examine the effect of a higher temperature at this later age. The results of tests on these cubes are given in Tables 5 and 6. In series B and C, immediate immersion at 38°C (regime Z) results in minimum strengths being obtained in a few days and so the cubes were tested at very early ages, the greatest age at test being 3 months.

EXPERIMENTAL PROCEDURES

The cements and aggregates were prepared as described on page 80; the quantities of the cement and the various fractions of aggregate were weighed to an accuracy of 1 part in 1000. The concrete was mixed in a rotating pan and paddle mixer of 37 l. or 85 l. capacity as appropriate, the water being added and measured to obtain the required workability or water/cement ratio. The concrete was mixed for 2 min, allowed to stand for 8 min to permit some of the absorption of the water by the aggregates to take place, and given a final mix for half a minute. When required, the workability was measured by the slump, Vebe and compacting factor tests in accordance with the requirements of BS 1881[17].

Various test specimens were made according to the requirements of each test series as follows.

Series A1. Thirty-two 101·6 mm cubes were cast from each batch; two cubes were tested at each age and curing condition (regimes X and Y). The 24 mixes were replicated and cast in random order so that the results given in Table 6 are the means of four cubes.

Series A2. Thirty-two 101·6 mm cubes were cast from each batch; these were divided for storage under two of the ten curing regimes A to K and the results given in Table 9 are the means of three cubes.

Series B. Thirty-three 100 mm cubes were cast from each batch; these were divided for storage under the three curing regimes X, Y and Z. The results given in Table 10 are the means of three cubes.

Series C. Thirty-three 100 mm cubes, nine 500 × 100 × 100 mm beams and nine 300 × 150 mm dia. cylinders were cast from two batches, and divided for storage under the three curing regimes X, Y and Z. The compressive strength and dynamic modulus results are the means from three test specimens but the static modulus results are the means of two test specimens.

The making and testing of the specimens to determine the compressive strength, and the dynamic and static modulus of elasticity were in accordance with the relevant parts of BS 1881[17,18].

In addition to these tests, thermocouples were embedded in two cubes from each mix cast in series A1 and A2 and connected to a multi-range electronic

TABLE 4: **Temperatures and times taken from continuous chart records for test series A1.**

Mix proportions by weight of cement:aggregate	Total water/cement ratio by weight		Stage indicated by ringed Nos. on Figure 1											
			1		2		3				4		5	
			Time from casting to start of reading (0) (min)		Time from 0 to start of rise of temperature (h)		Time from 0 to maximum temperature rise (h)		Maximum temperature attained (°C)		Time to return to normal temperature of 18°C (h)		Time for temperature to rise from 18°C to 38°C (min)	
	A*	B†	A*	B†	A	B	A	B	A	B	A	B	A	B
1:3	0·32	0·30	45	45	2·00	1·25	3·25	3·25	26·1	26·1	20·00	21·25	30	30
			55	40	2·00	1·75	4·00	3·50	26·1	26·1	21·25	18·50	30	30
	0·34	0·33	60	35	1·75	1·50	4·00	3·75	26·1	26·7	19·50	21·25	30	25
			30	30	1·75	1·75	4·00	4·00	26·1	25·6	18·75	20·75	30	30
	0·40	0·35	45	20	1·50	1·75	3·75	4·00	27·2	25·6	21·25	20·75	25	30
			35	35	2·00	1·75	4·50	4·00	26·1	26·1	16·50	20·00	25	30
1:4·5	0·35	0·35	50	35	1·25	1·75	3·00	3·75	23·9	23·9	18·00	15·00	15	25
			45	30	1·50	2·00	3·25	4·25	25·0	24·4	18·75	21·25	25	30
	0·44	0·42	50	50	2·00	1·75	4·50	4·25	24·4	23·9	20·25	13·25	30	30
			30	25	2·00	1·75	3·75	3·75	23·9	23·9	21·00	16·25	30	30
	0·50	0·48	45	35	2·00	1·75	5·50	4·25	22·8	25·6	14·75	19·25	30	30
			35	35	2·25	1·75	4·50	4·75	24·4	24·4	17·00	19·75	25	30
1:6	0·50	0·45	30	50	2·25	1·75	5·00	4·25	23·9	23·9	16·25	20·50	25	25
			45	45	1·75	2·00	3·75	4·25	23·9	23·9	21·75	21·00	30	30
	0·56	0·52	60	45	1·25	2·00	5·00	4·25	23·9	23·3	23·75	18·25	30	30
			35	45	2·50	2·00	5·25	4·75	23·3	22·2	18·00	21·75	30	25
	0·60	0·56	45	35	1·75	1·75	5·00	4·75	23·3	25·0	18·25	23·25	25	25
			40	30	2·25	2·25	5·00	4·75	21·1	27·1	15·50	21·75	30	30
1:7·5	0·65	0·59	35	35	2·00	2·25	4·50	4·75	21·7	23·3	14·50	19·00	30	20
			35	80	2·25	1·75	4·80	4·25	22·8	21·7	19·25	14·75	25	30
	0·73	0·68	30	60	2·50	1·75	5·50	5·25	23·3	22·8	21·75	14·50	30	25
			35	55	2·25	2·75	5·50	5·00	22·2	22·2	17·75	17·50	30	30
	0·80	0·73	50	35	2·50	2·25	6·75	5·00	25·0	22·2	23·00	19·00	30	30
			55	30	2·50	2·25	4·75	4·75	22·8	21·7	22·25	17·25	30	30
Mean			42·50	41·25	1·97	1·92	4·51	4·27	24·1	24·1	19·12	19·00	28	28
Grand mean and 95% confidence limits			42 ± 11		1·94 ± 0·68		4·39 ± 1·43		24·1 ± 3·0		19·00 ± 5·20		28 ± 3	

*A = Concrete mixes with Zone 2 sand
†B = Concrete mixes with Zone 4 sand

recorder to obtain a continuous record of the temperature rise during the first 24 h after casting. Data obtained from these recordings are given in Table 4 and shown in Figures 1 and 7.

Differential thermal analyses were made on samples from some of the crushed cubes at some ages and the degree of conversion was calculated as shown in equation 3. These results are given in Table 5.

Results and discussion

MAIN STRENGTH TEST SERIES A1

Workability

No attempt was made to fix the water/cement ratio to produce concretes having workabilities specified in terms of any of the three recognized test methods; the water content for the workability required was judged by visual assessment. The total water/cement ratios used are given in Table 6 and the free water/cement ratios are estimated by allowing for the absorption values of the aggregates given in Table 3. Unless otherwise stated, the water/cement ratios in this paper refer to these estimated values of free water/cement ratio.

The three levels of workability aimed for resulted in the following average measured characteristics:

Slump (mm)	4	23	93
Vebe time (s)	19	10	7
Compacting factor	0·78	0·90	0·93

The average water contents to achieve these levels of workability were about 125, 145 and 160 kg/m^2 respec-

TABLE 5: **Degree of conversion at different ages for test series A1.**

Mix proportions by weight of cement: aggregate	Details of sand used: Grading Zone	Details of sand used: Percentage of sand in total aggregate	Water/cement ratio by weight: Total	Water/cement ratio by weight: Estimated free	Degree of conversion (%) — Stored for 24 h in moist air at 18°C, then in water at 18°C: 28 days	3 months	1 year	5 years	8½ years	Stored for 24 h in moist air at 18°C, then in water at 38°C: 28 days	3 months	1 year	5 years	8½ years
1:3	2	40	0·32	0·27	20	20	25	25	40	50	80	85	85	90
			0·34	0·29	20	20	25	30	—	55	85	80	85	—
			0·40	0·35	20	20	30	30	—	65	85	80	85	—
	4	27	0·30	0·25	20	25	25	35	—	70	85	—	90	—
			0·33	0·28	20	25	—	25	—	70	80	—	80	—
			0·35	0·30	20	25	25	25	40	60	85	—	85	90
1:4·5	2	40	0·35	0·27	15	15	—	—	—	50	80	—	—	—
			0·44	0·36	20	20	30	30	—	50	85	80	85	—
			0·50	0·42	20	20	25	40	—	55	85	80	80	—
	4	27	0·35	0·28	20	20	25	25	35	40	85	80	85	90
			0·42	0·35	20	20	20	30	40	50	80	80	80	90
			0·48	0·41	20	20	25	30	—	60	80	80	85	—
1:6	2	40	0·50	0·40	15	15	25	30	—	50	80	90	85	—
			0·56	0·46	15	20	25	40	—	55	85	75	80	—
			0·60	0·50	20	20	25	40	—	65	85	80	85	—
	4	27	0·45	0·35	20	20	25	50	60	45	85	85	85	95
			0·52	0·42	20	20	25	40	—	55	80	60	—	90
			0·56	0·46	20	20	30	40	50	60	80	85	85	90
1:7·5	2	40	0·65	0·52	20	20	20	45	60	60	80	75	85	90
			0·73	0·60	20	25	25	40	70	70	80	85	85	90
			0·80	0·67	25	25	—	50	—	70	85	—	85	—
	4	27	0·59	0·47	15	20	35	35	55	55	80	85	80	85
			0·68	0·56	15	25	—	55	—	60	85	—	80	—
			0·73	0·61	25	25	25	50	—	65	80	85	—	—

tively. These are about 10 to 15 kg/m^3 less than those likely to be required for similar mixes made with ordinary Portland cement.

Table 6 shows that the mixes using the finer, Zone 4 sand generally require slightly lower water/cement ratios, although in some instances the workability, as measured by the various methods, is slightly reduced. As might be expected, there are greater reductions in water/cement ratios with the leaner mixes. Although the finer sands tend to reduce the harshness of HAC concrete mixes, this benefit is rather marginal and they could have an adverse effect if the sand content is not reduced as it was in this programme.

Temperature rise

The temperature rise in the middle of a 101·6 mm cube during the first 24 h storage in air at 18°C is represented diagrammatically in Figure 1, and the experimental results are given in Table 4. These show that the maximum temperature is reached in just under 4·5 h and that the cube returns to 18°C after about 19 h. After being demoulded at 24 h and placed in water at 38°C, the concrete takes only about 30 min to reach this temperature.

The average maximum temperature attained is 24°C, i.e. 6°C above the ambient laboratory temperature. As might be expected, the richer mixes reach a higher maximum temperature than the leaner mixes, about 26°C for the 1:3 mixes compared with about 22°C for the 1:7·5 mixes. Richer mixes also reach their maximum temperature earlier and take longer to cool down to the ambient temperature.

Degree of conversion

The degree of conversion calculated from equation 3 by using data obtained from DTA is given in Table 5 and its change with age is shown in Figure 6. The results given in Table 5 have been averaged into three groups according to the water/cement ratio, 0·27–0·40, 0·42–0·50 and 0·52–0·67, and these averages are plotted

TABLE 6: **Workability and compressive strength data for test series A1.**

Mix proportions by weight cement: aggregate	Details of sand used		Water/cement ratio by weight		Workability tests			Weight of fully compacted fresh concrete	Cement content	Compressive strength of 101·6 mm cubes (N/mm²)														
										Cured for 24 h in moist air (over 95% R.H.) at 18°C then in water														
										at 18°C (regime X)							at 38°C (regime Y)							at 18°C for about 8¾ years, then at 38°C for 3 months
	Grading zone	Percentage of sand in total aggregate	Total	Estimated free	Slump (mm)	Vebe time (s)	Compacting factor	(kg/m³)	(kg/m³)	24 h	7 days	28 days	3 mths	1 year	5 years	8½ years	7 days	28 days	3 mths	1 year	5 years	8½ years	9 years	
1:3	2	40	0·32	0·27	14	18·0	0·79	2400	556	62·0	75·7	83·6	88·2	89·8	83·0	87·3	80·0	57·1	40·7	44·2	47·8	49·8	—	—
			0·34	0·29	65	2·8	0·94	2370	546	57·2	74·9	80·6	87·0	78·2	81·0	—	77·5	53·4	33·9	34·7	39·6	—	—	—
			0·40	0·35	172	1·9	0·99	2355	535	52·5	69·3	76·5	84·8	85·2	77·5	—	70·5	41·0	26·8	29·7	32·5	—	—	—
	4	27	0·30	0·25	5	15·2	0·75	2395	557	67·9	83·6	88·5	87·9	91·5	81·7	—	81·8	63·6	46·1	49·5	52·8	—	52·5	46·3
			0·33	0·28	15	14·8	0·91	2415	558	63·0	74·5	82·5	86·5	87·2	79·7	—	77·0	52·1	41·1	43·5	45·6	—	—	—
			0·35	0·30	60	11·2	0·92	2395	550	58·0	72·2	72·8	87·2	85·3	84·9	87·0	74·0	49·8	35·2	37·3	42·2	44·5	—	—
1:4·5	2	40	0·35	0·27	0	24·0	0·68	2380	407	73·5	88·3	92·9	94·3	100·5	78·1	—	90·9	71·9	43·8	46·4	48·2	—	—	—
			0·44	0·36	30	5·5	0·92	2370	399	60·2	70·2	76·2	82·3	84·9	67·0	—	70·7	46·3	23·5	27·0	29·3	—	31·0	27·6
			0·50	0·42	120	2·2	0·96	2355	390	52·6	65·3	76·1	79·0	75·2	53·3	—	68·2	39·2	19·9	21·9	25·2	—	—	—
	4	27	0·35	0·28	2	17·0	0·76	2410	412	63·9	80·6	81·4	90·9	88·0	86·1	90·5	90·4	68·2	39·8	38·7	44·5	43·6	—	—
			0·42	0·35	18	8·8	0·85	2415	408	51·9	68·6	75·2	80·1	78·0	73·0	70·9	66·0	49·7	26·6	26·7	29·5	33·6	—	—
			0·48	0·41	110	5·0	0·92	2360	395	51·9	74·7	72·0	77·0	77·6	59·6	—	65·9	31·9	22·6	23·9	27·0	—	—	—
1:6	2	40	0·50	0·40	1	15·0	0·83	2365	315	52·1	62·6	73·9	75·3	76·1	63·1	—	65·8	39·2	22·7	24·5	28·1	—	—	—
			0·56	0·46	18	7·0	0·89	2360	312	48·0	57·4	65·5	67·0	63·9	40·9	—	59·3	23·2	17·6	19·8	19·4	—	25·3	23·0
			0·60	0·50	60	3·5	0·92	2350	309	44·8	54·5	60·4	62·4	58·2	46·0	—	53·4	21·9	15·0	15·4	18·4	—	20·2	17·2
	4	27	0·45	0·35	1	25·0	0·81	2395	321	60·4	70·5	82·2	81·0	81·8	75·6	76·8	73·7	55·6	27·2	30·0	33·7	36·0	—	—
			0·52	0·42	8	19·0	0·88	2370	315	53·0	64·9	74·0	75·5	72·4	54·6	—	62·7	36·6	21·0	21·0	25·0	—	—	—
			0·56	0·46	8	12·0	0·88	2370	313	47·0	57·7	68·8	71·1	64·0	45·2	40·2	54·0	31·7	18·5	19·6	23·3	23·8	—	—
1:7·5	2	40	0·65	0·52	4	10·0	0·84	2360	258	40·4	52·8	52·9	57·0	52·0	32·5	23·6	46·0	19·4	13·9	13·8	16·7	17·0	—	—
			0·73	0·60	25	3·5	0·88	2350	255	38·3	46·5	49·0	50·1	43·6	24·8	17·2	41·0	15·4	12·0	12·2	13·9	14·8	—	—
			0·80	0·67	90	2·5	0·93	2320	250	33·2	40·2	39·9	38·1	33·1	12·9	—	35·1	10·7	8·8	8·8	10·8	—	—	—
	4	27	0·59	0·47	8	25·8	0·77	2395	263	47·1	60·4	67·0	66·6	64·8	47·3	34·6	58·5	29·6	17·8	17·9	18·0	20·4	—	—
			0·68	0·56	8	23·0	0·89	2350	256	39·4	51·4	50·3	51·1	50·2	28·4	—	41·9	17·5	12·3	11·7	13·2	—	—	—
			0·73	0·61	125	18·5	0·90	2335	253	36·7	45·8	46·7	47·0	42·9	24·4	—	38·6	14·8	11·3	11·2	12·2	—	—	—

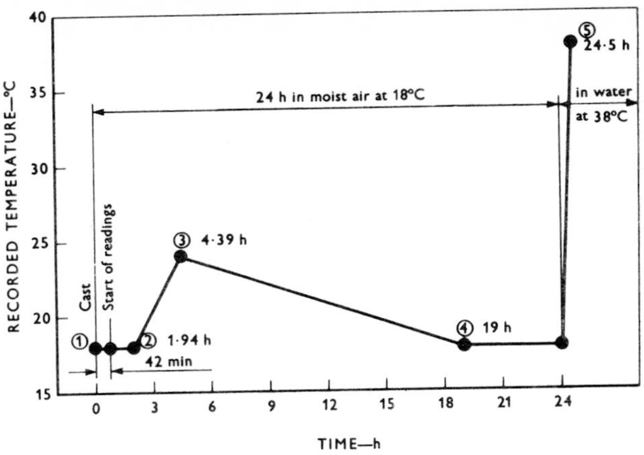

Figure 1: Series A1. Mean times and temperatures taken from continuous recording charts. The ringed numbers represent stages of development referred to in Table 4.

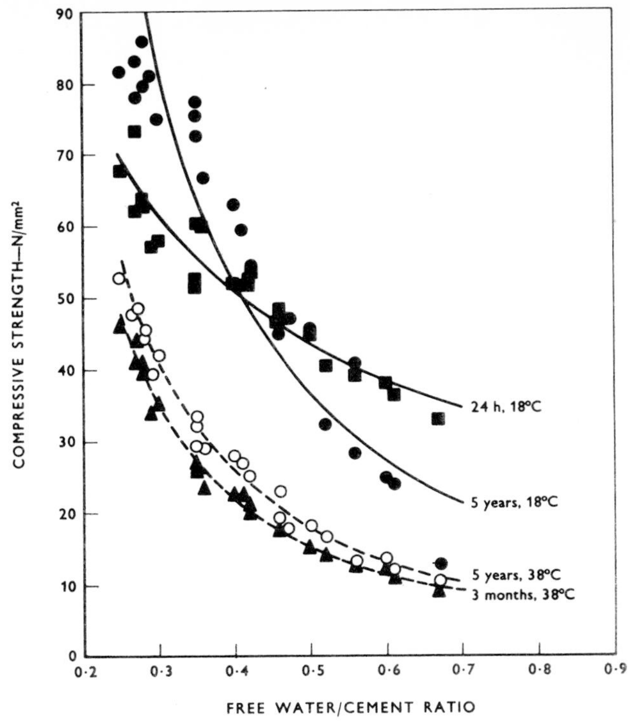

Figure 2: Series A1. Relationship between strength and water/cement ratio at different ages and storage temperatures.

in Figure 6. For the concrete stored in water at 38°C (regime Y), conversion proceeds rapidly, reaching a value of about 80% at 3 months and slowly increasing to about 90% at 8·5 years. Under this condition, the degree of conversion is not affected by the mix proportions although, as shown previously, richer mixes reach a higher temperature during initial curing. Figure 6 shows little difference in the development of conversion for the 24 mixes as grouped according to their water/cement ratio.

Concrete stored in water at 18°C (regime X) behaves differently. As expected, the concrete converts more slowly from about 20% at 28 days to about 25% at 1 year, the degree of conversion being independent of the mix proportions. However, at 5 years and upwards, leaner mixes (with higher water/cement ratios) convert more rapidly than richer mixes with lower water/cement ratios. At 5 years, the average conversion is 30%, 40% and 50% for water/cement ratios of 0·27–0·40, 0·42–0·50 and 0·52–0·67 respectively; at 8·5 years, it has increased to 45%, 50% and 65% respectively. Thus, for slow conversion, the degree of conversion at the greater ages depends upon the initial water/cement ratio.

Compressive strength development

Table 6 gives the mean compressive strengths for the 24 mixes at various ages up to 5 years, for the nine mixes tested at about 8·5 years and for the four mixes stored in water at 38°C instead of 18°C for the last 3 months before being tested at an age of about 9 years. These results show that the cement content of the mix has little effect upon the compressive strength of the concrete at any age. In some cases, leaner mixes are slightly stronger than richer mixes having the same water/cement ratio. The compressive strength at any age depends primarily upon the free water/cement ratio and the storage temperature.

From the experimental results given in Table 6, the best-fit relationships between compressive strength and water/cement ratio can be derived and some of these are shown in Figure 2. From such curves the average strength development for rounded values of the water/cement ratio have been obtained and these are given in Table 7 as percentages of the 24 h strength at different ages. The reduction in strength on conversion is sometimes expressed as a percentage of the 24 h strength, but this in turn depends upon the water/cement ratio, as can be seen from Table 7 and Figure 2. Figures 3 and 4 show the strength development curves for concrete having free water/cement ratios of 0·3, 0·4, 0·5 and 0·6 plotted from the derived data given in Table 7. For storage in water at 38°C, strength development follows a similar pattern, although the strength levels depend upon the water/cement ratio. There is an increase in strength above the 24 h value reaching a maximum at about 7 days, followed by a rapid reduction to a minimum value at 3 months of between 60 and 30% of the 24 h value, depending upon the water/cement ratio. This is followed by a very gradual increase in strength; at 8·5 years, the strength has increased, on average, by 17% of the 3 month minimum strength.

For storage in water at 18°C, strength development follows a different pattern and the water/cement ratio becomes an even more important factor. There is a longer period of increasing strength, the maximum value being reached between 3 months and 1 year, and

TABLE 7: **Strength development (expressed as a percentage of 24 h strength) of HAC concretes of different free water/cement ratios (calculated from data for test series A1).**

Free water/cement ratio by weight	Stored in water at 18°C							Stored in water at 38°C					
	24 h	7 days	28 days	3 months	1 year	5 years	8·5 years	7 days	28 days	3 months	1 year	5 years	8·5 years
0·25	100	123	136	146	150	—	—	131	109	68	73	79	73
0·30	100	123	135	144	144	133	137	127	90	57	61	66	68
0·35	100	124	135	142	139	116	121	124	77	49	52	57	62
0·40	100	124	134	140	135	102	104	121	67	43	45	51	55
0·45	100	124	133	138	131	92	85	118	59	38	40	45	49
0·50	100	124	133	137	128	83	71	116	53	35	36	41	39
0·55	100	124	133	136	126	76	57	114	48	32	33	38	39
0·60	100	124	133	135	123	70	44	112	44	29	30	34	35
0·65	100	125	133	134	121	66	—	111	41	27	28	32	—
0·70	100	125	132	133	119	61	—	109	37	26	26	30	—

followed by a slow reduction in strength, greater reductions being obtained at higher water/cement ratios. At the low free water/cement ratio of 0·30, a minimum strength is achieved at between 5 and 8·5 years (Figure 3) and the experimental results given in Table 6 show an increase in strength between 5 and 8·5 years for the mixes with very low water/cement ratios. It should also be noted that the minimum strength under these conditions is above the 24 h strength which is specified in Codes of Practice for structural design. At higher water/cement ratios, the strengths continue to decline for longer periods and, as can be seen from Figure 4, the 8·5 year results are below the 5 year results and an identifiable minimum has not yet been reached. Furthermore, the strength at these ages is well below the initial strength at 24 h and is approaching the

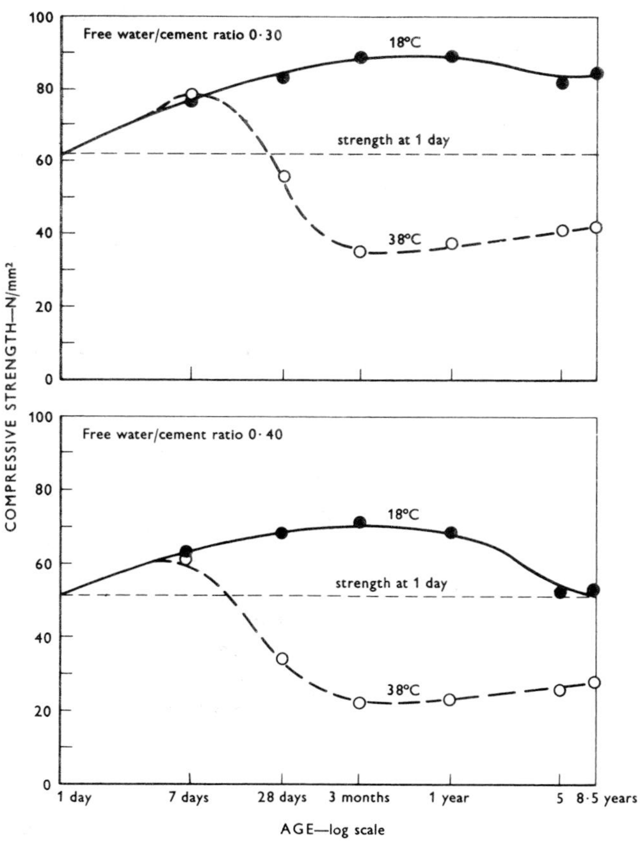

Figure 3: Series A1. Strength development at water/cement ratios of 0·30 and 0·40.

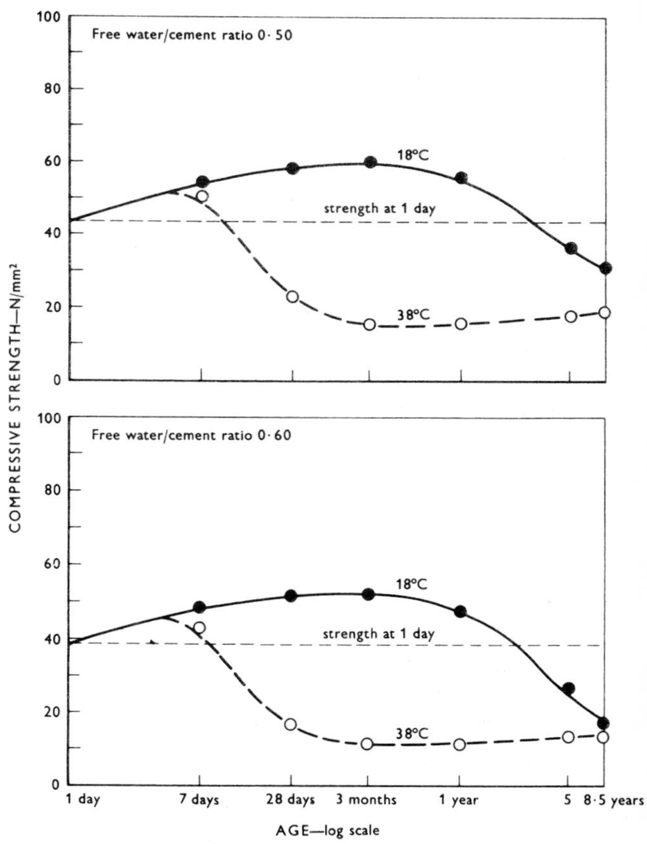

Figure 4: Series A1. Strength development at water/cement ratios of 0·50 and 0·60.

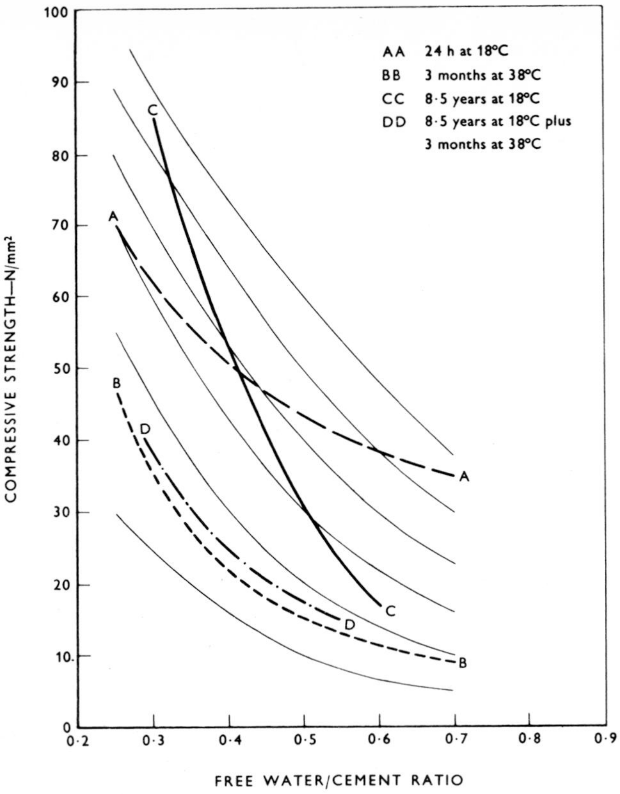

Figure 5: Series A1. Relationship between strength and water/cement ratio at different ages and storage conditions compared with relationships for ordinary Portland cement at different ages.

minimum strength obtained from storage in water at 38°C.

From Table 6 it can be seen that the effect of storing mature concrete for 3 months at 38°C after 8·5 years of continual storage at 18°C is to produce a reduction in strength. The strength is reduced to a value only slightly above the minimum strength obtained after 3 months initial storage at 38°C, and slightly below that of concrete stored for about 9 years at 38°C.

It is obvious that the water/cement ratio is an extremely important factor influencing the compressive strength and other characteristics of HAC concrete. Figure 5 shows four important relationships between the compressive strength and the free water/cement ratio obtained from this test series against a background grid of relationships for ordinary Portland cement concrete at different ages. Curve AA shows the relationship after 24 h storage as 18°C (see also Figure 2); this indicates that HAC concrete is less sensitive to changes in water/cement ratio than concrete made with ordinary Portland cement since the curve is less steep. Curve BB shows the minimum strength obtained at 3 months at 38°C for the 'fully converted' concrete, which corresponds to a degree of conversion of 80%. This shows a considerable reduction in strength, but a relationship similar to that obtained with ordinary Portland cement. The very low strength results could only be obtained with ordinary Portland cement concrete by testing it at very early ages. The strength requirements that were quoted in Section 12 of CP 110[19] before this Section was withdrawn, and the curves shown in Figure 12.1 of the handbook[20] to the Code, were based on these curves.

Curve CC shows the relationship after 8·5 years at 18°C; it is much steeper than the 24 h relationship and shows a greater sensitivity to water/cement ratio than ordinary Portland cement. An increase in water/cement ratio from 0·4 to 0·5 results in a reduction in strength of more than 20 N/mm². At water/cement ratios above about 0·40, the reduction in strength from the 24 h strength increases rapidly with increasing water/cement ratio. Curve DD shows the reduction in strength compared with curve CC, after concrete which has been stored in water at 18°C for 8·5 years has been subjected to 3 months' storage in water at 38°C.

Degree of conversion and strength

Since HAC concrete undergoes the process of conversion and as a result suffers a reduction in strength, it might be assumed that the degree of conversion and the strength were related. However, Table 5 shows similar levels of conversion at a given age (except for the greater ages at 18°C) over a wide range of mixes and water/cement ratios, whereas Table 6 shows a wide range of strengths at a given age. This range of strengths, of about 40 N/mm² at a given level of conversion, is due to the range of water/cement ratios used, and the range would be extended if other batches of cement were included.

Although there is no relationship between the absolute values of the degree of conversion and the strength, there appears to be a closer link between the degree of conversion and the compressive strength expressed as a percentage of the 24 h strength. It has been shown previously that the degree of conversion is independent of the mix proportions, except that at later ages the degree of conversion for storage in water at 18°C varies according to the water/cement ratio. Figure 6 shows the development of both conversion and strength (expressed as a percentage of the 24 h strength) with age for curing at 18°C (regime X) and at 38°C (regime Y). The conversion results given in Table 5 and the strengths expressed as percentages of the 24 h strength have been averaged for the three bands of free water/cement ratios shown in Figure 6. Since within these bands there is a considerable change in strength due to the water/cement ratio, Figure 6 shows the average percentage strength and the range of percentage strengths by the vertical bars at each age. The 8·5 year results relate to only some of the mixes (see Table 6) and thus the range shown in Figure 6 is reduced because the water/cement ratios do not cover the full range indicated. The degree of conversion is plotted on a descending scale and the similarity in behaviour of these two parameters with age is clearly

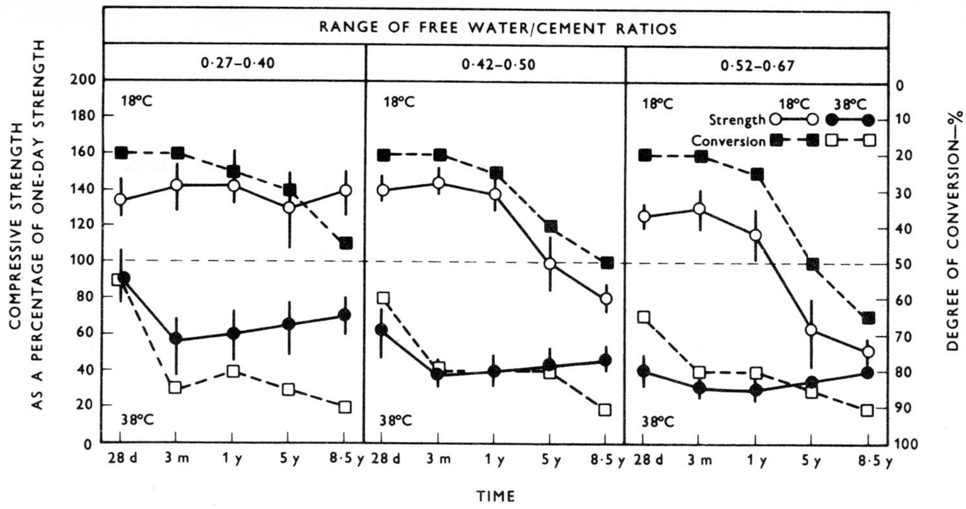

Figure 6: Development of conversion and strength expressed as a percentage of the 24 h strength from series A1.

seen. It should also be noted that, for storage at 38°C, the percentage strength increases slightly with age from the minimum at 3 months whereas the degree of conversion tends to increase.

TEST SERIES A2: EFFECT OF INITIAL CURING CONDITIONS

General

In series A2, ten different curing regimes were used during the first 24 h. They were called to A to K in order of increasing severity and details are given in Table 8. They comprised storage in water at the 'normal' temperature of 18°C and at a higher temperature, H (= 25, 35 or 50°C), for the consecutive periods shown in the Table. After the first 24 h, the cubes were stored in water at 18°C until tested.

As shown in Table 1 and described on page 80, there were fewer mix variables in this test series, but they still covered a range of water/cement ratios from 0·28 to 0·62. The strength results given in Table 9 represent 225 different combinations of mix proportions, curing regime and hot-water temperature.

Temperature rise

Temperature readings within cubes were recorded during the first 24 h. Richer mixes have a greater rise in temperature (cf. Table 4 for series A1), but this was only noticed with the curing regimes having a long delay at 18°C, i.e. regimes A to E. These produced a peak temperature at about 6 h varying from about 25°C for 1:3 mixes to 21°C for 1:7·5 mixes.

Figure 7 shows the type of temperature variations with time for the 10 curing regimes with the hot water at 35°C. Similar diagrams are obtained at 25°C and 50°C. As can be seen from Figure 7, when the concrete is placed in the hot water tank, the temperature of the concrete rises to exceed that of the hot water. This is most noticeable when the concrete is placed there immediately after casting, the concrete temperature reaching about 38°C for immersion in water at 35°C and about 56°C for immersion in water at 50°C. At these higher temperatures, there is a tendency for a 3 h delay (regime F) to result in slightly higher temperatures than immediate placing in hot water for 3 h (regime G). It takes about one hour for the temperature conditions within a cube to stabilize after changing from normal to hot water or vice versa.

Degree of conversion

In this series, the degree of conversion was calculated from DTA data measured only at 24 h and at 3 months and, since it was considered that only a low level of conversion would take place at 25°C, the tests were carried out on one 1:3 mix and one 1:7·5 mix at this temperature. As in series A1, it was found that, at these relatively early ages, the degree of conversion is independent of the mix proportions and water/cement ratio. The mix parameters can thus be ignored and the mean degree of conversion of all mixes used to examine

TABLE 8: **Test series A2. Curing regime during the first 24 h.**

Reference	Consecutive hours in		
	normal water at 18°C	hot water at H°C	normal water at 18°C
A	24	—	—
B	21	3	—
C	18	3	3
D	18	6	—
E	6	18	—
F	3	3	18
G	—	3	21
H	—	6	18
J	—	18	6
K	—	24	—

H = 25, 35 or 50°C

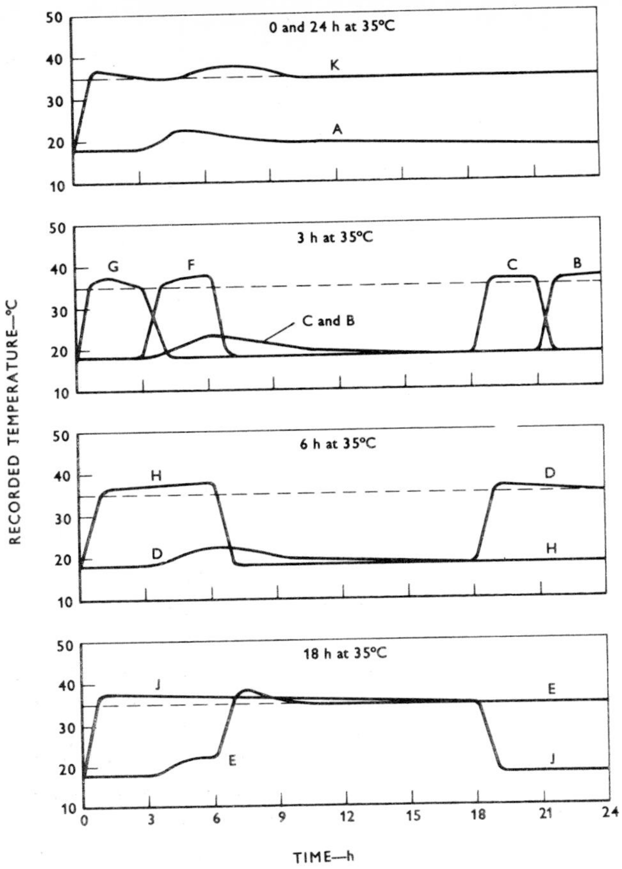

Figure 7: Typical temperature records during the first 24 h for the 10 curing regimes used in test series A2 when the high temperature was 35°C.

the effect of curing regimes and temperatures. The mean degrees of conversion at 24 h and at 3 months for these different conditions are shown in Figure 8. Unfortunately, DTA tests were not carried out at 3 months for any of the regimes involving storage at 35°C for 3 h. Figure 8 also shows the average level of conversion at 3 months from series A1 for concrete stored continuously in water at 18°C and at 38°C after the first 24 h in air (regimes X and Y)

With curing regime A (24 h at 18°C), the conversion was 10% at 24 h and 20% at 3 months (the 3 month value being the same as obtained in series A1). Similar levels of conversion were obtained with the water temperature at 25°C and, as can be seen from Figure 8, the length of time at this temperature or initial delays at 18°C have a negligible effect upon the degree of conversion. Similar low levels of conversion are also obtained with water temperatures of 35°C and 50°C for curing regimes A to D, i.e. provided these temperatures are not reached during the first 18 h after casting.

At storage temperatures of 35°C and 50°C occurring shortly after casting, the degree of conversion depends upon the actual curing regime. Regime F (3 h at 18°C before 3 h at 35 or 50°C) results in slightly higher conversion than regime G (immediate storage at high temperature for 3 h). Regimes J and K, which have long periods (18 h and 24 h) of immediate storage at high temperature, result in high conversions at 35°C and even higher at 50°C.

It is interesting to note that, where the degree of conversion at 24 h is less than 50%, it doubles by 3 months even though storage after the first 24 h is in water at 18°C. Where the 24 h conversion is 50% or more (three regimes at 50°C), the 3 month conversion remains about the same. These results show that, if the temperature rises considerably during the first few hours after casting, the conversion at 3 months is similar to that obtained from series A1, regime Y (24 h at 18°C followed by 3 months at 38°C). Thus 18–24 h at 35°C, and periods as short as 3 h at 50°C result in conversions of 70% and 75% compared with 82% after 3 months at 38°C from series A1.

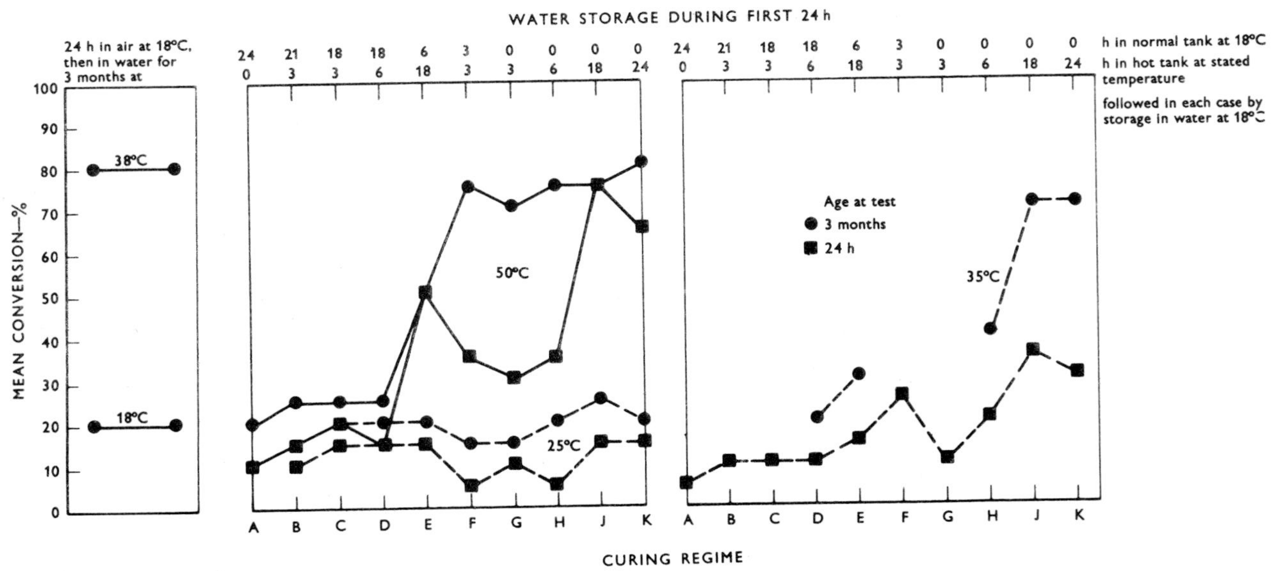

Figure 8: Degree of conversion (left) at 3 months from series A1 and (right) at 24 h and at 3 months from series A2.

Compressive strength development

The results from this test series show that the mix proportions have a small effect upon the temperature rise, and no effect upon the degree of conversion, but as usual they have a major effect upon the compressive strength. The compressive strength results at various ages up to 1 year are given in Table 9. As seen from series A1, the minimum strength at low storage temperatures is not reached for a number of years. It would therefore be misleading to refer to the lowest result for a particular mix and curing regime in Table 9 as the 'minimum' strength and such results within the limit of 1 year will be described as the 'lowest' strength.

It would be impracticable to plot all the 225 strength development curves resulting from the data given in Table 9, but the general form of the different developments should be noted, and a further examination made of the lowest strengths obtained. Figure 9 shows three very different types of strength development curves. This Figure relates to the two curing regimes A and K, i.e. 24 h continuously in water at 18°C or at the high temperatures. The curves shown are the average curves for all the water/cement ratios used with the strengths expressed as a percentage of the strength after 24 h storage in water at 18°C.

At 18°C the strength development follows a pattern similar to that in series A1, rising to a maximum value at 3 months and then declining, but at 1 year the strength is still above the original 24 h strength. At 25°C there is a slightly higher initial strength and a slightly more rapid decline at later ages. Raising the initial 24 h temperature to 35°C produces a considerable reduction in the initial strength followed by a slight gain in about 28 days, then a very rapid fall to about 50% of the normal 24 h strength at 3 months and 1 year.

Raising the initial 24 h temperature further to 50°C results in a completely different strength development curve. The 24 h strength is reduced to 50% of the normal 24 h strength, after which there is a gradual increase in strength with age. Various authors[2,3] have shown that the time to reach full conversion and minimum strength decreases with an increase in the temperature of continuous storage. With increasing temperature, the minimum strength also decreases. French, Montgomery and Robson[21] have shown that at 50°C the minimum strength occurs at 24 h. The present data given in Table 9 show a continued increase in strength for this regime at 50°C although, as shown in Figure 8, the degree of conversion is 65% at 24 h and increases to 80% at 3 months. The strength development curves for other regimes tend to follow one of the types of curve represented by those at 18°C and 25 or 35 or 50°C shown in Figure 9.

The actual strength at any age resulting from the regimes A and K depends upon how the water/cement ratio affects both the 24 h strength and the way it changes the average curves in Figure 9. It should be noted that placing the specimens immediately in water at 18°C, instead of in air at 18°C as in series A1, results in an increase in strength of about 10 N/mm².

Figure 10 shows the effect of curing temperature during the first 24 h upon the lowest strength (indicated by L against the appropriate value in Table 9) obtained during the first year for any curing regime A to K. Increasing the temperatures produces a drop in strength and the reduction is particularly severe be-

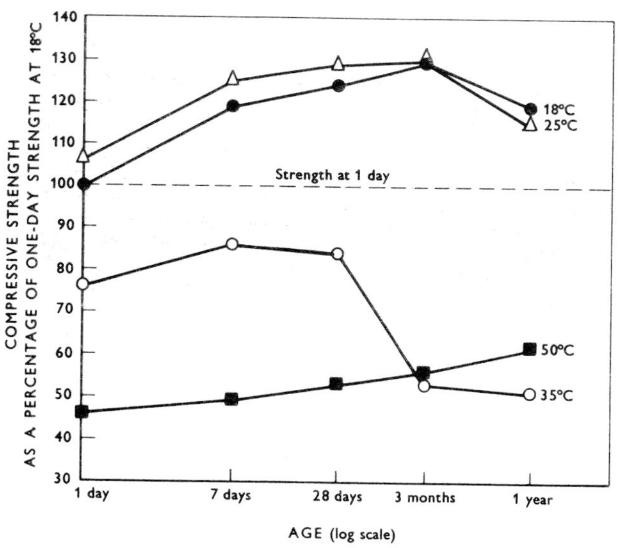

Figure 9: Average strength developments at 18°C after initial 24 h in water at the stated temperature.

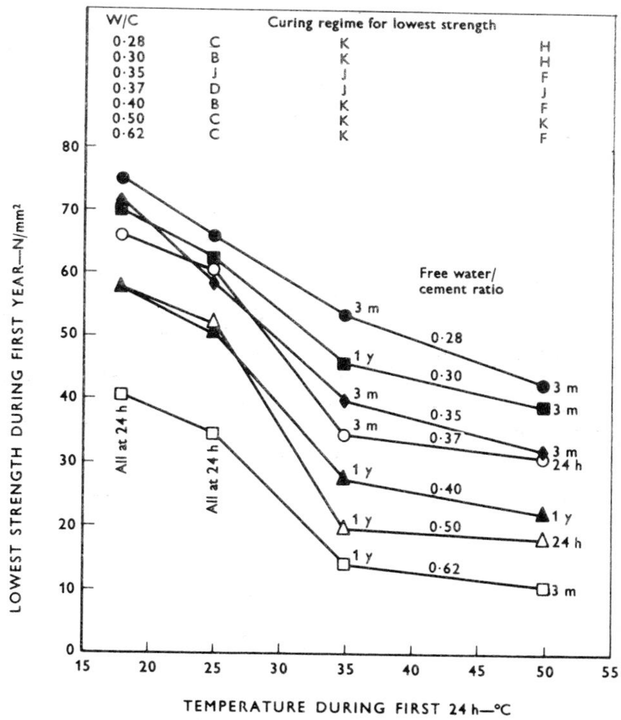

Figure 10: Effect of early temperature upon strength during the first year.

TABLE 9: Compressive strength data for test series A2.

Mix proportions by weight of cement: aggregate	Curing regime during first 24 h — Hours in			Water/cement ratio by weight		Compressive strength (N/mm²) of 101·6 mm cubes cured for the first 24 h under the stated conditions followed by storage in water at 18°C														
						Hot water at temperature H = 50°C					Hot water at temperature H = 35°C					Hot water at temperature H = 25°C				
	normal water at 18°C	hot water at H°C	normal water at 18°C	Total	Estimated free	24 h	7 days	28 days	3 months	1 year	24 h	7 days	28 days	3 months	1 year	24 h	7 days	28 days	3 months	1 year
	24	—	—			74·9	87·8	90·7	95·2	85·0	74·9	87·8	90·7	95·2	85·0	74·9	87·8	90·7	95·2	85·0
	21	3	—			76·3	87·5	89·0	95·3	88·1	77·9	92·2	93·1	92·3	96·0	68·5	79·6	89·1	93·9	90·1
	18	3	3			76·6	88·5	90·5	93·7	93·6	76·2	85·7	86·6	95·6	98·4	66·3L	75·3	90·7	96·2	96·5
	18	6	—			84·9	89·9	96·6	95·6	91·4	77·1	84·9	88·7	94·8	93·6	74·0	89·6	88·6	94·3	85·9
	6	18	—			78·5	82·8	83·5	83·0	82·4	84·7	84·5	86·0	89·6	88·9	71·2	82·6	90·1	98·3	89·9
1:3	3	3	18	0·33	0·28	75·3	67·8	75·3	58·6	53·8	81·4	91·5	96·8	101·5	76·2	78·2	89·1	94·2	100·2	95·2
	—	3	21			59·7	68·7	75·1	46·3	54·1	84·1	96·3	97·8	102·5	97·1	77·6	90·2	96·9	97·2	92·4
	—	6	18			60·0	65·2	64·7	42·5L	48·4	62·7	70·7	73·7	59·5	57·4	84·8	89·5	85·8	95·8	82·7
	—	18	6			50·0	51·6	58·1	62·7	60·7	62·5	77·8	86·0	79·8	55·8	70·9	86·3	88·8	94·0	90·0
	—	24	—			49·0	52·5	59·2	61·3	68·3	69·6	77·6	75·0	53·6L	54·7	82·5	94·0	98·1	104·3	88·0
	24	—	—			69·8	82·9	89·9	89·6	84·3	69·8	82·9	89·9	89·6	84·3	69·8	82·9	89·9	89·6	84·3
	21	3	—			78·8	85·4	93·2	90·0	91·8	73·8	81·2	69·6	75·5	73·7	62·3L	75·5	86·1	96·0	78·7
	18	3	3			78·3	84·3	92·7	94·3	88·8	75·6	81·6	68·5	79·1	84·7	65·8	80·8	88·8	84·1	81·6
	18	6	—			80·7	85·6	91·8	98·0	79·8	80·1	91·2	90·5	91·1	91·7	68·7	83·7	84·7	92·3	91·4
	6	18	—			72·1	74·3	76·4	79·6	78·6	79·8	73·2	80·5	88·8	84·3	62·9	79·6	88·0	91·9	89·9
1:3	3	3	18	0·35	0·30	56·5	67·5	67·2	49·2	45·6	70·1	85·7	84·3	83·9	76·3	78·7	86·5	99·5	97·1	98·8
	—	3	21			58·7	63·4	69·2	42·4	44·1	80·5	99·7	89·3	94·0	97·4	78·6	93·3	86·5	95·9	97·1
	—	6	18			52·2	55·8	49·8	38·8L	43·9	84·1	77·8	90·1	93·8	90·9	74·9	83·0	72·5	90·4	80·1
	—	18	6			43·3	45·6	51·7	54·7	51·6	69·8	73·4	79·8	56·9	54·1	74·7	93·1	91·6	100·0	93·6
	—	24	—			43·0	47·0	51·9	55·4	61·4	73·8	69·5	71·4	49·0	45·8L	79·0	90·5	84·1	90·6	90·0
	24	—	—			71·0	79·4	87·2	91·3	92·4	71·0	79·4	87·2	91·3	92·4	71·0	79·4	87·2	91·3	92·4
	21	3	—			73·0	82·1	86·6	97·8	93·5	69·4	82·4	76·5	89·4	97·4	62·5	77·0	77·2	90·6	88·3
	18	3	3			73·6	84·1	85·8	96·5	91·6	70·4	77·7	82·5	92·4	79·7	61·9	80·6	86·3	94·9	93·1
	18	6	—			71·6	86·5	92·4	96·5	90·5	71·0	77·6	83·7	90·1	75·7	59·0	73·2	81·6	83·0	94·5
	6	18	—			39·8	53·2	55·6	62·7	60·2	71·4	78·9	79·1	86·0	79·8	62·0	76·7	83·6	87·0	84·3
1:3	3	3	18	0·40	0·35	48·0	60·5	56·1	31·7L	39·0	65·8	79·4	79·5	75·6	60·2	66·9	77·4	85·4	97·8	84·5
	—	3	21			48·7	62·2	59·8	34·8	36·4	76·1	88·2	86·9	88·5	81·6	68·3	82·6	84·9	91·8	81·7
	—	6	18			49·4	58·3	49·3	35·7	43·2	75·8	86·1	88·2	87·6	82·1	72·6	81·6	77·5	94·1	84·5
	—	18	6			32·2	37·9	38·3	44·7	44·3	52·7	62·9	63·0	40·2L	40·7	58·5L	81·2	84·3	98·0	98·6
	—	24	—			33·4	37·1	39·2	44·3	46·3	51·0	63·5	71·0	41·6	43·0	72·3	87·0	87·3	94·7	100·0
	24	—	—			66·3	78·2	84·1	92·2	87·1	66·3	78·2	84·1	92·2	87·1	66·3	78·2	84·1	92·2	87·1
	21	3	—			67·8	79·5	83·2	88·3	84·3	69·6	80·5	85·0	90·8	77·8	64·2	79·7	84·5	92·3	97·9
	18	3	3			68·3	76·8	84·9	87·8	79·3	69·5	83·8	81·7	94·9	78·9	60·8	77·7	84·1	86·0	90·1
	18	6	—			72·6	82·0	84·3	86·7	81·6	74·2	81·3	85·5	91·3	91·2	60·5L	71·6	80·5	93·1	89·5
	6	18	—			47·1	54·7	60·1	62·8	51·6	69·9	78·1	82·8	72·9	77·6	68·5	83·3	89·6	85·2	97·5
1:4·5	3	3	18	0·45	0·37	44·5	53·0	45·0	35·1	38·0	67·0	75·2	82·5	73·6	61·7	71·6	82·9	78·5	96·4	85·6
	—	3	21			43·6	54·3	53·8	33·8	35·9	71·6	81·8	89·8	90·9	89·6	70·8	79·5	91·2	90·4	81·2
	—	6	18			41·9	49·0	44·0	33·2	34·8	60·4	76·9	79·1	72·2	65·3	73·3	84·3	85·5	100·0	100·0
	—	18	6			30·8L	32·8	35·9	39·6	41·9	49·6	53·3	53·4	35·2	39·0	70·5	79·6	78·8	92·7	63·4
	—	24	—			37·9	37·5	36·1	35·1	38·5	47·9	56·8	58·3	34·4L	37·9	70·1	81·3	90·3	89·5	76·1

TABLE 9 *continued*

1:6	24	—	—	0·50	0·40	57·7	69·3	79·2	78·1	72·9	57·7	69·3	79·2	78·1	72·9	57·7	69·3	79·2	78·1	72·9
	21	3	—			57·3	65·4	71·8	71·7	66·1	61·0	68·7	80·7	78·2	74·3	50·4L	70·9	80·5	82·9	79·6
	18	3	3			58·8	67·0	71·8	74·1	65·0	59·2	67·7	82·3	78·4	70·6	52·2	67·6	76·5	77·8	78·3
	18	6	—			61·5	69·4	74·0	79·7	69·2	56·2	66·7	76·8	69·9	72·1	51·8	66·0	71·7	81·4	67·4
	6	18	—			42·2	45·9	47·9	27·4	37·7	49·6	60·1	67·2	59·4	61·1	57·8	69·4	78·9	79·1	83·8
	3	3	18			39·4	48·2	43·2	29·7	22·0L	52·1	68·5	69·1	57·3	46·2	58·1	73·6	77·4	83·8	82·7
	—	3	21			35·6	42·0	41·9	26·1	27·0	60·5	78·1	78·9	83·1	73·6	60·7	74·3	76·5	84·0	71·6
	—	6	18			37·0	43·8	40·5	27·1	28·8	38·5	56·2	56·5	38·3	37·3	56·2	70·4	79·5	83·4	80·5
	—	18	6			26·0	27·9	27·9	28·5	32·5	43·1	49·2	49·2	30·2	29·1	60·1	66·7	74·7	78·7	72·4
	—	24	—			24·6	25·2	30·1	32·5	34·7	44·5	50·2	47·5	30·5	27·6L	66·5	80·2	78·7	78·8	61·4
1:6	24	—	—	0·55	0·45						56·1	68·2	69·8	79·6	69·2	56·1	68·2	69·3	79·6	69·2
	21	3	—								54·7	62·7	74·0	80·4	74·5	49·4L	70·8	78·7	77·3	72·5
	18	3	3								57·6	68·5	73·8	80·4	79·0	51·1	69·4	76·5	73·9	78·5
	18	6	—								59·0	68·5	78·5	82·8	77·6	57·1	69·0	75·1	78·7	74·3
	6	18	—								55·9	65·0	65·3	66·3	65·4	59·9	79·2	81·0	84·1	74·2
	3	3	18								52·7	70·7	69·2	63·2	52·3	62·1	73·2	80·2	84·7	78·4
	—	3	21								64·8	71·8	80·9	82·7	78·5	61·3	74·6	83·7	81·2	79·6
	—	6	18								52·4	67·2	65·8	62·3	60·5	72·6	85·2	92·1	90·9	94·2
	—	18	6								39·2	45·4	44·6	25·8	25·3L	59·0	71·2	74·7	78·5	70·1
	—	24	—								40·4	45·4	42·7	27·7	27·3	60·0	73·6	77·6	75·6	59·9
1:6	24	—	—	0·60	0·50	57·8	64·3	71·9	74·5	63·7	57·8	64·3	71·9	74·5	63·7	57·8	64·3	71·9	74·5	63·7
	21	3	—			57·6	65·0	69·2	66·0	60·3	53·4	66·5	64·4	71·9	58·4	53·0	64·5	75·1	70·6	68·7
	18	3	3			55·8	61·0	63·3	63·0	53·3	51·6	63·9	69·1	73·6	58·5	51·6L	63·8	71·5	69·9	74·7
	18	6	—			54·7	61·1	61·8	55·9	41·2	55·3	69·4	72·5	75·1	69·1	52·1	51·8	80·3	76·3	73·6
	6	18	—			33·9	38·1	35·9	31·9	23·3	48·5	55·2	59·9	53·3	49·6	52·9	68·5	68·3	72·3	67·6
	3	3	18			28·7	39·9	29·4	19·5	20·8	39·2	56·4	51·3	39·0	32·1	53·5	70·5	72·1	78·3	69·5
	—	3	21			28·7	37·6	31·5	18·5	19·2	56·5	70·3	73·5	73·2	72·6	55·0	73·7	76·3	85·4	79·2
	—	6	18			30·3	36·1	31·2	19·4	21·7	30·0	44·1	44·3	26·5	25·6	54·5	67·8	67·7	76·5	70·0
	—	18	6			19·0	21·6	19·8	21·2	23·6	31·2	39·5	37·4	20·7	21·2	52·0	72·6	73·4	72·0	62·0
	—	24	—			18·3L	19·1	21·1	23·1	25·4	34·8	41·2	37·2	20·5	19·6L	53·9	62·0	76·1	69·1	59·0
1:6	24	—	—	0·65	0·55	48·7	62·3	52·5	55·4	47·6										
	21	3	—			55·2	60·0	66·9	63·4	62·4										
	18	3	3			51·4	65·3	65·3	68·8	63·2										
	18	6	—			54·5	60·6	62·3	59·3	53·2										
	6	18	—			36·9	41·1	41·1	47·9	31·4										
	3	3	18			23·1	31·3	23·0	15·2	15·7										
	—	3	21			25·5	33·3	31·5	16·7	18·4										
	—	6	18			24·9	28·4	24·2	16·3	17·1										
	—	18	6			16·2	17·7	19·7	18·5	21·2										
	—	24	—			15·2L	19·1	17·5	20·3	22·1										
1:7·5	24	—	—	0·75	0·62	40·5	51·3	52·7	51·3	47·8	40·5	51·3	52·7	51·3	47·8	40·5	51·3	52·7	51·3	47·8
	21	3	—			44·2	49·9	48·8	54·1	50·5	38·5	50·6	48·6	49·0	46·3	38·7	53·2	55·7	54·1	46·3
	18	3	3			46·4	50·8	46·7	47·7	80·5	39·6	46·8	46·7	54·3	49·0	34·8L	52·5	50·1	50·6	52·0
	18	6	—			44·0	52·5	49·8	49·9	49·3	34·8	48·6	50·5	50·3	49·2	39·6	50·3	52·4	53·0	46·5
	6	18	—			25·7	28·7	24·5	24·8	18·3	40·3	39·9	37·0	40·0	32·8	40·6	52·8	55·1	54·1	50·3
	3	3	18			18·8	22·4	15·7	10·1L	10·8	28·2	35·1	27·9	25·3	17·2	41·6	58·1	55·9	54·9	52·7
	—	3	21			18·2	24·8	17·2	10·8	11·8	37·6	47·2	49·8	41·5	40·4	36·7	56·7	56·5	55·4	46·0
	—	6	18			20·2	23·0	19·0	12·2	13·7	29·5	38·3	37·0	33·8	27·9	35·4	53·7	54·5	50·1	44·9
	—	18	6			14·3	14·5	14·5	15·0	16·8	27·4	28·7	24·8	14·5	16·9	34·6	45·8	52·4	43·3	36·8
	—	24	—			12·8	14·2	14·6	14·6	17·4	24·3	29·2	22·9	14·5	14·1L	39·2	49·0	47·9	45·4	40·3

The letter L beside a strength value indicates the lowest strength value obtained for a particular mix.

tween 25 and 35°C. At 18°C and 25°C the lowest strengths occur at 24 h, and at 25°C they generally result from curing regimes B and C, i.e. 3 h at 25°C after at least 18 h at 18°C. At 35°C the lowest strength is attained at 3 months or 1 year, the difference in strengths at these ages being very small, and results from curing regimes J and K, i.e. immediate storage in water at 35°C for 18 or 24 h. At 50°C the conditions appear to be more variable, the lowest strength occurring at different ages and in five cases resulting from storage at this temperature for only 3 or 6 h (regimes F or H). However, the difference between the lowest value and the 24 h strength resulting from regimes J and K is usually very small (Table 9).

It must be emphasized that Figure 10 relates only to the lowest strength during the first year and not to the minimum strength that may be reached at later ages. However, series A1 has shown that, with water/cement ratios less than 0·40, the minimum strength for storage at 18°C does not fall below the 24 h value. Thus the lowest strengths at 24 h at 18°C shown in Figure 10 are likely to be minimum strengths for water/cement ratios less than 0·40, whereas for higher water/cement ratios these lowest strengths will be considerably higher than the minimum strengths eventually reached. Figure 10 illustrates the adverse effect upon the later strength of concrete with low water/cement ratios if the temperature rises considerably during the first 24 h.

Effect of curing regime upon strength

It is obvious that the curing conditions during the first 24 h have a most significant effect upon the subsequent strength of the concrete. It is convenient to compare the effect of the different curing regimes upon strength with the strength after 24 h at 18°C (regime A), which can be regarded as the 'normal' condition since it is the value used in specification for structural design. The comparison is made difficult by the fact that this strength depends upon the initial water/cement ratio. Figures 11 and 12 show the effect of curing regime and temperature upon the strengths at 24 h and at 1 year expressed as a percentage of the normal 24 h strength. The majority of results at intermediate ages lie between the strengths at these ages (Table 9).

It is convenient to divide the results into two groups according to the range of water/cement ratios. A maximum free water/cement ratio of 0·40 was specified in CP 110, and Figure 11 shows the strengths obtained from the different regimes expressed as the average percentage of the normal 24 h strength for mixes having free water/cement ratios between 0·27 and 0·40. Figure 12 is similar to Figure 11 except that it relates to concrete mixes having free water/cement ratios from 0·45 to 0·62. It is interesting to compare the pattern of these curves with those for conversion in Figure 8.

From Figures 11 and 12, the curing regimes can be divided into three groups depending upon their effect upon the strength at 24 h and at 1 year, viz. regimes A, B, C, D and E, regimes F, G and H, and regimes J and K. At 50°C regime E tends to lie between the first and second group of regimes. Regimes A, B, C and D have at least 18 h delay before the concrete is subjected to the high temperature and the 24 h strength only varies by about ±10% and the 1 year strength generally shows an increase of from 20 to 30%.

Regimes F, G and H, which involve 3 or 6 h at the high temperature during the first 6 h after casting, have a very marked effect upon the strength of concrete. For regime G (immediate storage for 3 h at 35°C), the 24 h strength and the 1 year strength are of the same order as those obtained from regimes A to D. However, if this period is extended to 6 h (regime H) or if there is a 3 h delay at 18°C (regime F), there is a considerable reduction in strength at both ages, particularly for the mixes having the higher water/cement ratios. At 50°C all three regimes result in a considerable reduction in strength. The 24 h strengths of the mixes with the low water/cement ratios are reduced to about 70% of the normal 24 h strength; furthermore, the strength at 1 year shows an even greater reduction to about 60% of the normal 24 h strength. For mixes with higher water/cement ratios, these reductions are increased to about 50% and 30% respectively. At 50°C these regimes produce the lowest strengths at 1 year.

The most severe regimes, J and K, in which the concrete is subjected to hot water for 18 h or 24 h immediately after casting, result in severe reductions in strength compared with the normal 24 h strength. These regimes at 35°C result in the lowest strengths at 24 h and at 1 year.

With low water/cement ratios, the 24 h strength is reduced to about 80% of the 24 h strength at 18°C and at high water/cement ratios to about 65% of the 24 h strength at 18°C. What is more disturbing is that the 1 year strength is reduced further still, to about 65% and 40% for low and high water/cement ratios respectively.

Regimes J and K at 50°C result in the lowest strengths at 24 h, only 55% and 35% of the normal 24 h strength for low and high water/cement ratios respectively. At this temperature, however, there is an increase in strength with these regimes between 24 h and 1 year.

From Figures 11 and 12 it is obvious that both the 24 h and the 1 year strengths vary greatly according to the curing conditions in the first 24 h. The range of strengths is similar for all water/cement ratios, but increasing the storage temperature increases the range of strength. The range resulting from the different curing regimes increases with age; at 25°C the average range increases from 14 N/mm^2 at 24 h to 23 N/mm^2 at 1 year, at 35°C the average range increases from 22 N/mm^2 at 24 h to 49 N/mm^2 at 1 year, and at 50°C the average range increases from 38 N/mm^2 at 24 h to 49 N/mm^2 at 1 year. This increase is due to the fact that with some curing regimes there is an increase in

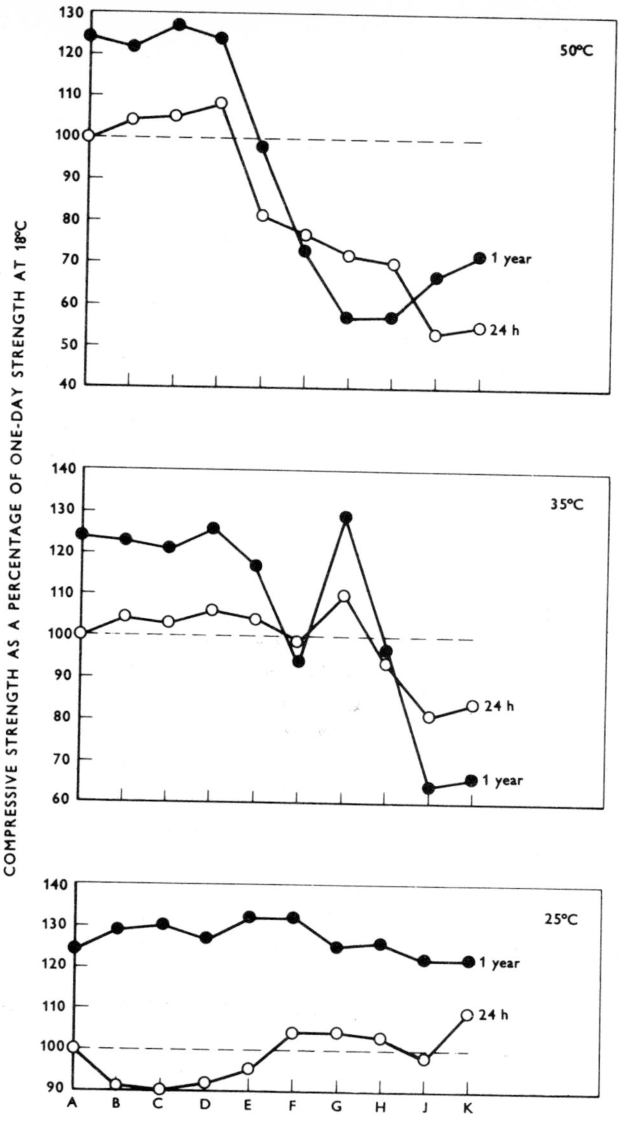

Figure 11: Effect of curing regime upon strength at 24 h and at 1 year for free water/cement ratios of 0·27 to 0·40.

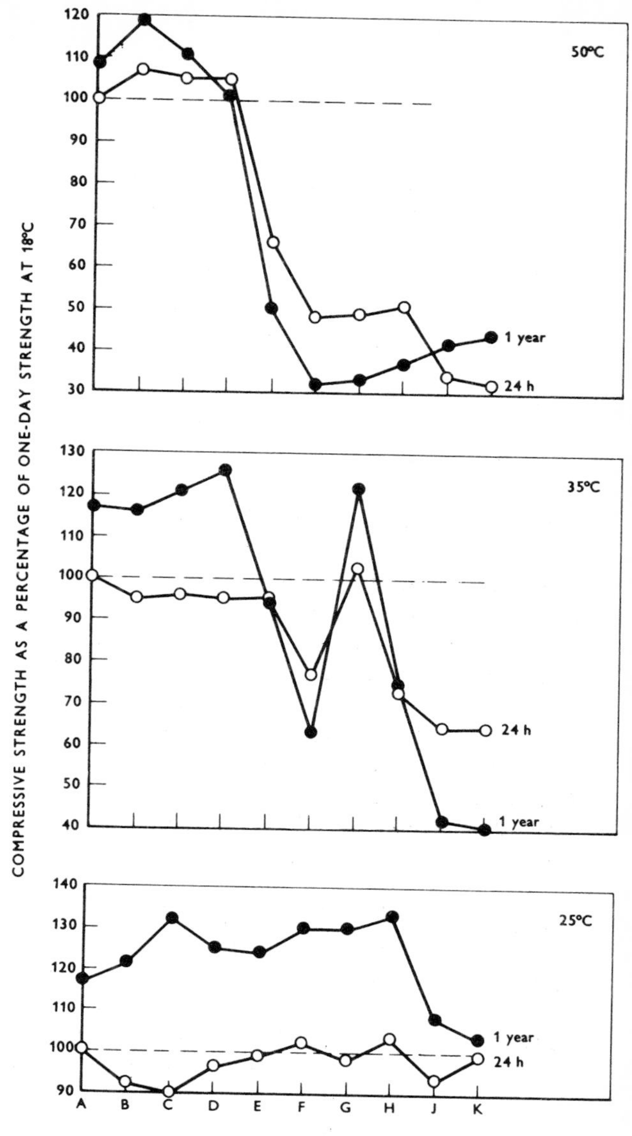

Figure 12: Effect of curing regime upon strength at 24 h and at 1 year for free water/cement ratios of 0·45 to 0·62.

strength between 24 h and 1 year, whilst with other curing regimes there is a decrease in strength.

Since structural concrete is based on the strength at 24 h, it is interesting to look at the reductions in strength due to the initial curing. The short early hot curing regimes F, G and H and the immediate longer hot curing regimes J and K result in the largest reductions in strength from the normal 24 h strength. Increasing the storage temperature increases the reduction in strength. At 25°C there is a negligible reduction at 24 h and an increase in strength at 1 year, but at 50°C the reduction in strength is in the order of 20 to 30 N/mm^2. With regimes F, G and H, there is generally a greater reduction in strength at 1 year than at 24 h. The importance of this increasing reduction in strength with age due to the initial curing conditions was discussed by Bate[22] in his report of the failure of roof beams at Stepney.

This test series shows that the strength at 1 year may be even more sensitive to slight changes in the early curing regime than is indicated by the strength measured at 24 h, particularly at higher water/cement ratios. Figure 13 shows the effect of small changes in the initial curing regime and the water/cement ratio upon the compressive strength at 1 year expressed as a percentage of the normal strength at 24 h at the different water/cement ratios used. The normal curing regime A, 24 h at 18°C, shows an average increase of about 20% in the 1 year strength. If the concrete is stored under regime F at 25°C (3 h at 18°C before 3 h at 25°C), the strength at 1 year shows an increase of about 30%. If, with this regime, the temperature is

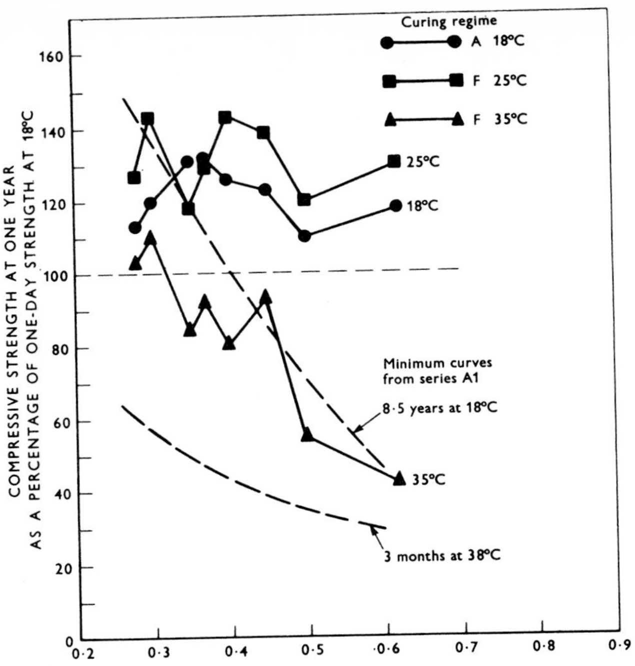

Figure 13: Combined effect of water/cement ratio and initial curing conditions upon the strength at 1 year.

raised to 35°C, the 1 year strength falls rapidly compared with the 24 h strength at increasing water/cement ratio, and the results are rather similar to those obtained from series A1 where the concrete is stored for 8·5 years in water at 18°C. It should also be noted that there is not such a serious reduction in strength if the water/cement ratio is kept below the recommended value of 0·40.

Degree of conversion and strength

The results from test series A1 (see Figure 6) indicated a link between the degree of conversion and the compressive strength expressed as a percentage of the normal 24 h strength, i.e. for curing in air at 18°C. In series A2, the degree of conversion and the compressive strength were measured at 24 h and at 3 months as shown in Figure 8 and Table 9. As reported previously, the degree of conversion measured in this series in independent of the water/cement ratio but this factor affects the compressive strength. The strengths at 24 h and at 3 months obtained with the different curing regimes and temperatures have been expressed as a percentage of the strength of the normally cured concrete (24 h at 18°C) and the averages for the same two groups of low and high water/cement ratios used in the previous section calculated. Figure 14 shows the relationship between the degree of conversion and the strength expressed as a percentage of the normal 24 h strength. This shows good relationships between these two parameters but they differ for different water/cement ratios and change with age, the water/cement ratio having a larger effect at 24 h than at 3 months. Low levels of conversion, say 20%, at 24 h resulting from the various curing regimes result in a reduction in strength, but the same value at 3 months is obtained from concrete 20 to 40% stronger than at 24 h. High levels of conversion, say 70%, result in serious reductions in strength both at 24 h and at 3 months but at the later age the reductions may be less.

TEST SERIES B: EFFECT OF ACCELERATING CONVERSION BY IMMEDIATE HEATING

In both the curing regimes X and Y used in the main strength series A1, the test specimens were kept for the

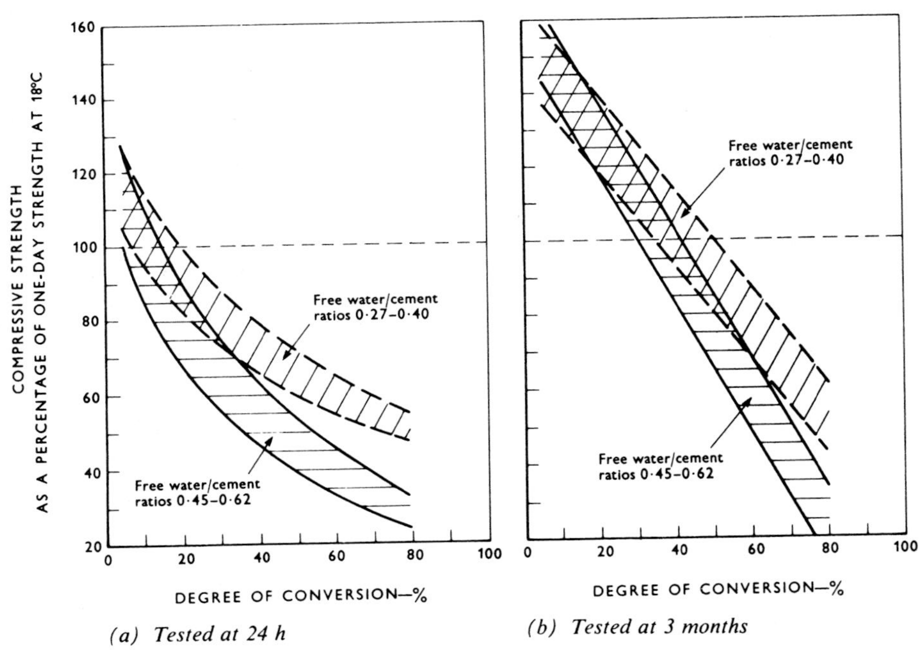

Figure 14: Relationship between conversion and strength expressed as a percentage of the strength after 24 h at 18°C (series A2).

first 24 h in air at 18°C. Subsequent storage in water at 18°C results in a minimum strength only being achieved after five or more years, but a minimum is achieved at 3 months by storage in water at 38°C. If the first 24 h delay is avoided by placing the cubes in water at 38°C immediately after casting, the minimum strength is achieved in about 5 days. This curing regime, Z, has been used by Lafarge as a method of obtaining the minimum strength on conversion at an early age.

In this test series, a 1:4·5 mix made with three different water/cement ratios, as recommended by the manufacturer, was used. The compressive strength results obtained from the three curing regimes X, Y and Z with six different batches of high alumina cement are given in Table 10. This shows that, for normal storage at 18°C (regime X), all the cements show an increase in strength up to an age of three months. With regime Y, the other batches of cement behave similarly to batch 685 (as used in series A1), a minimum strength generally being obtained at 3 months. In some instances, the lowest strength measured was at 28 days and in these cases the true minimum strength may be slightly less than either of these values and may occur between 28 days and 3 months.

The specimens that are placed at once in water at 38°C (regime Z) have a 24 h strength some 10 to 25 N/mm² below the normal 24 h strength. The strength continues to fall to a minimum value at 4 to 5 days and then shows signs of recovery, the concretes with the higher water/cement ratios having the greatest reduction. The minimum values obtained from the immediate heating at 38°C are similar to those obtained at 3 months when the heating is delayed by 24 h. The minimum values differ, particularly at the low water/cement ratio, depending upon the batch of cement. Figure 15 shows that the minimum strengths obtained from accelerated conversion are virtually the same as those obtained at 3 months from delayed conversion at 38°C.

Figure 16 shows the relationship between the compressive strength at 24 h at 18°C and at 3 months at 38°C, and the free water/cement ratio for the six batches of cement as given in Table 10. As with Portland cements, the strength of concrete at a given age depends upon the quality of the particular batch of cement. It has been claimed[2] that, when HAC concrete has fully converted, the strength to water/cement ratio relationship follows a very narrow band-width. Figure 16 shows that there is a considerable variation

TABLE 10: **Effect of delayed or immediate immersion at 38°C upon the strength development of 1:4·5 concrete made with different batches of high alumina cement (test series B).**

Cement batch No.	Water/cement ratio by weight		Workability test results			Compressive strength of 100 mm cubes (N/mm²)										
						Cured immediately in water at 38°C (regime Z)				Cured for 24 h in moulds at 18°C then in water at 18°C (regime X)				Cured for 24 h in moulds at 18°C then in water at 38°C (regime Y)		
	Total	Estimated free	Slump (mm)	Vebe time (s)	C.F.	24 h	4 days	5 days	6 days	24 h	7 days	28 days	3 months	7 days	28 days	3 months
685	0·35	0·27	0	15	0·77	59·2	47·6	46·1	45·1	—	86·9	86·2	86·3	70·6	45·6	39·0
	0·44	0·36	80	4	0·92	49·5	27·5	28·5	29·2	64·7	74·2	81·2	84·8	71·7	31·6	28·0
	0·50	0·42	Coll.	3	0·97	40·4	22·8	22·7	22·9*	60·4	75·1	79·7	76·3	68·5	22·0	22·6
759	0·35	0·27	0	13	0·78	69·1	57·5	61·1	60·3	81·6	91·5	95·9	100·9	79·6	61·3	54·2
	0·44	0·36	Coll.	1	0·93	51·2	36·3	35·7	36·6*	70·0	79·4	88·9	89·9	67·5	35·5	33·3
	0·50	0·42	Coll.	1	0·98	43·5	28·0	28·6	29·4	66·9	76·0	85·8	89·0	65·0	29·2	27·8
716	0·35	0·27	0	14	0·74	71·0	53·7	55·6	52·6*	80·7	90·5	92·4	88·6	84·3	64·2	44·2
	0·44	0·36	140	3	0·91	52·6	31·4	31·9	33·1	68·4	79·3	88·2	86·8	72·2	36·0	32·8
	0·50	0·42	Coll.	1	0·95	42·7	23·5	23·7	25·6*	62·5	73·7	78·4	83·7	64·2	26·0	26·1
765	0·35	0·27	0	13	0·77	45·6	48·3†	53·4‡	51·4	77·0	90·1	91·9	95·7	77·5	52·9	58·8
	0·44	0·36	35	4	0·97	45·3	31·7†	31·3‡	33·5	67·5	82·6	85·0	88·2	64·1	32·5	33·7
	0·50	0·42	Coll.	2	0·98	34·5	23·2†	23·2‡	24·2	55·2	73·9	83·1	81·0	39·8	24·6	25·4
794	0·35	0·27	0	22	0·72	57·1	54·3	54·4	54·3	76·0	85·0	91·3	83·1	88·7	65·0	50·2
	0·44	0·36	7	7	0·87	44·9	31·7	31·5	33·2	59·9	78·5	83·6	87·5	74·6	36·3	32·8
	0·50	0·42	80	2	0·95	32·3	22·2	24·0	24·0	51·2	69·3	78·9	79·1	64·2	24·1	22·8
796	0·35	0·27	0	20	0·69	54·0	49·0	50·9	48·5	69·2	85·5	89·7	89·1	75·8	61·8	54·8
	0·44	0·36	20	5	0·86	40·3	32·2	33·5	33·1	59·4	74·5	78·5	82·4	70·8	32·0	33·6
	0·50	0·42	Coll.	2	0·96	31·9	22·5	24·1	25·0	53·4	63·8	72·1	69·0	54·3	21·4	24·0

*Tested at 7 days †Tested at 3 days ‡Tested at 4 days

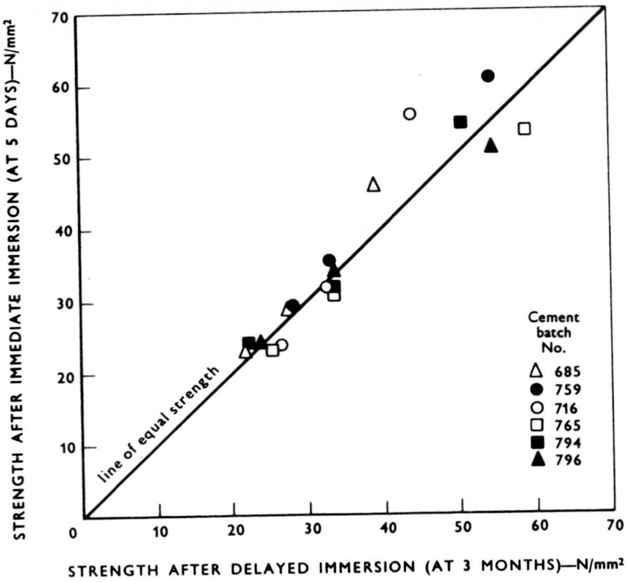

Figure 15: Relationship between minimum strengths obtained by immediate and delayed immersion in water at 38°C.

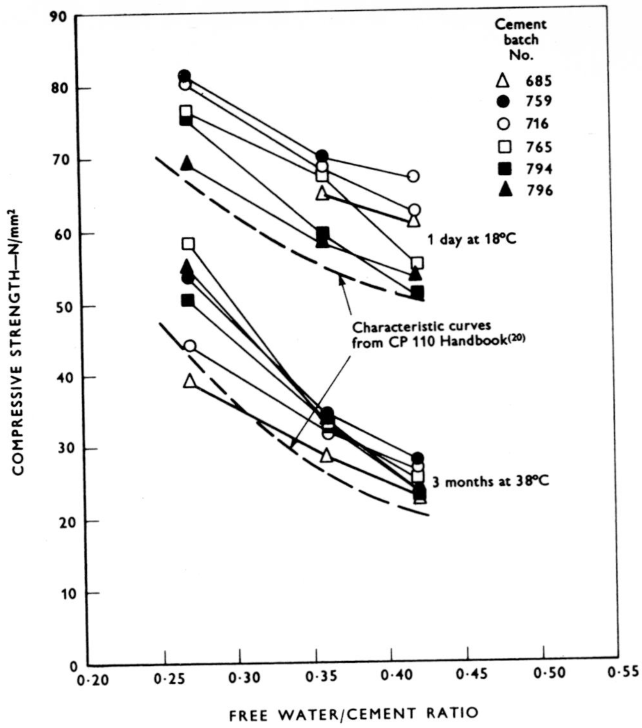

Figure 16: Relationship between 24 h strength, minimum converted strength and water/cement ratio for various batches of cement.

in the minimum converted strength, particularly at low water/cement ratios, depending upon the batch of cement, although all batches comply with BS 915.

Cement batch No. 685 was used in the main test series A1 and A2 and from Figure 16 it can be seen that this batch has a relatively high 24 h strength but the lowest converted strength at 3 months. Figure 16 also shows the relationship between the characteristic strength and water/cement ratio as given in the handbook to the Unified Code[20]; 95% of concrete should have strengths above the values indicated at the appropriate water/cement ratio. The 24 h strength of batch 685 is well above the characteristic strength, but when fully converted its strength was equivalent to the characteristic strength.

It should also be noted that it is not possible to estimate the minimum converted strength from the 24 h standard strength for a particular batch of cement, but the characteristic minimum converted strength can be related to the characteristic strength at 24 h.

TEST SERIES C: MODULUS OF ELASTICITY OF HAC CONCRETE

There are few published data on the modulus of elasticity of concrete made with high alumina cement, particularly when it has converted. For this short programme, cement batch No. 759 was used in the four mixes described in page 81 and shown in Table 1. The same curing regimes (X, Y and Z) as in series B were used in this series to study the effects of immediate or delayed heating at 38°C upon the modulus of elasticity compared with that of concrete stored at 18°C. Details of the experimental procedures are described on page 82.

The change in compressive strength with age under the different curing regimes follows a pattern similar to that previously obtained. The 1:6 mix with a total water/cement ratio of 0·50 has a higher compressive strength with all regimes than the richer 1:4·5 mix at the same total water/cement ratio. Although the total water/cement is the same, the estimated free water/cement ratio is slightly lower for the leaner 1:6 mix.

The modulus of elasticity is related to the compressive strength of the concrete as shown in Figure 17, which also shows the range of values to be expected using ordinary Portland cement derived from Table 64 of CP 110[19]. For a given compressive strength, the modulus of elasticity of HAC concrete is higher than that of ordinary Portland cement concrete. The values of the static modulus are more variable than those of the dynamic modulus.

As with the compressive strength, the elastic modulus is reduced on conversion, but not to the same extent, and the dynamic modulus is less affected than the static modulus. For example, HAC concrete with a 24 h compressive strength of 60 N/mm² is reduced to 30 N/mm² on full conversion (50% of 24 h strength), whereas the corresponding values of the static modulus are 40 kN/mm² to 30 kN/mm² (75%), and of the dynamic modulus 47 kN/mm² to 45 kN/mm² (95%).

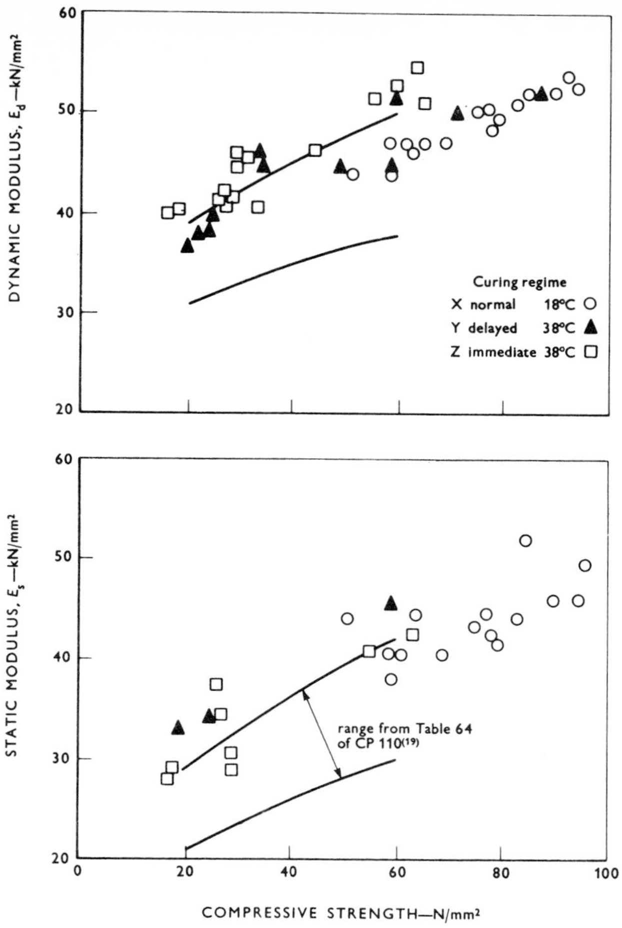

Figure 17: Relationship between modulus of elasticity and compressive strength.

The reduction in the modulus of elasticity is related to the water/cement ratio as shown below.

Free water/cement ratio	Property of fully converted concrete expressed as a percentage of the 24 h value		
	Compressive strength	Static modulus	Dynamic modulus
0·27	76	99	102
0·41	47	92	94
0·50	35	70	86

Neither the static modulus nor the dynamic modulus is affected by whether the heating at 38°C is immediate or delayed.

The static modulus of elasticity of HAC concrete can be estimated from measured values of the dynamic modulus of elasticity by using the same equation as for Portland cement concrete, as given in Appendix D of CP 110, i.e.

$$E_s = 1{\cdot}25E_d - 19 \text{ kN/mm}^2$$

The estimated values are generally correct within ± 4 kN/mm².

General discussion

This series of four test programmes provides new information on the characteristics of HAC concrete in the following areas:

(1) the variation in strength of concrete at 1 day at 18°C and fully converted at 38°C due to different batches of cement;
(2) the extreme importance of the water/cement ratio for the strength of concrete stored in water at 18°C for long periods;
(3) the effect of curing conditions during the first 24 h upon the strength at early and later ages;
(4) the quantitative measurement of conversion, the factors influencing it and its relationship with strength;
(5) the modulus of elasticity.

The variation in the 24 h strength at 18°C and the fully converted strengths at 38°C of concrete made with different batches of high alumina cement is clearly seen in Figure 16. Batch No. 685 complied with the requirements of BS 915 and has an average strength at 24 h. Its behaviour is similar to that of the other cements used in series B except that its minimum fully converted strength at 38°C is the lowest. There is thus no reason to suspect that high alumina cement in general should react differently from batch No. 685 to the various factors studied in test series A1 and A2 except that minimum values obtained on full conversion may generally be higher.

In future, it may be important to know the minimum strength likely to be achieved during the life of the concrete. An accelerated curing regime, such as that used by Lafarge (regime Z), could be used either as a requirement in BS 915 or as a production control test since, as shown in Figure 15, there is a good correlation between the results obtained at the early age of 5 days and the minimum results obtained at 3 months with a more delayed conversion. It should also be noted that the lowest strengths obtained from the harmful curing regimes in series A2 (Table 9) were above the minimum obtained at 3 months at 38°C from similar mixes used in series A1 (Table 6).

The degree of conversion is independent of mix proportions (including water/cement ratio) when conversion occurs rapidly as in series A1 at 38°C or as a result of the various hot regimes used in series A2. However, when the concrete converts slowly at 18°C, as in series A1, the degree of conversion at 5 years and more increases with increasing water/cement ratio. Series A2 shows that, provided temperatures as high as 50°C are not reached during the first 18 h after casting, the degree of conversion at 3 months is similar to that from continuous storage at 18°C. There is little effect upon the conversion at 3 months if the temperature during the first 24 h reaches 25°C, but at 35°C and particularly at 50°C the curing regime during the first 18 h may result in the conversion at 3 months being similar to that from continuous storage at 38°C (apart from the first 24 h) obtained in series A1. In his book, Robson[2] reports on the delay in the setting time and on the heat evolution at ambient temperatures of between about 20 to 35°C, and this may be related to the effect upon conversion and strength of changes in the curing regime during the first 6 h.

For most of the different initial curing regimes used in series A2, the degree of conversion doubled between 24 h and 3 months even though the storage between these ages was at 18°C, the same as in test series A1, which only resulted in 20% conversion at 3 months. If the degree of conversion in concrete several years old is low, say less than 30%, this would indicate that the concrete had a low water/cement ratio, had been satisfactorily cured during the first 24 h and had not been subjected to an adverse environment.

At the time of writing this paper, only some of the 10 year strength tests have been carried out, but these confirm the trends shown in Figures 3 and 4. At 38°C the strength continues to increase very slowly but at 18°C the strength development depends upon the water/cement ratio. With very low water/cement ratios the strength is increasing after reaching a minimum, but at high water/cement ratios the strength is continuing to decrease. It has been previously thought that the time to reach the minimum strength was independent of the mix proportions but these tests indicate that at 18°C it increases with the water/cement ratio. These tests (see Figure 5) also show the extreme importance of the water/cement ratio for the long-term strength of concrete stored at normal temperatures (18°C).

Work by Davey[23] and others has shown that delays in storage at higher temperatures result in a longer time being required to reach the minimum strength, but do not affect the minimum strength. This is confirmed by series B, where immediate storage in water at 38°C results in a minimum strength at about 5 days, whereas a delay of 24 h results in the same minimum (see Figure 15) being obtained in 3 months. It should also be noted from Table 6 that, in series A1, 3 months storage in water at 38°C after 8·5 years in water at 18°C produces minimum values similar to those obtained at 3 months after 24 h in air at 18°C followed by water storage at 38°C.

Considerable quantities of data have been published[2,3] on the adverse effect upon strength of high temperatures during the first 24 h. This current research shows the adverse effect that temperatures of 35°C and 50°C for periods as short as 3 or 6 h very shortly after casting can have upon the subsequent strength of the concrete. It should be noted that series A2 was not continued long enough to obtain true minimum strengths and that strength reductions were sometimes greater at 1 year than at 24 h, despite satisfactory curing between these two ages. Such temperature rises for short periods may occur in practice[2] and were considered as a contributory factor to be low-strength concrete found in Stepney[22].

It is generally considered that structures in the United Kingdom are not subjected to high temperatures. However, in the investigation following the collapse at Camden[15], it was thought that the temperature of the roof members could have risen to 25 or 30°C. It is unfortunate that no specimens in series A1 were stored at a temperature between 18°C and 38°C. From the results up to 1 year obtained from series A2, storage at 25°C produced a reduction in strength of about 10% compared with storage at 18°C, but more severe reductions may result at temperatures above 30°C. Neville[24] showed a reduction in strength with age at curing temperatures of 25, 30, 35 and 40°C but the tests were not continued long enough to reach minimum values; he reported[10] other work showing similar trends. Thomas[25] reported reductions in strengths at 20 years varying from 57 to 97% of the 1 year strength according to the water/cement ratio, for storage in water at 18°C, and from 67 to 87% for storage in air at 18°C.

There is thus considerable evidence supporting the results obtained in series A1. At storage temperatures between 18 and 38°C, the strength development curves will lie between those shown in Figures 3 and 4.

The value of the minimum strength and the time it is reached depend both upon the water/cement ratio and upon the temperature. At high water/cement ratios, virtually the same minimum strength is reached at temperatures of 38°C and 18°C; with the former this occurs at 3 months and with the latter at upwards of 8·5 years. It is reasonable to conclude that, at intermediate temperatures, a similar minimum strength will be attained at an intermediate age. However, at low water/cement ratios and the low temperature of 18°C, the minimum strength may remain above the 24 h value, but as the temperature is increased the minimum strength will be lower and occur at an earlier age.

The reductions in strength at the normal temperature of 18°C shown in this research were obtained from water-stored specimens, and it could be claimed that under practical conditions, where carbonation will occur and the concrete may dry out, such loss would not take place. Other research has shown that when concrete is oven-dried, conversion ceases and there is no further reduction in strength. However, Midgley[12] has shown that there is enough water present in most structural members for conversion to begin and that the conversion reaction liberates more water. Some of the research shows long-term reductions in strength for storage in air, and standard tests made at BRS on HAC concrete stored both in water at 18°C and in air at 18°C and 65% R.H. show similar strength reductions at 5 and 10 years.

The precise mechanism of the reduction in strength due to conversion is not known but cannot be explained by increased porosity alone. No relationship between the degree of conversion and the strength can be found from the results of test series A1 and A2. However, in both series, there are indications of relationships between the degree of conversion and the strength expressed as a percentage of the 24 h strength, but such apparent relationships change with age, curing conditions and water/cement ratio. It should also be noted that 100% conversion has not been achieved and that in series A1 the minimum strength at 38°C is reached at 3 months when the conversion is about 80%. At later ages the conversion increases slowly but so does the strength. At 18°C the minimum strength is attained at a lower degree of conversion; as the water/cement ratio is increased, the degree of conversion for minimum strength is increased and the minimum strength is decreased.

Conclusions

(1) High alumina cement requires slightly lower water contents than ordinary Portland cement to produce concretes having comparable workabilities. The use of fine sands, such as grading Zone 4, may be beneficial in reducing the harshness of some mixes but requires careful mix design and production control.

(2) The strength of concrete, converted or unconverted, at any age varies according to the batch of cement used.

(3) The rate of conversion depends upon the storage temperature. The degree of conversion, at all ages for continuous storage at 38°C, and when resulting from various initial curing regimes, is independent of the mix proportions. For continuous storage at 18°C, the degree of conversion at 5 years and later increases with increases in the water/cement ratio.

(4) The degree of conversion at 24 h resulting from various initial curing regimes, if less than 50%, doubles at 3 months even though the curing conditions between these ages do not normally result in this rate of conversion.

(5) There is no relationship between the degree of conversion and the compressive strength, but there appears to be a number of relationships between the degree of conversion and the compressive strength expressed as a percentage of the 24 h strength depending upon age and water/cement ratio. With rapid conversion at 38°C, the minimum strength occurs when the degree of conversion is about 80%. With slow conversion at 18°C, the minimum strength may occur at lower degrees of conversion.

(6) The strength development with age follows a characteristic pattern of an increase to a maximum strength, then a decrease to a minimum strength, followed by a very slow increase in strength with age. The maximum and minimum values and the time that they occur depend upon the storage temperature and the water/cement ratio.

(7) With continuous storage at 38°C, after 24 h at 18°C, the minimum strength is obtained at 3 months followed by a very slow increase in strength with age. With continuous storage at 18°C, the minimum strength appears to occur after 5 years at increasing ages with increasing water/cement ratio.

(8) Changing the storage at an age of 8·5 years from 18° to 38°C for 3 months results in a reduction in strength to about the minimum value obtained by rapid conversion at 38°C at an early age.

(9) At any age and storage condition, the strength depends upon the water/cement ratio. Very different relationships between these factors are obtained for the normal 24 h strength, the minimum strength obtained from rapid conversion at 38°C, and the strengths obtained at 8·5 years at 18°C. In the latter case, the strength is extremely sensitive to changes in the water/cement ratio and there is an increasing reduction in strength below the 24 h value for increasing water/cement ratios above 0·40.

(10) The minimum strength on conversion at 38°C cannot be estimated from the initial 24 h strength, but can be obtained at 5 days by an accelerated curing regime.

(11) The temperature conditions at different times during the first 24 h after casting are critical and can result in a serious reduction in strength at 24 h, and in some instances a greater reduction at 1 year. The tests were not continued long enough to obtain minimum values in all cases. For instance, if temperatures of 35 or 50°C are reached during the first 6 h after casting, considerable reductions in strength at 24 h result and these may increase at 1 year. Greater reductions are obtained at higher water/cement ratios.

(12) The modulus of elasticity of HAC concrete is higher than that of ordinary Portland cement concrete of the same compressive strength.

(13) The modulus of elasticity of HAC concrete is related to its compressive strength and is reduced on conversion. The reduction on conversion is much less than that of the compressive strength, but like the compressive strength it depends upon the water/cement ratio.

(14) The dynamic modulus is less affected by conversion than the static modulus. The static modulus can be estimated from the dynamic modulus by using the same relationship as applied to Portland cement concrete.

ACKNOWLEDGEMENTS

The author wishes to acknowledge the assistance and comments from his colleagues at the Building Research Station, particularly Dr H. G. Midgley for his assistance in providing data on conversion.

The work described has been carried out as part of the research programme of the Building Research Establishment of the Department of the Environment and this paper is published by permission of the Director. The work received financial support from the Lafarge Aluminous Cement Co. Ltd.

REFERENCES

1. BIED, J. *Recherches industrielles sur les chaux, ciments et mortiers*. (Industrial research on limes, cements and mortars.) Paris, Dunod, 1926. pp. 227.
2. ROBSON, T. D. *High alumina cements and concretes*. London, CR Books, 1926. pp. 263.
3. LEA, F. M. *The chemistry of cement and concrete*. Third edition. London, Edward Arnold, 1970. pp. 727.
4. BRITISH STANDARDS INSTITUTION. CP 114: 1948. *The structural use of reinforced concrete in buildings*. London. pp. 57.
5. NEWMAN, K. The design of concrete mixes with high alumina cement. *The Reinforced Concrete Review*. Vol. 5, No. 5. March 1960. pp. 269–301.
6. ROAD RESEARCH LABORATORY. *Design of concrete mixes*. 2nd edition. London, HMSO, 1950 (reprinted 1955). pp. 16. Road Research Road Note No. 4.
7. LEA, F. M. and WATKINS, C. M. *The durability of reinforced concrete in sea water*. London, HMSO, 1960. pp. 42. National Building Studies, Research Paper No. 30.
8. MASTERMAN, O. J. High alumina cement concrete with data concerning conversion. *Civil Engineering and Public Works Review*. Vol. 56, No. 657. April 1961. pp. 483–486.
9. NEVILLE, A. M. A study of deterioration of structural concrete made with high alumina cement. *Proceedings of the Institution of Civil Engineers*. Vol. 25. July 1963. pp. 287–324.
10. BRITISH STANDARDS INSTITUTION. CP 116: 1965. *The structural use of precast concrete*. London. pp. 153.
11. INSTITUTION OF STRUCTURAL ENGINEERS. *Report on the use of high alumina cement in structural engineering*. London, 1964. pp. 18.
12. MIDGLEY, H. G. The mineralogy of set high alumina cement. *Transactions of the British Ceramic Society*. Vol. 66, No. 4. June 1967. pp. 161–187. BRS Current Paper 19/68.
13. MIDGLEY, H. G. The conversion of high alumina cement. *Magazine of Concrete Research*. Vol. 27, No. 91. June 1975. pp. 59–77.
14. BRITISH STANDARDS INSTITUTION. BS 915: Part 1: 1947. *High alumina cement*. London. pp. 24.
15. BRITISH STANDARDS INSTITUTION. BS 882, 1201: 1965. *Aggregates from natural sources for concrete (including granolithic)*. London, pp. 24.
16. DEPARTMENT OF EDUCATION AND SCIENCE. *Report on the collapse of the roof of the Assembly Hall of the Camden School for Girls*. London, HMSO, 1973. pp. 12.
17. BRITISH STANDARDS INSTITUTION. BS 1881: 1952. *Methods of testing fresh concrete*. London. pp. 16.
18. BRITISH STANDARDS INSTITUTION. BS 1881: 1970. *Methods of testing concrete*. Parts 1 to 5. London. pp. 16, 36, 28, 28, 40.
19. BRITISH STANDARDS INSTITUTION. CP 110: Part 1: 1972. *The structural use of concrete. Part 1: Design, materials and workmanship*. London. pp. 154.
20. BATE, S. C. C. et al. *Handbook on the Unified Code for structural concrete (CP 110: 1972)*. London, Cement and Concrete Association, 1972. pp. 153.
21. FRENCH, P. J., MONTGOMERY, R. G. J. and ROBSON, T. D. High concrete strength within the hour. *Concrete*. Vol. 5, No. 8. August 1971. pp. 253–258.
22. BATE, S. C. C. *Report on the failure of roof beams at Sir John Cass's Foundation and Red Coat Church of England Secondary School, Stepney*. Garston, Building Research Establishment, 1974. pp. 18. CP 58/74.
23. DAVEY, N. *Influence of temperature upon the strength development of concrete*. London, HMSO, 1933. pp. 76. Building Research Station Technical Paper No. 14.
24. NEVILLE, A. M. The effect of warm storage conditions on the strength of concrete made with high alumina cement. *Proceedings of the Institution of Civil Engineers*. Vol. 10. June 1958. pp. 185–192.
25. THOMAS, F. G. Influence of time upon the strength of concrete. *RILEM Bulletin*. No. 9. December 1960. pp. 17–34.

Sprayed concrete: tunnel support requirements and the dry mix process (CP 18/77)

W.H. Ward and D.L. Hills

INTRODUCTION

In 1972-73 it was realised that there were likely to be many kilometres of tunnels to be built in the UK before the end of the century particularly for long-distance water transmission and sewerage disposal and that many of these tunnels would be in rock. It was also noted that there had been little innovation in rock tunnelling practice in the UK in recent decades compared with the developments that had taken place in Europe or with the significant improvements that had taken place in soft ground tunnelling in the UK.

Many of the rock tunnels would be in rather highly jointed, sedimentary rocks with mixed and changing face conditions and it seemed likely that in the immediate future many tunnels would be excavated by drilling and blasting rather than by tunnelling machines. Thus the shape of the tunnel would be irregular, being controlled to a large extent by the bedding and joint systems rather than by the excavation process.There would be a frequent need to provide immediate support of the ground to prevent falls of loosened rock and to protect the miners from accidents during the erection of the supports.

The traditional practice is to provide immediate support by means of expensive steel arches, lagging and rock packing where necessary and to complete the whole tunnel. This is followed by the construction of a cast in-situ concrete lining, often reinforced, and placed by continuous pumping behind collapsible steel shutters. The immediate supports have usually to be left in place and the numerous voids behind the final lining are grouted. Thus the tunnel is provided with two sets of supports which makes construction very expensive compared to the UK practice in soft ground. The accident rates are also high because the miners are exposed to rockfalls during erection of the immediate supports.

Significant economies could be made in the support of rock tunnels if only one support system is used. This would need to be placed as excavation proceeded so that it provided any necessary immediate support and it would need to be placed with the miners protected against rockfalls.

The use of prefabricated lining systems, such as precast concrete segments, does not appear to be economic because large quantities of grout and concrete are needed to fill the void between the irregular excavation and the lining. Indeed, cases were known in small sewer tunnels where the volume and the support value of the void filling material was greater than the lining itself.

Support by means of sprayed concrete which had been used extensively in Europe, but very little in the UK, seemed attractive. It has the advantages of moulding itself directly to the rock, can be placed quickly and remotely with rather simple equipment in layers of different thickness and might serve as both an immediate and permanent support.

A survey of the literature and visits to many tunnels in Europe showed the extent of the use of sprayed concrete. But detailed understanding of its supporting action or of the requisite properties of the deposited concrete were lacking. Indeed it appeared that the properties of the concrete were determined largely by the characteristics of the spraying machine and the method of deposition rather than by the rock supporting needs.

Two related experimental programmes of work were started:

1 to measure and compare the behaviour of sprayed concrete linings and other support systems at full scale in a special experimental tunnel built in connection with the construction of a major water transmission tunnel for the Kielder Water Scheme
2 to study the characteristics of existing concrete spraying equipment and the properties of the deposited concrete in laboratory and field trials.

General details of the first year's work in the experimental tunnel have already been published[1] and should be studied in relation to this paper.

There are two types of spraying processes in common use for depositing concrete on a surface, wet mix and dry mix. In the wet mix process, pre-mixed wet concrete is conveyed by pump or pneumatic placer along a pipe to a nozzle where compressed air is injected to accelerate and propel the materials to the surface. In the dry mix process a dry mix of concrete is pneumatically conveyed at high velocity to the nozzle, where water is injected. The dry mix is the more widely used and the first stage of our experimental work has been restricted to this process.

In this paper we give details of the structural performance of sprayed concrete used to support a tunnel in fissile shale which collapses rather rapidly without any support and we present the general characteristics and limitations of the dry mix spraying system and identify its problems.

STRUCTURAL REQUIREMENTS OF SPRAYED CONCRETE AS A SUPPORT IN ROCK TUNNELS

Comparative behaviour of different supports in the same rock

Our detailed knowledge of the structural behaviour of sprayed concrete as a support in rock tunnels is limited to our studies in the Kielder Water Scheme experimental tunnel[1]. In that project eight different support systems (including no support) are being monitored in the same rock – a highly fissile shale of Carboniferous age. Part of the tunnel (3.3 m diameter at a depth of about 100 m) was excavated by blasting and the other part by a roadheader machine. Four of the support systems used sprayed concrete and it is possible to compare the performance of all systems in the same rock conditions.

This is done in a simple way in Figure 1, which shows the typical downward rock displacements at points 0.3 m above the crown in each support system for a period of about 600 days since construction. The left hand diagram shows the results in the part excavated by blasting and the right hand one relates to the machine excavated part. Each of these measurements was commenced in a position 0.3 m behind the face soon after it had been advanced to that position.

The continuing falls of rock from the shoulders followed by the extensive and dramatic collapse of the roof in the unsupported length showed without any doubt the need for support.

The length supported with circular steel ribs at 1.0 m centres, partial lagging and rock packing showed large rock movements in the beginning on account of the loose packing and deformable lagging, but the rate of rock displacement decreased as the packing compressed and the ribs started to provide support. However, slow rock crushing continues behind the supports as small amounts of rock debris fall out between the lagging, allowing displacement to continue slowly, yet quite safely and with the ribs supporting only low loads – less than 1 m of rock overburden. The total roof displacement is now some 29 mm and the loading continues to increase slowly.

In the length supported by rows of seven fully resin-bonded rockbolts spaced at 0.9 m in the roof and shoulders, the roof continued to descend after a while at almost a constant rate and allowed rock crushing to continue. A large shoulder block fell out just before this length was re-supported about a year later by a complete ring of sprayed concrete. The rock bolts carried low loads and the new support of sprayed concrete is also quite lightly loaded, because of the large displacements (about 20 mm) that occurred prior to its construction.

Another length was supported with an arch of sprayed concrete placed in two layers each to a minimum thickness of 50 mm, with a layer of light steel mesh between. This allowed smaller movements than the rockbolted length, but again after a while the rock and the arch continued to descend at almost a constant speed, with rock crushing and concrete breaking away along the sides of the tunnel. After a year and some 11 mm of roof displacement it became necessary to re-support this length by completing the invert with sprayed concrete to form a closed ring support. The very instructive behaviour of this support length is presented in detail later.

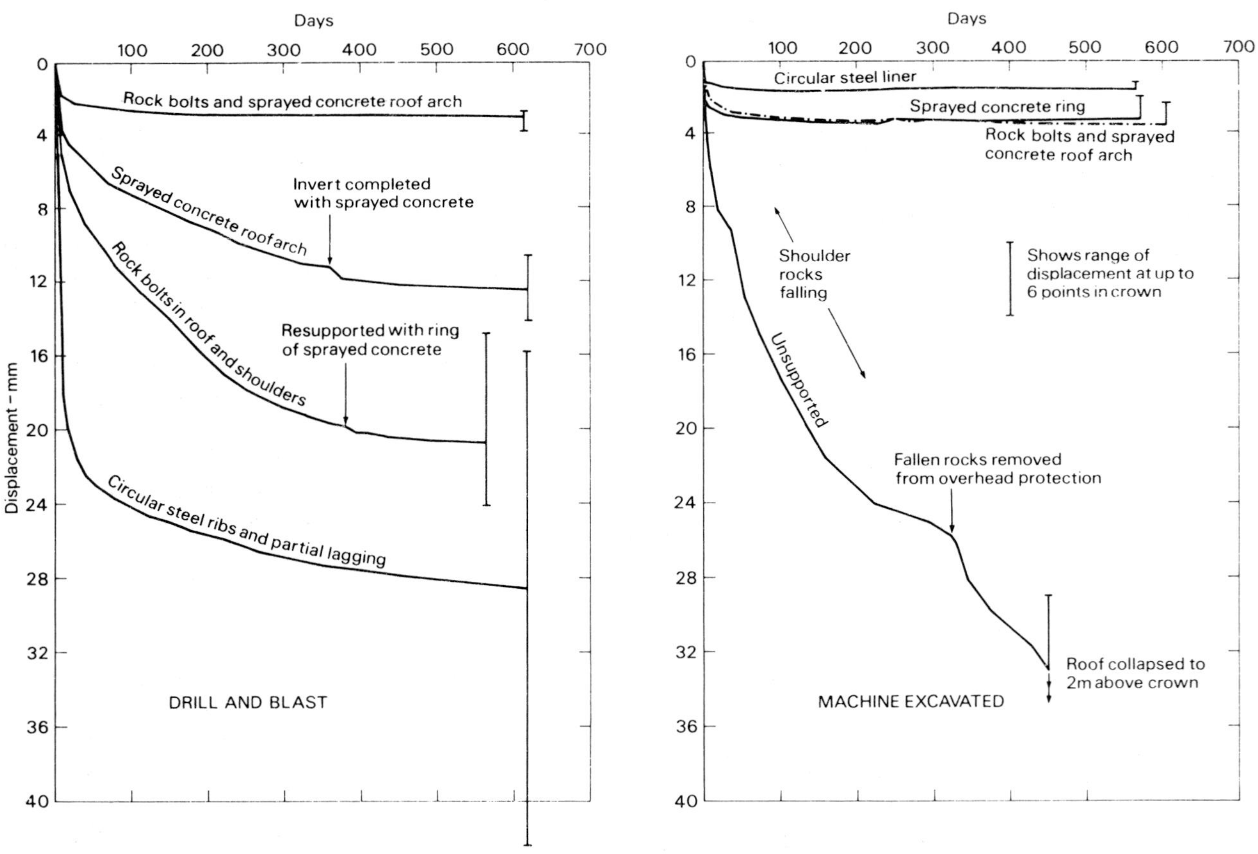

Figure 1

In two other lengths, one in the blasted part and another in the machine excavated part of the tunnel, the previous arch of sprayed concrete was combined with the earlier array of rockbolts, the rockbolts being inserted after the first layer of shotcrete had been applied. In both these lengths the supports behaved almost identically, only small displacements (3 mm to 4 mm) of the roof rock being recorded. However, displacements continue very slowly and the supports are not entirely satisfactory. In all three arches of sprayed concrete, with or without rockbolts, the early circumferential strains in the sprayed concrete were large and remained fairly high in the roof, but along the sides the concrete became hollow (detached from the rock) in places within a few days of placement. In the arch of sprayed concrete without bolts this was followed rather soon by extensive breaking away of the concrete and crushing of the rock behind. Just recently a similar failure had developed rapidly in a local area away from the instrumented part of the sprayed concrete with rockbolts length in the machine cut part of the tunnel. The reason for the localised rock crushing along the sides of the tunnel has been demonstrated by means of a simple model of the orientations and positions of the existing discontinuities in the rock mass in relation to the orientation and size of the tunnel[2].

The circular steel liner length, consisting of butt-welded 12.7 mm plate fitted tightly to the rock with a weak grout. and the length consisting of a complete ring of sprayed concrete, both in the machine excavated part of the tunnel, have performed in a similar way and are entirely satisfactory to date. The rock movements have been quite small, only 1 mm to 3 mm. Both of these supports carry quite high loadings equivalent to about 30 per cent of the total overburden of about 100 m of rock, which built up rapidly as the face advanced. During the second year after construction the loadings increased about a further 10 per cent.

It was made abundantly clear from the results of the first year's observations in the experimental tunnel[1] that the design requirements for a tunnel support involve a complex interaction between the dilation of the discontinuous rock mass and the deformation characteristics of the support. After a year there was at least a forty-fold variation in the loadings in the different supports, varying inversely with at least a twenty-fold variation of roof rock displacement.

Even at a particular section of one support system with given deformation characteristics the loading at a particular time after construction depends critically on the detailed timing of constructional operations. For example, on how fast or intermittently the face is advanced, where the section happens to be in relation to the face and what time has elapsed since the face was advanced. In other words the loading depends on the timing of the construction operations both before and after the particular section being considered. Many of these critical operational variables are often overlooked and can never be controlled strictly, except by very detailed monitoring measurements. They are among the hazards that need to be allowed for in design and construction and the miner should appreciate them even though he may have incomplete control over them.

Except for rocks which swell by absorbing water into their structure as they are unloaded, the more the surrounding rock is allowed to yield and dilate before the support is inserted, and the more the support yields under loading, the more lightly will the support be loaded and the greater will be the load carried by the rock. These features are well known to the old time miners in many countries, but they are sometimes forfotten today when the emphasis is on speed of construction and attempts are made, particularly in small tunnels, to place once and for all hard supports close up to the face. Nevertheless, even in deep tunnels where the deformation at the ground surface can be negligibly small despite large dilation of the rock surrounding the tunnel, too much dilation and too much yielding of the supports is unacceptable. Ultimate stability must be provided in the end.

The yielding limit does not appear to have been defined. It probably needs to be established in terms of the rock dilation. because this substantially augments the rock permeability. In water-bearing ground or around a leaky water tunnel the dilated rock mass becomes a significant conduit surrounding the tunnel. Internal erosion can develop and lead to instability and collapse of supports, unless the rock mass is grouted at considerable cost. How this limiting dilation should be defined remains an open question in the present state of knowledge, but it will be a function of the rock type and the way it disintegrates during dilation.

The deformation characteristics of the support system used in the experimental tunnel vary considerably, for example:

1 the supports consisting of steel ribs with lagging and rock packing have a stiffness which increases with rock displacement and is not well defined

2 the rockbolts themselves are of constant stiffness, but they can be displaced bodily with the rock surrounding them, or when their axial loading gradient exceeds a certain value may allow rock dilation; failure of the bolts themselves seems very unlikely in our particular rock

3 the steel liner is of constant stiffness and is fitted tight to the rock

4 a sprayed concrete support has a stiffness and strength which varies with time; additional layers may be added at different times so that the stiffness and strength of each layer will be different at the same time; normally an accelerator is added to the concrete to provide high early stiffness and strength, primarily for the purposes of preventing initial slumping and

peeling of the concrete from the rock; but this will also attract load to the concretes as the rock dilates; too rapid a gain in stiffness can lead to circumferential crushing failure of the concrete; equally if the rate of gain of stiffness and strength is too slow in relation to the rate at which the rock releases strain energy it will not support the rock adequately; clearly sprayed concrete is a versatile support, capable of having its time properties and thickness with time adjusted for optimum economy in relation to the behaviour of the dilating rock.

Detailed behaviour of a sprayed concrete arch, subsequently converted to a ring

We now examine in more detail the long-term behaviour of the simple arch of sprayed concrete built in the blasted part of the tunnel.

Near the centre of this experimental length in the morning of Day Minus Two the part to be instrumented was being approached and the face had been advanced 1 m to position A, see Figure 2. During the day a first array of rock extensometers had been installed, including extensometer 3, at a position 0.3 m behind the face and the first readings were taken later that day. Early in the morning on Day Zero (11 October 1974) the face was then advanced 2 m to position B.

Three to four hours later the first layer of concrete was sprayed between A and B over the roof and shoulders, and thinned out below axis level. This was covered with a layer of 200 x 200 x 6.4 mm wire mesh. At a position halfway between positions A and B vibrating-wire strain gauges (gauge length 14 cm) numbered 9 to 18 were tied lossely to the mesh and orientated circumferentially, see Figure 2. The second layer of sprayed concrete embedded the gauges and was completed ten hours after advancing to position B. The first readings were taken on the vibrating-wire gauges at this time. On Day One a second array of rock extensometers, including extensometer 9, were installed 0.3 m behind the face (ie 0.7 m in advance of the vibrating-wire gauges) and their first readings taken later that day. The face was then advanced 2 m to position C in the morning of Day Three. It was again advanced 1.9 m to position D near midday on Day Four. The sprayed concrete arch in each case was completed within seven hours of advancing the face.

The advance of the face in metres with time in relation to the position of the vibrating-wire gauges is plotted in the upper left corner of Figure 3 where it is labelled 'face ahead'.

The displacements of the rock 0.3 m above the roof, typified by extensometer 3 in the first array and 9 in the second array, are plotted with time across the lower half of Figure 3. Notice the sudden displacement of about 0.5 mm at extensometer 3 as the face is advanced from position A to B and again a sudden displacement of about 1 mm at gauge 9 when the face is advanced from B to C on Day Three. By Day Seven when the face had advanced some 10 m ahead the roof displacement at extensometer 9 has almost attained the value at extensometer 3. For the next two years, where Figure 3 has a compressed time scale, the rock displacements at both gauges 3 and 9 coincide with each other with differences of only a few tenths of a millimetre. Gauge 3 is, in fact, typical of all six measurements of rock displacement that were made in the roof of this experimental length.

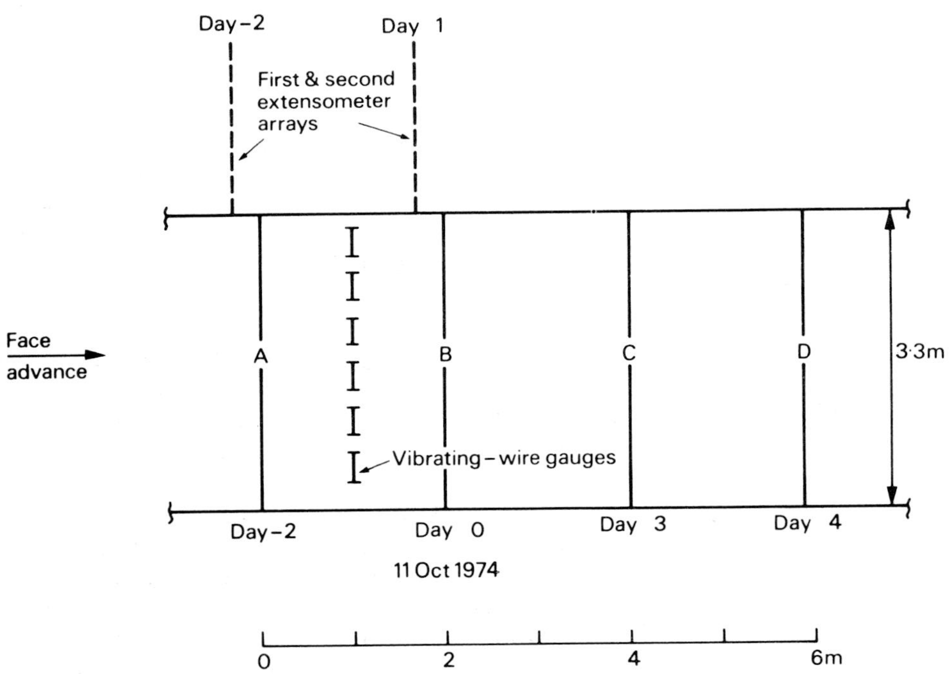

Figure 2

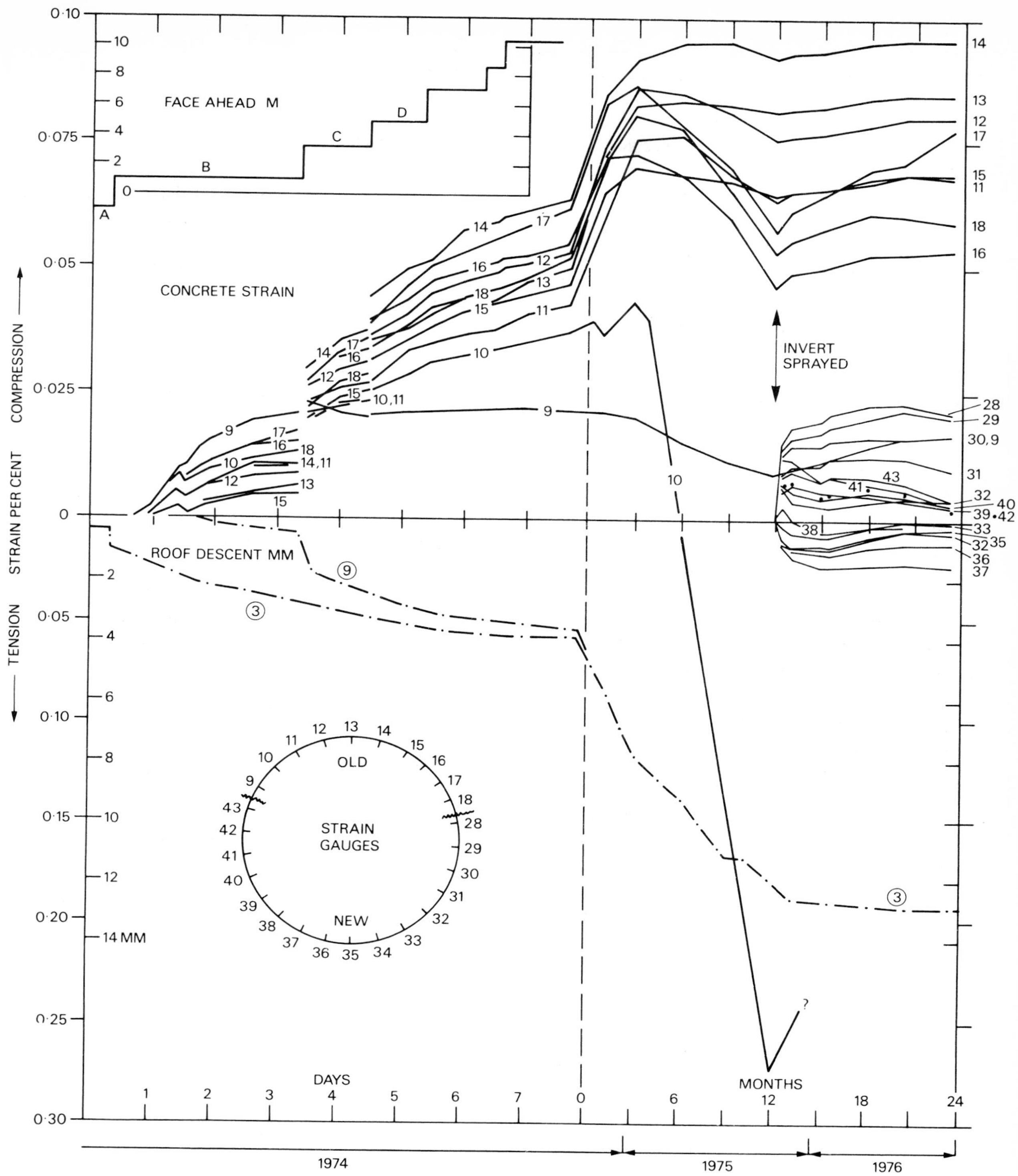

Figure 3

The circumferential compressive strains in vibrating-wire gauges 9 to 18 in the concrete are plotted with time in the upper part of Figure 3. Between Days One and Three when the face was stationary at B the strains in the shoulder concrete increased more than in the crown. The slight dip in the plot around the middle of Day One is associated with drilling the holes for the second extensometer array with rotary percussion drills. When the face was suddenly advanced from position B to C there was an abrupt increase in circumferential strain in all positions.

At gauge 9 the response to this effect was noticeably small. The strains continued to increase, though at gauge 9 there was a fall. Again when the face advanced beyond C there was a noticeable increase in the compressive strains, gauge 9 again being an exception.

During the week it was noted by tapping the sprayed concrete with a hammer that it had become hollow,ie detached from the rock, in places along the sides up to and somewhat above axis level, but remained tight everywhere in the crown. The strains continued to increase at all points, except at gauges 9 and 10, for some three months. Subsequently during the next nine months all compressive strains decreased, those in the crown only to a limited extent, but quite dramatically on the left shoulder at gauge 10. In the vicinity of this gauge a crack propagated from lower down the side, where the concrete had been disintegrating. During most of 1975 the rock displacements in the roof had become almost linear with time, and it was clear from our measurements that the sprayed concrete arch was descending as a whole and converging at about axis level.

It was necessary now to decide whether to allow this experiment to continue to a state of collapse or whether to start another experiment which would follow from an attempt to stabilise the support. We took the latter decision and the arch was stabilised after a year by carefully breaking away the loosened portions of sprayed concrete which had not already fallen away along the sides, and completing a ring of sprayed concrete by spraying two layers around the invert and sides.

In the new concrete further vibrating-wire gauges were embedded as before and numbered 28 to 43. Observations on all old and new instruments have now continued for a further year and are plotted in Figure 3. Except right in the invert where there are very slight tensile strains, there have been small compressive strains. At gauge 10 where the gauge went into high tension there has evidently been some yielding of the instrument wire, for after the strain was reversed to some extent it was no longer possible to obtain a reading – probably the wire has buckled into compression. On the right hand shoulder of the tunnel where the compression was falling most rapidly before resupporting (at gauges 16, 17 and 18) there was the most rapid increase in compression following completion of the ring.

The rock displacements after re-supporting have been exceedingly small, only a few tenths of a millimetre.

A number of vibrating-wire strain gauges orientated longitudinally were inserted in the sprayed concrete and also in panels of sprayed concrete kept in the tunnel. All these show very small strains and it is clear that the strains measured circumferentially in the lining are associated almost entirely with rock loading and not with thermal or moisture movement of the concrete.

It is useful to summarise the magnitudes of the strains and strain rates which occurred at different times in the experiment since these are important design parameters, which need to be quantified if the structural requirements of sprayed concrete as a support are to be specified to suit the rock conditions. Results from other rock conditions would be of considerable value also.

The mean and range of strains recorded in the concrete at different time intervals are summarised in Table 1.

The initial instantaneous strain associated with the first advance of the face may occur at any time in the early life of the sprayed concrete depending on the timing of the contractor's operations. Our instrumentation was partly responsible for the delay of 64 hours, otherwise the delay would have been only a few hours.

In a small 3 m tunnel the whole periphery of the tunnel can be sprayed in an hour or so to provide a closed ring support. In this case the early stiffness of the concrete is likely to be even more critical than in a large diameter tunnel in similar poor ground.

A 10 m tunnel in weak rock is usually excavated with two or three benches progressing behind each other, the concrete arch being extended towards the invert behind each bench. Thus many days, even some weeks, may pass before the sprayed concrete ring is closed. This construction procedure allows considerable dilation of the rock and much smaller structural demands on the sprayed concrete support than in a small tunnel where the ring is completed close behind the face. Even so, with multiple bench working in the poor rock zones of

Table 1 Compressive strains (per cent) in newly sprayed concrete arch

Time interval	Mean	Range
During first 64 hours when face stationary at B	0.012*	0.005 – 0.021
Instantaneous strain at 64 hours when face advanced from B to C	0.012	0.002+ – 0.020
During 64 to 91 hours when face stationary at C	0.005	0.002+ – 0.008
Instantaneous strain at 91 hours when face advanced from C to D	0.005	0.001+ – 0.009
Cumulative strain in first week	0.048	0.022 – 0.063
Cumulative strain in first month	0.064	0.021 – 0.085

*The strain gauges are likely to underestimate the actual strain in the very early life of the concrete on account of their stiffness.

+The lower values are associated with gauges on the left side where the concrete later fails in tension the large Tauern and Arlberg road tunnels the Austrians found it necessary to increase the circumferential compressibility of their sprayed concrete by systematically leaving several longitudinal gaps, around the periphery of the concrete. The steel continued across these gaps, buckled inwards and considerably reduced the hazards of falls of crushed concrete which occurred otherwise.

It is not possible at present to interpret the concrete strain observations in terms of stress, and no information is available on the early strength characteristics of the concrete in the mudstone tunnel. It is intended to measure the load-strain properties in the laboratory by straining newly sprayed concrete equipped with identical strain gauges at the rates measured in the field. The effects of high early strains on the long-term strength of the concrete will also be examined since they may be significant.

DRY MIX SPRAYING – LABORATORY TRIALS

Equipment and tests

A spraying machine of the widely used rotary chamber type was used, namely a Meynadier model GM57 with a two-speed electric drive and a nominal output of 2 m^3 to 6 m^3/hour, Figure 4. This machine consists of a multi-chamber rotor clamped, with its axis vertical, between rubber seals. The dry mix is stored in a hopper above the rotor and is gravity fed through a hole in the top seal into the rotor chambers. It is then carried round to the discharge point where compressed air is injected into the top of the chamber, forcing the mix into the outlet and hence the pipeline. Further compressed air is injected at the bend in the outlet to convey the mix away. At a point midway between the discharge and charge positions holes are provided through both top and bottom seals to vent the empty chambers to atmosphere.

A rubber hose 50 mm bore and 15 m or 20 m long was used for conveying the dry mix to the nozzle and an 8.9 m^3/min reciprocating compressor supplied the conveying air. For most of the trials a parallel steel nozzle was used with water injected 200 mm from the end, Figure 5. Water was supplied by a 0.7 MN/m^2 electric pump. The pressure, temperature and flow rate of the water and conveying air were measured in all trials.

Routine tests included the measurement of rebounded material, the wet and dry density of the deposited concrete and the compressive strength of 75 mm diameter cores cut normal to and in the direction of spraying. In selected trials the tensile strength and bond strength to the substrate were measured and the composition of both deposit and rebound determined by wet-mix analysis.

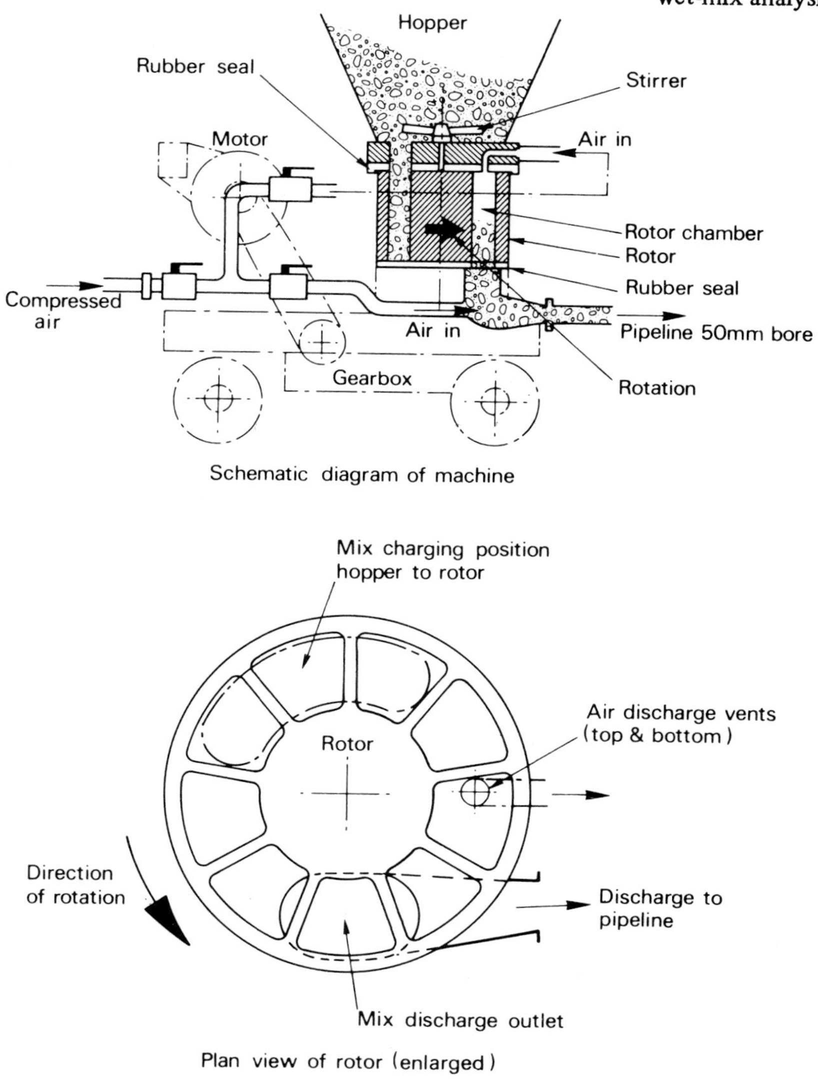

Figure 4

In addition, surfaces of test panels sawn parallel to the direction of spraying were examined to assess material distribution. Initially test panels 1 m^2 by 160 mm thick, mounted vertically in a large wall, were sprayed for subsequent testing; these were sprayed in four layers of equal thickness at approximately hourly intervals and will be referred to as large test panels. For the majority of trials, however, panels measuring 600 mm x 180 mm x 180 mm were sprayed in moulds with mesh tops and bottoms to allow for the escape of air and rebound; these small panels were sprayed to full thickness in one application.

All spraying, unless otherwise stated, was horizontal onto a smooth, vertical concrete surface with the nozzle held 1 m from the surface. No admixtures were used.

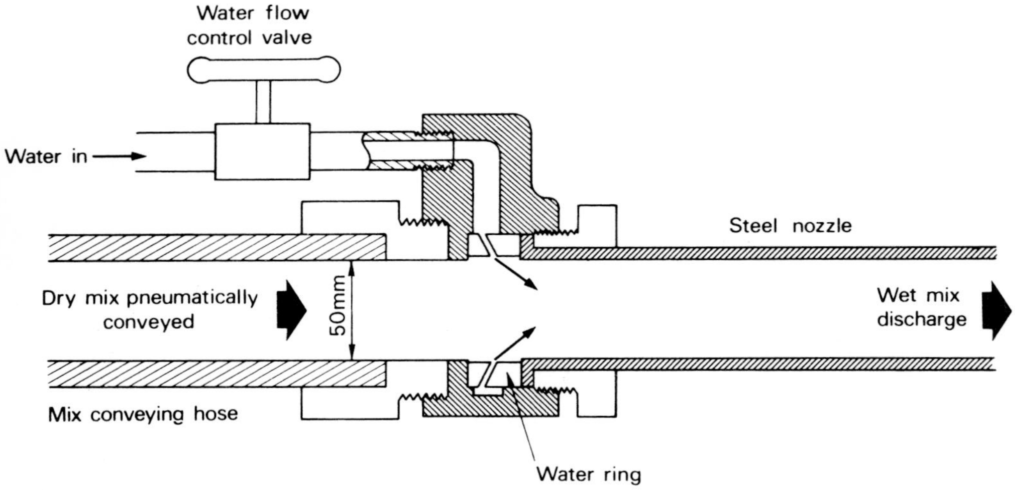

Figure 5 Nozzle and water injector

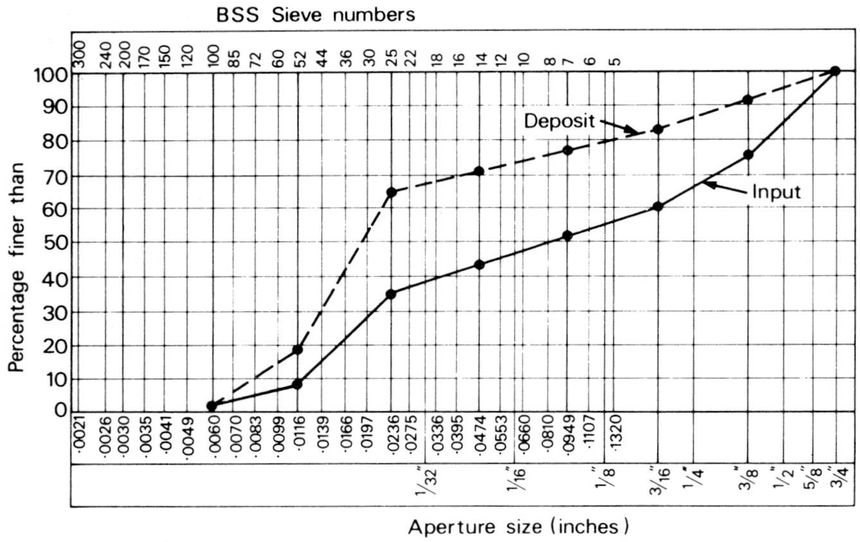

A Coarse aggregate mix

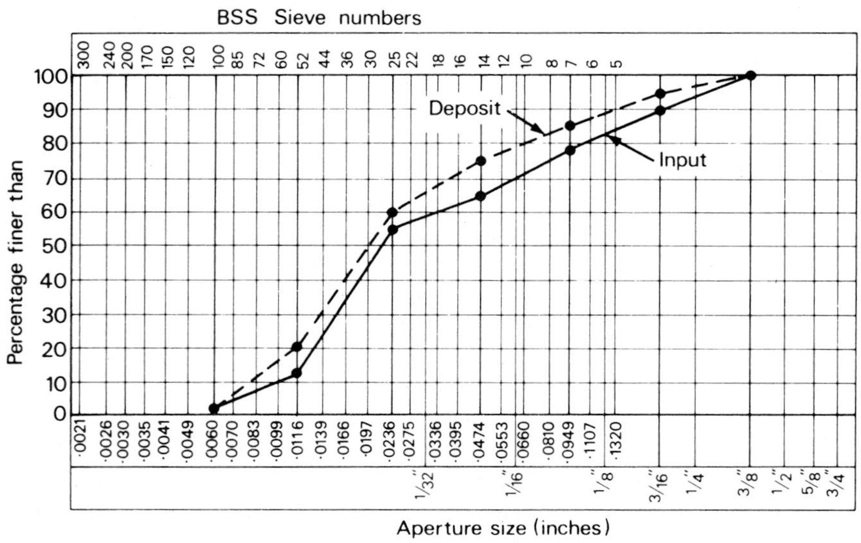

B Fine aggregate mix

Figure 6

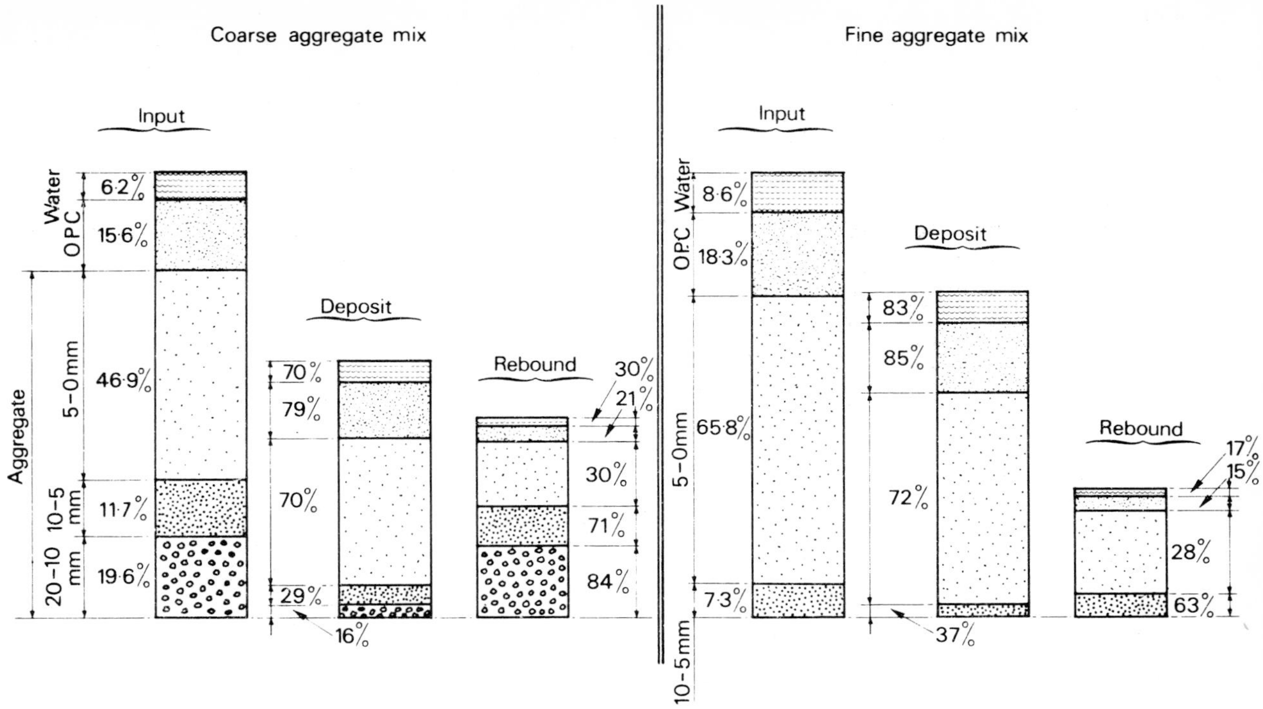

Figure 7 Mix compositions – input, deposit and rebound

These laboratory trials were designed to check the effects of mix and process variables on the deposited concrete and the results would not necessarily be expected to be the same as those obtained under practical site conditions.

Materials

Sprayed concrete mixes can be broadly divided into fine aggregate mixes, with a maximum size of 10 mm, and coarse mixes containing aggregates generally up to 20 m. Most of the work described here was carried out with the more widely used fine aggregate mix but a limited number of coarse mixes were also sprayed. The gradings used were intended to follow 'conventionally used' gradings[3]; the gradings are shown in Figure 6 and the total wet mix compositions, as discharged from the nozzle, in Figure 7.

General characteristics of the system

Let us make a general statement about what is required of concrete deposited by spraying and then see how far the dry-mix process goes towards meeting these requirements.

It is assumed the deposited material should be dense, homogeneous and isotropic with properties which can be predicted with sufficient accuracy for economic structural design and which are similar to those of well compacted cast concrete of similar composition. The laboratory trials have shown that a number of characteristics, apparently inherent to the process as it is generally used, are likely to detract from these properties. The more important ones are:

1 non-uniform distribution of the mix ingredients discharging from the nozzle, resulting in an inhomogenous deposit.

2 variations in the total water content of the mix, due to the quantity injected being at the discretion of the nozzleman and to time variations in the solids throughput, due to pulsatory delivery at the machine and to partial choking of the conveying system.

3 orientation of elongated particles in the deposit normal to the direction of spraying.

4 material losses due to rebound, resulting in a deposit of unpredictable composition.

These points will now be considered in detail.

Materials distribution in the spray

Non-uniform distribution of the mix at discharge from the nozzle is apparent in at least four distinct ways. The cyclic nature of the discharge of the dry mix from each rotor chamber into the conveying pipe results in 'packages' of material moving through the pipe and nozzle; since the water is injected at a constant rate close to the discharge the resulting spray and hence deposit tends to consist of alternating quantities of mix with varying water content, Figure 8a. The effect on the deposit is shown in Figure 9 (upper); here the mix was sprayed vertically downwards with the nozzle stationary and a section sawn through the resulting cone shows a number of layers consisting of alternate dense and porous bands. Each layer corresponds to a discharge from one chamber of the machine rotor. The pulses of materials can be seen frequently during spraying and appear more pronounced the larger the largest aggregate in the mix; even when the pulses cannot be seen their presence can be detected aurally by directing the spray against a suitable suface such as thin plywood or taut polythene sheet. The severity of pulsing is likely to depend on a number of factors, especially the geometry of the rotor chambers and their frequency of discharge.

The smaller the package size (ie chamber volume) and the faster the frequency of discharge, then the less severe the pulsing is likely to be. The spraying machine was driven so that the rotor chambers discharged at frequencies of either 65 per minute or 97 per minute, ie one discharge every 0.92 or 0.62 seconds. Examination of the spray at the nozzle by high-speed cine-photography showed the average coarse particle velocity to be around 25 m/sec so that at the highest rotor speed the 'pitch' of the pulses would be approximately 15 m at the nozzle although the pitch would be less close to the spraying machine. Each pulse contains 1.3 litres of dry mix (the capacity of each chamber) or about 2 kg. The pulses do not, of course, consist of discrete packages of material with large gaps between them; discharges from each chamber takes a finite time and the particles travel at different velocities along the pipe. There is also some evidence to suggest reduced pulsing severity the longer the conveying pipe, so some intermingling of the packages of material takes place and the distribution is probably like a series of overlapping skewed distribution curves.

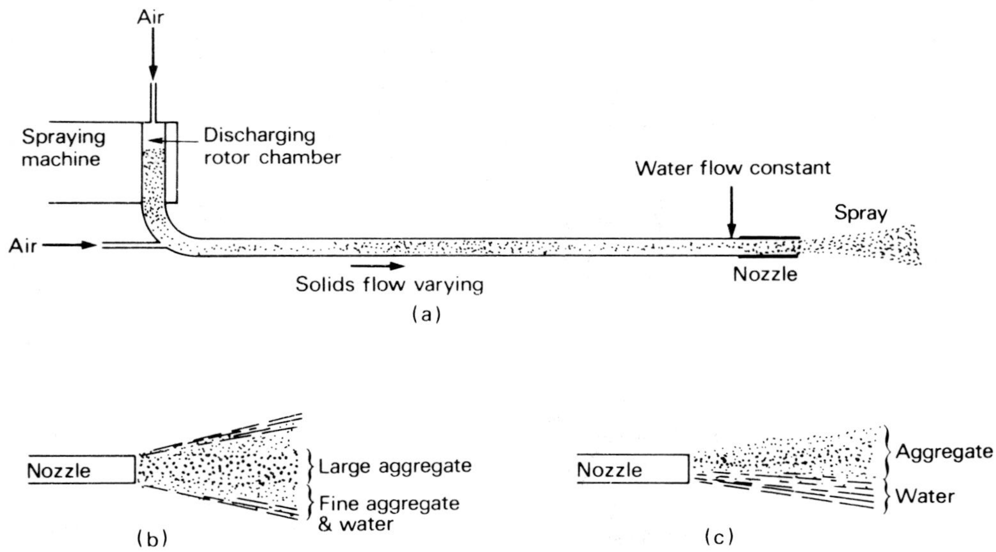

Figure 8 Distribution of materials in the pipeline and on discharge

Figure 9 Sections sprayed vertically downwards, nozzle stationary (upper), with three traverses of nozzle (lower)

A feature of the spray only apparent from the examination of cine film shot at high speed and described in more detail later is the irregular solids flow in the short term, ie in hundredths of a second. Apart from the longer term (around 1-second) cyclic variations due to the chambers discharging and described above, the discharge consists of alternating and irregular clusters of dense solids and sparse, virtually solid free regions. The resolution was not good enough to see the finest particles nor the distribution of the water in the spray.

Another form of non-uniform materials distribution is transverse segregation of materials on discharge from the nozzle. This is not always visually apparent and may be due to a combination of expansion of the conveying air and inadequate mixing of the water and finer particles, so that these are carried towards the edge of the spray cone, Figure 8b. As the nozzle is transversed across the work this effect again gives rise to layers of different density and composition. Figure 9 (lower) shows a section through a windrow of material sprayed vertically downwards with three traverses of the nozzle, that is in three layers, and three fairly distinct layers can be seen.

A fourth source of poor distribution is eccentric discharge, in which most of the solids appear to be confined to one side of the spray Figure 8c. The can be seen frequently and is illustrated in Figure 10. The reasons for this are not clear; it may be due to turbulence or some eccentric mixing effect at the water ring or due to the presence of a bend in the pipe close to the nozzle creating a sidewall effect. If the latter is the case then the effect lasts over considerable distances from the bend – the effect is still apparent with as much as 5 m of hose laid straight between the nozzle and the first bend. Also the stream of greater material concentration has been observed to suddenly change its position without changing the hose position and is not always associated with the region of the nozzle periphery associated with the outside of the nearest bend, as would be expected. The effect of this phenomenon is to produce transverse variations across the spray which can result in concrete of varying composition; it was observed during the spraying of the two specimens shown in Figure 9. The right hand side of both specimens appears weak and porous and may be due to insufficient water resulting in bulking and a region of non-plastic mix.

The combined effect of all these distribution problems is the layered product shown in Figure 11 (upper). The layering can be clearly seen and is surprisingly regular considering that the nozzle was moved in a fairly random way during the spraying of the panel. The density and composition through the depth of such a panel have not yet been checked in detail. The layering, although probably always present to a greater or lesser extent in dry-mix sprayed concrete, is not always apparent from broken pieces or even cores and a fairly large carefully sawn surface is necessary to reveal its presence clearly. In extreme cases, however, the layers can be seen in cores and the fracture planes in compression tests have been observed to lie along the layers when these are oriented towards the direction of loading.

Figure 10 Eccentric material discharge

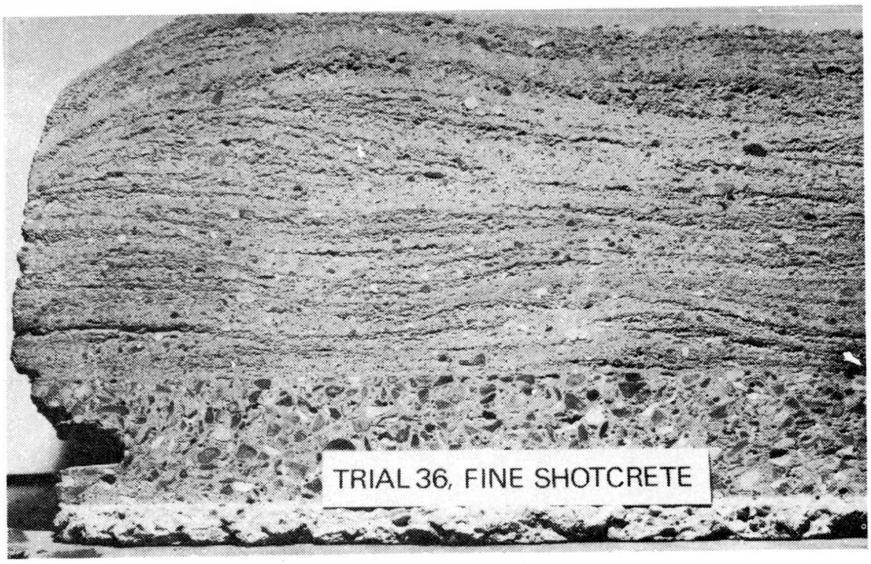

Figure 11 Sections through 160 mm thick panels sprayed with a short nozzle (upper) and a long nozzle (lower)

All the tests so far described were carried out with a parallel steel nozzle with the water injected 200 mm before the point of discharge; this will be referred to as the short nozzle.

The poor distribution of the solids as they arrive at the nozzle is inherent in the machine design. It is evidently not much influenced by the injection of water largely because of the very short time available for any mixing of the solids and water to occur before the stream disperses out of the nozzle. At a mean particle speed of about 25 m/sec, this being the order of speed determined by high-speed photography, the mixing time in the short nozzle is 0.008 sec. This is an incredibly short time for mixing and it is clear much of the wetting of the solids occurs beyond the nozzle.

The effect of increasing the mixing time by injecting the water into the materials stream 5 m back from the nozzle tip (this will be referred to as the long nozzle) led to a remarkable improvement. With the mixing time in the nozzle increased 25 times to 0.2 sec the mixing of water and solids was considerably improved and a large number of comparative tests showed that, whereas with the short nozzle layering was always present and became more pronounced with increasing water content, use of the long nozzle resulted consistently in a substantially uniform deposit with very little evidence of layering. Compare the sectioned panel shown in Figure 11 (upper), with that of Figure 11 (lower) produced with the same mix and spraying conditions but with the long nozzle. The effects of the position of the water injection on the appearance of the spray can be seen in Figure 12.

Figure 12 Discharge from short nozzle (upper) and long nozzle (lower)

The mechanism of water dispersion into a fast moving stream of dry mix has not been examined in detail and is an area which should be investigated further. The advantages of increasing the mixing time by the use of the long nozzle have been empirically established, however, and it is recommended the method should always be used for dry-mix spraying. There are no apparent practical disadvantages to the use of the long nozzle, such as blocking or plugging and in fact there are several advantages from the nozzleman's viewpoint, especially if combined with metered water control. It is not suggested 5 m is the optimum nozzle length. Injection of the water up to 10 m back from the hose end has been tried but without any apparent improvement in the mixing efficiency and placing the water ring less than 5 m from the hose end has a practical disadvantage because the nozzleman has to carry its weight. A higher air pressure is required at the spraying machine the longer the nozzle. A pressure about 13 per cent higher is required with a 5 m nozzle compared to the short nozzle and a limited amount of data suggests this may increase to as much as 60 per cent higher with a 10 m nozzle.

Variations in water : solids ratio

Two major causes of variations in the total water content of the mix, ie the ratio of water to solids, are variations in the solids throughput due to partial choking of the conveying system and variations in the amount of water injected imposed by the nozzleman.

Cause of irregular time changes in the solids throughput can be attributed to the moisture content of the 'dry' mix, machine design and operation, and pipeline layout. Sand moisture contents of 4 to 8 per cent are frequently quoted as being the optimum for the dry-mix spraying on the grounds that moisture contents lower than this lead to poor mixing and large amounts of airborne dust at the point of spraying, whilst higher moisture contents cause adhesion of the fines in the pipeline and ultimately blockages. Experience with both the long and short nozzle in the tunnel lining trials described later suggest better mixing and reduced dusting when the long nozzle is used so that specifying a minimum moisture content is probably not necessary. The sand used for the field trials described later varied in moisture content between three and six per cent with a mean of four per cent, below the maximum usually recommended for dry mix spraying, yet sufficient fines adhered in the outlet from the spraying machine to seriously affect the spraying performance of the equipment after about 45 minutes spraying and sometimes within 15 minutes of starting up with a clean machine. The question of the moisture content of the mix needs further investigation and the mix standing time between mixing and conveying also appears to have a significant bearing on the equipment performance; it is suggested, however, that sand for pneumatic conveying with the present machine should be specified as having a moisture content below four per cent.

Improvements could be made to the design of the spraying machine to reduce the adhesion of fines in the most critical part of the conveying system, namely the chamber beneath the discharging rotors into which the mix drops at low velocity and the attached taper in which the mix is accelerated into the line, by the main conveying air stream. The use of rubber liners, a more gradual change in the material flow direction, a longer taper or a different arrangement for the air inlet such as multiple or a tangential inlet, might reduce material build-up.

The second source of irregular variations in the water solids ratio is due to the nozzleman who in conventional practice has sole control of the quantity of water injected into the materials stream. The laboratory trials indicate the mix can be deposited with water/cement ratios in the range 0.45 to 0.65, that is a variation in the water of about 45 per cent. Such a large variation leads to significant variations in both the amount of material loss due to rebound and in the properties of the deposited concrete. The use of a water meter to ensure a steady flow at a predetermined rate would eliminate these variations and also simplify the nozzlemen's job.

Particle orientation

A characteristic of sprayed concrete is the tendency for elongated particles to lie with the long axis parallel to the plane of the sprayed surface. This is analogous to dropping a coin on a flat surface – it can only take up one orientation. This would give rise to anisotropy if the mix contains a high proportion of elongated material, particularly the larger aggregate.

Rebound

Probably the most obvious undesirable characteristic of sprayed concrete to anyone seeing it for the first time is the large amount of material rebounding from the surface and falling to the floor. This material is scrap and cannot be reused for anything aspiring to be a reasonable quality concrete for a number of reasons. Rebound has been variously reported[3,4,5] as being typically 20 to 50 per cent for overhead spraying, 10 to 36 per cent for horizontal spraying and zero when spraying more than 30° below the horizontal.

Many factors influence the amount of rebound, such as nozzle distance, nozzle orientation, particle size, shape and velocity, mix grading, cement content, water content, effects of admixtures and the nature of the surface being sprayed.

Particle velocity is of particular interest since this is likely to influence not only the amount of rebound but also the degree of compaction and hence strength and density of the deposit. Particle velocities as high as 150 m/sec have been reported[6]. However, analysis of film taken with a high-speed cine camera at speeds of 2500 to 4000 frames per second have shown that particles when discharging from the nozzle have mean velocities ranging from 15 m/s to 35 m/s, depending on the air pressure at the machine, with occasional particles travelling as slowly as 10 m/s and as fast as 56 m/s. Particle velocity increases with reducing particle size so that 20 mm stones are generally up to 20 per cent slower than the smallest sand particles resolvable on the film, about 3 mm diameter. A characteristic of the conveying system is the amount of spin imparted to many of the particles; some of the larger, elongated particles have been found to be spinning at up to 250 revolutions per second. These high spin rates could be an important factor in the rebound mechanism.

Another rebound factor is the nature of the surface onto which the concrete is sprayed. It is self-evident that only the finer particles coated with cement will adhere initially when spraying on a hard surface; as this layer increases in thickness a sufficient depth is reached at which progressively larger particles can be embedded, and the proportion of material rebounding would be expected to diminish. Most of the rebound measurements in these trials were made by spraying to a depth of 40 mm to 50 mm over about 1 m^2 onto smooth, vertical concrete panels and calulating the percentage loss be weighing the total rebound and comparing it either with the weight of the material deposited or the known total material throughput over the duration of the test. The maximum thickness that can be sprayed horizontally or vertically, even at low water contents, before the material slides or peels appears to be about 50 mm; greater thicknesses are likely to be possible with the aid of reinforcement or an uneven surface. This only applies when the deposit is relying on its cohesive strength and before any setting starts.

The measured average rebound for a large number of tests was about 40 per cent for coarse aggregate mixes with a water/cement ratio of 0.4 and 30 per cent for fine aggregate mixes with a water/cement ratio of approximately 0.45. Analyses of the mixes are summarised in Figure 7. It can be seen that the rebound loss increases with particle size so that very roughly 20 per cent of the water and cement are lost, 30 per cent of <5mm aggregates, 65 per cent of 5 to 10 mm aggregate and 85 per cent of 10 to 20 mm aggregate.

At increasing water contents the rebound loss is smaller. Figure 13 shows the relation between water content and rebound (fine aggregate mixes only) within the practical range of water contents ie water/cement ratio from 0.45 to 0.65. The relationship is almost linear with no significant difference between mixes sprayed with the long and short nozzles, varying from 25 to 30 per cent at a water/cement ratio of 0.45 to about 10 per cent at a water/cement ratio of 0.65. There is therefore a considerable benefit in material economy in spraying as wet as possible. The effects on the strength of the deposit will be described later.

Properties of the deposited concrete

The compressive and tensile strength and density of 75 mm diameter cores, cured in water, for replicate coarse and fine aggregate large (1 m^2) test panels are given in Table 2. Both compressive strength and density are lower than would be expected of well compacted concrete with similar cement and water contents. The coefficient of variation of compressive strength is high, being generally in the range normally associated with concrete made with a poor degree of control[7]. The direct tensile strength was determined by biaxial fluid test and the bond at the interface was found to be similar to the strength within the sprayed concrete, although the failure in the concrete probably occurred at one of the weak, porous layers. It was not possible to measure the tensile strength normal to the direction of spraying because of the escape of the compressing medium through the specimen ends via the porous layers.

It should be noted there is no significant difference in the transverse and longitudinal compressive strengths even though all the test panels exhibited the type of layering shown in Figure 9. Directional differences might be more apparent in, say, flexural testing, drying shrinkage or moisture movement tests; it is important to establish the effects of layering on the properties of the material but this would require testing on a more extensive and diverse scale than was possible in the trials reported here.

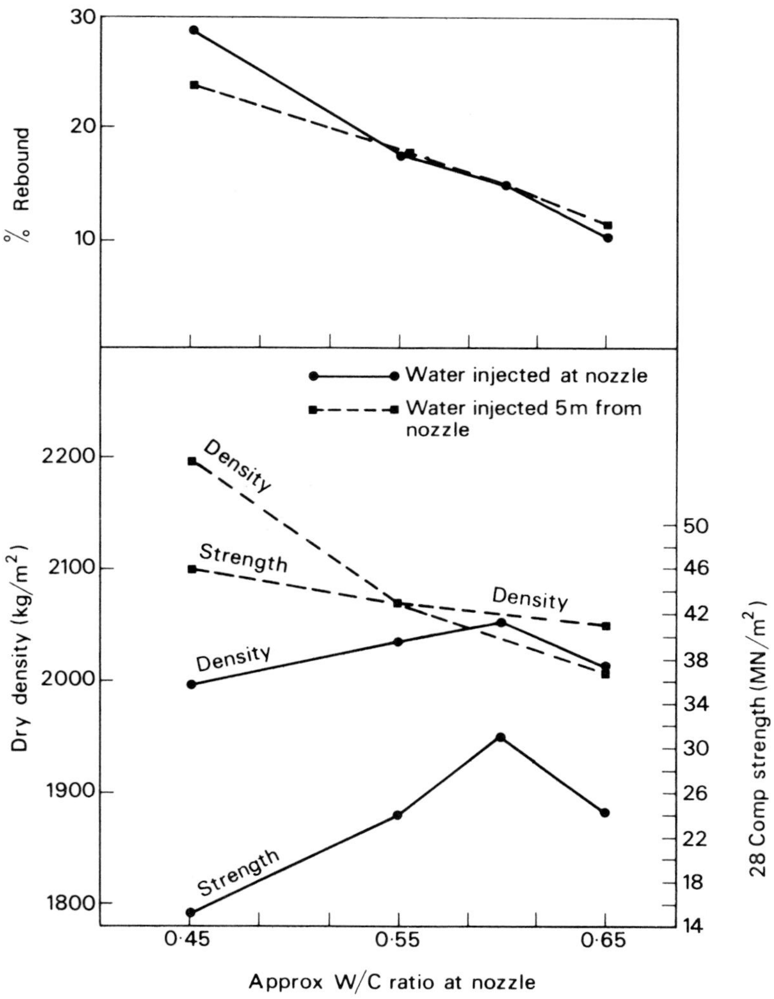

Figure 13 Effect of water content on rebound, strength and density

Table 2 Strength and density of cores, coarse and fine aggregate sprayed concrete – laboratory trials

		Mean 35-day equivalent cube strength compression – MN/m²					Mean 35-day tensile strength – MN/m² in direction of spraying only				Density kg/m³		
	Trial number number	In direction of spraying	Normal to direction of spraying	All specimens n		Coefficient of variation	In sprayed concrete n		At interface n		n	Oven dry	Saturated surface dry
Coarse aggregate mix	32*	15.4	17.3	16	16.3	20%	9	0.9	4	1.1	4	2005	2188
	33*	22.3	23.4	16	22.9	21%	11	1.5	2	1.3	4	2018	2203
	34	22.5	21.6	17	22.0	15%	9	1.0	3	1.4	7	1986	2185
Fine aggregate mix	36	33.1	29.7	16	31.4	23%	12	0.9	3	1.4	6	2028	2227
	37	29.7	32.3	16	31.0	28%	10	1.1	4	1.0	7	2027	2222
	38	27.6	34.0	16	30.8	28%	12	1.0	3	1.2	7	2017	2217

Note: n = number of specimens tested

* – test cores not capped

All subsequent spraying tests were carried out with smaller test pieces measuring 600 x 180 x 180 mm and the fine aggregate mix only in order to accumulate data more rapidly in the face of limited resources and a large number of variables. It was apppreciated that the properties of such small specimens would be even less representative than the large test panels of what could be attained in practice on site, and they were intended only to show any significant changes due to changing variables.

A large number of trials were conducted and such factors as nozzle distance, nozzle tip design, position of water injection, water content and air pressure were examined.

It was found that the smaller test panels yielded generally lower compressive strengths than the 1 m² panels and also a considerable directional difference; mean strengths were as much as 50 per cent lower at low water contents and in all specimens sprayed with the short nozzle the longitudinal strength was higher than the transverse, on average 35 per cent higher. These differences due to specimen size are attributed in part to the manner in which they were made. The large panels were sprayed to their 160 mm thickness in four equal layers at 45- to 60-minute intervals so that each new layer was deposited on a partially set layer. The small panels, being well supported by the mould, were sprayed to the full 160 mm thickness in one application and it is thought incipient vertical cracks formed in the material due to slight slumping in the mould. At higher water contents, very large lens-shaped cracks were clearly visible in sawn sections, and were probably due to poor mixing or partial slumping.

Small panels sprayed with the 5 m nozzle did not show such defects; the appearance of sawn sections showed fairly uniform distribution of the materials with no marked layering or cracking and no directional strength differences.

The advantage of high water content on reduced rebound has been described and the strength and density of concrete sprayed with both long and short nozzles over the range of water contents used for the rebound tests are shown in Figure 13. The density and strength of material sprayed with the short nozzle both increase with increasing water content up to a water/cement ratio of about 0.6, then apparently start to fall. This fall could be due to a chance effect, such as trapped rebound, or the shape of the curve might be indicating perhaps increasing density due to a more plastic mix with increasing water content up to a certain optimum, followed by a reduction due to the normal trend in well designed and compacted concrete to reduced density and strength with increasing water. The material sprayed with the 5 m nozzle, however, shows a steady reduction in strength and density with increasing water as might be expected. The strength, though, is very much higher at all water contents than that of the mix sprayed with the short nozzle. The difference in the results for the two nozzle types may be attributed to differences in the water content of the deposited concrete. The graphs are plotted against the water content as discharged from the nozzle and it is likely that better wetting occurs with the long nozzle and therefore less water is lost to the surrounding atmosphere; photographs of the spray tend to support this. Additionally since a higher air pressure is required with the long nozzle the discharge velocity and hence the compaction effort may be greater, resulting in a higher density and strength. Further wet mix analysis and measurements of particle velocities with the long nozzle are needed before a true comparison can be made of the effects of the nozzle types on the deposited concrete.

Particle velocity has been mentioned and the effects of increasing the air flow and hence particle velocity on the deposit are shown in Figure 14. These tests were carried out at low water contents using the short nozzle. Particle velocity increases in direct proportion to the air pressure at the spraying machine, as does the density and compressive strength of the fine aggregate concrete. The density and strength of the coarse aggregate concrete falls at the higher particle velocities and this may be due to excessive rebound interference especially in the confines of the mould forming the small test panels; materials sprayed on an effectively boundless surface might not show such a fall in strength. It is important to remember that not only are the placing conditions changing with increasing spray velocity, ie the degree of compaction, but also the composition of the mix is changing because of the probability of higher rebound. The relation between spray velocity and rebound has not yet been examined in detail but some indication of the scale of change can be had from two tests with the coarse aggregate mix; in one test at an air pressure of 0.15 MN/m^2 the rebound was approximately 35 per cent and the compressive strength 22 MN/m^2 whereas in a second test at an air pressure of 0.43 MN/m^2 the rebound increased to 60 per cent and the strength to 40 MN/m^2.

A useful area for further research would be to extend the information of the type shown in Figure 14 to a wider range of air pressures and to examine the effects of varying the water content and the position of water injection. Rebound should also be measured. Optimum process conditions could then be established for a given mix.

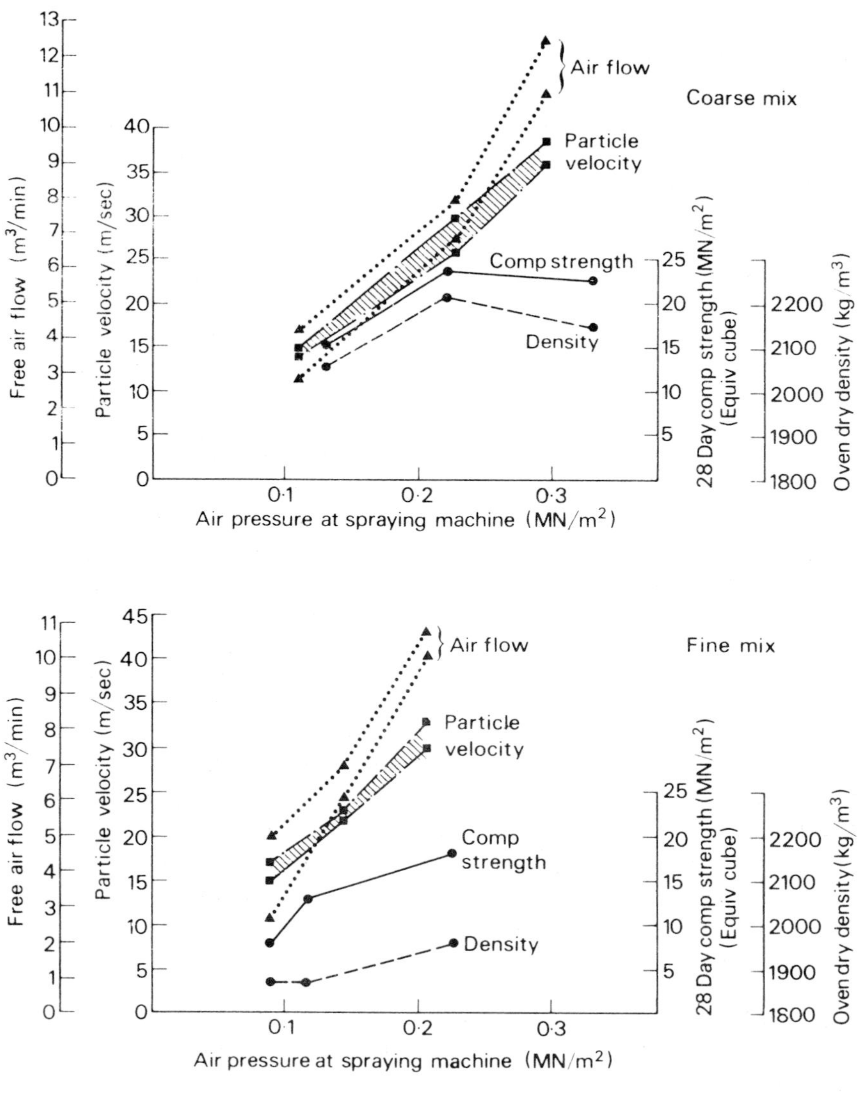

Figure 14 Effect of air pressure on particle velocity, density and strength – coarse and fine mixes

DRY MIX SPRAYING – FIELD TRIALS

Site, materials and equipment

The opportunity to apply some of the results of the laboratory trials arose with the need to provide additional support to two lengths of the Kielder experimental tunnel. These lengths, each approximately 12 m long and 3.3 diameter, were in the drill and blast length of the mudstone heading and were originally supported during construction one year previously by (a) rock bolts and (b) sprayed concrete. Additional sprayed concrete 100 mm to 150 mm thick was required in both lengths to form a complete circular support.

The rock bolt length had blasted in places to a section more square than circular, particularly above the axis. Since it was required to bring the length to an approximately circular profile the application of sprayed concrete up to 1 m thick would be necessary at the shoulders. In the roof the rock bolts were linked with 50 x 50 x 3.2 mm diameter steel mesh. Loose rock had collected on the mesh with hollows behind; for safety reasons it was decided not to attempt to remove the loose pieces but to spray as much concrete as possible through the mesh before building up to the required thickness.

In the sprayed concrete length the concrete had been applied during the construction of the tunnel (about one year earlier) in the roof and down to within 0.5 m of the floor; the average thickness above axis level was 160 mm, reinforced with mesh, and below axis level thinning down to 25 mm. The sprayed concrete below the axis had broken away in places and become hollow elsewhere, ie detached from the rock and it was proposed to break out this old material and spray the bottom half of the length thus completing a full circle of concrete.

The materials used were ordinary Portland cement, a Zone 3 (BS882) marine sand and 10 mm single-size, crushed limestone coarse aggregate. The mix proportions were:

OPC/aggregate ratio 1:5,
sand/coarse aggregate ratio 7:3.

This mix was similar to that used orignally in the sprayed concrete lengths except that the marine sand had to be substituted for a crushed sand which was no longer readily available.

It was intended to use a liquid set-accelerating admixture to facilitate spraying in the tunnel and a commercial sodium silicate solution was chosen on the basis of the extensive use of silicates in sprayed concrete in Europe and its relatively low cost compared with proprietary concrete admixtures.

The same equipment used for the laboratory trials was used at the tunnel. The spraying machine was located above ground adjacent to the concrete mixer and aggregate supply in order to centralise as many operations as possible. The mix was conveyed some 150 m to the nozzle through both steel and rubber pipe of 50 mm bore. The equipment layout is shown in Figure 15 (opposite). A field telephone was used for communication between the nozzleman and the surface team.

Over a period of nine working days 130 tonnes of mix was sprayed in the tunnel, the actual spraying time being approximately 20 hours, ie a spraying rate of 6.5 tonnes/hour. The machine was run at low speed to minimise conveying problems over the comparatively long distance.

Spraying

The spraying was in two phases. The first phase consisted of filling in the overbreak in the rock bolt length to bring the tunnel to a circular profile, and this consumed 25 per cent of the total material used. The average rebound was 57 per cent. The short nozzle was used with the water controlled by the nozzleman. No admixtures were used and the material was sprayed with an average water/cement ratio of approximately 0.40 although the water content did vary considerably as the nozzleman changed the flow rate seeking an optimum.

The second phase consisted of placing the lining proper in a complete circle in the rock bolt section and in the lower half of the sprayed concrete length with a lap joint to the old concrete at axis level. Two layers of 200 mm mesh were incorporated in the rock bolt section and one layer in the sprayed concrete section. The long nozzle was used for this work with the nozzle tip removed and the water, injected 5 m back, was preset to a constant flow rate over which the nozzleman had no direct control. An admixture was incorporated. The sodium silicate originally proposed proved unsuitable in above-ground trials and a proprietary aluminate-based liquid admixture, Sigunite liquid, was brought in and used at short notice. The water/cement ratio was increased to 0.53 average and maintained at this level fairly consistently. The average rebound in this phase was 40 per cent. The total volume of placed concrete amounted to about 40 m^3.

The first phase spraying was arduous. The varying angles of the rock surfaces made spraying at 90° difficult. The absence of a set-accelerating admixture meant the mix had to be sprayed with a low water content to produce the firmest possible deposit to minimise sloughing; only about 30 mm could be built up so the nozzleman had to keep moving over the whole 10 m of the experimental length to give the material time to set before applying a further layer. Visibility was very poor due to airborne dust and water. All these factors contributed to the very high rebound loss.

In the second phase the advantages of the long nozzle and the accelerating admixture were quickly apparent. With the water ring moved back 5 m and the steel nozzle tip removed the nozzleman merely held the end of the conveying hose. There also appeared to be less dusting with the long nozzle. In addition the nozzleman's task

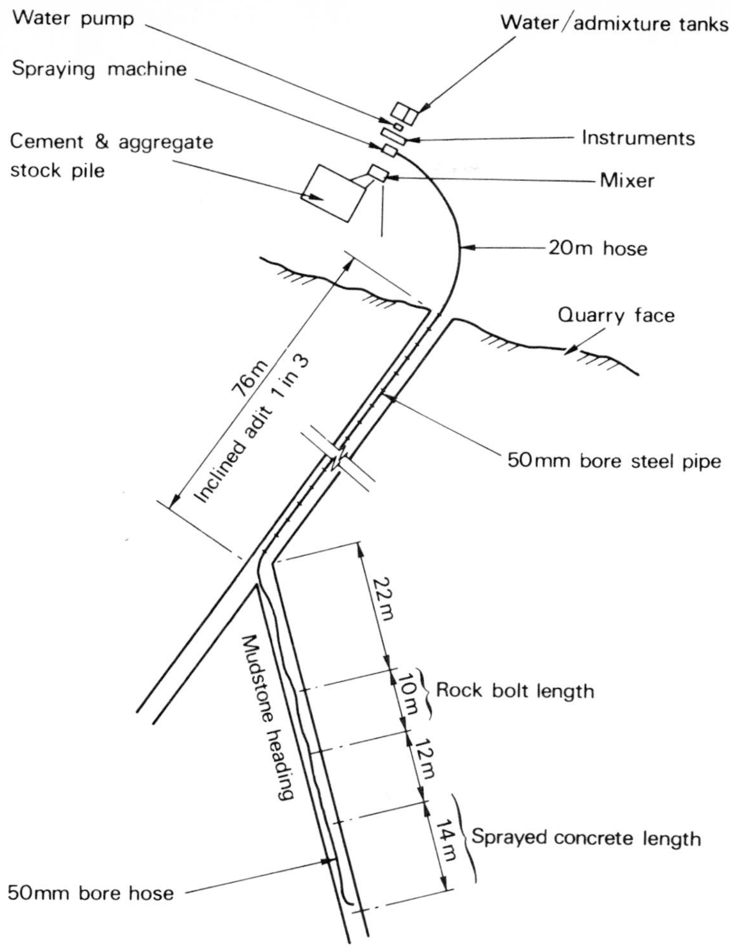

Figure 15 Equipment layout – field trials

was eased by remotely controlling the water. It was found that with the equipment running smoothly the water flow could be kept constant for any spraying orientation; small corrections to the remote valve were made by an observer in the tunnel who was in communication with the surface team where the flowmeter was located. Furthermore with the liquid accelerating admixture added to the water the mix could be sprayed much wetter and the deposit could be built up over quite a small area to a considerable thickness without any danger of sloughing. Under the right conditions the mix was hardening within about 5 minutes of spraying (the assessment of hardening was subjective; when the deposit could not be deformed by prodding with the finger it was deemed 'hard').

Problems were encountered with cement-admixture compatibility. Several different cement batches from the factory nearby were used in the course of the job and each required a different admixture dose. Temperature changes probably also contributed since not only were the weather conditions above ground changing, and hence the temperature of the aggregate and water, but also the cement temperature varied over a wide range; new cement batches arrived at high temperature – one batch was measured at 110°C. Time did not permit proper compatibility testing and the dose was varied in the mixing tanks on the surface according to visual assessments of the deposit in the tunnel. The admixture-water dilution had to be varied between 1:6 and 1:12 to produce the same hardening rate. Some form of metering device by which the admixture dose could be quickly and accurately changed according to a quick compatibility test would have been an asset.

It is worth noting that the admixture accounted for 40 per cent of the in-place material cost of £26/m^3.

The outstanding feature of the conveying or delivery side of the operation was the frequency of line and machine blocking. The reasons for this have already been discussed. Spraying was stopped 47 times in 21 hours because of blocking in the line or, more often, the machine outlet, and cleaning out accounted for 15 per cent of the total project time (see below). No blockages occurred at all in the 5 m nozzle, nor did the water ring show any serious signs of clogging at the end of each day. Under steady conditions the air requirements for conveying were 10 m^3/minute of free air at a pressure of 0.38 to 0.44 MN/m^2 on the inlet side of the machine. A ten per cent variation in pressure below or above this range usually resulted in either a blocked line or leaks at the machine rotor seals and an unnecessarily high spray velocity.

One cause of a number of early blockages was a very small difference in the internal diameters at the junction of the steel pipe and rubber hose at the bottom of the adit. Although this was only a 2 mm step it was sufficient to cause a solid plug of mix in the line after a short time. It was cured by chamfering the step.

The pipe couplings allowed a small degree of axial misalignment in the steel pipes used in the adit which was sufficient to cause a very high wear rate just downstream of the coupling. Two pipe lengths were completely perforated by abrasion, one after only 18 hours use. Most of the steel pipe sections showed similar wear. The rubber hoses in the conveying line showed signs of abrasion but no measurable reduction in wall thickness.

An analysis of the project time directly attributable to the spraying operation is given in Table 4.

Allowing for the uniqueness and experimental nature of the project and the comparative inexperience of the team it would be interesting to know if the 25 per cent efficiency is typical in 'real' tunnel spraying. Nothing appears to have been published on this. When rebound loss is taken into account, the useful spraying time drops to 15 per cent, with 10 per cent for 'spraying rebound'.

Strength of cores

A month after spraying 42 cores 75 mm diameter were cut, 21 from each tunnel length and a number of these were compression tested at age three months. Density was also measured. The results are given in Table 3. The bulk of the tested cores were from the lining proper, which was sprayed with the long nozzle and incorporated the admixture, rather than from the overbreak infill.

Table 3 Strength and density of cores – field trials

Core position	Mean 3 month compressive strength (equivalent cube)			Density kg/m³		
	MN/m²	Standard deviation	n	Oven dry	Saturated surface dry	n
Test beam - in direction of spraying	31.2	8.9	6	2118	2285	7
Test beam 90° to direction of spraying	30.8	8.4	6			
Tunnel lining 0 new concrete						
Cores 1, 2	–	–	0	2077	2287	8
Cores 3, 4	31.6		9	2085	2281	1
Cores 5, 6	29.6		8	2027	2232	2
Cores 7, 8	32.1		2	2015	2219	1
Cores 9, 10	29.4		4	–	–	0
Mean of all	30.6	3.9	23	2064	2272	12
Tunnel lining – old concrete	(Age approximately 14 months)					
Cores 7, 8	20.6		3	2064	2273	1
Cores 9, 10	22.5		3	–	–	0
Mean of all	21.6	4.0	6	2064	2273	1

Core positions:
1, 2 vertically down
3, 4 45° below horizontal
5, 6 horizontal
7, 8 45° above horizontal
9, 10 vertically up.

Table 4

Activity	Approximate hours	Approximate % of time
Spraying	20	25
Unloading and positioning of consumable materials	6	8
Setting up, changing position in tunnel and other preparations	8	10
Removal of rebound, rock, etc	8	10
Reinforcement fixing	6	8
Setting up instrumentation	4	5
Blockages, cleaning line and spraying machine	12	15
Delays in delivery of consumables	6	8
Support equipment breakdowns	10	10

The composition of the deposited concrete is not known. The composition as sprayed was approximately:

Water	182 litres/m^3
Cement	345 kg/m^3
Sand	1206 kg/m^3
10 mm aggregate	517 kg/m^3

w/c ratio 0.53

a/c ratio 5.0

At about 40 per cent average rebound, if it is assumed the amount of each ingredient rebounding was approximately as determined in the laboratory test described earlier then the composition of the deposited concrete would be:

Water	230 litres/m^3
Cement	440 kg/m^3
Sand	1270 kg/m^3
10 mm aggregate	310 kg/m^3

w/c ratio 0.53

a/c ratio 3.6

The magnitude of the strengths and densities shown in Table 3 are unremarkable at mean values of 30.6 MN/m^2 and 2272 kg/m^3, particularly at the probable cement and water contents given above, although the reduction in strength due to the admixture is not known. What is particularly interesting is the very low variation in these values. With a standard deviation of 3.9 MN/m^2 (coefficient of variation 12.6 per cent) this could be considered concrete with a high degree of control. Cores cut from the floor were not tested since these were of poor quality and broke into small pieces during coring; this is attributed to the large quantity of rebound trapped. Spraying vertically downwards presents a problem because of the difficulty of removing the rebound and it may be preferable to gravity place and consolidate by ramming, perhaps using the scoop extensions available with this type of equipment for decelerating the spray – in which case a wetter mix with reduced or no admixture incorporated would probably be necessary.

Small test panels of the type used in the laboratory trials were made during the spraying of the second phase ie with the long nozzle, admixture incorporated and a 0.53 water/cement ratio. The strength of cores cut from these panels, although rather more scattered, had a mean value virtually the same as those cut from the lining and also showed an insignificant directional difference. This is somewhat at variance with the earlier laboratory findings at lower water contents and suggests, as previously mentioned, that the minimum test panel size for results representative of spraying on an effective boundless surface becomes less critical the wetter the mix.

Cores cut from the old lining in the roof of the sprayed concrete section, about 14 months old, had a compressive strength some 30 per cent lower than the newer lining at 3 months, at a mean strength of 21.6 MN/m^2 although sprayed with a similar machine and mix. A short nozzle was used for the earlier spraying and the water content is not known.

In addition a powder set-accelerating admixture, Sigunite, was hand fed into the machine at a high and probably very variable rate.

Machine operator and nozzleman

The experience of the machine operator is just as important as that of the nozzleman; he should be able to detect impending blockages or reduced output and performance due to partial clogging in the machine by observing the air pressure at the machine, pulsations in the hose, the volume of material venting from the chamber exhausts and the flow of materials from the hopper, and take fast corrective action to prevent uneven delivery to the nozzle or a total blockage in the pipeline.

Correct operation and maintenance of the spraying machine are of great importance, such as ensuring the rubber rotor seals are not unduly worn and are evenly compressed to prevent loss of air and fines, ensuring the machine hopper is kept full and the mix flows freely into the rotor chambers, and ensuring the walls of the chambers, outlet, taper and chamber exhaust vents are free from any major accumulation of material.

On long pipe runs partial blockages may occur which are self clearing so that spraying can continue, but if these occur persistently or the material flow becomes erratic or markedly reduced for longer than about 10 seconds, the machine should be stopped, stripped, cleaned and checked. At the end of a shift or for any prolonged stop the machine should be cleaned thoroughly. The importance of all this increases with greater conveying distances. The time lost in locating and clearing a pipe blockage in perhaps poor light and in the confined and congested space in a tunnel can be considerable.

The aim of the men and equipment concerned with the supply and transmission side of the process is to ensure a steady, constant stream of materials to the nozzle so that the nozzleman can concentrate on the placing side of the process without constantly having to correct the water flow to compensate for uneven delivery of solids.

A great deal had been written about the skill required of the nozzleman and how the success or failure of the job depends on his skill. The experience gained from both the laboratory and field trials suggests that one of his major tasks, which has a direct bearing on the quality of the deposit, should be eliminated, that is sole control over the amount of water injected into the mix. It is suggested that the optimum amount of water should be predetermined by mix design considerations and trial spraying, and that this amount of water only is used. All that is needed to control the water delivered to the nozzle is a simple water flow meter and a flow control valve, which would be located at the spraying machine. This type of control would result in a more uniform, predictable deposit and possibly less material loss due to rebound.

CONCLUSIONS

Sprayed concrete acted in the experimental tunnel as a structural arch or ring in circumferential compression.

The concrete was subjected to high early strain rates particularly when placed close to the face soon after it had been advanced to that position and again when the face was advanced soon after the concrete was placed. Depending on the timing of constructional operations a peak instantaneous compressive strain of about 0.02 per cent could occur at an early age, and peak values of 0.06 per cent in the first week and of 0.09 per cent in the first month were measured.

The loading attracted by a concrete arch or ring of given thickness depends critically on the amount of dilation that has occurred in the rock before the concrete is placed, and, where little dilation has been allowed, on the rate at which the concrete stiffens in early age.

For economy combined with early safe support the stiffening/strengthening rates of the concrete need to be optimised in relation to the dilating properties of the rock and the timing and positions of constructional operations.

The versatility of sprayed concrete makes it potentially suitable to meet a wide range of constructional and ground requirements in rock tunnels, but more development and better control is necessary to achieve these objectives.

The limited laboratory and field trials show that there are a number of features in dry-mix spraying, as it is generally used, which increase the variability of the deposited concrete and which are inherent in the system even under good control and placement conditions.

It appears that the simple procedures of increasing the mixing time by means of a long nozzle, and predetermining and presetting the amount of water injected can improve the deposited concrete and reduce the demands on the nozzleman.

The use of an accelerating admixture, correctly dosed and of predictable compatibility with the cement, has considerable practical advantages in allowing the mix to be sprayed at a higher water content for maximum strength and minimum rebound loss.

The effects of admixtures on the long-term strength and particularly on the early time-dependent properties required for the optimum performance of a tunnel support system have yet to be investigated but are likely to be of major importance in small tunnels.

ACKNOWLEDGEMENTS

The construction and the work carried out in the Kielder Water Scheme experimental tunnel is the responsibility of a Steering Committee comprising representatives of the Northumbrian Water Authority and their consulting engineers, Babtie, Shaw and Morton, The Water Research Centre, the Director-General of Water Engineering and the Building Research Establishment of the Department of the Environment. Thyssen(GB) Limited, as contractors to the Northumbrian Water Authority, built the tunnel and provided attendance on the spraying trials.

REFERENCES

1 **Ward W H, Coats D J and Tedd P.** Performance of tunnel support systems in the Four Fathom Mudstone, *Tunnelling '76*, Institution of Mining and Metallurgy, 329–340. Also Building Research Establishment Current Paper CP25/76.

2 **Ward W H.** Rigid-body model of rock discontinuities shows tunnel collapses and support needs. Proceedings 6th European Conference on Soil Mech and Foundation Engg, Vienna, 1977, **2.2**, 133–135

3 **Kobler H G.** Dry-mix coarse-aggregate shotcrete as underground support, Table 3-1, in *Shotcreting,* American Concrete Institute, 1966, Publication SP-14.

4 **Ryan T.** Steel fibres in gunite, Tunnels and Tunnelling, 1975, **7** (4)

5 **Tynes W O and McCleese W F.** Investigation of Shotcrete, US Army Waterways Experiment Station, Vicksburg, 1974, Technical Report C–74–5

6 **Lorman W R.** Engineering properties of Shotcrete, American Concrete Institute, 1968, Publication SP–14A.

7 **Neville A M.** *Properties of Concrete,* Chapter 10, Pitman, London, 1973

Index

Where a topic is mentioned several times under the same heading only the first occurrence has normally been cited.

EDITOR'S NOTE

In addition to the Current Papers printed in full in the main body of this book, which covers the years 1973-1977, there are others in this broad subject area which for various reasons (such as, for example, their more localised appeal) have not been included. Their titles are, however, listed below for the benefit of any reader whose interests may extend to these areas.

7/73	The economics of lightweight aggregate structural concrete
8/73	Notes on the CIB Draft Practice Manual on lightweight aggregate concrete (1972)
22/73	Research in support of the 1971 revision of BS 368: precast concrete flags
31/73	Report of aggregates and waste materials Working Group
3/74	Admixtures for concrete
58/74	Report on the failure of roof beams at St. John Cass's Foundation and Red Coat Church of England Secondary School, Stepney
76/74	Results of exposure tests on various types of concrete blocks
78/74	A survey of possible sources in Wales of raw materials for the manufacture of a lightweight expanded slate aggregate
100/74	Synthetic aggregate sources and resources
41/75	The economic and environmental benefits of increased use of pfa and granulated slag
54/75	Field measurements of the sound insulation of cavity party walls of aerated concrete blockwork in local authority housing
59/75	High alumina cement concrete: research and results
72/75	The conversion of high alumina cement
99/75	Accuracy of in-situ concrete
17/76	Estimates of the requirements for aggregates
23/77	Chemical resistance of concrete
25/77	A simple pull-out test to assess the strength of in-situ concrete